New Trends in Catalysis for Sustainable CO2 Conversion

New Trends in Catalysis for Sustainable CO2 Conversion

Editors

Javier Ereña
Ainara Ateka

MDPI • Basel • Beijing • Wuhan • Barcelona • Belgrade • Manchester • Tokyo • Cluj • Tianjin

Editors
Javier Ereña
University of the Basque
Country UPV/EHU
Spain

Ainara Ateka
University of the Basque
Country UPV/EHU
Spain

Editorial Office
MDPI
St. Alban-Anlage 66
4052 Basel, Switzerland

This is a reprint of articles from the Special Issue published online in the open access journal *Catalysts* (ISSN 2073-4344) (available at: http://www.mdpi.com).

For citation purposes, cite each article independently as indicated on the article page online and as indicated below:

LastName, A.A.; LastName, B.B.; LastName, C.C. Article Title. *Journal Name* **Year**, *Volume Number*, Page Range.

ISBN 978-3-0365-5911-7 (Hbk)
ISBN 978-3-0365-5912-4 (PDF)

© 2022 by the authors. Articles in this book are Open Access and distributed under the Creative Commons Attribution (CC BY) license, which allows users to download, copy and build upon published articles, as long as the author and publisher are properly credited, which ensures maximum dissemination and a wider impact of our publications.

The book as a whole is distributed by MDPI under the terms and conditions of the Creative Commons license CC BY-NC-ND.

Contents

About the Editors vii

Preface to "New Trends in Catalysis for Sustainable CO2 Conversion" ix

Javier Ereña and Ainara Ateka
New Trends in Catalysis for Sustainable CO_2 Conversion
Reprinted from: *Catalysts* 2022, *12*, 1300, doi:10.3390/catal12111300 1

Muhammad Amin, Saleem Munir, Naseem Iqbal, Saikh Mohammad Wabaidur and Amjad Iqbal
The Conversion of Waste Biomass into Carbon-Supported Iron Catalyst for Syngas to Clean Liquid Fuel Production
Reprinted from: *Catalysts* 2022, *12*, 1234, doi:10.3390/catal12101234 5

Karolína Simkovičová, Muhammad I. Qadir, Naděžda Žilková, Joanna E. Olszówka, Pavel Sialini, Libor Kvítek and Štefan Vajda
Hydrogenation of CO_2 on Nanostructured Cu/FeO_x Catalysts: The Effect of Morphology and Cu Load on Selectivity
Reprinted from: *Catalysts* 2022, *12*, 516, doi:10.3390/catal12050516 19

Xinye Liu, Gad Licht, Xirui Wang and Stuart Licht
Controlled Transition Metal Nucleated Growth of Carbon Nanotubes by Molten Electrolysis of CO_2
Reprinted from: *Catalysts* 2022, *12*, 137, doi:10.3390/catal12020137 33

Xinye Liu, Gad Licht, Xirui Wang and Stuart Licht
Controlled Growth of Unusual Nanocarbon Allotropes by Molten Electrolysis of CO_2
Reprinted from: *Catalysts* 2022, *12*, 125, doi:10.3390/catal12020125 55

Antoni Waldemar Morawski, Katarzyna Ćmielewska, Kordian Witkowski, Ewelina Kusiak-Nejman, Iwona Pełech, Piotr Staciwa, Ewa Ekiert, Daniel Sibera, Agnieszka Wanag, Marcin Gano and Urszula Narkiewicz
CO_2 Reduction to Valuable Chemicals on TiO_2-Carbon Photocatalysts Deposited on Silica Cloth
Reprinted from: *Catalysts* 2022, *12*, 31, doi:10.3390/catal12010031 83

Li Li, Ye Wang, Qing Zhao and Changwei Hu
The Effect of Si on CO_2 Methanation over Ni-xSi/ZrO_2 Catalysts at Low Temperature
Reprinted from: *Catalysts* 2021, *11*, 67, doi:10.3390/catal11010067 99

Jaeyong Sim, Sang-Hyeok Kim, Jin-Yong Kim, Ki Bong Lee, Sung-Chan Nam and Chan Young Park
Enhanced Carbon Dioxide Decomposition Using Activated $SrFeO_{3-\delta}$
Reprinted from: *Catalysts* 2020, *10*, 1278, doi:10.3390/catal10111278 113

Fengyang Ju, Jinjin Zhang and Weiwei Lu
Efficient Electrochemical Reduction of CO_2 to CO in Ionic Liquid/Propylene Carbonate Electrolyte on Ag Electrode
Reprinted from: *Catalysts* 2020, *10*, 1102, doi:10.3390/catal10101102 127

Davide M. S. Marcolongo, Francesco Nocito, Nicoletta Ditaranto, Michele Aresta and Angela Dibenedetto
Synthesis and Characterization of *p-n* Junction Ternary Mixed Oxides for Photocatalytic Coprocessing of CO_2 and H_2O
Reprinted from: *Catalysts* 2020, *10*, 980, doi:10.3390/catal10090980 141

Heather D. Willauer, Matthew J. Bradley, Jeffrey W. Baldwin, Joseph J. Hartvigsen, Lyman Frost, James R. Morse, Felice DiMascio, Dennis R. Hardy and David J. Hasler
Evaluation of CO_2 Hydrogenation in a Modular Fixed-Bed Reactor Prototype
Reprinted from: Catalysts **2020**, *10*, 970, doi:10.3390/catal10090970 . . . . **161**

Tomáš Weidlich and Barbora Kamenická
Utilization of CO_2-Available Organocatalysts for Reactions with Industrially Important Epoxides
Reprinted from: Catalysts **2022**, *12*, 298, doi:10.3390/catal12030298 . . . . **175**

Daniel Weber, Tina He, Matthew Wong, Christian Moon, Axel Zhang, Nicole Foley, Nicholas J. Ramer and Cheng Zhang
Recent Advances in the Mitigation of the Catalyst Deactivation of CO_2 Hydrogenation to Light Olefins
Reprinted from: Catalysts **2021**, *11*, 1447, doi:10.3390/catal11121447 . . . . **217**

Sheikh Tareq Rahman, Jang-Rak Choi, Jong-Hoon Lee and Soo-Jin Park
The Role of CO_2 as a Mild Oxidant in Oxidation and Dehydrogenation over Catalysts: A Review
Reprinted from: Catalysts **2020**, *10*, 1075, doi:10.3390/catal10091075 . . . . **257**

About the Editors

Javier Ereña

Javier Ereña Loizaga graduated in Chemistry (1991, Speciality: Industrial Chemistry) at the University of the Basque Country (UPV/EHU, Leioa, Spain). Since 1990, the year he joined the laboratories of the Chemical Engineering Department from the University of the Basque Country, his research field has been focused on the development of catalytic processes for obtaining fuels and raw materials from alternative sources to petroleum.

In 1996 he defended his doctoral thesis within a collaboration agreement between "Catalytic Processes and Waste Valorization" research group (well established/consolidated high-performance group in the Basque Country) and the Center for Chemical Reactors from the University of Western Ontario (Canada). The objective was to obtain gasoline from synthesis gas and CO_2.

Since 2000, his research work has been mainly focused on the study of the following research lines (keys for the industrial development of the concept of Bio-refinery):
- H_2 synthesis by use of steam to reform DME and ethanol.
- Direct DME synthesis (STD process). The interest of this integrated process is based on both the product (clean fuel and H_2 source for fuel cells) and the raw materials (synthesis gas and CO_2).
- The development of catalytic processes of interest from the perspective of sustainability.

Since 1996, he has been a professor at the University of the Basque Country, where he integrates his research activities and teaching in subjects related to chemical engineering.

Ainara Ateka

Ainara Ateka Bilbao graduated in Chemical Engineering (2009) at the University of the Basque Country (UPV/EHU) and is member of the PROCAT-VARES group since 2010. The group focuses on the developing and intensifying catalytic processes, pursuing to valorize CO_2 and waste of the society into value added products and/or raw materials.

Within the group, her research line is related to valorizing CO_2, seeking for the production of oxygenated compounds or a varied range of hydrocarbons (olefins, gasoline, aromatics). Her work involves catalyst preparation and characterization, thermodynamic analysis, addressing deactivation and developing kinetic models. Specifically in the following integrated process:
- Synthesis of dimethyl ether (DME) from CO_2 and syngas mixtures (implying methanol formation from CO_2 and conversion into DME). Topic of her Doctoral Thesis (2014)
- Conversion of oxygenates (methanol/DME) into olefins (MTO and DTO processes)
- Direct production of hydrocarbons integrating two processes: oxygenates production and conversion into hydrocarbons in different ranges (light olefins, gasoline, aromatics).

Since 2022 she works as assistant professor at the University of the Basque Country (UPV/EHU), combining her teaching activities with the mentioned research activities.

Preface to "New Trends in Catalysis for Sustainable CO2 Conversion"

The challenge of overcoming climate change and limiting global temperature increase has raised interest in carbon dioxide storage (CCS) and utilization (CCU) technologies. CCS technologies pursue the long-term storage of large amounts of CO_2 of anthropogenic nature (emitted by the industry, combustion processes for energy production and vehicles) after its capture.

On the other hand, CCU technologies are more intriguing as methods. In this approach, CO_2 is not considered a waste, but is used as a feed for producing value-added products. Among the CCU routes, the catalytic conversion stands out. Nonetheless, there is still significant demand for research into this field, as for example in the following areas:

- Development of new catalysts and routes for the catalytic conversion/valorization of CO_2.
- Studying new processes for value-added product synthesis from CO_2, mainly seeking for fuels and chemicals.
- Intensification of existing processes and optimization of operating conditions and catalysts.

Javier Ereña and Ainara Ateka
Editors

Editorial

New Trends in Catalysis for Sustainable CO_2 Conversion

Javier Ereña * and Ainara Ateka *

Department of Chemical Engineering, University of the Basque Country UPV/EHU, P.O. Box 644, 48080 Bilbao, Spain
* Correspondence: javier.erena@ehu.eus (J.E.); ainara.ateka@ehu.eus (A.A.)

Citation: Ereña, J.; Ateka, A. New Trends in Catalysis for Sustainable CO_2 Conversion. *Catalysts* **2022**, *12*, 1300. https://doi.org/10.3390/catal12111300

Received: 19 October 2022
Accepted: 19 October 2022
Published: 23 October 2022

Publisher's Note: MDPI stays neutral with regard to jurisdictional claims in published maps and institutional affiliations.

Copyright: © 2022 by the authors. Licensee MDPI, Basel, Switzerland. This article is an open access article distributed under the terms and conditions of the Creative Commons Attribution (CC BY) license (https://creativecommons.org/licenses/by/4.0/).

Over the past few decades, there have been many advances in the world, leading to improvements in quality of life. Due to demographic and industrial growth, consumption has increased, as well as the amount of waste and pollution. Today, global warming and climate change are mainly attributed to the emission of anthropogenic greenhouse gases, with carbon dioxide (CO_2) being the most relevant one, due to the huge emissions of this gas into the atmosphere (mainly derived from the consumption of fossil fuels).

Carbon capture and storage (CCS) is a physical process consisting of separating the CO_2 (emitted by industry and the combustion processes for energy generation) and transporting it to geological storage to isolate it from the atmosphere in the long term. However, the most promising routes for CO_2 mitigation are those pursuing its catalytic valorization. By applying specific catalysts and suitable operating conditions, CO_2 molecules react with other components to form longer chains (i.e., hydrocarbons). Accordingly, effort should be made to catalytically valorize CO_2 (alone or co-fed with syngas) as an alternative way for reducing greenhouse gas emissions and obtaining high-value fuels and chemicals.

Carbon capture and utilization (CCU) is a developing field with significant demand for research in the following aspects:

- The development of new catalysts, catalytic routes, and technologies for CO_2 conversion;
- The study of new processes for obtaining fuels and chemicals from CO_2;
- Optimization of the catalysts and the reaction conditions for these processes;
- Further steps in advanced processes using CO_2-rich feeds ($H_2 + CO_2$ or CO_2 mixed with syngas), increasing product yields.

This Special Issue on "New Trends in Catalysis for Sustainable CO_2 Conversion" shows new research on the development of catalysts and catalytic routes for CO_2 valorization, and the optimization of the reaction conditions for the process.

This Special Issue includes ten articles and three reviews. In the paper by Amin et al., waste biomass is converted into activated carbon, and then a carbon-supported iron-based catalyst is prepared (Fe–C) [1]. The catalyst is used to assess the influence of temperature on the subsequent transformation of syngas to liquid fuels. Potassium is used as a structural promoter in the iron carbon-supported catalyst to boost catalyst activity and structural stability. Potassium promoter increases gasoline conversion from 36.4% (Fe–C) to 72.5% (Fe–C–K), and diesel conversion from 60.8% (Fe–C) to 80.0% (Fe–C–K). The influence of copper content and particle morphology on the performance of Cu/FeO_x catalysts in the hydrogenation of CO_2 is analyzed by Simkovičová et al. [2]. All the catalysts, with a copper content between 0 and 5 wt%, are found to be highly efficient, with CO_2 conversion reaching 36.8%. Their selectivity towards C_1 (versus C_2–C_4, C_2–$C_{4=}$, and C_{5+} products) is dependent on the catalyst's composition, and morphology, and on the temperature. The results indicate new potential methods of altering the morphology and composition of iron-oxide-based particles. The electrolysis of CO_2 in molten carbonate as an alternative mechanism to synthesize carbon nanomaterials is introduced in the paper by Liu et al. [3]. This study focuses on controlled electrochemical conditions in molten lithium carbonate to split CO_2 absorbed from the atmosphere into carbon nanotubes (CNT), and into various macroscopic assemblies of CNT, which may be useful for nano-filtration. Different CNT morphologies are prepared electrochemically by varying the anode's and cathode's composition and architecture, varying the electrolyte's composition pre-electrolysis processing, and varying the application and density of currents. The paper by Liu et al. shows that CO_2 can be transformed into distinct nano-bamboo, nano-pearl, nano-dragon, solid and hollow nano-onion, nano-tree, nano-rod, nano-belt, and nano-flower morphologies [4]. The capability to produce these unusual nanocarbon allotropes at high purity by a straightforward electrolysis process opens an array of inexpensive unique materials with high strength, electrical and thermal conductivity, flexibility, charge storage, lubricant, and robustness properties. A new photocatalyst for CO_2 reduction is presented by Morawski et al. [5]. The photocatalyst is prepared from a combination of a commercial TiO_2 with a mesopore structure and carbon spheres with a microporous structure with high CO_2 adsorption capacity. The combined TiO_2–carbon spheres/silica cloth photocatalysts show higher efficiency in the two-electron CO_2 reduction towards CO than in the eight-electron reaction to methane. The 0.5 g graphitic carbon spheres combined with 1 g of TiO_2 results in almost 100% selectivity to CO. In the article by Li et al., a series of Ni-xSi/ZrO_2 catalysts (x = 0, 0.1, 0.5, 1 wt% of Si) are prepared by co-impregnation with ZrO_2 as support and Si as a promoter [6]. The effect of different amounts of Si on the catalytic performance is investigated for CO_2 methanation. It is found that adding the appropriate amount of Si improves the catalytic performance of the Ni/ZrO_2 catalyst at a low reaction temperature (250 °C). It is observed that Si enhances the interaction between Ni and ZrO_2 and increases Ni dispersion, the number of active sites (including surface Ni^0), the number of oxygen vacancies, and the number of strong basic sites on the catalyst surface. Sim et al. report efficient CO_2 decomposition results using a nonperovskite metal oxide, $SrFeCo_{0.5}O_x$, in a continuous-flow system [7]. In this study, the authors obtain enhanced efficiency, reliability under isothermal conditions, and catalytic reproducibility through cyclic tests using $SrFeO_{3-\delta}$. Activated oxygen-deficient $SrFeO_{3-\delta}$ decomposes CO_2 into carbon monoxide (CO) and carbon (C). In the work by Ju et al., a new type of electrolyte solution constituted by ionic liquids and propylene carbonate is used as the cathodic solution to study the conversion of CO_2 on an Ag electrode [8]. Linear sweep voltammetry and Tafel characterization indicate that the chain length of 1-alkyl-3-methyl imidazolium cation has strong influences on the catalytic performance for CO_2 conversion. Electrochemical impedance spectroscopy shows that the imidazolium cation absorbed on the Ag electrode surface stabilizes the anion radical ($CO_2^{\bullet-}$), leading to the enhanced efficiency of CO_2 conversion. The synthesis and characterization of both binary (Cu_2O, Fe_2O_3, and In_2O_3) and ternary (Cu_2O–Fe_2O_3 and Cu_2O–In_2O_3) transition metal mixed-oxides as photocatalysts for solar-driven CO_2 conversion into energy-rich species is reported by Marcolongo et al. [9]. Two different preparation techniques (high-energy milling and coprecipitation) are compared. The composition and synthetic methodologies of mixed-oxides, the reactor geometry, and the method of dispersing the photocatalyst sample play key roles in the light-driven reaction of CO_2–H_2O. Willauer et al. study the behavior on CO_2 hydrogenation of an iron-based catalyst in a commercial-scale fixed-bed reactor under different feed rates and product recycling conditions [10]. CO_2 conversion increases from 26% to as high as 69% by recycling a portion of the product stream, and CO selectivity is greatly reduced from 45% to 9% in favor of hydrocarbon production. In addition, the catalyst is successfully regenerated for optimum performance. The review by Weidlich and Kamenická summarizes new developments in the ring opening of epoxides with a subsequent CO_2-based formation of cyclic carbonates [11]. The review highlights recent and major developments for sustainable CO_2 conversion from 2000 to the end of 2021. The syntheses of epoxides, especially from bio-based raw materials, are described, such as the types of raw materials used (vegetable oils or their esters) and the reaction conditions. The aim of this review is also to compare the types of homogeneous non-metallic catalysts. Weber et al. review recent advances in the mitigation of catalyst deactivation of CO_2 hydrogenation to light olefins [12]. The authors provide a brief summary of the two dominant reaction pathways (CO_2–Fischer–Tropsch and methanol-mediated pathways), mechanistic insights, and catalytic materials for CO_2 hydrogenation to light olefins. Then, they describe the main deactivation mechanisms caused by carbon deposition, water formation, phase transformation, and metal sintering/agglomeration. The main focus of this review is to provide a useful resource for researchers to correlate catalyst deactivation and the recent research effort on catalyst development for enhanced olefin yields and catalyst stability. CO_2 oxidative dehydrogenation is a greener alternative to the classical dehydrogenation method. The availability, cost, safety, and soft oxidizing properties of CO_2, with the assistance of appropriate catalysts at an industrial scale, can lead to breakthroughs in the pharmaceutical, polymer, and fuel industries. The review by Rahman et al. focuses on several applications of CO_2 in oxidation and oxidative dehydrogenation systems [13]. These processes and catalytic technologies reduce the cost of utilizing CO_2 in chemical and fuel production, which may lead to commercial applications in the imminent future.

In summary, these manuscripts clearly show the relevance of CO_2 conversion for the production of fuels and raw materials, avoiding CO_2 emission into the atmosphere and reducing global warming. Nowadays, efforts are being made on the co-feeding of CO_2 with syngas and on the use of new catalytic processes for CO_2 conversion under mild reaction conditions.

We would like to thank all the authors for their works and the reviewers for improving the quality of the papers with their comments. We are honored to be the Guest Editors of this Special Issue. We are also grateful to all the staff of the *Catalysts* Editorial Office.

Author Contributions: Conceptualization, J.E. and A.A.; writing—original draft preparation, J.E. and A.A.; writing—review and editing, J.E. and A.A.; visualization, J.E. and A.A.; supervision, J.E. and A.A. All authors have read and agreed to the published version of the manuscript.

Funding: This research was funded by the Ministry of Science, Innovation and Universities of the Spanish Government (PID2019-108448RB-100), the Basque Government (Project IT1645-22), the European Regional Development Funds (ERDF) and the European Commission (HORIZON H2020-MSCA RISE-2018. Contract No. 823745).

Conflicts of Interest: The authors declare no conflict of interest.

References

1. Amin, M.; Munir, S.; Iqbal, N.; Wabaidur, S.M.; Iqbal, A. The Conversion of Waste Biomass into Carbon-Supported Iron Catalyst for Syngas to Clean Liquid Fuel Production. *Catalysts* **2022**, *12*, 1234. [CrossRef]
2. Simkovičová, K.; Qadir, M.I.; Žilková, N.; Olszówka, J.E.; Sialini, P.; Kvítek, L.; Vajda, S. Hydrogenation of CO_2 on Nanostructured Cu/FeO$_x$ Catalysts: The Effect of Morphology and Cu Load on Selectivity. *Catalysts* **2022**, *12*, 516. [CrossRef]
3. Liu, X.; Licht, G.; Wang, X.; Licht, S. Controlled Transition Metal Nucleated Growth of Carbon Nanotubes by Molten Electrolysis of CO_2. *Catalysts* **2022**, *12*, 137. [CrossRef]
4. Liu, X.; Licht, G.; Wang, X.; Licht, S. Controlled Growth of Unusual Nanocarbon Allotropes by Molten Electrolysis of CO_2. *Catalysts* **2022**, *12*, 125. [CrossRef]
5. Morawski, A.W.; Ćmielewska, K.; Witkowski, K.; Kusiak-Nejman, E.; Pełech, I.; Staciwa, P.; Ekiert, E.; Sibera, D.; Wanag, A.; Gano, M.; et al. CO_2 Reduction to Valuable Chemicals on TiO$_2$-Carbon Photocatalysts Deposited on Silica Cloth. *Catalysts* **2022**, *12*, 31. [CrossRef]
6. Li, L.; Wang, Y.; Zhao, Q.; Hu, C. The Effect of Si on CO_2 Methanation over Ni-xSi/ZrO$_2$ Catalysts at Low Temperature. *Catalysts* **2021**, *11*, 67. [CrossRef]
7. Sim, J.; Kim, S.H.; Kim, J.Y.; Lee, K.B.; Nam, S.C.; Park, C.Y. Enhanced Carbon Dioxide Decomposition Using Activated SrFeO$_{3-\delta}$. *Catalysts* **2020**, *10*, 1278. [CrossRef]
8. Ju, F.; Zhang, J.; Lu, W. Efficient Electrochemical Reduction of CO_2 to CO in Ionic Liquid/Propylene Carbonate Electrolyte on Ag Electrode. *Catalysts* **2020**, *10*, 1102. [CrossRef]
9. Marcolongo, D.M.S.; Nocito, F.; Ditaranto, N.; Aresta, M.; Dibenedetto, A. Synthesis and Characterization of p-n Junction Ternary Mixed Oxides for Photocatalytic Coprocessing of CO_2 and H_2O. *Catalysts* **2020**, *10*, 980. [CrossRef]
10. Willauer, H.D.; Bradley, M.J.; Baldwin, J.W.; Hartvigsen, J.J.; Frost, L.; Morse, J.R.; DiMascio, F.; Hardy, D.R.; Hasler, D.J. Evaluation of CO_2 Hydrogenation in a Modular Fixed-Bed Reactor Prototype. *Catalysts* **2020**, *10*, 970. [CrossRef]
11. Weidlich, T.; Kamenická, B. Utilization of CO_2-Available Organocatalysts for Reactions with Industrially Important Epoxides. *Catalysts* **2022**, *12*, 298. [CrossRef]

12. Weber, D.; He, T.; Wong, M.; Moon, C.; Zhang, A.; Foley, N.; Ramer, N.J.; Zhang, C. Recent Advances in the Mitigation of the Catalyst Deactivation of CO_2 Hydrogenation to Light Olefins. *Catalysts* **2021**, *11*, 1447. [CrossRef]
13. Rahman, S.T.; Choi, J.R.; Lee, J.H.; Park, S.J. The Role of CO_2 as a Mild Oxidant in Oxidation and Dehydrogenation over Catalysts: A Review. *Catalysts* **2020**, *10*, 1075. [CrossRef]

Article

The Conversion of Waste Biomass into Carbon-Supported Iron Catalyst for Syngas to Clean Liquid Fuel Production

Muhammad Amin [1,2], Saleem Munir [1,3], Naseem Iqbal [1,*], Saikh Mohammad Wabaidur [4] and Amjad Iqbal [5,6,*]

1. U.S.-Pakistan Centre for Advanced Studies in Energy (USPCAS-E), Department of Energy Systems Engineering, National University of Sciences and Technology, Islamabad 44000, Pakistan
2. Department of Energy Systems Engineering, Seoul National University, Seoul 08826, Korea
3. DEQ, ETSEQ, Universitat Rovira i Virgili, Avinguda dels Països Catalans 26, 43007 Tarragona, Spain
4. Department of Chemistry, College of Science, King Saud University, Riyadh 11451, Saudi Arabia
5. Department of Materials Technologies, Faculty of Materials Engineering, Silesian University of Technology, 44-100 Gliwice, Poland
6. CEMMPRE-Center for Mechanical Engineering Materials and Processes, Department of Mechanical Engineering, University of Coimbra, Rua Luí's Reis Santos, 3004-531 Coimbra, Portugal
* Correspondence: naseem@uspcase.nust.edu.pk (N.I.); amjadiqbalfalak@gmail.com (A.I.); Tel.: +92-51-9085-5281 (N.I.)

Citation: Amin, M.; Munir, S.; Iqbal, N.; Wabaidur, S.M.; Iqbal, A. The Conversion of Waste Biomass into Carbon-Supported Iron Catalyst for Syngas to Clean Liquid Fuel Production. *Catalysts* **2022**, *12*, 1234. https://doi.org/10.3390/catal12101234

Academic Editors: Ainara Ateka and Javier Ereña

Received: 26 August 2022
Accepted: 8 October 2022
Published: 14 October 2022

Publisher's Note: MDPI stays neutral with regard to jurisdictional claims in published maps and institutional affiliations.

Copyright: © 2022 by the authors. Licensee MDPI, Basel, Switzerland. This article is an open access article distributed under the terms and conditions of the Creative Commons Attribution (CC BY) license (https://creativecommons.org/licenses/by/4.0/).

Abstract: Syngas has been utilized in the production of chemicals and fuels, as well as in the creation of electricity. Feedstock impurities, such as nitrogen, sulfur, chlorine, and ash, in syngas have a negative impact on downstream processes. Fischer–Tropsch synthesis is a process that relies heavily on temperature to increase the production of liquid fuels (FTS). In this study, waste biomass converted into activated carbon and then a carbon-supported iron-based catalyst was prepared. The catalyst at 200 °C and 350 °C was used to investigate the influence of temperature on the subsequent application of syngas to liquid fuels. Potassium (K) was used as a structural promoter in the Fe-C catalyst to boost catalyst activity and structural stability (Fe-C-K). Low temperatures (200 °C) cause 60% and 80% of diesel generation, respectively, without and with potassium promoter. At high temperatures (350 °C), the amount of gasoline produced is 36% without potassium promoter, and 72% with promoter. Iron carbon-supported catalysts with potassium promoter increase gasoline conversion from 36.4% (Fe-C) to 72.5% (Fe-C-K), and diesel conversion from 60.8% (Fe-C) to 80.0% (Fe-C-K). As seen by SEM pictures, iron particles with potassium promoter were found to be equally distributed on the surface of activated carbon.

Keywords: biomass; Fischer–Tropsch synthesis; carbon-supported iron catalyst; gasoline; diesel

1. Introduction

Researchers throughout the world are striving to develop alternate technologies for manufacturing chemicals and hydrocarbons from coal as crude oil sources become depleted [1,2]. Fischer–Tropsch Synthesis was invented by a German chemist (FTS). Syngas, a mixture of hydrogen and carbon monoxide produced from coal, natural gas, or biomass, is used in the FTS process [3]. There are two stages to the FT reaction. The catalyst is employed in the first stage to produce straight-chain hydrocarbons, as well as water and/or carbon dioxide [4]. Hydrocracking is employed in the second stage to create the products and split them into liquid fuel. To boost production, various catalysts (Fe, Co, Ru, and Ni) have been utilized [3]. Due to their better conversion and selectivity, as well as their ease of CO dissociation, metal catalysts are favored in the FTS process [5]. Due to their inexpensive cost, favorable reactional conditions, and high catalytic activity, iron (Fe) and cobalt catalysts are frequently selected and utilized in commercial FTS reactors [6].

Fe catalysts are less expensive than other catalysts, and provide more product distribution flexibility than other catalysts. The Fe-based catalyst is more appealing, since it has a better water gas shift reaction with a lower H_2/CO ratio [3]. For iron-based catalysts, there are two forms: supported and bulk, both of which are manufactured using the wet impregnation method. To prevent particle agglomeration, a supported iron catalyst requires the inclusion of metal support. In comparison to bulk iron catalysts, they can also have a higher mechanical strength [7]. Some support materials, such as SiO_2, Al_2O_3, TiO_2, and carbon, have been widely used in recent research to influence catalyst performance in the FTS process [8,9]. The FTS activity and selectivity are influenced by the physical and chemical features of the support. They have been used for a long time to help with the FTS process. Carbon compounds, such as carbon black and activated carbon, have been used for this purpose [10]. Supported iron FT catalysts provide a larger iron dispersion; therefore, research has shifted away from bulk iron FT catalysts in recent years. Supported iron catalysts improve the performance of iron species by optimizing the interaction of iron species with support catalysts [11].

An iron metal catalyst's support serves as a macromolecular ligand that either directly or indirectly increases the reactivity of the active site [7]. Carbon materials' catalytic performance is influenced by their electrical characteristics, surface chemistry, and porous structure [12]. Because activated carbons have a porous morphology, they havebeen used as a support material. In this study, activated carbon derived from "Lantana Camara" is used as a support. Selecting the right carbon support could affect the overall FTS process [13,14].

Lower activity is one of the key problems of iron-based catalysts, which can be improved by using the correct promoters, such as potassium, sodium, and copper, which are regarded as advantageous for Fe-catalysts. To improve selectivity and optimize the spatial organization of active iron species, a variety of promotors are utilized [3]. Researchers have recently focused their attention on promoters such as potassium, copper, sodium, and manganese [6]. Potassium (K) promoter has been widely used in carbon-supported iron catalysts to improve the selectivity of hydrocarbons. Potassium promoter reduces methane formation, and enhances the production of short olefins. The use of potassium promoter with carbon support showed a marginal effect on the reactivates of CO and CO_2. The addition of potassium as a promoter creates effects on an optimized iron-carbide by donating electrons to an active surface of iron. That is why potassium is used as a promoter in this study. For this purpose, the incipient wetness impregnation method was used to produce the carbon-supported iron catalyst with potassium promoter for Fischer–Tropsch synthesis [14].

FTS reactions are classified into two types: low-temperature FTS reactions (200–280 °C) and high-temperature FTS reactions (300–350 °C). The low temperature FTS reactions have been used to make specific C_{5+} hydrocarbons. For the coproduction of liquid hydrocarbons, however, high temperature FTS reactions have been used. At a temperature of 270 °C, the synergistic effect of active co-sites and acidic sites has been deemed the best reaction temperature for gasoline synthesis [15]. Given the foregoing, the goal of this research was to look at FTS reactions at low temperatures (LT-FTS) of 200 °C and high temperatures (HT-FTS) of 350 °C to see how reaction temperature and catalyst type affected direct syngas conversion via Fischer–Tropsch synthesis. Potassium (K) was used as a "structural promoter" (Fe-C-K) to make the Fe-C catalyst more active and stable, and its effect on the next process will be looked into as well.

To the best of our knowledge, no study on the development of a carbon-supported iron catalyst by using *Lantana Camara* has been conducted. The oil nature of *Lantana Camara* enhanced the production of gasoline and diesel. Therefore, this article could be beneficial for researchers and industrialists for the preparation of a carbon-supported catalyst by using *Lantana Camara* as a carbon support.

2. Results and Discussions

2.1. X-ray Diffraction

Figure 1 depicts changes in the crystal structure of various catalysts (Figure 1A–C). A carbonaceous structure peak at angle 23 can be observed in the XRD pattern of activated carbon derived from *"Lantana Camara"* in Figure 1A [16]. Three processes are involved in the chemical activation using H_3PO_4: dehydration, degradation, and coagulation. Research revealed that the phosphoric acid forms pores with raw precursor during the carbonization phase [17,18].

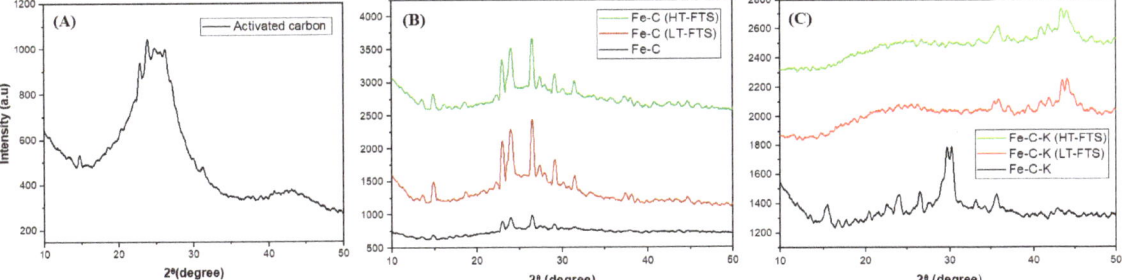

Figure 1. XRD pattern: (**A**) activated carbon; (**B**) Fe-C catalyst, Fe-C LT-FTS, and Fe-C HT-FTS; (**C**) Fe-C-K catalyst, Fe-C-K LT-FTS, and Fe-C-K HT-FTS.

Figure 1B shows the XRD patterns of the carbon-supported iron catalyst, Fe-C. Because activated carbon was utilized as a support, the large peak at 23° can be attributed to the amorphous structure of the carbon. The reflection at a value of 26.5° becomes more discernible during FTS reactions, and is attributed to the graphite carbon [19]. The peak at angles of 24° and 32° in both low- and high-temperature catalysts (Fe-C (LT-FTS) and Fe-C (HT-FTS)) demonstrates the creation of Fe_2O_3 [7].

Figure 1C shows the XRD patterns of the carbon-supported iron catalyst with potassium promoter (Fe-C-K). Iron carbide (Fe_3C) synthesis in the Fe-C-K catalyst at an angle of 37° at low and high temperatures is responsible for the formation of gasoline [20]. During the activation procedure, Fe_2O_3 were converted to Fe_3O_4 and Fe_5C_2 [21]. Researchers suggested that the creation of Fe_5C_2 at all times boosts diesel formation [22,23]. The Fe_5C_2 is considered as an active phase, and is critical for obtaining simultaneous improved selectivity and activity in iron-based Fisher–Tropsch synthesis. Thermal reduction and carburization methods have been frequently used for this purpose. The creation of Fe_5C_2 occurs during the FTS reaction, as shown by the post Fe-C catalyst, and this is observed at an angle of 44° [24]. It was thought that iron carbide (Fe_3C) to Fe_5C_2, which has a better ability to pick out C_{5+}, was the most active iron phase for making the FTs [25]. The results show that changing the catalyst qualities by changing the chemical and physical properties of carbon improves metal oxide reduction [7].

2.2. Scanning Electron Microscopy

Scanning electron microscopy (SEM) images of activated carbon are shown in Figure 2a,b, demonstrating that its interior surface is riddled with voids. These chambers are upright to allow for the development and provision of iron particles, resulting in long-lasting contact between carbon and iron particles.

Figure 3a shows the precise development of iron particles on the surface of activated carbon. On the surface of activated carbon, the production of needle formations is uniformly distributed. The Fe-C (Figure 3b) iron particles are absorbed and agglomerated during the low-temperature (LT-FTS) reaction due to an increase in the Fe crystallite size. The Fe-C (Figure 3c) marginally alters the activation of iron oxides in a high-temperature (HT-FTS) reaction. This demonstrates that temperature has a significant impact on the structure and properties of the catalyst.

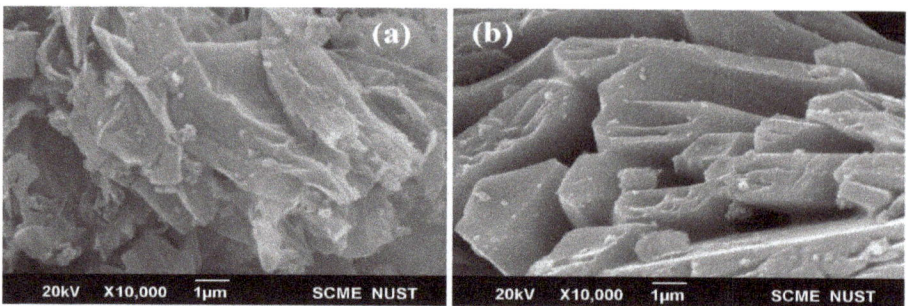

Figure 2. Scanning electron microscopy of prepared activated carbon (**a**,**b**) *Lantana Camara*.

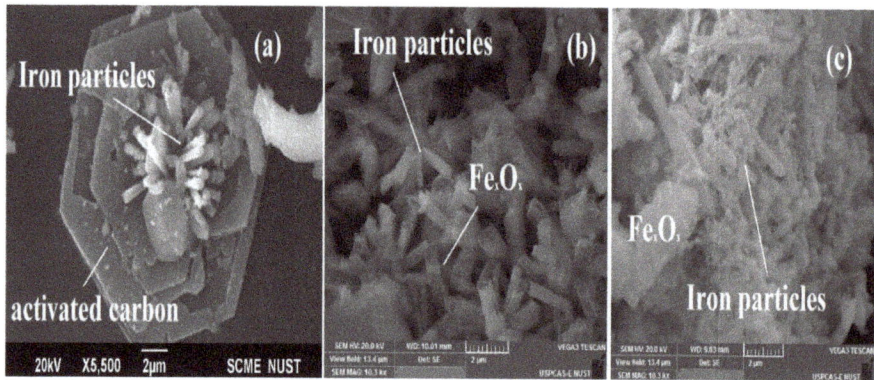

Figure 3. Scanning electron microscopy of Fe-C catalyst (**a**), Fe-C LT-FTS (**b**), and Fe-C HT-FTS (**c**).

In Figure 4a, the potassium particles are dispersed into the carbon surface to create the circular structure. This is because only 3% of the promoter was used. The needle and spherical Fe-C-K particles (Figure 4b) react inside the FT reactor to generate carbide particles when a low-temperature (LT-HTS) reaction is carried out. Figure 4c shows that Fe-C-K particles bonded each other to form active carbide phase at high temperatures (HT-FTS).

Figure 4. Scanning electron microscopy of Fe-C-K catalyst (**a**), Fe-C-K LT-FTS (**b**), and Fe-C-K HT-FTS (**c**).

2.3. Energy-Dispersive X-ray Analysis

The EDS results of the catalyst show that the increase in the carbon percentage in the catalysts are responsible for the development of the active carbide phase. Additionally, the presence of oxygen is responsible for the development of Fe_2O_3 and Fe_3O_4[26]. Table 1 shows the EDS analysis of the prepared catalysts.

Table 1. EDS elemental composition of FTS catalyst and activated carbon.

Compound Name	EDS Analysis (Wt %)				
	Phosphorus (%)	Carbon (%)	Oxygen (%)	Iron(%)	Potassium (%)
Activated Carbon	17.9	77.5	4.6	0.0	0.0
Fe-C	10.5	15.0	55.5	17.0	0.0
Fe-C (LT-FTS)	6.7	28.1	42.5	22.6	0.0
Fe-C (HT-FTS)	4.5	51.3	40.0	4.0	0.0
Fe-C-K	12.2	13.3	48.9	21.3	3.8
Fe-C-K (LT-FTS)	9.1	10.3	46.9	24.9	4.1
Fe-C-K (HT-FTS)	1.0	69.8	17.0	8.4	3.5

2.4. Thermogravimetric Analysis

Figure 5 shows the thermal stability of activated carbon (a), as well as the catalysts, Fe-C (b) and Fe-C-K (e). At a temperature of 110–250 °C, the water molecules are evaporated [17]. In Figure 5, the evaporation of water molecules occurred between 110–250 °C, as shown by Equations (1)–(4). At roughly 450 °C, partial thermal degradation of the carbon-supported iron-based catalyst with and without potassium promoters occurs at low and high temperatures, as illustrated in Figure 5b–d,f,g, and shown by Equations (5) and (6). Above 550 °C, however, full thermal damage occurs, as shown by Equations (7)–(10). The following steps describe the overall process of thermal degradation of the Fe-C catalyst with and without potassium promoter (Equations (1)–(10)). First, goethite decomposes (Equation (1)) between 200 and 250 degrees Celsius [27,28]. The shape of Fe_2O_3 (hematite) is then reduced to FeO. In addition, the reaction of iron oxide with CO at 250–300 °C produces Fe_2O_3(magnetite) (Equations (2)–(6)). In the XRD peaks, the production of magnetite was also plainly visible. The creation of CO_2 is linked to the reduction of iron oxides to generate the iron carbide carbonaceous material at temperatures above 350 °C, as shown by Equations (7)–(10) [29].

$$2FeO(OH) \rightarrow Fe_3O_3 + H_2O \quad (1)$$

$$3FeCO_3 \rightarrow Fe_3O_4 + 2CO_2 + CO \quad (2)$$

$$FeCO_3 \rightarrow FeO + CO_2 \quad (3)$$

$$FeC_2O_4 \cdot 2H_2O \rightarrow FeC_2O_4 + 2H_2O \quad (4)$$

$$3FeC_2O_4 \rightarrow Fe_3O_4 + 4CO + 2CO_2 \quad (5)$$

$$3Fe_2O_3 + CO \rightarrow 2Fe_3O_4 + CO_2 \quad (6)$$

$$6FeO(OH) + CO \rightarrow 2Fe_3O_4 + CO_2 + 3H_2O \quad (7)$$

$$FeC0_3 \rightarrow FeO + CO_2 \quad (8)$$

$$xFe_3O_4 + xCO \rightarrow 3Fe_xC + xCO_2 \quad (9)$$

$$xFeO + xCO \rightarrow Fe_xC + xCO_2 \quad (10)$$

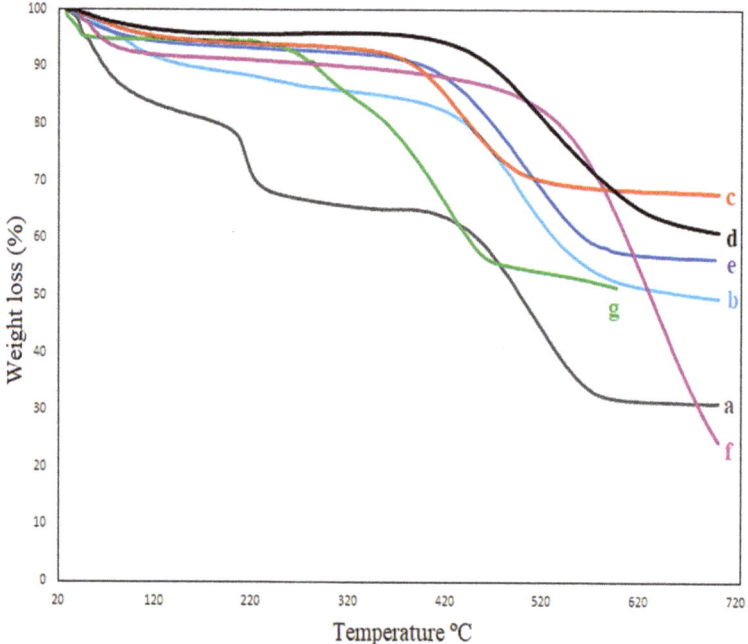

Figure 5. TGA analysis of activated carbon (**a**), Fe-C catalyst (**b**), Fe-C LT-FTS (**c**), Fe-C HT-FTS (**d**), Fe-C-K catalyst (**e**), Fe-C-K LT-FTS (**f**), and Fe-C-K HT-FTS (**g**).

2.5. Brunauer–Emmett–Teller Analysis

The BET surface area of activated carbon and the produced catalyst is shown in Table 2. The results reveal that the surface area of activated carbon is 167.4 m^2/g, and the surface area of the Fe-C catalyst is 78.7 m^2/g. The surface area of the potassium-based promoter catalyst was reduced by 11.1 m^2/g because its particles occluded tiny holes of the active sites with their own particles. The high surface area of activated carbon, as compared to the iron catalyst in the iron phase, is greatly disseminated over the surface, forming minute particles of iron oxides. When Fe_2O_3 is calcined, it can transform into Fe_3O_4. Surface carbon species may be able to aid this process by acting as an active support [30].

Table 2. BET surface area of activated carbon, Fe-C, and Fe-C-K catalyst.

Sample ID	BET Surface Area (m^2/g)
Activated Carbon	167.4
Fe-C	78.7
Fe-C-K	11.1

2.6. Effect of Temperature and Promoter on Downstream Syngas Applications

2.6.1. Effect of Promoter and Carbon Support

Table 3 shows that adding potassium promoter in an iron carbon-supported catalyst has a substantial impact on the conversion of gasoline, from 36.4 percent (Fe-C) to 72.5 percent (Fe-C-K). A low potassium concentration is thought to be beneficial for promoting the FTS reaction and converting iron carbides to the inactive iron oxide phase. Potassium offers a number of benefits, including increasing the quantity of alkenes in hydrocarbon synthesis, and inhibiting the creation of methane [31]. In general, using a potassium promoter in conjunction with an Fe/Ac-based catalyst reduces the quantity of n-paraffins in FTS products while significantly increasing branching paraffin [32]. When promoters are added to an iron-based catalyst, the catalyst's physical and chemical properties change, as well as the conversion and selectivity of CO_2 and C_{5+} [33]. The support of carbon helps for the formation of the iron-carbide phase. Due to the iron-carbide phase, the conversion rate of gasoline and diesel was increased. The Figure 6 shows the mechanism of a carbon-supported iron catalyst containing potassium as a promoter.

Table 3. Distribution of carbon chain with oxygenates and non-oxygenates.

Sample ID	Distribution of Carbon Chain (%)					Oxygenates (%)	Non-Oxygenates (%)
	C_2–C_4 Light Hydrocarbons	C_5–C_{11} Gasoline	C_{12}–C_{22} Diesel	>C_{23} Lubricating Oil	C_{8+}		
Fe-C (LT-FTS)	9.67	26.41	60.84	3.08	81.34	46.40	53.60
Fe-C-K (LT-FTS)	0.78	16.99	80.05	2.18	96.59	18.30	81.70
Fe-C (HT-FTS)	9.60	36.49	47.75	6.16	80.23	33.50	66.50
Fe-C-K (HT-FTS)	0.03	72.51	27.28	0.18	91.43	17.20	82.80

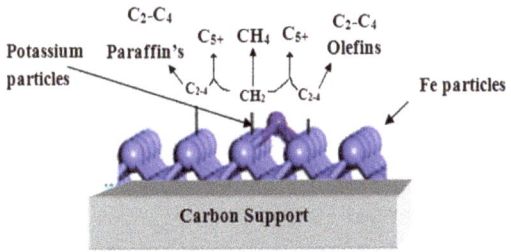

Figure 6. Mechanism of a carbon-supported iron catalyst containing potassium as a promoter.

Potassium was found to be a good promoter that enhanced the water–gas shift reaction, enhancing the selectivity toward olefins, and suppressing the formation of methane. Potassium promoter improves the performance of an iron catalyst by affecting the catalyst phase. In a catalyst, potassium acts as an electron donor when absorbed on the surface of an iron carbide-oxide phase. It inhibits the reduction from α-Fe_2O_3 to α-Fe_3O_4, and promotes CO dissociation. In general, a small amount of potassium as a promoter is sufficient to stimulate FTS activity [31–34]. In this study, the results revealed that the use of potassium as a promoter suppressed the formation of methane, and enhanced the C_{5+} formation.

2.6.2. Effect of Temperature

The active phase of iron-based catalysts is affected by the operating temperature of FTS. The iron oxide FeO becomes hematite Fe_2O_3, and then magnetite Fe_3O_4 when the temperature of the reactor rises [35].The iron particles of the promoted catalyst stabilize on the surface of the carbon matrix during the FTS reaction, preventing aggregation [36]. When using a carbon-supported iron catalyst without a promoter at higher temperatures, the number of non-oxygenates increased from 53% to 66%. (Fe-C). In a carbon-supported iron catalyst system with the promoter, the number of non-oxygenates increased from 81.7 percent to 82.8 percent at a higher temperature (Fe-C-K). Diesel generation increased from 60% to 80% at low temperatures (200 °C), with and without the potassium promoter. In the absence of a potassium promoter, however, the synthesis of gasoline increased from 36% to 72% at high temperatures (350 °C). The results of carbon chain distribution with and without oxygenates are shown in Table 3.

2.6.3. Carbon Chain Distribution

The FTS method may create C_5–C_{11} (gasoline) and C_{12}–C_{22} (diesel) hydrocarbon chains, making it a desirable transportation fuel. Various characteristics, such as the active phase and the quality of the support utilized, were used to evaluate the performance of FTS catalysts [37]. The FTS test was carried out with and without potassium promoters on a carbon-supported iron catalyst. After a reduction reaction at 450 degrees Celsius, water molecules were formed as a result of pre-adsorbed water molecules on the catalyst surface. When hydrogen and oxygen that have been adhered to the catalyst mix and begin to break down, they can produce water.

As a result, there were two main sources of adsorbed oxygen: first, polluted molecular oxygen from the gas phase, and second, the CO dissociation product. Fe_5C_2 was formed during the reduction reaction, which is considered the most active phase of the process [38,39]. Furthermore, the use of activated carbon increased the catalytically active phase's anchoring. The samples from the separators were collected in glass vials and sealed to prevent evaporation after the reaction. The lower white layer in Figure 7, which is oxygenated, and the higher brown layer, which contains C_{5+} hydrocarbons, were separated in the collected samples. Figure 8 depicts the results, which show various peaks and all potential chemicals included in the product. The overall percentage of different hydrocarbons in the mixture is shown in Table 3.

Figure 7. Sample collection.

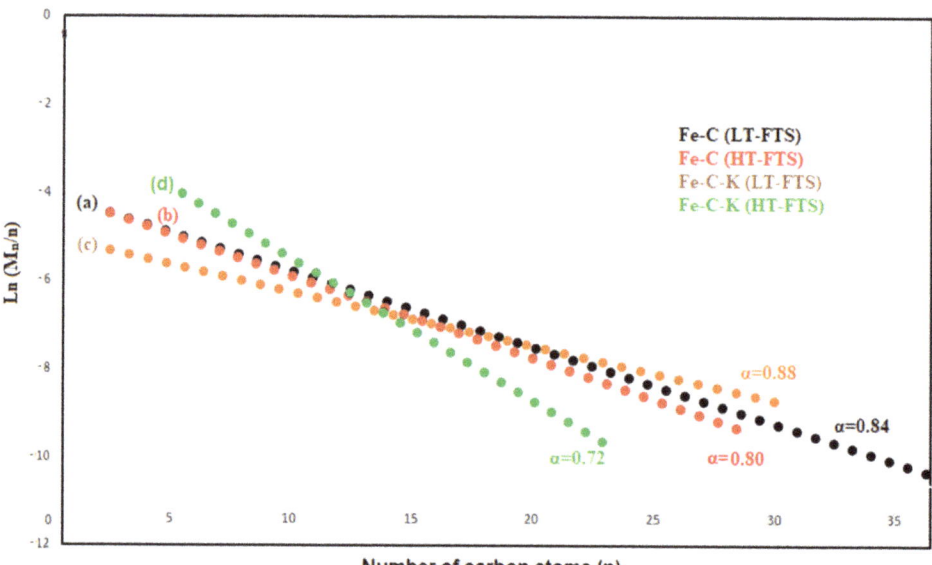

Figure 8. Product composition: (a) Fe-C (LT-FTS), (b) Fe-C (HT-FTS), (c) Fe-C-K (LT-FTS), and (d) Fe-C-K (HT-FTS).

2.6.4. P's Influence

Because phosphorus causes choking during the reaction in a fixed bed reactor, and affects reactor performance, it is not considered as good for the FTS process. The GCMS data revealed the creation of phosphorus derived during the FTS reaction ($C_7H_{18}NO_3P$, $C_8H_{18}NO_3P$, and $C_{18}H_{15}OP$). The yield of the FT product will increase if other chemical activation agents are utilized in the synthesis of activated carbon. For making activated carbon in the future, activation agents that are KOH, K_2CO_3, or $FeCl_3$ will be good. These activation agents will nothave an effect on the performance of a fixed bed reactor.

2.6.5. Distribution of Oxygen

FTS produces a significant amount of oxygenate. The amount of oxygenated compounds produced is determined by the catalyst utilized and the process parameters. The aqueous phase typically contains oxygenates in the range of 3–25 wt% [40]. According to researchers [15], increased metal loading on the activated carbon-supported Fe/K catalyst resulted in greater selectivity for oxygenates. Because Lantana Camara has a lot of sesquiterpene hydrocarbon essential oil components, the amount of oxygenates, such as $C_3H_4O_2$, $C_4H_8O_4$, C_6H_6O, $C_7H_{14}O$, $C_{10}H_{20}O$, $C_{16}H_{34}O$, and many more found in this study, was the highest (46%) out of all the plants.

According to Anderson–Schulz–Flory (ASF) Distribution, diesel is produced as the major component, with a value of roughly 0.9 using LTFT. Gasoline requires slightly lower values of 0.7–0.8 under HTFT circumstances [41].

$$\alpha = \left(0.2332 \cdot \left(\frac{yCo}{yCo + yH_2}\right) + 0.633\right) \cdot (1 - 0.0039 \cdot ((T(°C) + 273) - 573)) \quad (11)$$

The ASF model is illustrated in Equation (11) to determine the "α". The "α" is determined using the gradient of the linearized expression in the log Mn/n against the n plot, which is given as Equations (12) and (13) [42]. Figure 6 shows the computed value "α" using the ASF model.

$$\frac{Mn}{n} = (1-\alpha)^2 \cdot \alpha^{(n-1)} \quad (12)$$

$$\ln \alpha = n \ln \alpha + \ln\left[\frac{(1-\alpha)^2}{\alpha}\right] \quad (13)$$

Some of the recent studies about the Fe/AC catalyst demonstrate that the use of activated carbon (3–5 wt%) provides the majority of the C_1–C_6 hydrocarbon composition. In this study, as the weight percentage of activated carbons (10–15%) was increased, the formation of C_1–C_{20} hydrocarbons was the main product of FTs. Furthermore, the role of activated carbon in Fe/Ac-based catalysts has to improve the anchoring of the catalytically active phase.

The addition of carbon as a support provides a great advantage in terms of high surface area and thermal and chemical stability. The carbon support made from biomass would result in an additional advantage from an economic perspective. The carbon support has a remarkable feature which allows the bonding of extra heteroatoms on its surface. The carbon support in the FTS process seems to be beneficial for the formation of iron-carbide species [5]. Lantana Camara is composed of isocaryophyllene (16.7%), germacrene D (12.3%), bicyclogermacrene (19.4%), andvalecene (12.9%), and caryophyllene isomers were detected in *Lantana Camara's* essential oil composition [43]. Table 4 shows the carbon-supported iron-based catalyst results. Recently, some research revealed that the addition of carbon as support enhanced the C_{5+}. In this research work, the addition of carbon support made from Lantana Camara enhanced the formation of C_{5+} hydrocarbons by upto 90%.

Table 4. Hydrocarbon selectivity of different Fe-carbon-supported catalysts.

Sample Name	Temperature (°C)	Pressure (MPa)	H_2/CO Ratio	Hydrocarbon Selectivity		Ref
				C_{1-4}	C_{5+}	
Fe (carbon-based iron catalyst)	220	2	2:1	67.16	31.20	[28]
Fe/15C (carbon support)	340	2	1:1	61.0	39.0	[44]
Fe-AC (AC=activated carbon)	300	2	2:1	46.3	53.7	[8]
Fe/N-CNT-h (CNT=carbon nanotube)	275	0.8	2:1	39.1	60.9	[9]
Fe/AC (AC=activated carbon)	340	1	1:1	37.7	62.3	[23]
FeO_x (carbon-supported)	260	1	1:1	35.6	64.4	[5]
Fe-1200 (carbon-encapsulated)	300	2	1:1	31.2	68.8	[10]
Fe(4700)(MOF-based Fe-catalyst)	300	2	1:1	24.3	75.7	[39]
Fe/AC (activated carbon)	200	2	1:1	9.67	90.33	This work
Fe/AC (activated carbon)	350	2	1:1	9.60	90.40	This work

3. Materials and Method

3.1. Preparation of Activated Carbon

The leftover leaves of "*Lantana camara*" (24 g) were collected from SMME Department, NUST Islamabad, (latitude, 33.6360821; longitude, 72.9897282) Pakistan. The surfaces of the leaves were cleaned three times with distilled water to remove dust. Before being crushed, the leaves were dried out in the open atmosphere for 3 to 4 days. The dried leaves were crushed and sieved. After that, an H_3PO_4 (wt 50%) solution was used to activate the fine biomass powder. The initial concentration of the H_3PO_4 solution was 85%. The required H_3PO_4 volumes per (17.38 g) dry raw biomass material were 12.13 mL [45]. The slurry-based material was calcined in a box resistance furnace for two hours at 500–600 degrees Celsius. Then, the obtained char was rinsed with distilled water to remove extra H_3PO_4 from the char [46]. After that, it was dried for an hour at 110 °C to yield 4 g of activated carbon [45–47].

3.2. Catalyst Preparation

By incorporating 10% iron nitrate (3.2 g Fe $(NO_3)_3 \cdot 9H_2O$) and 4 g of "*Lantana Camara*"-activated carbon, the Fe-C sample was prepared. The sample was then calcined for three h at 350 °C in a box resistance furnace [48]. The prepared Fe-C catalyst was mixed with 3 wt% potassium carbonate (0.2675 g K_2CO_3) by adding 10 mL of deionized water dropwise and letting it sit at room temperature for a day to allow the potassium particles to properly soak into the Fe-C catalyst. In a box resistance furnace, the sample was calcined for three h at 400 °C [31]. The sample was maintained in an airtight sample bottle after cooling to room temperature. The procedures for making carbon-supported iron catalysts with and without potassium promoters are shown in Figure 9.

3.3. Catalyst Characterization

Various techniques were used to characterize the Fe-C and Fe-C-K catalysts. The surface area was examined using the Brunauer–Emmett–Teller method (SBET) (Micromeritics Gemini VII 2390t VI.03, Norcross, GA, USA). Energy dispersive X-ray analysis and a scanning electron microscope were used to assess changes in surface morphology (JEOL JSM 6490 LA, Akishima, Japan). The crystal structure was determined using the X-ray diffraction method (XRD-D8 advanced by Bruker, Bermen, Germany). The thermal deterioration was revealed by the TGA investigation (Model: DTG-60 H, TGA 5500 TA Instrument, New Castle, DE, USA).

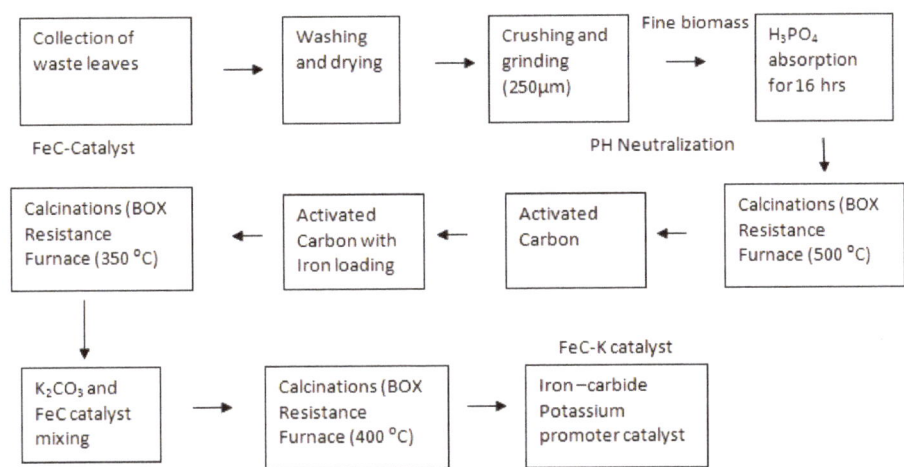

Figure 9. Synthesis of activated carbon-based catalyst by wet impregnation process.

3.4. Catalyst Performance Test

A Fischer–Tropsch synthesis (FTS) system with a fixed-bed reactor was used to assess the catalytic performance. In the reactor, one gram of the prepared catalyst (Fe-C or Fe-C-K) was loaded. The catalyst was reduced for 5 h in a fixed bed reactor at 450 °C with a hydrogen gas flow of 30 sccm and a nitrogen gas flow of 10 sccm. The reactor was cooled to 350 °C to promote the reaction at 20 bar in the presence of a H_2/CO (1:1 ratio). The pressure was controlled using a back-pressure regulator. A liquid product sample was taken from the two separators. Gas chromatography was used to evaluate the tail gases in real time (Shimadzu GC-2010 Plus, Kyoto, Japan). Gas chromatography–mass spectrometry was used to assess the liquid sample taken from the separator (GCMS-QP2020, Shimadzu Japan). The product selectivity was calculated using the peak area from the GCMS results. Figure 10 depicts the schematic diagram of the FTS plant that was employed in this experiment.

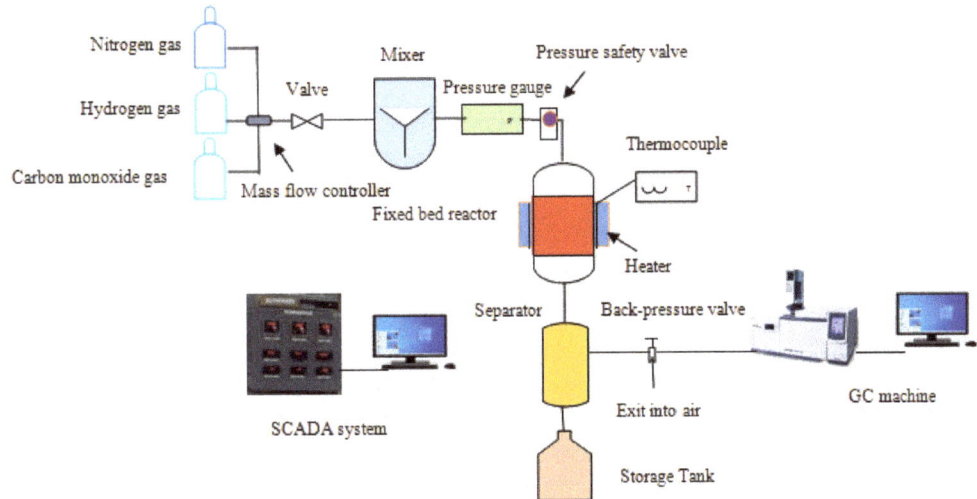

Figure 10. Schematic diagram of the FTS setup.

4. Conclusions

In this study, the effects of temperature and catalyst promoters on the formation of liquid hydrocarbons were studied. The Fe-C-K-supported catalyst with potassium promoter had 72 percent gasoline (C_5–C_{11}) selectivity at high temperatures. The GCMS results show the presence of C_{8+} hydrocarbons with selectivities of 96 percent and 91 percent when using a Fe-C-K catalyst at low and high temperatures. As a result, this research concludes that these catalysts may have a significant impact on the generation of gasoline and diesel in the future. Lantana Camara is composed of isocaryophyllene (16.7%), germacrene D (12.3%), bicyclogermacrene (19.4%), and valecene (12.9%), and caryophyllene isomers were detected in *Lantana Camara's* essential oil composition. Because of the oily nature of "Lantana Camara," the activated carbon utilized in the catalyst preparation had a significant impact on the increased hydrocarbon production. This research concludes that the use of activated carbon as a support made from Lantana Camara in FTS catalysts has a significant influence in improving hydrocarbon selectivity.

Author Contributions: Conceptualization, writing—original draft, formal analysis, M.A.; funding acquisition, S.M.W.; investigation, methodology, S.M.; project administration, resources, visualization, supervision, writing—review and editing, N.I., A.I. All authors have read and agreed to the published version of the manuscript.

Funding: Authors are grateful to the Researchers Supporting Project Number (RSP2022R448), King Saud University, Riyadh, Saudi Arabia.

Institutional Review Board Statement: Not applicable.

Informed Consent Statement: Not applicable.

Data Availability Statement: The data are presented in the present research article.

Acknowledgments: The work was conducted at the U.S.-Pakistan Center for Advanced Studies in Energy (USPCASE), NUST. The authors appreciate the technical support by the Energy Storage and Conservation lab staff and the Advanced Energy Materials lab. Authors are grateful to the Researchers Supporting Project Number (RSP2022R448), King Saud University, Riyadh, Saudi Arabia.

Conflicts of Interest: The authors declare no conflict of interest.

References

1. Sharif, M.F.; Arslan, M.; Iqbal, N.; Ahmad, N.; Noor, T. Development of hydrotalcite based cobalt catalyst by hydrothermal and co-precipitation method for Fischer-Tropsch synthesis. *Bull. Chem. React. Eng. Catal.* **2017**, *12*, 357–363. [CrossRef]
2. Amin, M.; Shah, H.H.; Fareed, A.G.; Khan, W.U.; Chung, E.; Zia, A.; Farooqi, Z.U.R.; Lee, C. Science Direct Hydrogen production through renewable and non- renewable energy processes and their impact on climate change. *Int. J. Hydrogen Energy* **2022**, *47*, 33112–33134. [CrossRef]
3. Ma, G.; Wang, X.; Xu, Y.; Wang, Q.; Wang, J.; Lin, J.; Wang, H.; Dong, C.; Zhang, C.; Ding, M. Enhanced Conversion of Syngas to Gasoline-Range Hydrocarbons over Carbon Encapsulated Bimetallic FeMn Nanoparticles. *ACS Appl. Energy Mater.* **2018**, *1*, 4304–4312. [CrossRef]
4. Martínez-Vargas, D.X.; Sandoval-Rangel, L.; Campuzano-Calderon, O.; Romero-Flores, M.; Lozano, F.J.; Nigam, K.D.P.; Mendoza, A.; Montesinos-Castellanos, A. Recent Advances in Bifunctional Catalysts for the Fischer-Tropsch Process: One-Stage Production of Liquid Hydrocarbons from Syngas. *Ind. Eng. Chem. Res.* **2019**, *58*, 15872–15901. [CrossRef]
5. Dos Santos, R.G.; Alencar, A.C. Biomass-derived syngas production via gasification process and its catalytic conversion into fuels by Fischer Tropsch synthesis: A review. *Int. J. Hydrogen Energy* **2020**, *45*, 18114–18132. [CrossRef]
6. Li, Y.; Lu, W.; Zhao, Z.; Zhao, M.; Lyu, Y.; Gong, L.; Zhu, H.; Ding, Y. Tuning surface oxygen group concentration of carbon supports to promote Fischer-Tropsch synthesis. *Appl. Catal. A Gen.* **2021**, *613*, 118017. [CrossRef]
7. Xiong, H.; Jewell, L.L.; Coville, N.J. Shaped Carbons As Supports for the Catalytic Conversion of Syngas to Clean Fuels. *ACS Catal.* **2015**, *5*, 2640–2658. [CrossRef]
8. Valero-Romero, M.J.; Rodríguez-Cano, M.Á.; Palomo, J.; Rodríguez-Mirasol, J.; Cordero, T. Carbon-Based Materials as Catalyst Supports for Fischer–Tropsch Synthesis: A Review. *Front. Mater.* **2021**, *7*, 617432. [CrossRef]
9. Xiong, H.; Motchelaho, M.A.; Moyo, M.; Jewell, L.L.; Coville, N.J. Fischer-Tropsch synthesis: Iron-based catalysts supported on nitrogen-doped carbon nanotubes synthesized by post-doping. *Appl. Catal. A Gen.* **2014**, *482*, 377–386. [CrossRef]

10. Chernyak, S.A.; Ivanov, A.S.; Maksimov, S.V.; Maslakov, K.I.; Isaikina, O.Y.; Chernavskii, P.A.; Kazantsev, R.V.; Eliseev, O.L.; Savilov, S.S. Fischer-Tropsch synthesis over carbon-encapsulated cobalt and iron nanoparticles embedded in 3D-framework of carbon nanotubes. *J. Catal.* **2020**, *389*, 270–284. [CrossRef]
11. Barrios, A.J.; Gu, B.; Luo, Y.; Peron, D.V.; Chernavskii, P.A.; Virginie, M.; Wojcieszak, R.; Thybaut, J.W.; Ordomsky, V.V.; Khodakov, A.Y. Identification of efficient promoters and selectivity trends in high temperature Fischer-Tropsch synthesis over supported iron catalysts. *Appl. Catal. B Environ.* **2020**, *273*, 119028. [CrossRef]
12. Gerber, I.C.; Serp, P. A Theory/Experience Description of Support Effects in Carbon-Supported Catalysts. *Chem. Rev.* **2020**, *120*, 1250–1349. [CrossRef] [PubMed]
13. Yan, L.; Liu, J.; Wang, X.; Ma, C.; Zhang, C.; Wang, H.; Wei, Y.; Wen, X.; Yang, Y.; Li, Y. Ru catalysts supported by Si3N4 for Fischer-Tropsch synthesis. *Appl. Surf. Sci.* **2020**, *526*, 146631. [CrossRef]
14. Liu, G.; Chen, Q.; Oyunkhand, E.; Ding, S.; Yamane, N.; Yang, G.; Yoneyama, Y.; Tsubaki, N. Nitrogen-rich mesoporous carbon supported iron catalyst with superior activity for Fischer-Tropsch synthesis. *Carbon N. Y.* **2018**, *130*, 304–314. [CrossRef]
15. Chun, D.H.; Rhim, G.B.; Youn, M.H.; Deviana, D.; Lee, J.E.; Park, J.C.; Jeong, H. Brief Review of Precipitated Iron-Based Catalysts for Low-Temperature Fischer–Tropsch Synthesis. *Top. Catal.* **2020**, *63*, 793–809. [CrossRef]
16. Xu, J.; Chen, L.; Qu, H.; Jiao, Y.; Xie, J.; Xing, G. Preparation and characterization of activated carbon from reedy grass leaves by chemical activation with H3PO4. *Appl. Surf. Sci.* **2014**, *320*, 674–680. [CrossRef]
17. Liu, Y.; Yao, X.; Wang, Z.; Li, H.; Shen, X.; Yao, Z.; Qian, F. Synthesis of Activated Carbon from Citric Acid Residue by Phosphoric Acid Activation for the Removal of Chemical Oxygen Demand from Sugar-Containing Wastewater. *Environ. Eng. Sci.* **2019**, *36*, 656–666. [CrossRef]
18. Gao, Y.; Yue, Q.; Gao, B.; Li, A. Insight into activated carbon from different kinds of chemical activating agents: A review. *Sci. Total Environ.* **2020**, *746*, 141094. [CrossRef]
19. Teng, X.; Huang, S.; Wang, J.; Wang, H.; Zhao, Q.; Yuan, Y.; Ma, X. Fabrication of Fe2C Embedded in Hollow Carbon Spheres: A High-Performance and Stable Catalyst for Fischer-Tropsch Synthesis. *ChemCatChem* **2018**, *10*, 3883–3891. [CrossRef]
20. Yang, X.; Zhang, H.; Liu, Y.; Ning, W.; Han, W.; Liu, H.; Huo, C. Preparation of iron carbides formed by iron oxalate carburization for fischer-tropsch synthesis. *Catalysts* **2019**, *9*, 347. [CrossRef]
21. Chun, D.H.; Park, J.C.; Hong, S.Y.; Lim, J.T.; Kim, C.S.; Lee, H.; Yang, J.; Hong, S.J.; Jung, H. Highly selective iron-based Fischer-Tropsch catalysts activated by CO 2-containing syngas. *J. Catal.* **2014**, *317*, 135–143. [CrossRef]
22. Lu, Y.; Yan, Q.; Han, J.; Cao, B.; Street, J.; Yu, F. Fischer–Tropsch synthesis of olefin-rich liquid hydrocarbons from biomass-derived syngas over carbon-encapsulated iron carbide/iron nanoparticles catalyst. *Fuel* **2017**, *193*, 369–384. [CrossRef]
23. Zhao, Q.; Huang, S.; Han, X.; Chen, J.; Wang, J.; Rykov, A.; Wang, Y.; Wang, M.; Lv, J.; Ma, X. Highly active and controllable MOF-derived carbon nanosheets supported iron catalysts for Fischer-Tropsch synthesis. *Carbon N. Y.* **2021**, *173*, 364–375. [CrossRef]
24. Tang, L.; He, L.; Wang, Y.; Chen, B.; Xu, W.; Duan, X.; Lu, A. Selective fabrication of χ-Fe5C2 by interfering surface reactions as a highly efficient and stable Fischer-Tropsch synthesis catalyst. *Appl. Catal. B Environ.* **2021**, *284*, 119753. [CrossRef]
25. Lyu, S.; Wang, L.; Li, Z.; Yin, S.; Chen, J.; Zhang, Y.; Li, J.; Wang, Y. Stabilization of ε-iron carbide as high-temperature catalyst under realistic Fischer–Tropsch synthesis conditions. *Nat. Commun.* **2020**, *11*, 1–8. [CrossRef]
26. Sun, B.; Xu, K.; Nguyen, L.; Qiao, M.; Tao, F.F. Preparation and Catalysis of Carbon-Supported Iron Catalysts for Fischer-Tropsch Synthesis. *ChemCatChem* **2012**, *4*, 1498–1511. [CrossRef]
27. Einemann, M.; Neumann, J.; Thomé, A.G.; Wabo, S.G.; Roessner, F. Quantitative study of the oxidation state of iron-based catalysts by inverse temperature-programmed reduction and its consequences for catalyst activation and performance in Fischer-Tropsch reaction. *Appl. Catal. A Gen.* **2020**, *602*, 117718. [CrossRef]
28. Shakeri, J.; Joshaghani, M.; Hadadzadeh, H.; Shaterzadeh, M.J. Methane carbonylation to light olefins and alcohols over carbon–based iron–and cobalt–oxide catalysts. *J. Taiwan Inst. Chem. Eng.* **2021**, *122*, 127–135. [CrossRef]
29. Pérez, S.; Mondragón, F.; Moreno, A. Iron ore as precursor for preparation of highly active χ-Fe5C2 core-shell catalyst for Fischer-Tropsch synthesis. *Appl. Catal. A Gen.* **2019**, *587*, 117264. [CrossRef]
30. Jiang, F.; Zhang, M.; Liu, B.; Xu, Y.; Liu, X. Insights into the influence of support and potassium or sulfur promoter on iron-based Fischer-Tropsch synthesis: Understanding the control of catalytic activity, selectivity to lower olefins, and catalyst deactivation. *Catal. Sci. Technol.* **2017**, *7*, 1245–1265. [CrossRef]
31. Niu, L.; Liu, X.; Wen, Y.; Yang, Y.; Xu, J.; Li, Y. Effect of potassium promoter on phase transformation during H2 pretreatment of a Fe2O3 Fischer Tropsch synthesis catalyst precursor. *Catal. Today* **2020**, *343*, 101–111. [CrossRef]
32. Chen, Y.; Wei, J.; Duyar, M.S.; Ordomsky, V.V.; Khodakov, A.Y.; Liu, J. Carbon-based catalysts for Fischer-Tropsch synthesis. *Chem. Soc. Rev.* **2021**, *50*, 2337–2366. [CrossRef]
33. Witoon, T.; Chaipraditgul, N.; Numpilai, T.; Lapkeatseree, V.; Ayodele, B.V.; Cheng, C.K.; Siri-Nguan, N.; Sornchamni, T.; Limtrakul, J. Highly active Fe-Co-Zn/K-Al2O3 catalysts for CO2 hydrogenation to light olefins. *Chem. Eng. Sci.* **2021**, *233*, 116428. [CrossRef]
34. Amoyal, M.; Vidruk-Nehemya, R.; Landau, M.V.; Herskowitz, M. Effect of potassium on the active phases of Fe catalysts for carbon dioxide conversion to liquid fuels through hydrogenation. *J. Catal.* **2017**, *348*, 29–39. [CrossRef]
35. Di, Z.; Feng, X.; Yang, Z.; Luo, M. Effect of Iron Precursor on Catalytic Performance of Precipitated Iron Catalyst for Fischer–Tropsch Synthesis Reaction. *Catal. Letters* **2020**, *150*, 2640–2647. [CrossRef]

36. Wang, A.; Luo, M.; Lü, B.; Song, Y.; Li, M.; Yang, Z. Effect of Na, Cu and Ru on metal-organic framework-derived porous carbon supported iron catalyst for Fischer-Tropsch synthesis. *Mol. Catal.* **2021**, *509*, 111601. [CrossRef]
37. Lu, F.; Chen, X.; Lei, Z.; Wen, L.; Zhang, Y. Revealing the activity of different iron carbides for Fischer-Tropsch synthesis. *Appl. Catal. B Environ.* **2021**, *281*, 119521. [CrossRef]
38. Zhao, H.; Liu, J.; Yang, C.; Yao, S.; Su, H.; Gao, Z.; Dong, M.; Wang, J.; Rykov, A.I.; Wang, J.; et al. Synthesis of iron-carbide nanoparticles: Identification of the active phase and mechanism of fe-based fischer-tropsch synthesis. *CCS Chem.* **2021**, *3*, 2712–2724. [CrossRef]
39. Cho, J.M.; Kim, B.G.; Han, G.Y.; Sun, J.; Jeong, H.K.; Bae, J.W. Effects of metal-organic framework-derived iron carbide phases for CO hydrogenation activity to hydrocarbons. *Fuel* **2020**, *281*, 118779. [CrossRef]
40. Ma, W.; Kugler, E.L.; Dadyburjor, D.B. Promotional effect of copper on activity and selectivity to hydrocarbons and oxygenates for Fischer-Tropsch synthesis over potassium-promoted iron catalysts supported on activated carbon. *Energy Fuels* **2011**, *25*, 1931–1938. [CrossRef]
41. Claeys, M.; van Steen, E. Basic studies. *Stud. Surf. Sci. Catal.* **2004**, *152*, 601–680. [CrossRef]
42. Aluha, J.; Hu, Y.; Abatzoglou, N. Effect of CO concentration on the α-value of plasma-synthesized Co/C catalyst in Fischer-Tropsch synthesis. *Catalysts* **2017**, *7*, 69. [CrossRef]
43. Sousa, E.O.; Colares, A.V.; Rodrigues, F.F.G.; Campos, A.R.; Lima, S.G.; Costa, J.M.G. Effect of Collection Time on Essential Oil Composition of Lantana camara Linn (Verbenaceae) Growing in Brazil Northeastern. 2010. Available online: http://www.acgpubs.org/RNP (accessed on 5 October 2022).
44. Tafjord, J.; Rytter, E.; Holmen, A.; Myrstad, R.; Svenum, I.; Christensen, B.E.; Yang, J. Transition-Metal Nanoparticle Catalysts Anchored on Carbon Supports via Short-Chain Alginate Linkers. *ACS Appl. Nano Mater.* **2021**, *4*, 3900–3910. [CrossRef]
45. Amin, M.; Shah, H.H.; Iqbal, A.; Farooqi, Z.U.R.; Krawczuk, M.; Zia, A. Conversion of Waste Biomass into Activated Carbon and Evaluation of Environmental Consequences Using Life Cycle Assessment. *Appl. Sci.* **2022**, *12*, 5741. [CrossRef]
46. Kumar, A.; Jena, H.M. Preparation and characterization of high surface area activated carbon from Fox nut (Euryale ferox) shell by chemical activation with H3PO4. *Results Phys.* **2016**, *6*, 651–658. [CrossRef]
47. Amin, M.; Chung, E.; Shah, H.H. Effect of different activation agents for activated carbon preparation through characterization and life cycle assessment. *Int. J. Environ. Sci. Technol.* **2022**, 1–12. [CrossRef]
48. Zhao, X.; Lv, S.; Wang, L.; Li, L.; Wang, G.; Zhang, Y.; Li, J. Comparison of preparation methods of iron-based catalysts for enhancing Fischer-Tropsch synthesis performance. *Mol. Catal.* **2018**, *449*, 99–105. [CrossRef]

Article

Hydrogenation of CO_2 on Nanostructured Cu/FeO$_x$ Catalysts: The Effect of Morphology and Cu Load on Selectivity

Karolína Simkovičová [1,2], Muhammad I. Qadir [1], Naděžda Žilková [1], Joanna E. Olszówka [1], Pavel Sialini [3], Libor Kvítek [2,*] and Štefan Vajda [1,*]

[1] Department of Nanocatalysis, J. Heyrovský Institute of Physical Chemistry v.v.i., Czech Academy of Sciences, Dolejškova 2155/3, 18223 Prague, Czech Republic; karolina.simkovicova@jh-inst.cas.cz (K.S.); muhammad.qadir@jh-inst.cas.cz (M.I.Q.); nadezda.zilkova@jh-inst.cas.cz (N.Ž.); joanna.olszowka@jh-inst.cas.cz (J.E.O.)
[2] Department of Physical Chemistry, Faculty of Science, Palacký University Olomouc, 17. listopadu 12, 77900 Olomouc, Czech Republic
[3] Laboratory of Surface Analysis, University of Chemistry and Technology, Technická 3, 16628 Prague, Czech Republic; sialinip@vscht.cz
* Correspondence: libor.kvitek@upol.cz (L.K.); stefan.vajda@jh-inst.cas.cz (Š.V.)

Abstract: The aim of this work was to study the influence of copper content and particle morphology on the performance of Cu/FeO$_x$ catalysts in the gas-phase conversion of CO_2 with hydrogen. All four investigated catalysts with a copper content between 0 and 5 wt% were found highly efficient, with CO_2 conversion reaching 36.8%, and their selectivity towards C_1 versus C_2-C_4, C_2-C_4=, and C_{5+} products was dependent on catalyst composition, morphology, and temperature. The observed range of products is different from those observed for catalysts with similar composition but synthesized using other precursors and chemistries, which yield different morphologies. The findings presented in this paper indicate potential new ways of tuning the morphology and composition of iron-oxide-based particles, ultimately yielding catalyst compositions and morphologies with variable catalytic performances.

Keywords: heterogeneous catalysis; CO_2 hydrogenation; CO_2 conversion; methane; hydrocarbons; iron oxide; copper nanoparticles

1. Introduction

Ever since the industrial revolution, human activity has emitted massive amounts of CO_2 into the atmosphere by burning fossil fuels. For the last two decades, the annual emissions of CO_2 increased to nearly 37 billion tons. The current global CO_2 concentration in the atmosphere exceeded 415 ppm, which is expected to rise to 500 ppm by the end of 2030 [1,2]. In order to mitigate global warming, a 70–80% reduction in CO_2 production should be reached by 2050. This, of course, creates a major challenge for modern science due to the fundamental contribution of CO_2 to global warming via the greenhouse effect [3]. Hence, one of the research topics of modern science is CO_2 capture and/or its conversion to hydrocarbons [4,5], which can be utilized as energy sources or precursors in the chemical industry. In particular, CO_2 can be used as a feedstock in many organic reactions, catalytically converting CO_2 into alcohols (methanol); hydrocarbons, such as methane, ethane, or benzene; CO or carbonates; and even derivates of hydrocarbons (e.g., carboxylic acids, aldehydes, amides, and esters) [6–11]. The topic of CO_2 hydrogenation has been researched intensively for the last decades, however, there is still plenty of room to develop new routes toward catalysts with high conversion, durability, and desired selectivity. The synthesis of C1 products by CO_2 hydrogenation may follow three different paths, according to the open literature. One is a process known as reverse water gas shift (RWGS), the second is called the formate pathway, and the third pathway is the direct C–O split of CO_2 [12]. The highly positive Gibbs free energy change for CO_2 conversion to CH_4 (i.e., 1135 kJ mol^{-1}) makes the CO_2 methanation reaction thermodynamically adverse [13], and the extremely stable and unreactive nature of CO_2 molecules, its conversion to value-added fuels, is a challenging problem that requires high energy input [14,15]. Currently, the process in which CO_2 and H_2 are thermally activated on the catalyst's surface, resulting in the formation of hydrocarbons, seems to be extremely interesting for industrial implementations. During this process, CO_2 and H_2 are first transformed into CO and water (RWGS), then CO can enter a reaction with excess H_2 to generate hydrocarbons through the Fisher–Tropsch (FT) synthesis process [16]. CO_2 hydrogenation usually produces lower molecular weight hydrocarbons [16], and the conversion rate and selectivity to desired products depend on catalyst composition as one of the factors which determine performance. Iron is one of the most studied catalysts for both FT and CO_2 hydrogenation, as it can adsorb and activate CO_2 [17], which is a prerequisite for the conversion of CO_2 to short-chain olefins, thanks to its intrinsic RWGS and FT activity. It is accepted that during CO_2 hydrogenation, highly active Fe species (Fe(0) and Fe_5C_2) are generated, which activate the formed CO to subsequently undergo sequential hydrogenation steps with the generation of CH, CH_2, and CH_3 reactive intermediates that can polymerize to higher hydrocarbons or are fully hydrogenated to form methane [18,19] Fe catalysts with alkali metal promoters for CO_2 conversion to light olefins were reported in several papers; these catalysts typically require operating temperatures over 300 °C and pre-treatment under hydrogen or a CO atmosphere for over 12 h [20–29]. Bimetallic catalysts, such as Co-Fe [30], were reported with improved activity towards the production of methane in CO_2 hydrogenation. Fe catalysts with Cu as promoters have also been reported, possessing low selectivity for light olefins while dominantly producing methane [31,32]. Combinations of Cu and Fe catalysts have already been reported with improved selectivity to olefins [33,34] and copper-based catalysts were investigated for their RWGS performance related to CO_2 activation [35]. Numerous Cu-based catalysts have been reported with high selectivity toward methanol formation [36–46], including copper tetramer (Cu_4) clusters [47]. Copper nanoparticles have been reported to be able to generate C_2-C_3 products with high selectivity [48]. In this paper, we focus on the design of efficient catalysts for CO_2 hydrogenation based on copper-iron oxide (Cu/FeO_x) to leverage both Cu's and FeO_x's inherent abilities, with Cu serving as the RWGS catalyst and hydrogen activator and FeO_x yielding hydrocarbons by FT.

2. Experimental Section

All chemicals, oxalic acid ($C_2H_2O_4$), N,N-dimethylacetamide (C_4H_9NO, anhydrous), iron (II) chloride tetrahydrate ($FeCl_4 \cdot 4H_2O$), copper sulfate pentahydrate ($CuSO_4 \cdot 5H_2O$) and hydrazine hydrate ($N_2H_4 \cdot H_2O$) were purchased from Sigma Aldrich; the gases used: CO_2 (99.99%), H_2 (99.99%) and He (99.99%) were acquired from Airgas. Deionized water (purity 0.05 µS·cm^{-1}, AQUAL 29, Merci) was used for the preparation of the solutions for the synthesis of the catalysts. A Sonicator SONOPULS HD 4400 Ultrasonic homogenizer and an Eppendorf Centrifuge 5702 were used for mixing the solution and for improving the dispersion of FeOx in it during synthesis, and for the separation of the solid products, respectively.

2.1. Preparation of the Catalysts

The FeO_x and Cu/FeO_x catalysts were fabricated using the wet impregnation method [37]. To prepare FeO_x, 1 mmol of oxalic acid was dissolved in 10 mL of N,N-dimethylacetamide. Then, a solution of 1 mmol of iron (II) chloride in 12 mL of deionized water was added at room temperature. The reaction was completed after 5 min, and iron (II) oxalate was separated by centrifugation, washed with deionized water and ethanol, and dried in a vacuum at 60 °C for 2 h. The obtained yellow powder of iron (II) oxalate was spread in a crucible in a thin layer and treated at the temperature of 175 °C in air for 12 h to obtain FeO_x [49,50].

Cu/FeO$_x$ were prepared as follows. Typically, 1 g of the already prepared FeO$_x$ was dispersed in 188 mL of deionized water. Then, a certain volume of 15.7 mmol/L of an aqueous solution of copper sulfate pentahydrate, calculated to the desired final load of Cu (2.55 mL for 1 wt%, 7.65 mL for 3 wt% and 12.75 mL for 5 wt%) was added. After 10 min of sonication, 50 mL of 4.95 mmol/L of the solution of hydrazine hydrate was poured into the reaction mixture and was sonicated for an additional 10 min. The resulting reddish-brown solid was isolated by centrifugation, washed with water and ethanol, and dried in a flow box under an inert nitrogen atmosphere at room temperature for 12 h.

The reference copper-free catalyst (FeO$_x$) was prepared using the same procedure, however, instead of using a copper sulfate solution in the first synthesis step, deionized water was added to the solution. The samples, according to their nominal Cu content of 1%, 3%, and 5%, are named as 1%-Cu/FeO$_x$, 3%-Cu/FeO$_x$, and 5%-Cu/FeO$_x$, respectively; the pure iron oxide sample is named as FeO$_x$ thorough the manuscript.

2.2. Characterization of the Catalysts

Transmission electron microscopy (TEM) characterization was conducted with a JEOL TEM-2100 multipurpose electron microscope. For scanning electron microscopy (SEM) analysis, a HITACHI SU6600 scanning electron microscope was used. Scanning transmission spectroscopy with high-angle annular dark-field scanning transmission electron microscopy (STEM/HAADF), equipped with an energy-dispersive X-ray spectroscope (EDX), was used for the elemental mapping and obtaining the STEM/HAADF images of the prepared catalysts.

Powder X-ray diffraction (XRD) analysis was performed in a Malvern Panalytical Empyrean diffractometer and the quantification of the individual Fe components in the fresh and used catalysts was completed by using Rietveld analysis of the obtained XRD data, with the High Score Plus (Malvern Panalytical) software utilizing the PDF-4+ and ICSD databases.

Atomic absorption spectroscopy (AAS) was conducted with an Analytik Jena AG ContrAA 300 spectrometer with flame ionization. The sorption of gas was measured on the surface area analyzer Autosorb iQ-C-MP (Quantachrome Anton Paar), using ASiQWin software at 77 K up to the saturation pressure of N$_2$. For calculating the specific surface area, the Brunauer-Emmett-Teller (BET) model was used. The points for multipoint BET were determined using Roquerol's method and are within the standard range of $p/p_0 = 0.05$ to 0.3. Before the surface analysis, all catalysts were treated at 130 °C for 12 h under vacuum.

X-ray photoelectron spectroscopy (XPS) was conducted on a PHI 5000 VersaProbe II spectrometer with monochromatic AlK$_\alpha$ radiation.

UV-visible spectroscopy was performed using a UV-Vis-NIR spectrometer Perkin-Elmer Lambda 950.

2.3. Catalytic Testing

Tests of the catalysts were performed in a Microactivity Reactor System, (PID Eng&Tech/Micromeritics) coupled to a quartz tube reactor of 320 mm in length and 10 mm inner diameter. A total of 200 mg of a catalyst was placed onto 20 mg of quartz wool in the middle of the reactor and the reactor was conditioned at 250 °C in He with a flow of 30 mL/min for 40 min. No other pre-treatment of the catalysts was performed. The reaction mixture used contained CO$_2$, H$_2$, and He at the ratio of 1:5:4 giving 11% and 49% of CO$_2$ and H$_2$ in He, respectively. A total flow of 25 mL/min was used at a pressure of 1 bar. The reaction products were analyzed on an Agilent 6890 gas chromatograph equipped with TCD (column HP-PLOT/Q) and FID (column Al$_2$O$_3$/KCl) detectors, using an injection after 20 min of reaching the given temperature. In the range of 250 °C to 410 °C, the temperature was raised in steps of 30 °C, at a rate of 5 °C/min (see Figure S1 for the double temperature ramp applied). After reaching the highest temperature in the first ramp, the reactor was cooled down to 250 °C under He. Next, the catalytic test was repeated using an identical heating ramp as the first one.

3. Results and Discussion

SEM micrographs of the as-prepared catalysts are shown in Figure 1, revealing two distinctly different morphologies. FeO$_x$ (Figure 1a) and 1%-Cu/FeO$_x$ (Figure 1b) show up as rods, with up to about 5 μm in length and thinner (ca. 500 nm) 1%-Cu/FeO$_x$ than for FeO$_x$ (ca. 1.3 μm), both with a coarse surface structure. The morphology of 3%-Cu/FeO$_x$ and 5%-Cu/FeO$_x$ (Figure 1c,d, respectively) is very different, reminding us of a structure of sponge or wool. We hypothesize that the observed differences in morphology reflect the differences in the composition of the reaction mixtures used in the synthesis of this set of catalysts. One parameter is the presence of hydrazine, which has reducing and basic properties; the other cause affecting morphology is the variable concentration of $SO_4{}^{2-}$, which is introduced to the system by adding CuSO$_4$ solution in increasing amounts with a growing Cu loading. With CuSO$_4$ being acidic, the low pH could impact the level of etching of FeO$_x$, [51] as well as the chemical composition.

Figure 1. SEM images of fresh (**a**) FeO$_x$, (**b**) 1%-Cu/FeO$_x$, and (**c**) 3%-Cu/FeO$_x$, (**d**) 5%-Cu/FeO$_x$. For additional images, see Figure S9.

SEM images with EDX analysis of the highlighted part are depicted in Figure S2. These images show the elemental composition on the surface of the fresh catalysts. The chemical composition and elemental distribution of FeO$_x$ and Cu/FeO$_x$ catalysts were characterized by EDX mapping and their EDX spectra. The STEM/HAADF images are depicted for FeOx in Figure S3a, 1%-Cu/FeOx in Figure S4a, 3%-Cu/FeOx in Figure S5a, and 5%-Cu/FeOx in Figure S6a. The images are depicting the same nanorod morphology as shown by the TEM images. As shown, the two dominant elements observed were Fe and O, which were distributed evenly in the catalysts. EDX elemental mapping is shown in Figures S3–S6. Elemental maps, including copper, iron, and carbon, shown in Figures S4–S6, for the copper-containing catalysts, present uniformly dispersed copper in these catalysts. The detected copper particulates appear to be around 2 nm in diameter. In these maps, carbon is shown to be present as the common contamination from exposing the samples to air or detected on the sample grid.

The effect of the composition of the synthesis solution on the composition of the iron component was also confirmed by XRD (Figure S7), with compositions listed for the individual catalysts in Table 1.

Table 1. Fractions of various forms of iron oxide components in the fresh catalysts, obtained from XRD data using Rietveld refinement of analysis and specific surface area of fresh catalysts, as determined from adsorption/desorption of nitrogen on the surface of the fresh catalysts. Multipoint BET was assessed using Roquerol's method.

Catalyst	Iron Oxide Composition [a]			Specific Surface Area	Actual Cu load [b]
	α-Fe$_2$O$_3$ [%]	α-FeO(OH) [%]	Fe$_3$O$_4$ [%]	[m^2/g]	[%]
FeO$_x$	43.5	54.4	2.1	114	0
1%-Cu/FeO$_x$	81.2	16.8		114	0.7
3%-Cu/FeO$_x$	43.6	56.4		93	2.7
5%-Cu/FeO$_x$	48.3	51.7		92	5.3

[a] Obtained from XRD. [b] Obtained from atomic absorption spectroscopy.

It is important to note, that the mean size of coherent domains obtained from XRD data can underestimate the particle size, because it does not take into account the possibly present non-crystalline surface layer, and the mean X-ray coherence length (MLC) in the case of a multiphase system can be less accurate as well. The data are summarized in Table S1. FeO$_x$ consists of a coherent domain length of 9 nm for α-Fe$_2$O$_3$, 12 nm for α-FeO(OH), and 11 nm for Fe$_3$O$_4$. Whereas, 1%-Cu/FeO$_x$ has 12 nm for α-Fe$_2$O$_3$ and 11 nm for α-FeO(OH) and (not in the table) 56 nm for the copper phase. The 3%-Cu/FeO$_x$ catalyst has 19 nm for α-Fe$_2$O$_3$ and 9 nm for α-FeO(OH). The 5%-Cu/FeO$_x$ catalyst has 19 nm α-Fe$_2$O$_3$ and 10 nm α-FeO(OH). Diffraction peaks of copper were detected only in 1%-Cu/FeO$_x$, at 50.3°, 58.8°, and 87.9°, due to too small a particle size and/or their amorphicity in the other samples [52]. The chemical composition of iron and copper in the Cu/FeO$_x$ samples was determined from the analysis of XPS spectra, as shown in Figures S8–S10, including the deconvoluted spectra and the results from fitting the spectra. The broad Fe 2p$_{3/2}$ peak indicates the presence of Fe in different valence states. The deconvolution of this peak shows the presence of a mixture of Fe^{2+} and Fe^{3+} in the fresh catalysts, where all spectra can be deconvoluted by two main components and one satellite feature. The spectral features above 933.0 eV are related to Cu^{2+} in copper oxides [53,54]. See also Table S2 for a summary of the XPS analysis.

Adsorption–desorption isotherms (Figure S11) were recorded at 77 K up to the saturation pressure of nitrogen to determine the surface area of the prepared catalysts using the Brunauer–Emmett–Teller (BET) theory. The hysteresis of the two curves shows that the prepared catalysts are porous. The specific surface area of the catalysts was calculated to be 114 m^2/g for FeO$_x$, 114 m^2/g for 1%-Cu/FeO$_x$, 93 m^2/g for 3%-Cu/FeO$_x$, and 92 m^2/g for 5%-Cu/FeO$_x$. See Table 1 for the data summary, reflecting on the difference in the morphologies of the catalysts. The UV-vis spectra of the fresh catalysts (Figure S12) show an absorption peak at 275 nm, associated with that of iron oxides [55].

3.1. Catalyst Testing

The catalysts did not undergo any reduction pre-treatment before the tests. Thus, the first temperature ramp can be considered as a pre-treatment step of the catalyst directly under the reactants which have a reducing character due to the excess hydrogen in it. Figure 2 shows the evolution of the CO$_2$ conversion (left column) and the product selectivities (right column) of the studied catalysts during the applied double temperature ramp. The drop in conversion between the individual ramps (Figure 2c) and the constantly evolving selectivity observed for FeO$_x$ (see Figure 2b and Table S3) indicate that this particular catalyst did not converge to its final state during the applied double temperature ramp. On the contrary, the identical CO$_2$ conversion observed for the Cu/FeO$_x$ catalysts at the highest temperatures of the first and the second temperature ramp, along with the experimental accuracy comparable product selectivities, provide a hint about the completed annealing of 1%-Cu/FeO$_x$, 3%-Cu/FeO$_x$, and 5%-Cu/FeO$_x$ during the first temperature ramp.

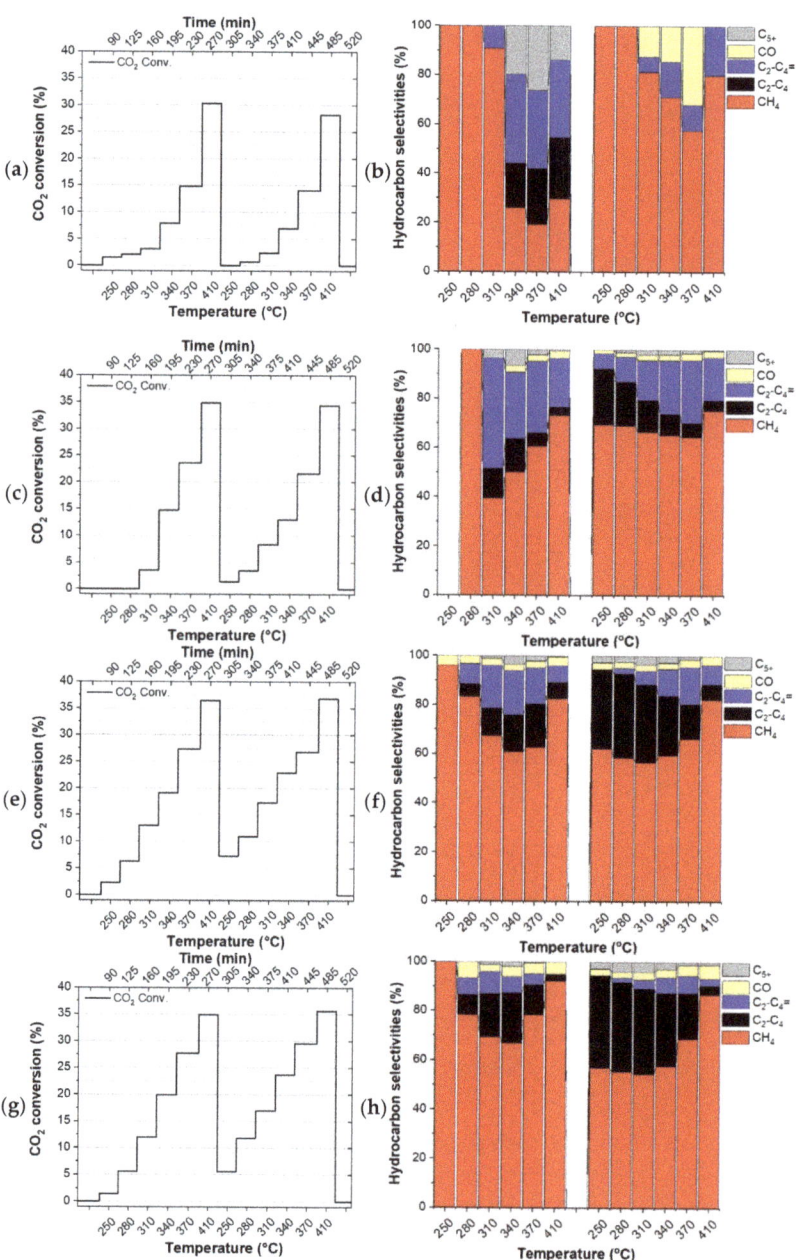

Figure 2. Hydrogenation of CO_2. CO_2 conversion and selectivities by (**a**,**b**) FeO_x, (**c**,**d**) 1%-Cu/FeO_x, (**e**,**f**) 3%-Cu/FeO_x, (**g**,**h**) 5%-Cu/FeO_x. A total of 200 mg of catalyst and $CO_2/H_2/He$ (1:5:4, total flow 25 mL/min).

Since the catalysts evolve during the first temperature ramp, their performance during the first temperature ramp is not discussed here. The comparison of CO_2 conversion shows that the addition of copper boosts conversion from around 28% up to about 37%, with respect to the copper-free reference catalyst. While all three interrogated Cu/FeO_x catalysts primarily produce methane along with a small amount of carbon monoxide, depending on the temperature and copper content, these catalysts produce up to about a total of 43% of C_2-C_5 olefins and paraffins. As reported in the literature, copper plays a dual role, activating hydrogen for the reaction with CO_2 itself and also for the reduction of iron oxide via hydrogen spillover, increasing the overall activity of iron-based catalysts.

Using the 1%-Cu/FeO$_x$ catalyst (Figure 2c,d), a relatively steady fraction of CH_4 is produced in the applied temperature range, with methane fraction amounting to around 75% at the highest temperature of 410 °C, and about 66% at lower temperatures. With an increasing temperature, this catalyst exhibits primarily a temperature-dependent switch from paraffins to olefins. In the case of the 3%-Cu/FeO$_x$ (Figure 2e,f) and 5%-Cu/FeO$_x$ (Figure 2g,h) catalysts, methane formation dominates at 410 °C with a methane selectivity of 82% and 86%, respectively. At lower temperatures in the presence of a small amount of C_{5+} (in the order of about ~2%) up to 37% and 39% C_2-C_4 olefins and paraffins are produced in the temperature region of 250–370 °C. The formation of the relative fractions of C_2-C_4 olefins and paraffins is dependent on the copper contents in Cu/FeOx catalysts. As the concentration of copper increases from 1% to 5%, the selectivity of paraffin hydrocarbons also increases, accompanied by the decrease in the olefins fraction. Significantly fewer olefins are produced on 3%-Cu/FeO$_x$ in comparison with 1%-Cu/FeO$_x$, and an even smaller amount is produced on 5%-Cu/FeO$_x$. In terms of selectivity, we hypothesized the central role of the morphology of the catalyst because the catalysts of comparable composition synthesized differently turned out to convert CO_2 primarily into methanol or benzene [37], some others produce methanol and small hydrocarbons [56], while the currently presented catalysts solely produce hydrocarbons. The presence of Cu plays an important role in both enhancing CO_2 conversion and the selectivity toward higher hydrocarbons. Cu not only acts as an RWGS catalyst but also reduces the iron oxides into metallic iron ($Fe_2O_3 \rightarrow Fe_3O_4 \rightarrow FeO \rightarrow Fe$), which is converted into Fe_5C_2 due to the carbonation process [18] to the active phase that produces higher hydrocarbons [57]. As the concentration of Cu increases, the fraction of Fe_5C_2 also increases, as confirmed by the XRD of the used catalysts (Table 2).

Table 2. Fractions of various forms of iron oxide and iron carbide components in the spent catalysts, obtained from XRD data using Rietveld refinement of analysis and specific surface area of spent catalysts, as determined from adsorption/desorption of nitrogen on the surface of the spent catalysts. Multipoint BET was assessed using Roquerol's method.

Catalyst	Iron Oxide/Carbide Composition			Specific Surface Area
	Fe_3O_4 [%]	Fe_5C_2 [%]	Fe_3C [%]	[m^2/g]
FeO$_x$	100.0			18.6
1%-Cu/FeO$_x$	95.0	5.0		18.6
3%-Cu/FeO$_x$	71.3	28.7		18.0
5%-Cu/FeO$_x$	59.8	37.2	3.0	17.4

3.2. Characterization of the Catalysts after Catalytic Tests

Spent catalysts were characterized by scanning electron microscopy (SEM), transmission electron microscopy (TEM), high-angle annular dark-field scanning transmission electron microscopy (STEM/HAADF), energy-dispersive X-ray spectroscopy (EDX), X-ray diffraction (XRD), sorption of N_2 with the BET model and X-ray photoelectron spectroscopy (XPS).

SEM micrographs of spent catalysts (Figure 3) show a massive restructuring of all the catalysts, featuring intertwined polymorphs which indicate the presence of Fe_5C_2 or Fe_3C [37,58]. Figure 4 shows TEM images of the spent catalysts, which for FeO$_x$, reveal particles with a mean diameter of 68 ± 26 nm, and for the Cu/FeO$_x$ catalysts, particle sizes increase with copper content: 1%-Cu/FeO$_x$—61 ± 19 nm, 3%-Cu/FeO$_x$—75 ± 25 nm, and 5%-Cu/FeO$_x$—81 ± 35 nm to 150 nm. Additional TEM images can be found in Figure S13.

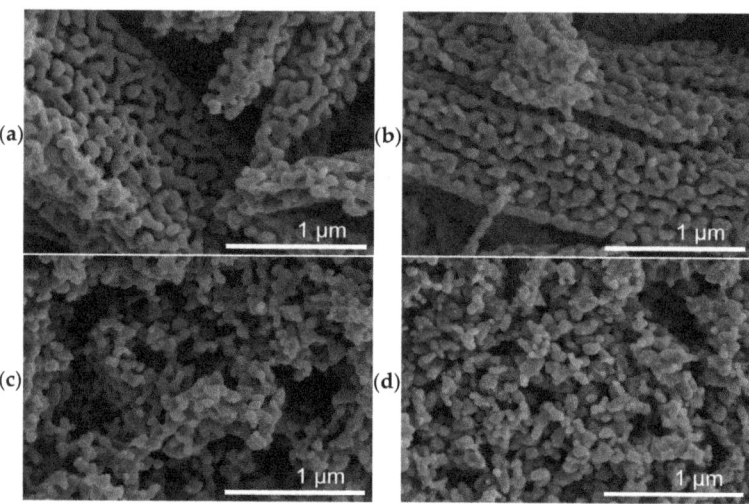

Figure 3. SEM images of (**a**) spent FeO$_x$, (**b**) spent 1%-Cu/FeO$_x$, (**c**) spent 3%-Cu/FeO$_x$, and (**d**) spent 5%-Cu/FeO$_x$ showing major restructuring at the micrometer and sub-micrometer scale. Additional images are shown in Figure S10.

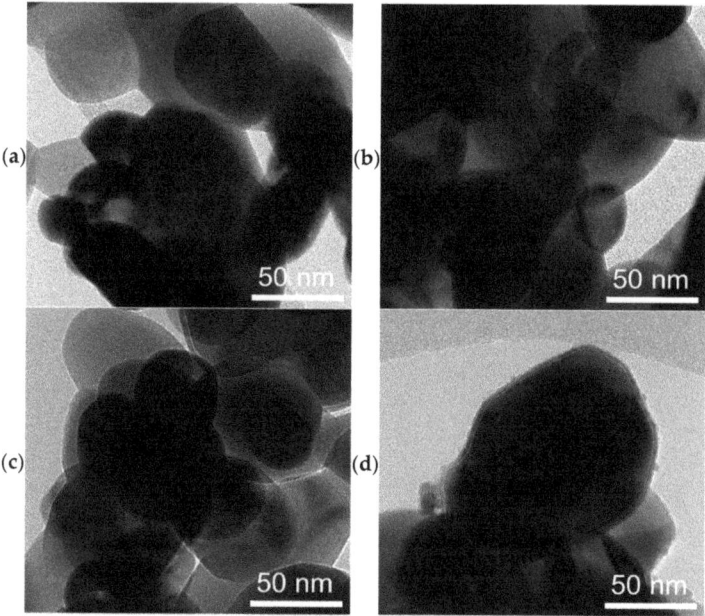

Figure 4. TEM images of spent catalysts after CO$_2$ hydrogenation of (**a**) FeO$_x$, (**b**) 1% Cu/FeO$_x$, (**c**) 3% Cu/FeO$_x$, and (**d**) 5% Cu/FeO$_x$ showing the transformation of the rod-like particles of the fresh catalysts into roughly spherical particles.

SEM images with EDX analysis are depicted in Figure S14. STEM/HAADF imagining of spent catalysts are depicted in Figures S15a, S16a and S17a, and Figure 5a for FeO$_x$,

1%-Cu/FeO$_x$, 3%-Cu/FeO$_x$, and 5%-Cu/FeO$_x$, respectively. These figures show the presence of nanoscale Fe species of about 150 nm in diameter with Cu agglomerates of 10 nm in diameter. From the EDX analysis, Fe and O appear to be dominant elements (Figures S15b, S16b and S17b and Figure 5b), forming the core of the spent FeO$_x$ particles. There is a recognizable layer of carbon (13 nm) seen on the surface of the particles, indicating the deposition of carbon during the reaction. This layer is most clearly visible in the spent 5%-Cu/FeO$_x$ (Figure 5c), and the least in the reference FeO$_x$ catalysts (Figure S15). In addition, as seen in Figures S16c and S17c and Figure 5c, copper became highly dispersed in all copper-containing spent catalysts.

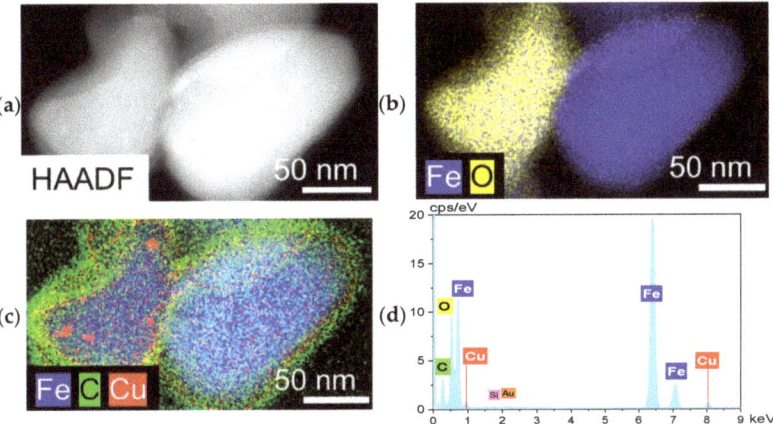

Figure 5. (a) STEM-HAADF of spent 5%-Cu/FeO$_x$, and (b–d) EDX elemental mapping of Fe, O, C and Cu.

The fractions of different iron oxides and carbides present in the used samples obtained by XRD (Figure S18) are summarized in Table 2. Spent FeO$_x$ consists of 100% Fe$_3$O$_4$, spent 1%-Cu/FeO$_x$ has 95% Fe$_3$O$_4$ and 5% Fe$_5$C$_3$; the 3%-Cu/FeO$_x$ catalyst contains 71.3% Fe$_3$O$_4$ and 28.7% Fe$_5$C$_3$; and 5%-Cu/FeO$_x$ has 59.8% Fe$_3$O$_4$, 37.2% Fe$_5$C$_3$ and 3% Fe$_3$C. The XPS of the spent catalysts reveals the presence of metallic Fe along with Fe^{2+} and Fe^{3+}. The components at the binding energy of around 932.6 eV are related to Cu(0).

The XPS of the spent catalysts similarly shows the presence of metallic Fe along with Fe^{2+} and Fe^{3+}. The components at the binding energy of around 932.6 eV are related to Cu(0) (see Figure S8). Table S2 summarizes the chemical constituents of the spent catalysts.

With the increasing copper load, an increase can be seen in the iron carbides content, which is consistent with the role of copper in reducing iron oxide. We note that copper was not detected by XRD on the spent catalysts either, which indicates that copper is present in either an amorphous form or the size of the majority of the copper particles is below the detection limit of XRD, as indicated by STEM-HAADF (Figure 5).

XRD further confirmed the formation of iron carbides in the copper-containing catalysts. From the XRD data, we obtained the mean X-ray coherence length (MLC), which is summarized in Table S4. Spent FeO$_x$ consists of a coherent domain length of 63 nm for the Fe$_3$O$_4$ phase. Whereas, spent 1%-Cu/FeO$_x$ has 63 nm for Fe$_3$O$_4$ and 29 nm for Fe$_5$C$_2$, and spent 3%-Cu/FeO$_x$ has 47 nm for Fe$_3$O$_4$ and 40 nm for Fe$_5$C$_2$. Spent 5%-Cu/FeO$_x$ has 36 nm for Fe$_3$O$_4$, 35 nm for Fe$_5$C$_2$, and 23 nm for Fe$_3$C.

Adsorption–desorption isotherms of spent catalysts (Figure S19) were obtained at 77 K up to the saturation pressure of nitrogen to determine the surface area of the spent catalysts. As opposed to the fresh catalyst, the adsorption/desorption curves have a course that suggests a smaller surface area, perhaps due to the inaccessibility of pores due to the carbon deposition. The specific surface area of the catalysts was calculated to be 18.6 m^2/g for FeO_x, 18.6 m^2/g for 1%-Cu/FeO_x, 18.0 m^2/g for 3%-Cu/FeO_x, and 17.4 m^2/g for 5%-Cu/FeO_x. Thus, this is just a mere fraction of the as-made catalyst prior to the catalytic test, with the specific surface areas of the spent catalysts being about 10 times smaller (see Table 2) than that of the fresh catalysts (see Table 1).

For the convenience of the reader, Figure S20 summarizes the CO_2 conversion and product selectivity during the second temperature ramp for the investigated catalysts, compared to the individual reaction temperatures. In order to test the stability of the 5%-CuFeOx catalysts, four more temperature ramps were performed (Figure S21). After the second temperature ramp, the CO_2 conversion decreased with each ramp, accompanied by an increase in CO generation and a decrease in the fraction of CH_4 and higher hydrocarbons. After the fourth ramp, the CO_2 conversion decreased from the initial 35% to 19%. This change could be caused by the accumulation of carbon on the surface of the catalysts, as confirmed by STEM-HAADF and elemental mapping (Figure 5), where a layer of 13 nm thick of carbon is seen on the surface of the particles (Figure 5c).

A review of the related literature, including this study, is summarized for the convenience of the reader in Table S5 and Figure S22.

4. Conclusions

Cu/FeO_x catalysts with a Cu content of 0%, 1%, 3%, and 5% were found to be highly active in CO_2 hydrogenation with the conversion reaching 36.8%. The observed selectivity to C_1 versus C_2-C_4, C_2-C_4 =, and C_5+ hydrocarbons is dependent on the catalyst's composition, morphology, and temperature. The observed range of products is diametrically different from those observed for catalysts with similar compositions but synthesized using other precursors. Depending on their copper content, the as-made particles possess rather distinct morphologies, most likely attributable to the changing ratio of the hydrazine and copper sulfate used during the synthesis of the individual samples, which affects the extent of the etching of the iron-oxide particles, as well as their crystallinity, chemical composition, and morphology at various scales. The findings presented in this paper indicate potential new ways of tuning the morphology and composition of iron-oxide-based particles, for example, by balancing the relative ratio of hydrazine and SO_4^{2-} ions or by using different precursors in the synthesis, ultimately yielding catalyst compositions and morphologies with variable catalytic performance. We note that, while some alkali (K, Na) doped Fe-based catalysts have a higher CO_2 conversion, these require harsh conditions for pre-treatment under hydrogen or a CO atmosphere at temperatures exceeding 350 °C for extended times up to 12 h. Our catalysts have very low selectivity towards carbon monoxide as compared to other reported catalysts.

Supplementary Materials: The following supporting information can be downloaded at: https://www.mdpi.com/article/10.3390/catal12050516/s1. Figure S1: Double temperature ramp applied during catalytic testing; Figure S2: SEM/EDX of fresh catalysts; Figure S3: STEM/HAADF of fresh FeO_x; Figure S4: STEM/HAADF of fresh 1%-Cu/FeO_x; Figure S5: STEM/HAADF of fresh 3%-Cu/FeO_x; Figure S6: STEM-HAADF of fresh 5%-Cu/FeO_x; Figure S7: XRD patterns of fresh catalysts; Table S1: The mean X-ray coherence length of fresh catalysts; Figure S8: XPS of fresh and spent catalysts; Figure S9: XPS of FeO_x; Figure S10: Wide scan XPS spectra of fresh and spent catalysts; Table S2: Chemical composition of Fe and Cu in catalysts; Figure S11: Adsorption and desorption isotherms of fresh catalysts; Figure S12: UV-vis spectra of fresh catalysts; Table S3: Conversion and selectivities of FeO_x and Cu/FeO_x catalysts; Figure S13: TEM images of fresh and spent catalysts; Figure S14: SEM/EDX of spent catalysts; Figure S15: STEM/HAADF images of spent FeO_x; Figure S16: STEM/HAADF images of spent 1%-Cu/FeO_x; Figure S17: STEM/HAADF images of spent 3%-Cu/FeO_x; Figure S18: XRD patterns of spent catalysts; Table S4: The mean X-ray coherence length of spent catalysts; Figure S19: Adsorption and desorption isotherms of spent catalysts; Figure S20: Comparison of the performance of the catalysts; Figure S21: CO_2 conversion and selectivity over the course of 6 consecutive temperature ramps; Table S5: Literature comparison of CO_2 hydrogenation by different iron-based catalysts; Figure S22: Literature comparison of CH_4/CO selectivity against CO_2 conversion.

Author Contributions: K.S. synthesized the catalysts and performed their tests, analyzed data, and drafted the manuscript; M.I.Q. contributed to the catalyst testing and discussed the results; N.Ž. participated in the catalyst testing, discussed the catalytic results, and contributed to the writing of the manuscript; J.E.O. measured and interpreted the UV-vis spectra; P.S. collected and evaluated the XPS spectra; L.K. contributed to the characterization of the catalysts, the interpretation of the results and correcting the manuscript; Š.V. planned and correlated the effort, discussed and interpreted the results, and wrote and finalized the manuscript. All authors have read and agreed to the published version of the manuscript.

Funding: K.S., M.I.Q., N.Z. and S.V. gratefully acknowledge the support from the European Union's Horizon 2020 research and innovation program under grant agreement No 810310, which corresponds to the J. Heyrovsky Chair project ("ERA Chair at J. Heyrovský Institute of Physical Chemistry AS CR—The institutional approach towards ERA"). The funders had no role in the preparation of the article. The work of L.K was supported by the ERDF project "Development of pre-applied research in nanotechnology and biotechnology" (No. CZ.02.1.01/0.0/0.0/17_048/0007323) and an internal grant of Palacky University IGA_PrF_2021_032.

Institutional Review Board Statement: Not applicable.

Informed Consent Statement: Not applicable.

Data Availability Statement: The data presented in this study are available on reasonable request from the corresponding authors.

Conflicts of Interest: The authors declare no conflict of interest.

References

1. Garba, M.D.; Usman, M.; Khan, S.; Shehzad, F.; Galadima, A.; Ehsan, M.F.; Ghanem, A.S.; Humayun, M. CO_2 towards fuels: A review of catalytic conversion of carbon dioxide to hydrocarbons. *J. Environ. Chem. Eng.* **2021**, *9*, 104756. [CrossRef]
2. Humayun, M.; Ullah, H.; Usman, M.; Habibi-Yangjeh, A.; Tahir, A.A.; Wang, C.; Luo, W. Perovskite-type lanthanum ferrite based photocatalysts: Preparation, properties, and applications. *J. Energy Chem.* **2022**, *66*, 314–338. [CrossRef]
3. Anderson, T.R.; Hawkins, E.; Jones, P.D. CO_2, the greenhouse effect and global warming: From the pioneering work of Arrhenius and Callendar to today's Earth System Models. *Endeavour* **2016**, *40*, 178–187. [CrossRef] [PubMed]
4. Mac Dowell, N.; Fennell, P.S.; Shah, N.; Maitland, G.C. The role of CO_2 capture and utilization in mitigating climate change. *Nat. Clim. Chang.* **2017**, *7*, 243–249. [CrossRef]
5. Centi, G.; Perathoner, S. Opportunities and prospects in the chemical recycling of carbon dioxide to fuels. *Catal. Today* **2009**, *148*, 191–205. [CrossRef]
6. Dupont, J. Across the Board: Jairton Dupont. *ChemSusChem* **2015**, *8*, 586–587. [CrossRef] [PubMed]
7. Peters, M.; Köhler, B.; Kuckshinrichs, W.; Leitner, W.; Markewitz, P.; Müller, T.E. Chemical Technologies for Exploiting and Recycling Carbon Dioxide into the Value Chain. *ChemSusChem* **2011**, *4*, 1216–1240. [CrossRef]
8. North, M.; Pasquale, R.; Young, C. Synthesis of cyclic carbonates from epoxides and CO_2. *Green Chem.* **2010**, *12*, 1514–1539. [CrossRef]
9. De, S.; Dokania, A.; Ramirez, A.; Gascon, J. Advances in the Design of Heterogeneous Catalysts and Thermocatalytic Processes for CO_2 Utilization. *ACS Catal.* **2020**, *10*, 14147–14185. [CrossRef]
10. Ra, E.C.; Kim, K.Y.; Kim, E.H.; Lee, H.; An, K.; Lee, J.S. Recycling Carbon Dioxide through Catalytic Hydrogenation: Recent Key Developments and Perspectives. *ACS Catal.* **2020**, *10*, 11318–11345. [CrossRef]
11. Valenti, G.; Melchionna, M.; Montini, T.; Boni, A.; Nasi, L.; Fonda, E.; Criado, A.; Zitolo, A.; Voci, S.; Bertoni, G.; et al. Water-Mediated ElectroHydrogenation of CO_2 at Near-Equilibrium Potential by Carbon Nanotubes/Cerium Dioxide Nanohybrids. *ACS Appl. Energy Mater.* **2020**, *3*, 8509–8518. [CrossRef]
12. Roy, S.; Cherevotan, A.; Peter, S.C. Thermochemical CO_2 Hydrogenation to Single Carbon Products: Scientific and Technological Challenges. *ACS Energy Lett.* **2018**, *3*, 1938–1966. [CrossRef]
13. Zhang, Y.; Xia, B.; Ran, J.; Davey, K.; Qiao, S.Z. Atomic-Level Reactive Sites for Semiconductor-Based Photocatalytic CO_2 Reduction. *Adv. Energy Mater.* **2020**, *10*, 1903879. [CrossRef]
14. Humayun, M.; Ullah, H.; Shu, L.; Ao, X.; Tahir, A.A.; Wang, C.; Luo, W. Plasmon Assisted Highly Efficient Visible Light Catalytic CO_2 Reduction over the Noble Metal Decorated Sr-Incorporated g-C3N4. *Nanomicro. Lett.* **2021**, *13*, 209. [CrossRef]
15. Gao, G.; Jiao, Y.; Waclawik, E.R.; Du, A. Single Atom (Pd/Pt) Supported on Graphitic Carbon Nitride as an Efficient Photocatalyst for Visible-Light Reduction of Carbon Dioxide. *J. Am. Chem Soc.* **2016**, *138*, 6292–6297. [CrossRef] [PubMed]
16. Choi, Y.H.; Jang, Y.J.; Park, H.; Kim, W.Y.; Lee, Y.H.; Choi, S.H.; Lee, J.S. Carbon dioxide Fischer-Tropsch synthesis: A new path to carbon-neutral fuels. *Appl. Catal. B* **2017**, *202*, 605–610. [CrossRef]
17. Riedel, T.; Shulz, H.; Schaub, G.; Jun, K.-W. Fischer–Tropsch on iron with H_2/CO and H_2/CO_2 as synthesis gases: The episodes of formation of the Fischer–Tropsch regime and construction of the catalyst. *Top. Catal.* **2003**, *26*, 41–54. [CrossRef]

18. Lopez Luna, M.; Timoshenko, J.; Kordus, D.; Rettenmaier, C.; Chee, S.W.; Hoffman, A.S.; Bare, S.R.; Shaikhutdinov, S.; Roldan Cuenya, B. Role of the Oxide Support on the Structural and Chemical Evolution of Fe Catalysts during the Hydrogenation of CO_2. *ACS Catal.* **2021**, *11*, 6175–6185. [CrossRef]
19. Qadir, M.I.; Weilhard, A.; Fernandes, J.A.; de Pedro, I.; Vieira, B.J.C.; Waerenborgh, J.C.; Dupont, J. Selective Carbon Dioxide Hydrogenation Driven by Ferromagnetic RuFe Nanoparticles in Ionic Liquids. *ACS Catal.* **2018**, *8*, 1621–1627. [CrossRef]
20. Hwang, J.S.; Jun, K.-W.; Lee, K.-W. Deactivation and regeneration of Fe-K/alumina catalyst in CO_2 hydrogenation. *Appl. Catal A-Gen.* **2001**, *208*, 217–222. [CrossRef]
21. Hong, J.-S.; Hwang, J.S.; Jun, K.-W.; Sur, J.C.; Lee, K.-W. Deactivation study on a coprecipitated Fe-Cu-K-Al catalyst in CO_2 hydrogenation. *Appl. Catal. A-Gen.* **2001**, *218*, 53–59. [CrossRef]
22. Pérez-Alonso, F.J.; Ojeda, M.; Herranz, T.; Rojas, S.; González-Carballo, J.M.; Terreros, P.; Fierro, J.L.G. Carbon dioxide hydrogenation over Fe–Ce catalysts. *Catal. Commun.* **2008**, *9*, 1945–1948. [CrossRef]
23. Rodemerck, U.; Holeňa, M.; Wagner, E.; Smejkal, Q.; Barkschat, A.; Baerns, M. Catalyst Development for CO_2 Hydrogenation to Fuels. *ChemCatChem* **2013**, *5*, 1948–1955. [CrossRef]
24. Aitbekova, A.; Goodman, E.D.; Wu, L.; Boubnov, A.; Hoffman, A.S.; Genc, A.; Cheng, H.; Casalena, L.; Bare, S.R.; Cargnello, M. Engineering of Ruthenium–Iron Oxide Colloidal Heterostructures: Improved Yields in CO_2 Hydrogenation to Hydrocarbons. *Angew. Chem. Int. Ed.* **2019**, *58*, 17451–17457. [CrossRef] [PubMed]
25. Al-Dossary, M.; Ismail, A.A.; Fierro, J.L.G.; Bouzid, H.; Al-Sayari, S.A. Effect of Mn loading onto MnFeO nanocomposites for the CO_2 hydrogenation reaction. *Appl. Catal. B* **2015**, *165*, 651–660. [CrossRef]
26. Liang, B.; Duan, H.; Sun, T.; Ma, J.; Liu, X.; Xu, J.; Su, X.; Huang, Y.; Zhang, T. Effect of Na Promoter on Fe-Based Catalyst for CO_2 Hydrogenation to Alkenes. *ACS Sustain. Chem. Eng.* **2019**, *7*, 925–932. [CrossRef]
27. Wei, J.; Yao, R.; Ge, Q.; Wen, Z.; Ji, X.; Fang, C.; Zhang, J.; Xu, H.; Sun, J. Catalytic Hydrogenation of CO_2 to Isoparaffins over Fe-Based Multifunctional Catalysts. *ACS Catal.* **2018**, *8*, 9958–9967. [CrossRef]
28. Choi, Y.H.; Ra, E.C.; Kim, E.H.; Kim, K.Y.; Jang, Y.J.; Kang, K.-N.; Choi, S.H.; Jang, J.-H.; Lee, J.S. Sodium-Containing Spinel Zinc Ferrite as a Catalyst Precursor for the Selective Synthesis of Liquid Hydrocarbon Fuels. *ChemSusChem* **2017**, *10*, 4764–4770. [CrossRef]
29. Kim, K.Y.; Lee, H.; Noh, W.Y.; Shin, J.; Han, S.J.; Kim, S.K.; An, K.; Lee, J.S. Cobalt Ferrite Nanoparticles to Form a Catalytic Co–Fe Alloy Carbide Phase for Selective CO_2 Hydrogenation to Light Olefins. *ACS Catal.* **2020**, *10*, 8660–8671. [CrossRef]
30. Gnanamani, M.K.; Jacobs, G.; Hamdeh, H.H.; Shafer, W.D.; Liu, F.; Hopps, S.D.; Thomas, G.A.; Davis, B.H. Hydrogenation of Carbon Dioxide over Co–Fe Bimetallic Catalysts. *ACS Catal.* **2016**, *6*, 913–927. [CrossRef]
31. Ronda-Lloret, M.; Rothenberg, G.; Shiju, N.R. A Critical Look at Direct Catalytic Hydrogenation of Carbon Dioxide to Olefins. *ChemSusChem* **2019**, *12*, 3896–3914. [CrossRef] [PubMed]
32. Bradley, M.J.; Ananth, R.; Willauer, H.D.; Baldwin, J.W.; Hardy, D.R.; Williams, F.W. The Effect of Copper Addition on the Activity and Stability of Iron-Based CO_2 Hydrogenation Catalysts. *Molecules* **2017**, *22*, 1579. [CrossRef] [PubMed]
33. Wang, S.-G.; Liao, X.-Y.; Cao, D.-B.; Huo, C.-F.; Li, Y.-W.; Wang, J.; Jiao, H. Factors Controlling the Interaction of CO_2 with Transition Metal Surfaces. *J. Phys. Chem. C* **2007**, *111*, 16934–16940. [CrossRef]
34. Wang, W.; Jiang, X.; Wang, X.; Song, C. Fe–Cu Bimetallic Catalysts for Selective CO_2 Hydrogenation to Olefin-Rich C_2^+ Hydrocarbons. *Ind. Eng. Chem. Res.* **2018**, *57*, 4535–4542. [CrossRef]
35. Jurković, D.L.; Pohar, A.; Dasireddy, V.D.B.C.; Likozar, B. Effect of Copper-based Catalyst Support on Reverse Water-Gas Shift Reaction (RWGS) Activity for CO_2 Reduction. *Chem. Eng. Technol.* **2017**, *40*, 973–980. [CrossRef]
36. Previtali, D.; Longhi, M.; Galli, F.; Di Michele, A.; Manenti, F.; Signoretto, M.; Menegazzo, F.; Pirola, C. Low pressure conversion of CO_2 to methanol over Cu/Zn/Al catalysts. The effect of Mg, Ca and Sr as basic promoters. *Fuel* **2020**, *274*, 117804. [CrossRef]
37. Halder, A.; Kilianová, M.; Yang, B.; Tyo, E.C.; Seifert, S.; Prucek, R.; Panáček, A.; Suchomel, P.; Tomanec, O.; Gosztola, D.J.; et al. Highly efficient Cu-decorated iron oxide nanocatalyst for low pressure CO_2 conversion. *Appl. Catal. B* **2018**, *225*, 128–138. [CrossRef]
38. Zhang, P.; Araki, Y.; Feng, X.; Li, H.; Fang, Y.; Chen, F.; Shi, L.; Peng, X.; Yoneyama, Y.; Yang, G.; et al. Urea-derived Cu/ZnO catalyst being dried by supercritical CO_2 for low-temperature methanol synthesis. *Fuel* **2020**, *268*, 117213. [CrossRef]
39. Xiong, S.; Lian, Y.; Xie, H.; Liu, B. Hydrogenation of CO_2 to methanol over Cu/ZnCr catalyst. *Fuel* **2019**, *256*, 115975. [CrossRef]
40. Dong, X.; Li, F.; Zhao, N.; Xiao, F.; Wang, J.; Tan, Y. CO_2 hydrogenation to methanol over Cu/ZnO/ZrO_2 catalysts prepared by precipitation-reduction method. *Appl. Catal. B* **2016**, *191*, 8–17. [CrossRef]
41. Wang, G.; Mao, D.; Guo, X.; Yu, J. Enhanced performance of the CuO-ZnO-ZrO_2 catalyst for CO_2 hydrogenation to methanol by WO_3 modification. *Appl. Surf. Sci.* **2018**, *456*, 403–409. [CrossRef]
42. Li, S.; Wang, Y.; Guo, L. A highly active and selective mesostructured Cu/AlCeO catalyst for CO2 hydrogenation to methanol. *Appl. Catal. A-Gen.* **2019**, *571*, 51–60. [CrossRef]
43. Sedighi, M.; Mohammadi, M. CO_2 hydrogenation to light olefins over Cu-CeO_2/SAPO-34 catalysts: Product distribution and optimization. *J. CO2 Util.* **2020**, *35*, 236–244. [CrossRef]
44. Gao, P.; Xie, R.; Wang, H.; Zhong, L.; Xia, L.; Zhang, Z.; Wei, W.; Sun, Y. Cu/Zn/Al/Zr catalysts via phase-pure hydrotalcite-like compounds for methanol synthesis from carbon dioxide. *J. CO2 Util.* **2015**, *11*, 41–48. [CrossRef]
45. Yang, B.; Liu, C.; Halder, A.; Tyo, E.C.; Martinson, A.B.F.; Seifert, S.; Zapol, P.; Curtiss, L.A.; Vajda, S. Copper Cluster Size Effect in Methanol Synthesis from CO_2. *J. Phys. Chem. C* **2017**, *121*, 10406–10412. [CrossRef]

46. Dasireddy, V.D.B.C.; Likozar, B. The role of copper oxidation state in Cu/ZnO/Al$_2$O$_3$ catalysts in CO$_2$ hydrogenation and methanol productivity. *Renew. Energy* **2019**, *140*, 452–460. [CrossRef]
47. Liu, C.; Yang, B.; Tyo, E.; Seifert, S.; DeBartolo, J.; von Issendorff, B.; Zapol, P.; Vajda, S.; Curtiss, L.A. Carbon Dioxide Conversion to Methanol over Size-Selected Cu$_4$ Clusters at Low Pressures. *J. Am. Chem Soc.* **2015**, *137*, 8676–8679. [CrossRef]
48. Khdary, N.H.; Alayyar, A.S.; Alsarhan, L.M.; Alshihri, S.; Mokhtar, M. Metal Oxides as Catalyst/Supporter for CO$_2$ Capture and Conversion, Review. *Catalysts* **2022**, *12*, 300. [CrossRef]
49. Zboril, R.; Machala, L.; Mashlan, M.; Hermanek, M.; Miglierini, M.; Fojtik, A. Structural, magnetic and size transformations induced by isothermal treatment of ferrous oxalate dihydrate in static air conditions. *Phys. Status Solidi C* **2004**, *1*, 3583–3588. [CrossRef]
50. Datta, K.J.; Gawande, M.B.; Datta, K.K.R.; Ranc, V.; Pechousek, J.; Krizek, M.; Tucek, J.; Kale, R.; Pospisil, P.; Varma, R.S.; et al. Micro–mesoporous iron oxides with record efficiency for the decomposition of hydrogen peroxide: Morphology driven catalysis for the degradation of organic contaminants. *J. Mater. Chem. A* **2016**, *4*, 596–604. [CrossRef]
51. Kim, E.; Kim, H.; Park, B.J.; Han, Y.H.; Park, J.H.; Cho, J.; Lee, S.S.; Son, J.G. Etching-Assisted Crumpled Graphene Wrapped Spiky Iron Oxide Particles for High-Performance Li-Ion Hybrid Supercapacitor. *Small* **2018**, *14*, e1704209. [CrossRef] [PubMed]
52. Zhu, D.; Wang, L.; Yu, W.; Xie, H. Intriguingly high thermal conductivity increment for CuO nanowires contained nanofluids with low viscosity. *Sci. Rep.* **2018**, *8*, 5282. [CrossRef] [PubMed]
53. Halder, A.; Lenardi, C.; Timoshenko, J.; Mravak, A.; Yang, B.; Kolipaka, L.K.; Piazzoni, C.; Seifert, S.; Bonačić-Koutecký, V.; Frenkel, A.I.; et al. CO$_2$ Methanation on Cu-Cluster Decorated Zirconia Supports with Different Morphology: A Combined Experimental In Situ GIXANES/GISAXS, Ex Situ XPS and Theoretical DFT Study. *ACS Catal.* **2021**, *11*, 6210–6224. [CrossRef]
54. Pauly, N.; Tougaard, S.; Yubero, F. Determination of the Cu 2p primary excitation spectra for Cu, Cu$_2$O and CuO. *Surf. Sci* **2014**, *620*, 17–22. [CrossRef]
55. Bouafia, A.; Laouini, S.E.; Khelef, A.; Tedjani, M.L.; Guemari, F. Effect of Ferric Chloride Concentration on the Type of Magnetite (Fe3O4) Nanoparticles Biosynthesized by Aqueous Leaves Extract of Artemisia and Assessment of Their Antioxidant Activities. *J. Clust. Sci.* **2020**, *32*, 1033–1041. [CrossRef]
56. Yang, B.; Yu, X.; Halder, A.; Zhang, X.; Zhou, X.; Mannie, G.J.A.; Tyo, E.; Pellin, M.J.; Seifert, S.; Su, D.; et al. Dynamic Interplay between Copper Tetramers and Iron Oxide Boosting CO$_2$ Conversion to Methanol and Hydrocarbons under Mild Conditions. *ACS Sustain. Chem. Eng.* **2019**, *7*, 14435–14442. [CrossRef]
57. Yao, B.; Xiao, T.; Makgae, O.A.; Jie, X.; Gonzalez-Cortes, S.; Guan, S.; Kirkland, A.I.; Dilworth, J.R.; Al-Megren, H.A.; Alshihri, S.M.; et al. Transforming carbon dioxide into jet fuel using an organic combustion-synthesized Fe-Mn-K catalyst. *Nat. Commun.* **2020**, *11*, 6395. [CrossRef]
58. Sayed, F.N.; Polshettiwar, V. Facile and sustainable synthesis of shaped iron oxide nanoparticles: Effect of iron precursor salts on the shapes of iron oxides. *Sci. Rep.* **2015**, *5*, 9733. [CrossRef]

Article

Controlled Transition Metal Nucleated Growth of Carbon Nanotubes by Molten Electrolysis of CO_2

Xinye Liu [1], Gad Licht [2], Xirui Wang [1] and Stuart Licht [1,2,*]

[1] Department of Chemistry, George Washington University, Washington, DC 20052, USA; xyl@gwmail.gwu.edu (X.L.); xiruiwang@email.gwu.edu (X.W.)
[2] C2CNT, Carbon Corp, 1035 26 St NE, Calgary, AB T2A 6K8, Canada; glicht@wesleyan.edu
* Correspondence: slicht@gwu.edu

Abstract: The electrolysis of CO_2 in molten carbonate has been introduced as an alternative mechanism to synthesize carbon nanomaterials inexpensively at high yield. Until recently, CO_2 was thought to be unreactive, making its removal a challenge. CO_2 is the main cause of anthropogenic global warming and its utilization and transformation into a stable, valuable material provides an incentivized pathway to mitigate climate change. This study focuses on controlled electrochemical conditions in molten lithium carbonate to split CO_2 absorbed from the atmosphere into carbon nanotubes (CNTs), and into various macroscopic assemblies of CNTs, which may be useful for nanofiltration. Different CNT morphologies were prepared electrochemically by variation of the anode and cathode composition and architecture, variation of the electrolyte composition pre-electrolysis processing, and variation of the current application and current density. Individual CNT morphologies' structures and the CNT molten carbonate growth mechanisms are explored using SEM (scanning electron microscopy), TEM (transmission electron micrsocopy), HAADF (high angle annular dark field), EDX (energy dispersive xray), X-ray diffraction, and Raman methods. The principle commercial technology for CNT production had been chemical vapor deposition, which is an order of magnitude more expensive, generally requires metallo-organics, rather than CO_2 as reactants, and can be highly energy and CO_2 emission intensive (carries a high carbon positive, rather than negative, footprint).

Keywords: nanocarbon; carbon nanotubes; carbon dioxide electrolysis; molten carbonate; greenhouse gas mitigation

Citation: Liu, X.; Licht, G.; Wang, X.; Licht, S. Controlled Transition Metal Nucleated Growth of Carbon Nanotubes by Molten Electrolysis of CO_2. Catalysts 2022, 12, 137. https://doi.org/10.3390/catal12020137

Academic Editor: Javier Ereña

Received: 17 December 2021
Accepted: 20 January 2022
Published: 22 January 2022

Publisher's Note: MDPI stays neutral with regard to jurisdictional claims in published maps and institutional affiliations.

Copyright: © 2022 by the authors. Licensee MDPI, Basel, Switzerland. This article is an open access article distributed under the terms and conditions of the Creative Commons Attribution (CC BY) license (https://creativecommons.org/licenses/by/4.0/).

1. Introduction

Global CO_2 has risen rapidly, accelerating extinction risk [1–4]. CO_2 is a highly stable molecule and is difficult to remove from the environment [5]. One means to mitigate CO_2 under consideration is its low-energy chemical transformation to a (i) stable, (ii) useful, and (iii) valuable product, with a low cost and low carbon footprint of production. The transformed CO_2's product stability prevents the captured CO_2 from re-emission, the product's usefulness provides a buffer to store the captured carbon, and its high-value (ideally higher than the cost of CO_2 transformation) provides an economic incentive to remove the greenhouse gas. Graphitic nanocarbons, such as carbon nanotubes made from CO_2, may meet several of these transformed CO_2 product requirements. For example, its basic structure of layered graphene retains the durability of graphite, whose hundreds-of-millions-year-old mineral deposits attest to its long term stability, while CNTs' market value of $100,000 to $400,000/tonne can provide a revenue, rather than a cost, while removing CO_2.

Multiwalled CNTs (Carbon NanoTubes) are comprised of concentric cylindrical graphene sheets. CNTs have a measured tensile strength of 93,900 MPa, which is the highest tensile strength of any material [6,7]. Other useful properties of CNTs include high electrical capacity, high thermal conductivity, high flexibility, high capacity for charge storage, and catalysis. CNTs applications range from stronger, lighter structural materials including cement, aluminum, and steel admixtures [8], to medical applications [9,10], electronics, batteries and supercapacitors [11,12], sensors and analysis [13–15], plastics and polymers [16–20], textiles [21], hydrogen storage [22], and water treatment [23,24].

The deliberate thought of this study is that the superior physical chemical properties of CNTs, and in particular CNTs made by consuming CO_2, will cause a demand for its application, thereby incentivizing CO_2 consumption and driving climate change mitigation by decreasing emissions of the greenhouse gas CO_2.

Increased pathways for the use of CO_2 as a molten carbonate electrolysis reactant to synthesize value-added CNTs will provide the effect of opening a path to consume this greenhouse gas to mitigate climate change, and its transformation to CNTs will provide a stable material to store carbon removed from the environment.

To date, the CNT market has been limited due to a high cost of production. Commercially, CNTs are mainly produced by chemical vapor deposition (CVD) and not from CO_2 [25,26]. CVD production of CNTs is chemical and energy intensive and expensive, leading to current costs of $100 K to $400 K per tonne of CNT, and CVD production has a high carbon footprint [27].

Prior attempts to transform CO_2 to carbon nanotubes or graphene have been low yield and energy intensive, such as the production of graphitic flakes using high-pressure CO_2 or by electrolysis in molten $CaCl_2$ electrolytes. Undesired byproducts included Al_2O_3, hydrogen, and hydrocarbons, and from the electrolysis, carbon monoxide byproducts from an 850 °C electrolysis splitting in molten $CaCl_2$ electrolytes [28,29].

However, CO_2 has a strong affinity for certain molten carbonates. In 2009, a process was introduced to mitigate the greenhouse gas CO_2 through molten electrolytic splitting and transformation at elevated temperatures. Pathways were opened to the high-purity, renewable energy electrolytic splitting of CO_2 to solid carbon by demonstrating that in molten lithium carbonate (melting point 723 °C) electrolytes, the four-electron molten electrolysis reduction of tetravalent carbon to solid products dominates below 800 °C. With rising electrolysis temperature between 800 and 900 °C, the two-electron reduction to a CO product increasingly dominates, and by 950 °C, the transition to the alternative CO byproduct is complete [30].

The solid carbon product of CO_2 electrolysis was further refined to graphitic nanocarbons through the discovery of the catalyzed molten electrolysis transition metal nucleated growth of carbon nanotubes and carbon nanofibers in lithium carbonate electrolytes [31–33]. The process was given the acronym C2CNT (Carbon dioxide to Carbon NanoTubes). The ^{13}C isotope of CO_2 was used to track carbon through the C2CNT process from its origin (CO_2 as a gas phase reactant) through its transformation to a CNT or carbon nanofiber product, and the CO_2 originating from the gas phase served as the renewable C building blocks in the observed CNT product [32]. The net reaction is:

$$\text{Dissolution: } CO_2(gas) + Li_2O(soluble) \rightleftharpoons Li_2CO_3(molten) \tag{1}$$

$$\text{Electrolysis: } Li_2CO_3(molten) \rightarrow C(CNT) + Li_2O \text{ (soluble)} + O_2(gas) \tag{2}$$

$$\text{Net: } CO_2(gas) \rightarrow C(CNT) + O_2(gas) \tag{3}$$

The electrolytic CO_2 splitting in molten carbonates could occur at electrolysis potentials of less than 1 volt [33]. Due to the high affinity of CO_2 towards reaction 1, that is, for the Li_2O present in the electrolyte, in the C2CNT process, the electrolytic splitting could occur as a direct capture of carbon in the air without CO_2 pre-concentration, or with exhaust gas, or with concentrated CO_2. As it directly captured CO_2 from the air, no further introduction of CO_2 was needed throughout the group's experiments. By means of variation of the electrolysis setup, the process could produce, in addition to conventional morphology CNT: doped CNTs, helical CNTs, and magnetic CNTs, as illustrated in Figure 1 [31–44]. Studied applications of electrolytic CNTs from CO_2 include batteries [34], CO_2 transformation from power station flue gas [45], and the substantial decrease in the carbon footprint of structural materials as CNT composites, including CNT-cement, CNT-steel, and CNT-aluminum [8,46], as well as modification of the CO$_2$ splitting process to yield other CNMs including carbon nano-onions, carbon platelets, and graphene [47–51].

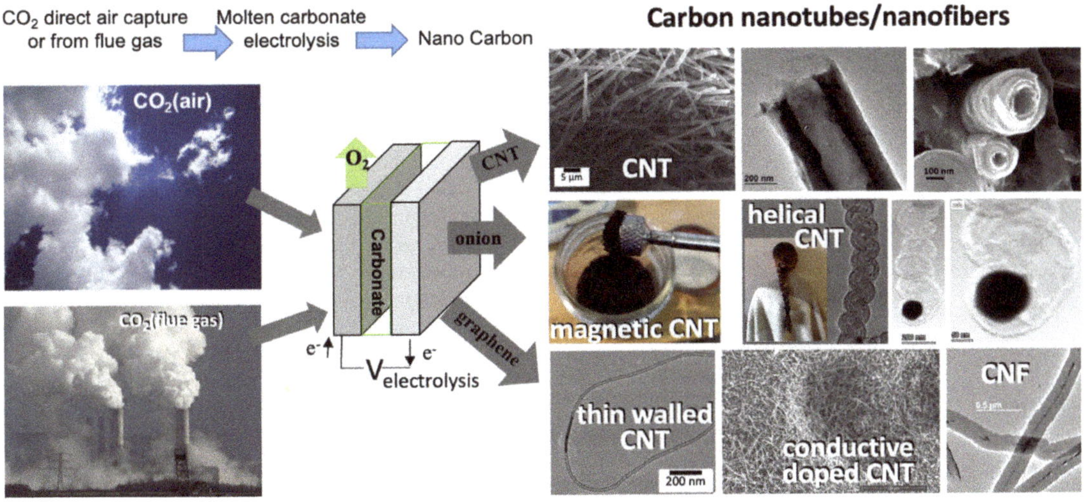

Figure 1. High-yield electrolytic synthesis of carbon nanotubes from CO$_2$, either directly from the air or from smokestack CO$_2$, in molten carbonate.

The rise of this greenhouse gas is causing extensive climate change and damage to the planet's ecosphere, and its mitigation is one of the most pressing challenges of our time [1–5]. Technical, catalyst-driven solutions to mitigate climate change are of the highest significance to the catalyst, not only due to their probes of a new chemistry to catalyze nanocarbon formation, but also by galvanizing the community with action towards mitigation of the existential climate change threat facing the planet. This study provides four contributions to a catalyst-driven solution to climate change: (i) Ten distinct, new electrochemical procedures are presented to transform CO$_2$ to CNTs at high purity. The procedures produce a variety of distinct CNT morphologies ranging from curled to straight, short to long, and thin to thick. (ii) This study explores the transition metal nucleation that catalyzes the process to produce high-purity carbon nanotubes. (iii) The study demonstrates new syntheses of macroscopic assemblies of CNTs using the C2CNT process, with structural implications towards their potential applications for nano-filtration and neural nets. (iv) This study provides an extensive carbon nanotube baseline to a companion study in which the same electrochemical components are utilized in new configurations to generate entire new classes of non-carbon nanotube graphitic nanocarbons.

2. Results and Discussion

2.1. Electrolytic Conditions to Synthesize High-Purity, High-Yield CNTs from CO$_2$

The first part of this study systematically explores electrochemical parameters to reveal a wide variety of conditions that yield a high-purity, high-yield CNT product by means of electrolysis of CO$_2$ in 770 °C lithium carbonate. An in depth look at the material composition and morphologies of the products is conducted, particularly around the transition metal nucleation zone of CNT growth. The latter part of this study reveals molten electrochemical conditions that produce macroscopic assemblies of CNTs. This study also serves as a sister study [51] in which small electrolytic changes in the 770 °C molten Li$_2$CO$_3$ yield major changes to the product consisting of new, non-CNT nanocarbon allotropes.

Previously, we showed that the high production rate (using a high electrolysis current density, J, of 0.6 A/cm^2) electrolytic splitting of CO$_2$ in molten Li$_2$CO$_3$ electrolyte using a Muntz Brass cathode (60% Cu and 40% Zn) and a Nichrome C (60% Ni, 24% Fe and

16% Cr) anode produced a high-quality (97% purity), high-aspect-ratio carbon nanotube (CNT) product with the addition, to the electrolyte, of either 0.1 wt% Fe_2O_3 [43] or 2.0 wt% Li_2O [41,44]. The addition of higher concentrations of either iron or lithium oxide to the electrolyte increased the formation of defects in the CNTs, as measured by Raman spectroscopy, which at this higher current density induced spiraling of the CNT during growth and the controlled growth of a variety of helical carbon nano-allotropes including single- and double-braided helices was observed, as well as flat, spiral morphologies [43,44].

Here, conditions related to the high-purity CNT synthesis were systematically varied to determine other electrochemical conditions that support the high-purity, low-defect formation of straight (non-helical) CNTs. Examples of the conditions that were varied include the composition of the cathode and anode, the additives to the lithium carbonate electrolyte, and the current density and time of the electrolysis. Variations of the electrodes include the use of cathode metal electrodes such as Muntz brass Monel, or Nichrome alloys. Anode variations include noble anodes such as iridium, various nickel-containing anodes including nickel, Nichrome A or C, Inconel 600, 625, or 718, or specific layered combinations of these metals. Electrolyte additives that were varied include Fe_2O_3, and nickel or chromium powder, and electrolyses were varied over a wide range of electrolysis current densities. Several electrolyses studied here, which yielded high-purity, high-yield carbon nanotubes, are described in Table 1. Scanning electron microscopy (SEM) of the products of a variety of those CNT syntheses was conducted by CO_2 electrolysis in molten Li_2CO_3 at 770 °C, and the results are presented in Figure 2.

Table 1. A variety of Electrolytic CO_2 splitting conditions in 770 °C Li_2CO_3 producing a high yield of carbon nanotubes.

Electrolysis	Cathode	Anode	Additives (wt% Powder)	Electr Time	Current Density A/cm^2	Product Description
A	Muntz Brass	Nichrome C	0.1% Fe_2O_3	0.5 h	0.6	97% Straight 50–100 µm CNT
B	Muntz Brass	Nichrome A	0.1% Fe_2O_3	4 h	0.15	94% Straight 20–80 µm CNT
C	Muntz Brass	Inconel 718	0.1% Fe_2O_3 0.1% Ni	4 h	0.15	96% curled CNT
D	Muntz Brass	Nichrome C	0.1% Fe_2O_3	15 h	0.08	70% 10–30 µm CNT
E	Muntz Brass	Inconel 625 3 layers Inconel 600	0.1% Fe_2O_3	15 h	0.08	97% 20–50 µm CNT
F	Muntz Brass	Inconel 718 2 layers Inconel 600	0.1% Fe_2O_3	4 h	0.15	98% Straight 100–500 µm CNT
G	Muntz Brass	Inconel 718 3 layers Inconel 600	0.1% Fe_2O_3 0.1% Ni	15 h	0.08	90% Curled CNT or fibers
H	Muntz Brass	Nichrome C	0.1% Fe_2O_3	1 h	0.4	96% Straight 100–200 µm CNT
I	Monel	Nichrome C	0.1% Fe_2O_3	1 h	0.4	97% Straight 20–50 µm CNT
J	Monel	Nickel	/	2 h	0.2	70% thin 10–20 µm CNT Rest: Onions
K	Monel	Nichrome C	0.1% Fe_2O_3	2 h	0.1	97% 30–60 µm straight CNT
L	Monel	Nichrome C	0.5% Fe_2O_3	15 h	0.08	~25% curled CNT ~70% straight CNT
M	Monel	Iridium	0.81% Cr	18 h	0.08	97% thin 50–100 µm CNT

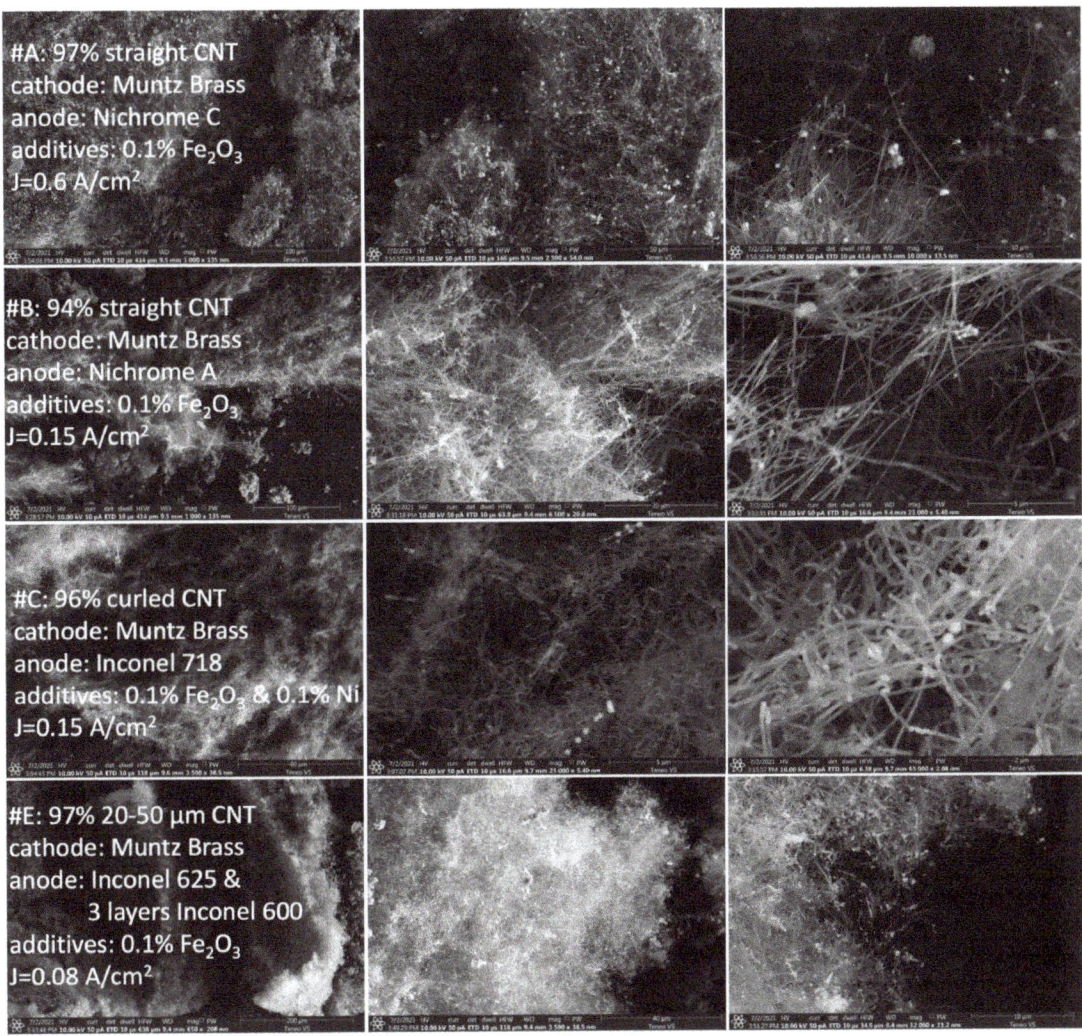

Figure 2. SEM of the synthesis product of high-purity, high-yield carbon nanotubes under a variety of electrochemical conditions by electrolytic splitting of CO_2 in 770 °C Li_2CO_3. The washed product was collected from the cathode subsequent to the electrolysis described in Table 1. Moving left to right in the panels, the product was analyzed by SEM with increasing magnification. Scale bars in panels (starting from left) are for panels A: 100, 50, and 10 µm; for panels B: 100, 20, and 5 µm; for panels C: 40, 5, and 2 µm; for panels E: 200, 40, and 10 µm.

For Electrolysis A, the top row of Table 1 presents the electrochemical conditions, and the top row of Figure 2 presents the SEM results of the product of a repeat of the electrochemical conditions of the described 0.1 wt% Fe_2O_3 electrolysis (the same lithium carbonate electrolyte, the same Muntz Brass cathode and Nichrome C anode, and the same 0.6 A/cm^2 current density and 30 min electrolysis duration), but used a simpler (from a material perspective) alumina (ceramic Al_2O_3), rather than stainless steel 304 electrolysis cell casing. Use of the alumina casing in this study limited the pathways for metals to enter, thereby the reducing parameters to evaluate, and possibly effect, the electrolytic system. Note, however, that the stainless steel 304 was not observed to corrode, and the switch from stainless to alumina was not observed to materially affect the electrolysis product. The CNT product was again 97% purity, and the coulombic efficiency was 99%, which quantified the measured available charge (current multiplied by the electrolysis time) to the measured number of four electrons per equivalent of C in the product, and the carbon nanotube length was 50 to 100 µm.

The second row of Figure 2 (panels B) represents a change only in the current density, which was lowered to 0.15 A/cm^2, and the electrolysis time, which was increased to 4 h, and the result was a decrease in product purity to 94%, a decrease in CNT length to 20–80 µm, and a modest decrease in coulombic efficiency to 98%. At this current density, as observed in the third row of Figure 2, panels C, the addition of 0.1 wt% Ni along with the 0.1 wt% Fe$_2$O$_3$ resulted in 96% purity, zigzag-patterned and twisted, rather than straight CNTs. These twists could be induced by over-nucleation decreasing control of the CNT linear growth. In the most magnified of these product images (right side of the Figure, 2 µm bar resolution), evidence of the over-nucleation can be observed in the larger nodules visible at the CNT tips and joints.

At a low current density of 0.08 A/cm^2, with an electrolyte additive of 0.1 wt% Fe$_2$O$_3$, the conventional Muntz Brass and Nichrome electrodes exhibited a significant drop in CNT product purity to 70%. Coulombic efficiency tended to drop off with current density, and, in this case, the coulombic efficiency of the synthesis was 82%. Product purity could be increased by refining the mix of transition metals available to the electrolytes or increasing the surface area. The alloy compositions of the metals used as electrodes are presented in Table 2. The metal variation was further refined by combining the metals in Table 2 as anodes; for example, using a solid sheet of one Inconel alloy, layered with a screen or screens of another Inconel alloy. This approach was utilized to obtain the results shown in the lowest row of Figure 2 (panels E); an anode of Inconel 625 with three layers of (spot welded) 100-mesh Inconel 600 screen, a return to a single electrolyte additive (0.1 wt% Fe$_2$O$_3$), and a very low current density of 0.08 A/cm^2 were utilized. As seen in panels E of the figure, the product was high purity (97%) and consisted of 20–50 µm length CNTs, and the coulombic efficiency was 75%. Not shown, but included in Table 1 (Electrolysis G), are the results obtained under the same electrode, and the same 0.08 A/cm^2 electrolysis conditions. However, with the electrolyte addition of both 0.1 wt% Fe$_2$O$_3$ and 0.1 wt% Ni at J = 0.15 A/cm^2, the product was twisted CNTs as shown in Figure 2 panels C, the purity was 96%, and the coulombic efficiency was 80%.

Table 2. Compositions of various alloys used (weight percentage).

Alloy	Ni %	Fe %	Cu %	Zn %	Cr %	Mo %	Nb & Ta %
Nichrome C	60	24			16		
Nichrome A	80				20		
Inconel 600	52.5	18.5			19.0	3.0	3.6
Inconel 718	72% min	6–10			14–17		
Inconel 625	58	5 max			20–23	8–10	4.15–3.15
Monel	67		31.5				
Muntz Brass			60	40			

2.2. Electrolytic Conditions to Synthesize High-Purity, High-Yield CNTs from CO$_2$

The syntheses listed in Table 1 delineated the electrochemical growth conditions for the high-purity growth of carbon nanotubes, each exhibiting the characteristic concentric multiple-graphene cylindrical walls. This can be observed in Figure 3, which presents Transmission Electron Microscopy (TEM) and High Angle Annular Dark-Field TEM (HAADF) results of a typical example (the product of Electrolysis E, as further described in Table 1 and Figure 2), and which provides general structural and mechanistic information of carbon nanotubes synthesized by molten electrolysis. As seen in the top row of the figure, the carbon nanotubes were formed by successive, concentric layers of cylindrical graphene. The graphene can be identified by its characteristic inter-graphene layer separation of

0.33 to 0.34 nm as measured in the figure by the spacing between the dark layers of uniform blocked electron transmission on the magnified top right side of the figure. This CNT had an outer diameter of 74 nm, and inner diameter of 46 nm, and by counting dark rows, it can be determined that the number of graphene layers in this CNT was 41. The right side of the third row of the figure measures the carbon elemental profile of the CNT. This profile was swept laterally from the tube's exterior (no carbon) through the left wall (carbon), then through the void of interior of the tube (low carbon from the exterior backside wall), then through the right wall (carbon), and finally to the exterior of the tube on the outer left side (no carbon). Additionally, the integrated elemental profile of area 1 of this panel is shown, which exhibits 100.0% carbon (fit error 1.3%).

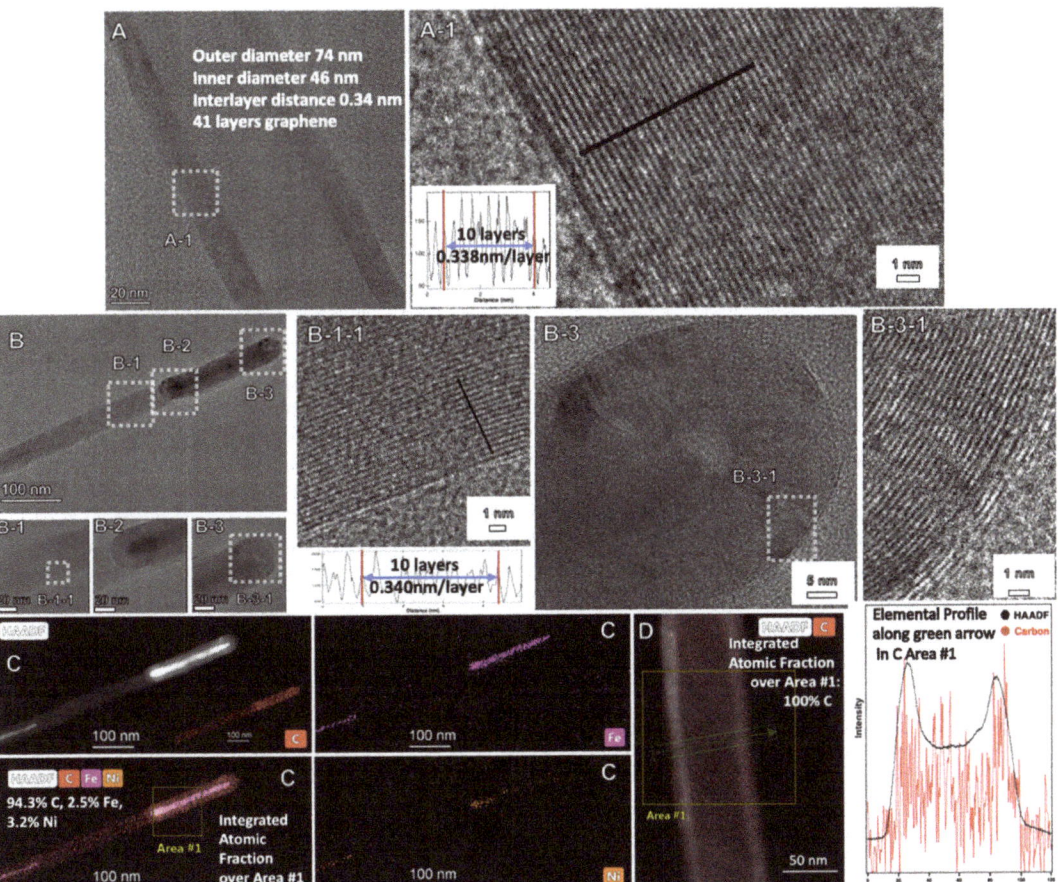

Figure 3. TEM and HAADF of the synthesis product of high-purity, high-yield carbon nanotubes under the Electrolysis E (Table 1) electrochemical conditions by electrolytic splitting of CO_2 in 770 °C Li_2CO_3. In the top row, the product is analyzed by TEM with scale bars of 20 nm (left panel) or 1 nm (right). Moving left to right in the second row, there are scale bars of 100, 5, 5, and 1 nm. The third row's scale bars are 100 or 50 nm. The bottom row scale bars are 20, 1, and 1 nm. Panels: (**A,B**), B-1, B-2, B-3, B-3-1 TEM; A-1, B-1-1 TEM, with measured graphene layer thickness. (**C**): Elemental HAADF elemental analysis; (**D**): HAADF element analysis with (right side) elemental profile.

In Figure 3 on the right side of row 2, the parallel 0.34 nm spacing for the graphene layers in the CNT walls is again observable. This panel also includes dark areas of metal trapped within the CNT, and serves as a snapshot in time of the growth of the CNT. In the third row of the figure, it can be seen that HAADF analysis of Area 1 revealed an elemental composition for this area, including the walls with the trapped interior metal, of 94.4% carbon, 2.5% Fe, and 3.2% Ni, as distributed according to the individual C, Fe, and Ni HAADF maps included in the figure. The second row of the figure also shows the tip of the CNT, which included trapped metal. The transition metal served as a nucleating agent, which supported the formation of the curved graphene layers shown at the tip of the CNT, which is a major component of the CNT growth mechanism. While occurring in an entirely different physical chemical environment than chemical vapor deposition (CVD), this molten carbonate electrolysis process of transition metal nucleated growth of CNTs appeared to be similar to those noted to occur for CVD CNT growth. This was the case despite the fact that CVD is a chemical/rather than electrochemical process, and occurs at the gas/solid, rather than liquid/solid interface.

2.3. Tailored Electrochemical Growth Conditions Producing High-Aspect-Ratio CNTs from CO_2

Figure 4 presents the product's SEM of the electrochemical configuration that yielded the longest (100 to 500 μm long) and highest purity (98%) CNTs at high coulombic efficiency (99.5%) of those studied here (described as Electrolysis F in Table 1). As with the previous configuration that yielded nearly as high-purity, but shorter, CNTs. The synthesis used an 0.1 wt% Fe_2O_3 additive to the Li_2CO_3 electrolyte, a Muntz Brass cathode, and an Inconel 718 anode with a layered Inconel 600 screen. However, this synthesis found an optimization in CNT purity and length using two, rather than three layers of Inconel 600, and using a higher current density (0.4, rather than 0.08, A/cm^2) and shorter electrolysis time (4, rather than 15, hours). With a diameter of <0.2 μm, these CNTs could have an aspect ratio of >1000. As correlated with the alloy composition in Table 1, the smaller number of Inconel 600 layers reflected the need for the inclusion of anodic molybdenum available in that alloy, but at a controlled, lower concentration, to achieve the resultant high-purity, high-aspect-ratio CNTs. As seen in the figure, the CNTs were densely packed and largely parallel, and as discussed in sect. 2.5, would be a useful candidate for use in nano-filtration.

Figure 5 presents TEM and HAADF probes of the high-aspect-ratio CNT product of Electrolysis F (as described in Table 1, and by SEM in Figure 4). As seen on the right side of the middle row of the figure, the CNT walls consisted of parallel carbon (layers separated by the characteristic 0.33–0.24 nm graphene layer spacing). As seen in the elemental analysis of Area 1 in the lowest row, the areas consisted of hollow tubes composed of 100% carbon. However, as seen in the TEM of the top two rows and in the bottom row as the HAADF elemental profiles, there were also extensive portions of the tubes that were intermittently filled with metal. In the bottom row of the figure, a lateral cross sectional elemental CNT profile scanned through Area 2 from the outside, through the CNT and then out the opposite wall, shows that the wall was composed of carbon, while the inner region also contained iron as the dominant metal coexisting with some nickel.

2.4. Tailored Electrochemical Growth Conditions Producing Thinner CNTs from CO_2

Figure 6 demonstrates additional electrochemical conditions that yielded high-purity carbon nanotubes by CO_2 molten electrolysis. In the top row, panels H, as with high current density (Figure 2 panels A), a moderate current density of 0.4 A/cm^2 (with the same electrolyte, a Muntz Brass cathode and a Nichrome C anode) yielded high-purity (96%) CNTs that were somewhat longer (100–200 μm) at a coulombic efficiency approaching 100%. Switching the cathode material to Monel, as shown in the second row (Figure 6, panels a), yielded shorter 20–50 μm CNTs with 97% purity and coulombic efficiency again approaching 100%. Not shown in the figure, but included in Table 1 (Electrolysis D), is the fact that a switch from Nichrome C to a pure nickel anode (while retaining the Monel Cathode, and with electrolyte additives at J = 0.2 A/cm^2) led to a substantial drop in CNT purity to 70% with the remainder of the product consisting of nano-onions. A drop of current density from 0.4 A to 0.1 A/cm^2, as shown in in Figure 6 panels K, yielded 97%

purity CNTs of 30–60 μm length with only a small drop of coulombic efficiency to 97%. In a single panel of L located in the lower left corner of Figure 6, it can be seen that an overabundance of Fe_2O_3 was added, which was previously observed to lead to a loss of control of the synthesis specificity [45]. In this case, the total purity of the CNTs remained high at ~95%, but this consisted of two distinct morphologies of CNT in the product. The majority product, at ~75%, was twisted CNTs, and the minority product, at ~20%, was straight CNTs. Finally, in panels M on the middle and right lowest row of Figure 6, a noble metal, iridium, was used as the anode (along with the Monel Cathode) at a low 0.08 A/cm² current density. Transition metals released from the anode, during its formation of a stable oxide over layer, can contribute to the transition metals ions that are reduced at the cathode and serve as nucleation points for the CNTs. This was not the case here due to the high stability of the iridium. Instead, as a single, high concentration transition metal, 0.81 wt% Cr was made as the electrolyte additive. The product was highly pure (97%) CNTs that were the thinnest shown (<50 nm diameter), were 50–100 μm long for an aspect ratio > 1000, and formed at a coulombic efficiency of 80%.

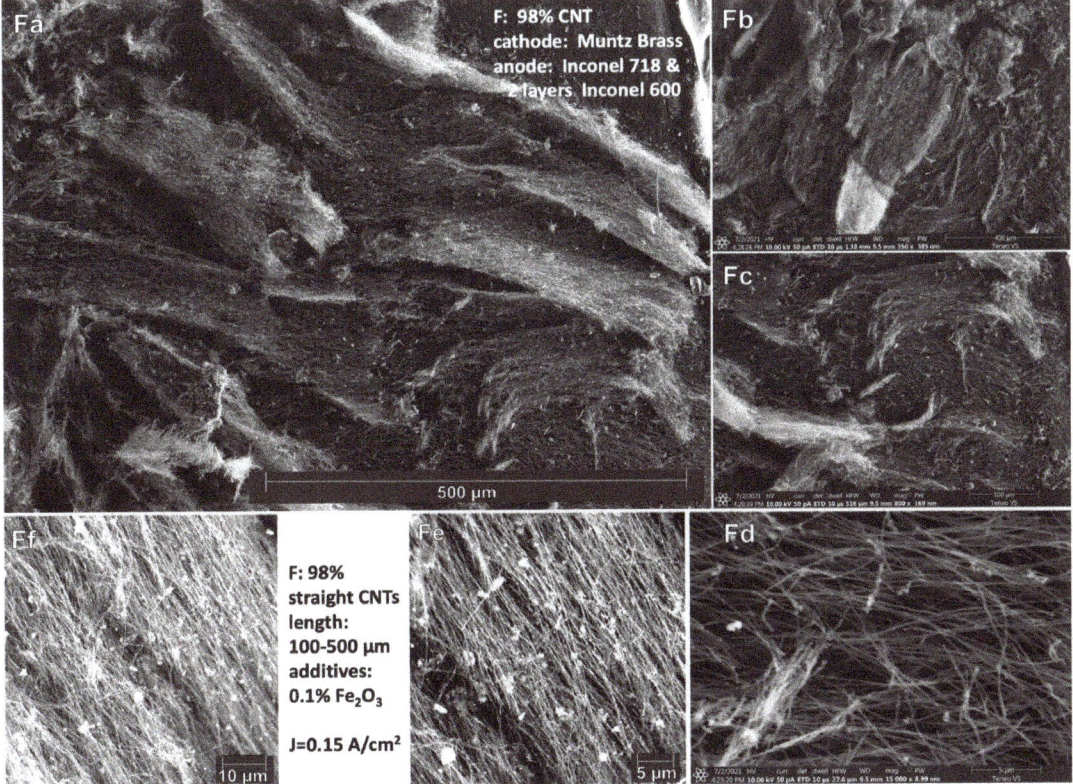

Figure 4. SEM of the synthesis product of high-aspect-ratio (and high-purity and high-yield) carbon nanotubes prepared by electrolysis F in Table 1, splitting CO_2 in 770 °C Li_2CO_3. Moving left to right in the panels, the product is analyzed by SEM with increasing magnification. Scale bars in panels Fa-Ff (clockwise from top) are 500, 400, 100, 5, 5, and 10 μm.

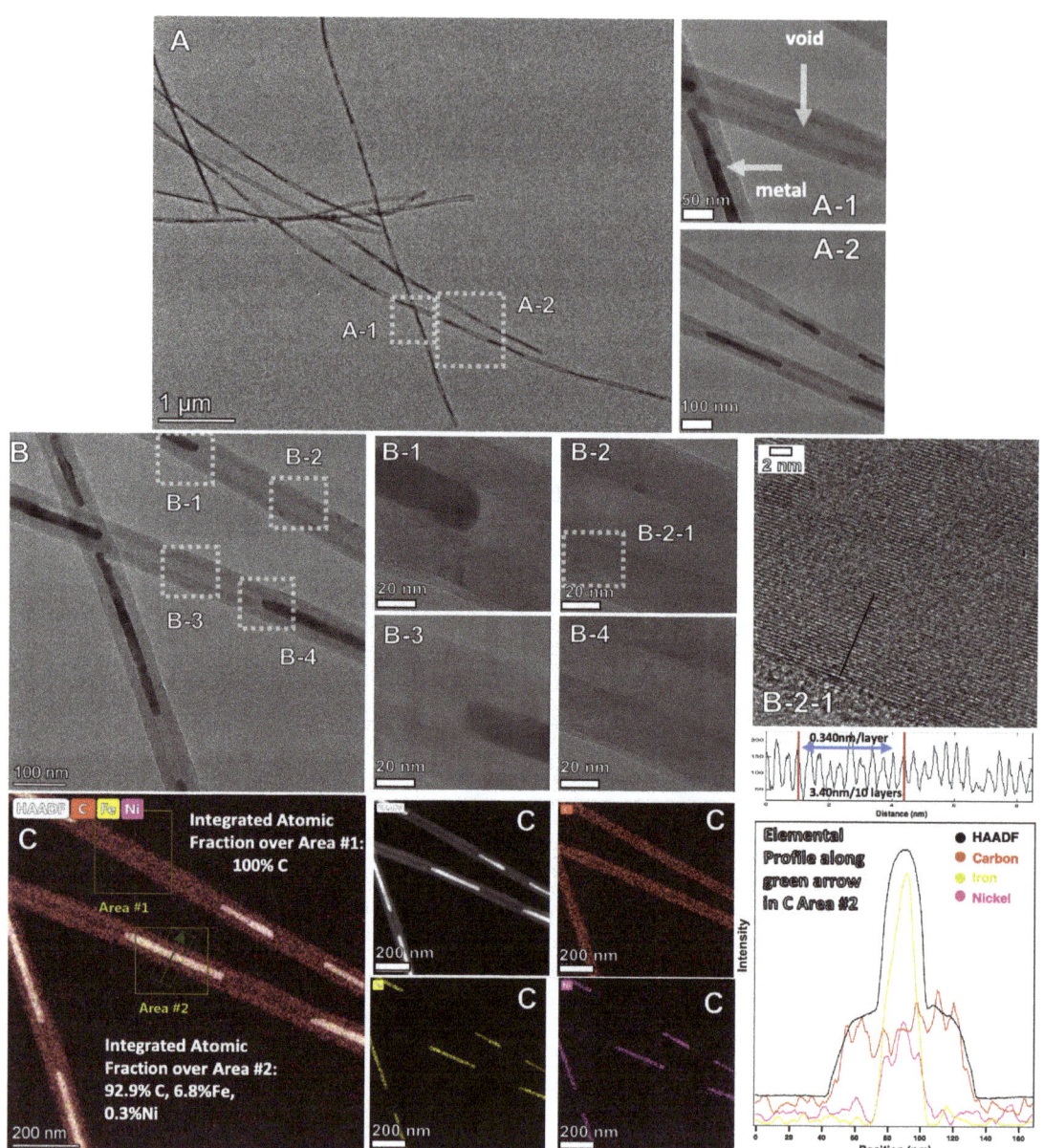

Figure 5. TEM and HAADF of the synthesis product of high-purity, high-yield carbon nanotubes under the Electrolysis F (Table 1) electrochemical conditions by electrolytic splitting of CO_2 in 770 °C Li_2CO_3. In the top row, the product is analyzed by TEM with scale bars of 1 μm (left panel) or 100 nm (right). Scale bars in the middle right moving left to right are of 50, 20, and 1 nm. HAADF measurements in the bottom panel each have scale bars of 200 nm. Panels: (**A**), A-1, A-2, (**B**), B-1, B-2, B-3, B-4: TEM; B-2-1 TEM with measured graphene layer thickness. (**C**): Elemental HAADF elemental analysis with (right side) elemental profile.

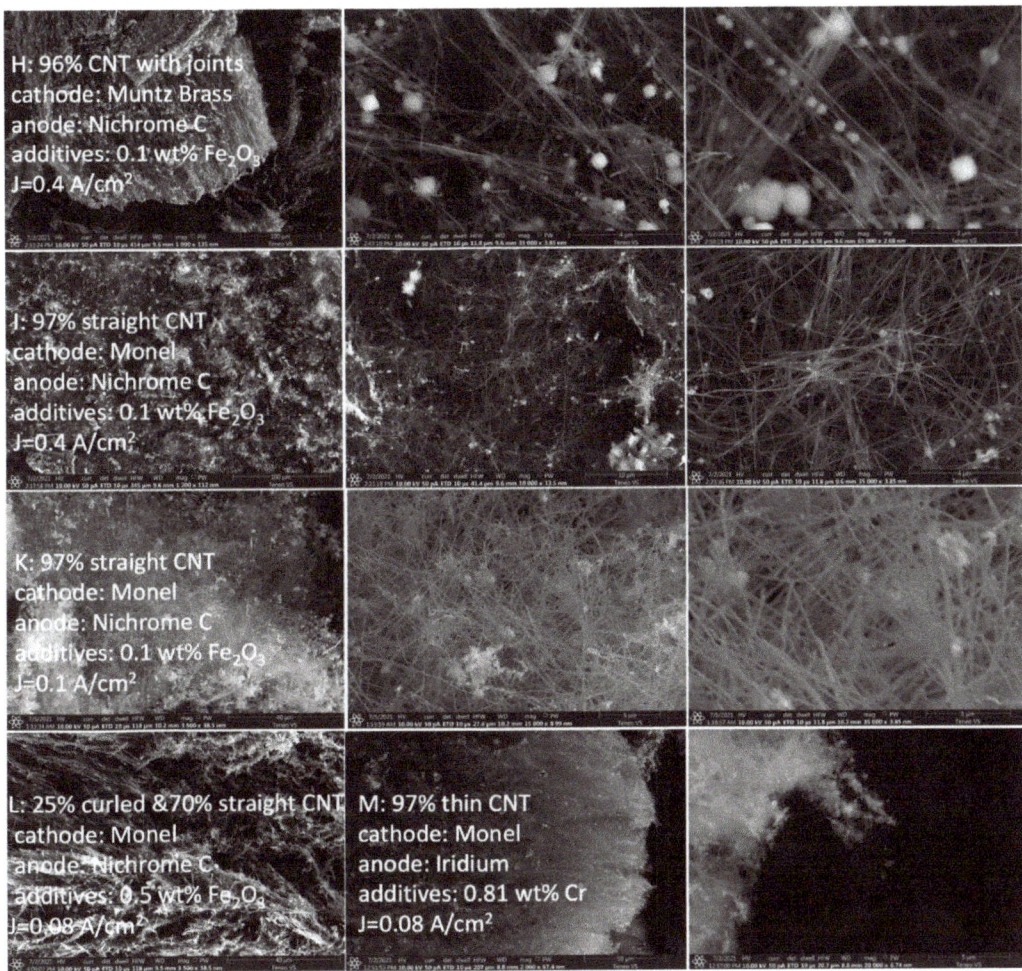

Figure 6. SEM of the synthesis product of high-purity, high-yield carbon nanotubes under a variety of electrochemical conditions by electrolytic splitting of CO_2 in 770 °C Li_2CO_3. The washed product was collected from the cathode subsequent to the electrolysis described in Table 1. Scale bars (starting from left) are for panels H: 100, 4, and 2 µm; for panels I: 100, 10, and 4 µm; for panels K: 40, 5, and 3 µm; for panels L: 40, 50, and 5 µm.

SEM analysis of several of the synthesis products, specifically Electrolyses H, B, and C, revealed evidence of nodules that appeared as "buds" attached to the CNTs. These buds were the most consistent in Electrolysis H and were explored by TEM and HAADF, as shown in Figure 7. It is fascinating, as seen in the top row of the figure, that the buds generally had a spherical symmetry, and while not prevalent in the structure, appeared in a structure comparable to grape bunches growing on a vine. The buds generally contained a low level of the transition metal nucleating metal, such as the 0.3% Fe shown, and the rest of the structure was generally pure carbon, with an occasional metal core. This low level of metal used was easily removed by an acid wash. Previously introduced higher levels of Ni or Fe could lead to magnetic carbon nanotubes with useful properties of recyclability, filtration, and shape-shifting materials among other applications [41]. As seen in the left side of the second row, the carbon nanotube walls continued to exhibit the regular 0.33 to

0.34 graphene interwall separation, as seen on the right side of the row, and the adjacent CNTs may have had merged or distinct graphene structures. Similarly, as seen in the third row of Figure 2, the adjacent buds on the CNTs could have graphene walls that bended to join, and were shared, or, as seen in the fourth row, appeared instead to be distinct (intertwined, not merged) structures.

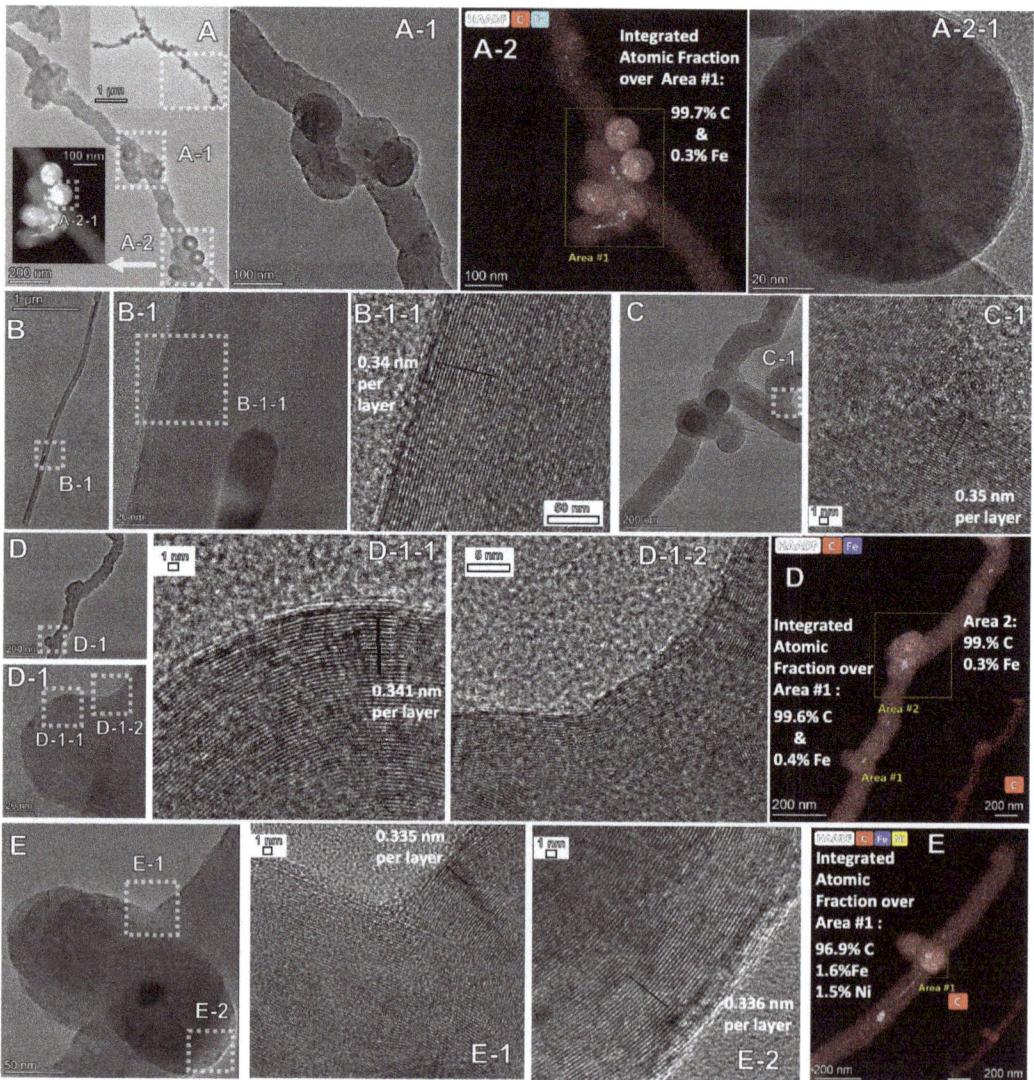

Figure 7. TEM and HAADF of the synthesis product of carbon nanotubes that exhibited nodules or buds under the Electrolysis H (Table 1 and SEM on top row of Figure 6) electrochemical conditions by electrolytic splitting of CO_2 in 770 °C Li_2CO_3. In the top row, the product is analyzed with scale bars from left to right of 200, 100, 20 and 100 nm. Scale bars in the second right row are of 1μm, then 20, 5, 200, and 1 nm. The third row scale bars are 200, 20, 1, 5, and 200 nm. The bottom row scale bars are 50, 1, 1, and 200 nm. TEM Panels: (**A**) A-1, A-2-1, (**B**), B-1, B-1-1, (**C**), C-1, C-2, D-1-1, D-1-2, (**E**), E-1, E2: TEM; A-2, (**D**,**E**) also include Elemental HAADF elemental analyses.

2.5. Electrochemical Conditions to Synthesize Macroscopic Assemblies of CNTs

In addition to individual CNTs, the final series of electrolyses generated useful macroscopic assemblies of CNTs. There has been interest in densely packed CNTs for nanofiltration, and also, due to their high density of conductive wires, as an artificial neural net [52–62]. CNTs' aerogels have been reported as being formed by CVD and/or also reported as being formed within molds. Their sorbent properties have been investigated for applications such as the cleanup of chemical leakages under harsh conditions. Those studies noted that such aerogel matrices, consisting of highly porous, intermingled CNTs, can be repeatedly compressed to a small fraction of their initial volume without damaging the structure of the carbon nanomaterials [52–56].

The term "Nanofiltration" was proposed in 1984 to solve the terminology problem for a selective reverse osmosis process that allows ionic solutes in a feed water to permeate through a membrane [57]. In addition to low energy consumption, with respect to alternative unit operations such as distillation and evaporation, thermal damage of heat-sensitive molecules can be minimized during the separation due to the potential for low operating temperatures of nanofiltration [58]. Small-diameter CNTs were demonstrated due to their ability to be used as a nanofiltration or molecular sieve to selectively remove larger size molecules from smaller size molecules, such as the selective removal of cyclohexane from n-hexane [57]. There is a need to control the fabrication of macroscopic assemblies of CNTs to optimize their capabilities for nanofiltration, and CNT assemblies synthesized by CVD and laser ablation have been investigated [52–56,58–61].

An artificial neural network is a collection of interconnected nodes that loosely model the neurons in a biological brain. Estimates of biologic neuronal density (rat) are in the range of 100 in a 100 µm cube (10^5/mm^3) on each side. The fabrication of an artificial neural network with a structure that mimics that number of nanowires and nodes presents a challenge. However, this is in the same size domain as macroscopic assemblies of CNTs. For example, Gabay and co-workers explored the engineered self-organization of neural networks using carbon nanotube clusters [38], emphasizing the need for improved pathways to fabricate and control macroscopic assemblies of CNTs.

The macroscopic assemblies observed in this study are referred to as nano-sponge, densely packed parallel CNTs, and nano-web CNTs, in Table 3 and Figure 8. The Nano-sponge assembly was formed by Electrolyses N with Nichrome C serving as both the cathode and the anode, with 0.81% Ni powder added to the 770 °C Li_2CO_3 electrolyte, the initial current ramped upwards (5 min each at 0.008, 0.016, 0.033 and 0.067 A/cm^2), then a 4h current density of 0.2 A/cm^2 generating a 97% purity nanosponge at 99% coulombic efficiency. As previously described, long, densely packed, parallel carbon nanotubes were produced in Electrolysis F with a 0.1 wt% Fe_2O_3 additive to the Li_2CO_3 electrolyte, a Muntz Brass cathode and an Inconel 718 anode, and two layers of Inconel 600 screen at 0.15 A/cm^2.

Table 3. Systematic variation of CO_2 splitting conditions in 770 °C Li_2CO_3 to optimize the formation of macroscopic assemblies of nanocarbons with densely packed carbon nanotubes.

Electrolysis	Cathode	Anode	Additives (wt% Powder)	Electrolysis Time	Current Density (A/cm^2)	Product Description
N	Nichrome C	Nichrome C	0.81% Ni	4 h	0.2	97% nano-sponge CNT
F	Muntz Brass	Inconel 718 2 layers Inconel 600	0.1% Fe_2O_3	4 h	0.15	98% dense packed straight 100–500 µm CNT
P	Muntz Brass	Nichrome C 3 layers Inconel 600	0.1% Fe_2O_3	15 h	0.08	97% 50–100 µm nano-web CNT
Q	Monel	Nichrome C	0.81% Ni	3 h	0.2	92% 5–30 µm nano-web CNT Rest: onions

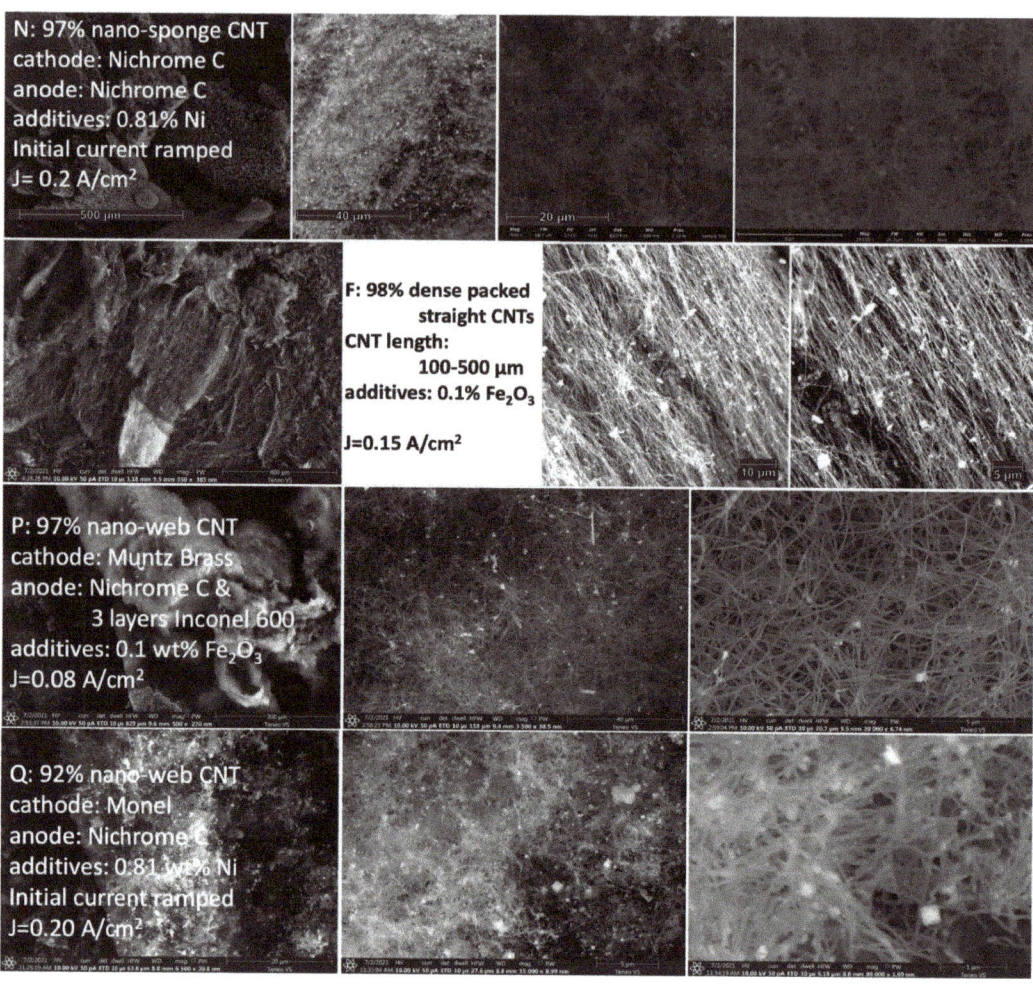

Figure 8. SEM of the synthesis product consisting of carbon nanotubes arranged in various packed macroscopic structures that are amenable to nano-filtration. The washed product was collected from the cathode subsequent to the electrolysis described in Table 3. Moving from left to right in the panels, the product is analyzed by SEM with increasing magnification. Scale bars in panels These include nano-sponge, dense packed straight, and nano-web CNTs. Moving from left to right in the panels, the product is analyzed by SEM in the top two rows (N), and subsequent rows P and Q, with increasing magnification. Scale bars in panels (starting from left) are for top panels N: 500, 40, 20, and 8 μm; for lower panels N: 400, 10, and 5 μm; for panels P: 300, 40, and 5 μm; for panels Q: 500, 40, 20, and 8 μm.

As opposed to the parallel assembly produced in Electrolysis F, nano-web aptly describes the interwoven carbon nanotubes from Electrolyses P and Q, presented in the lower rows of Table 3 and Figure 8. Two different routes to the nano-web assembly are summarized. The first uses an 0.1% Fe_2O_3 additive, a Muntz Brass cathode, and an Inconel 718 anode with three layers of Inconel 600 screen, at 0.08 A/cm², generating a nano-web with a purity of 97% at a coulombic efficiency of 79%. The second pathway uses an 0.81% Ni powder additive, a Monel cathode, and Nichrome C anode, at 0.28 A/cm², generating a nano-web with a purity of 92% at a coulombic efficiency of 93%.

The densely packed straight CNTs had an inter-CNT spacing ranging from 50 to 300 nm; moreover, the CNTs were highly aligned, providing unusual nanofiltration opportunities for both this size domain, and for an opportunity to filter 1D from 3D morphologies. The nano sponge did not have this alignment feature, and as shown in Figure 8, provided nanofiltration pore sizes of 100 to 500 nm, while the nano-web product provided nanofiltration with pore sizes of 200 nm to 1 µm. Future studies will investigate the effectiveness of this portfolio of macroscopic CNT assemblies for nanofiltration.

2.6. Raman and XRD Characterization of the CNTs and Their Macro-Assemblies

Figure 9 presents the Raman spectra effect of variation of the CNT electrolysis conditions on the CNT assembly products from CO_2 electrolysis in 770 °C Li_2CO_3. The Raman spectrum exhibits two sharp peaks ~1350 and ~1580 cm^{-1}, which correspond to the disorder-induced mode (D band) and the high-frequency E_{2G} first order mode (G band), respectively, and an additional peak, the 2D band, at 2700 cm^{-1}. In the spectra, the graphitic fingerprints lie in the 1880–2300 cm^{-1} and are related to different collective vibrations of sp-hybridized C-C bonds.

Table 4. Raman spectra of a diverse range of carbon allotropes and macro-assemblies formed by molten electrolysis.

CO_2 Molten Electrolysis Product Description	ν_D (cm^{-1})	ν_G (cm^{-1})	ν_{2D} (cm^{-1})	I_D/I_G	I_{2D}/I_G
Nano-web	1342.5	1577	2689.6	0.28	0.50
Dense packed CNTs	1342.5	1577.4	2694.8	0.46	0.49
Nano-sponge	1352.5	1580.6	2687.3	0.67	0.62

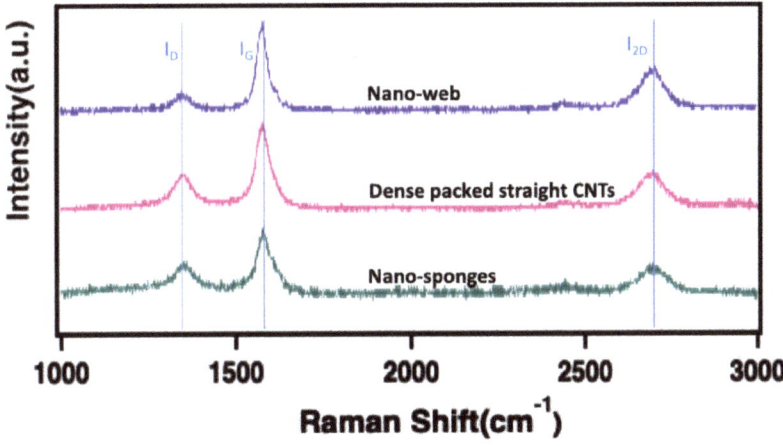

Figure 9. Raman spectra of the synthesis product consisting of various labeled carbon nanotube assemblies synthesized by the electrolytic splitting of CO_2 in 770 °C Li_2CO_3 with a variety of systematically varied electrochemical conditions described in Table 4.

Interpretation of the Raman spectra provides insights into the potential applications of the various carbon allotropes. As shown in Figure 9, the intensity ratio between the D band and the G band (I_D/I_G) was calculated; this ratio, and the observed shift in the I_G frequency, are useful parameters to evaluate the relative number of defects and degree of graphitization, and are presented in Table 4. Note in particular that, of the nano-sponge, nano-web, and densely packed CNT assemblies described in Figure 8 and Table 3, the nano-web CNT assembly exhibited low disorder, with I_D/I_G = 0.36, as shown in Table 4, the densely packed CNT assembly exhibited intermediate disorder with I_D/I_G = 0.49, and the nano-sponge exhibited the highest disorder with $I_D/I_G = 0.62$ while accompanied by a shift in I_G frequency.

That is, for the assemblies with increasing I_D/I_G ratio: CNT nano-web < Dense packed CNT < CNT nano-sponge.

Previously, increased concentrations of iron oxide added to the Li_2CO_3 electrolyte were correlated with an increasing degree of disorder in the graphitic structure [38]. It should be noted that these defect levels each remain relatively low as the literature is replete with reports of multiwalled carbon nanotubes made by other synthetic processes with $I_D/I_G > 1$. Lower defects are associated with applications that require high electrical power and strength, while high defects are associated with applications that permit high diffusivity through the structure, such as those associated with increased intercalation and higher anodic capacity in Li-ion batteries and higher charge supercapacitors.

Along with the XRD library of relevant compound spectra, XRD results of the CNT assembly products are presented in Figure 10, prepared as described in Figure 8 and Table 3. Each of the spectra exhibited strong diffraction peaks at $2\theta = 27°$, characteristic of graphitic structures. The nano-sponge XRD spectra was distinct from the others, with a dominant peak at $2\theta = 43°$, indicating the presence of iron as $Li_2Ni_8O_{10}$ and chromium as $LiCrO_2$ by XRD spectra match. The XRD result of this nano-sponge exhibited little or no iron carbide. On the other hand, both the nano-web and densely packed straight CNTs exhibited additional significant peaks at $2\theta = 42$ and $44°$, indicative of the presence of iron carbide, Fe_3C. The diminished presence of defects previously noted by the Raman spectra for the other densely packed CNTs, along with the XRD presence of $Li_2Ni_8O_{10}$, $LiCrO_2$, and Fe_3C, provide evidence that the co-presence of Ni, Cr, and Fe as nucleating agents can diminish defects in the CNT structure compared to Ni and Cr alone.

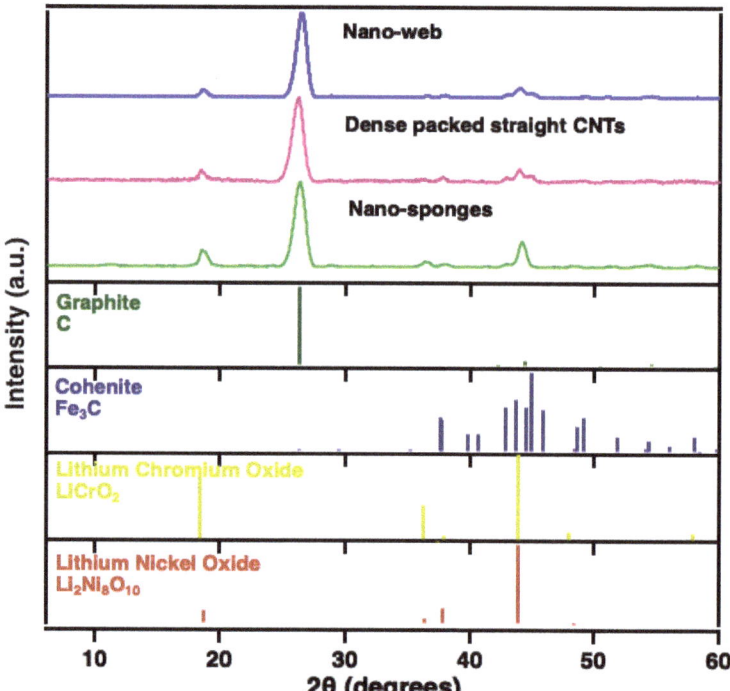

Figure 10. XRD of the synthesis product consisting of various labeled carbon nanotube assemblies synthesized by the electrolytic splitting of CO_2 in 770 °C Li_2CO_3 with a variety of systematically varied electrochemical conditions described in Table 4.

3. Materials and Methods

3.1. Materials

Lithium carbonate (Li_2CO_3, 99.5%), lithium oxide (Li_2O, 99.5%), lithium phosphate Li_3PO_4 (Li_3PO_4, 99.5%), iron oxide (Fe_2O_3, 99.9%, Alfa Aesar), and boric acid (H_3BO_3, Alfa Aesar 99+%) were used as electrolyte components in this study. For electrodes, Nichrome A (0.04-inch-thick), Nichrome C (0.04-inch-thick), Inconel 600 (0.25-in thick), Inconel 625 (0.25-in thick), Monel 400, Stainless Steel 304 (0.25-in thick), Muntz Brass (0.25-in thick), were purchased from onlinemetals.com (accessed on 14 December 2021). Ni powder was 3–7 µm (99.9%, Alfa Aesar). Cr powder was <10 µm (99.2%, Alfa Aesar). Co powder was 1.6 µm (99.8%, Alfa Aesar). Iron oxide was 99.9% Fe_2O_3 (Alfa Aesar). Co powder was 1.6 µm (99.8%, Alfa Aesar). Inconel 600 (100 mesh) was purchased from Cleveland Cloth. The electrolysis was a conducted in a high form crucible >99.6% alumina (Advalue).

3.2. Electrolysis and Purification

Specific electrolyte compositions of each electrolyte are described in the text. The electrolyte was pre-mixed by weight in the noted ratios then metal or metal oxide additives were added if used. The cathode was mounted vertically across from the anode and immersed in the electrolyte. Generally, the electrodes were immersed subsequent to electrolyte melting. For several noted electrolyses, once melted, the electrolyte was maintained at 770 °C ("aging" the electrolyte) prior to immersion of the electrolytes, followed by immediate electrolysis. Generally, the electrolysis was driven with a described constant current density. As noted, for some electrolyses, the current density was ramped in several steps building to the applied electrolysis current, which was then maintained at a constant current density. Instead, most of the electrolyses were initiated, and held, at a single constant current. The electrolysis temperature was 770 °C using CO_2 directly from the air. In the C2CNT process, the electrolytic splitting can occur as direct air carbon capture without CO_2 pre-concentration [31–34,36–44,47–51], or with concentrated CO_2, or CO_2 exhaust gas including during the scaling up of this process, in which the CO_2 transformation to CNTs was awarded the 2021 Carbon XPrize XFactor award for producing the most valuable product from CO_2 [32,45,46,62–64]. In this study, as the electrolysis cell directly captured CO_2 from the air via Equation (1), no additional introduction of CO_2 was needed. A simple measure of sufficient CO_2 uptake is whether the electrolyte level falls during the course of the electrolyte. As mentioned, the ^{13}C isotope of CO_2 was previously used to track carbon through the C2CNT process from its origin (CO_2 as a gas phase reactant) through its transformation to a CNT or carbon nanofiber product, and the CO_2 originating from the gas phase served as the renewable C building blocks in the observed CNT product [32]. If CO_2 uptake was insufficient, then the carbonate electrolyte was instead consumed in accord Equation (2), rather than renewed in accord with Equations (1) and (2) in tandem. For example, when conducting at high electrolysis rates of 1 A/cm^2 or greater (not the situation of this study), then gas containing CO_2 needed to be bubbled into the electrolyte, otherwise the electrolyte was consumed and the level of electrolyte visibly fell [40].

3.3. Product Characterization

The raw product was collected from the cathode after the experiment and cool down, followed by an aqueous wash procedure that removed electrolyte congealed with the product as the cathode cooled. The washed carbon product was separated by vacuum filtration. The washed carbon product was dried overnight in a 60 °C oven, yielding a black powder product.

The coulombic efficiency of electrolysis is the percent of applied, constant current charge that was converted to carbon, determined as:

$$100\% \times C_{experimental}/C_{theoretical} \quad (4)$$

This was measured by the mass of washed carbon product removed from the cathode, $C_{experimental}$, and calculated from the theoretical mass, $C_{theoretical} = (Q/nF) \times (12.01 \text{ g C mol}^{-1})$, which was determined from Q, the time-integrated charged passed during the electrolysis, F, the Faraday (96485 As mol^{-1} e^{-}), and the n = 4 e^{-} mol^{-1} reduction of tetravalent carbon, consistent with Equation (2).

Characterization: The carbon product was washed, and analyzed by PHENOM Pro Pro-X SEM (with EDX), FEI Teneo LV SEM, and by FEI Teneo Talos F200X TEM (with EDX). XRD powder diffraction analyses were conducted with a Rigaku D = Max 2200 XRD diffractometer and analyzed with the Jade software package. Raman spectra were collected with a LabRAM HR800 Raman microscope (HORIBA). This Raman spectrometer/microscope used an incident laser light with a high resolution of 0.6 cm^{-1} at 532.14 nm wavelength.

4. Conclusions

Molten carbonate electrolysis of CO_2 provides an effective path for the C2CNT synthesis of CNTs and macroscopic CNT assemblies. This study explored a variety of electrochemical configurations, systematically varying the electrode composition, electrode current density, electrolysis time, current ramping initiation, and electrolyte additives and their concentrations. The highest observed CNT purity synthesis (97%) utilized a specialized anode consisting of two layers of high-surface-area Inconel 600 (screen) on Inconel 718, a Muntz Brass cathode, with an 0.1 wt% Fe_2CO_3 additive to the 770 °C Li_2CO_3 electrolyte, and with the electrolysis current conducted for 4 h at an intermediate current density (without current ramping) of 0.15 mA/cm^2. The product, as analyzed by SEM, was aligned CNTs with a length of 100 to 500 μm and an aspect ratio of over 1000. The anode, cathode, and electrolyte additive choice were found to be effective for controlling the transition metal nucleation, which is critical to high-purity electrolytic CNT growth.

In a sister paper [51], slight variations of the synthesis parameters led to the formation of a variety of new, high-purity, non-CNT nanocarbon allotropes. All syntheses in the present study produced a majority of CNTs, but the morphology of the CNT product changed widely with the synthesis conditions. Depending on the synthesis conditions, alternate CNT products that were as short as 10 to 30 μm, and curled, rather than straight, or mixed with carbon nano-onions were observed. The high-purity product exhibited a sharp XRD graphic peak, and a low Raman I_D/I_G ratio, which was indicative of low defects in the carbon structure. The XRD also contained iron carbide, and nickel and chromium lithium oxides, which, based on TEM, were found to be located within the CNT.

TEM HAADF showed that the inner core of the CNT length was generally free of metals (void, with 100% carbon walls), but in other areas, the void was filled with transition metal. As they were produced by molten carbonate electrolysis, the CNT walls were conclusively shown to be comprised of highly uniform concentric, cylindrical graphene layers with graphene characteristics and inter-layer spacing of 0.33 to 0.34 nm. When the internal transition metal was within the CNT tip, the layered CNT graphene walls were observed to bend in a highly spherical fashion around the metal supporting the transition metal nucleated CNT growth mechanism. Several syntheses had unusual nodules, many of them highly spherical, on the CNT, generally comprising carbon and containing a low level of internal transition metal. Generally, intersecting CNTs did not merge, but in a few cases, graphene layers bent to become part of the CNT intersection, which was consistent with the occurrence of the occasional, related growth of intersecting CNTs, such as branching.

The study also demonstrates new syntheses of assemblies of CNTs by means of the C2CNT process, with structural implications towards their potential applications for nanofiltration and neural nets and demonstrated pores sizes ranging from 50 nm to 1 μm.

Author Contributions: Conceptualization, S.L. and G.L.; methodology, S.L., X.L., G.L. and X.W.; writing S.L. and G.L. All authors have read and agreed to the published version of the manuscript.

Funding: C2CNT LLC funded this research through the C2CNT LLC XPrize support funding.

Data Availability Statement: The authors confirm that the data supporting the findings of this study are available within the article.

Conflicts of Interest: The authors declare no conflict of interest.

References

1. CO$_2$-Earth: Daily CO$_2$ Values. Available online: https://www.co2.earth/daily-co2 (accessed on 7 December 2021).
2. NASA: Global Climate Change: The Relentless Rise of Carbon Dioxide. Available online: https://climate.nasa.gov/climate_resources/24/ (accessed on 7 December 2021).
3. Urban, M.C. Accelerating extinction risk from climate change. *Science* **2015**, *348*, 571–573. [CrossRef]
4. Pimm, S.L. Climate disruption and biodiversity. *Curr. Biol.* **2009**, *19*, R595–R601. [CrossRef]
5. Praksh, G.K.; Olah, G.A.; Licht, S.; Jackson, N.B. Reversing Global Warming: Chemical Recycling and Utilization of CO$_2$. In *Report of the 2008 National Science Foundation-Sponsored Workshop*; University of Southern California: Los Angeles, CA, USA, 2008. Available online: https://loker.usc.edu/ReversingGlobalWarming.pdf (accessed on 7 December 2021).
6. Yu, M.-F.; Lourie, O.; Dyer, M.; Moloni, K.; Kelly, T.; Ruoff, R. Strength and Breaking Mechanism of Multiwalled Carbon Nanotubes Under Tensile Load. *Science* **2000**, *287*, 637–640. [CrossRef] [PubMed]
7. Chang, C.-C.; Hsu, H.-K.; Aykol, M.; Hung, W.; Chen, C.; Cronin, S. Strain-induced D band observed in carbon nanotubes. *ACS Nano* **2012**, *5*, 854–862. [CrossRef]
8. Licht, S.; Liu, X.; Licht, G.; Wang, X.; Swesi, A.; Chan, Y. Amplified CO$_2$ reduction of greenhouse gas emissions with C2CNT carbon nanotube composites. *Mater. Today Sustain.* **2019**, *6*, 100023. [CrossRef]
9. Kaur, J.; Gill, G.S.; Jeet, K. Chapter 5 Applications of Carbon Nanotubes in Drug Delivery: A Comprehensive Review, in Characterization and Biology of Nanomaterials for Drug Delivery. In *Characterization and Biology of Nanomaterials for Drug Delivery*; Mohapatra, S.S., Ranjan, S., Dasgupta, N., Thomas, S., Eds.; Elsevier: Amsterdam, The Netherlands, 2019; Volume 1, pp. 113–135.
10. Yang, H.; Xu, W.; Liang, X.; Yang, Y.; Zhou, Y. Carbon nanotubes in electrochemical, colorimetric, and fluorimetric immunosensors and immunoassays: A review. *Microchim. Acta* **2020**, *187*, 206. [CrossRef] [PubMed]
11. Moghaddam, H.K.; Maraki, M.R.; Rajaei, A. Application of carbon nanotubes(CNT) on the computer science and electrical engineering: A review. *Int. J. Reconfig. Embed. Syst. (IJRES)* **2020**, *9*, 61–82. [CrossRef]
12. Sehrawat, P.; Julien, C.; Islam, S.S. Carbon nanotubes in Li-ion batteries: A review. *Mater. Sci. Eng. B* **2016**, *213*, 12–40. [CrossRef]
13. Han, T.; Nag, A.; Mukhopadhyay, S.C.; Xu, Y. Carbon nanotubes and its gas-sensing applications: A review. *Sens. Actuators A* **2019**, *291*, 107–143. [CrossRef]
14. Kumar, S.; Pavelyev, V.; Tripathi, N.; Platonov, V.; Sharma, P.; Ahmad, R.; Mishra, P.; Khosla, A. Review—Recent Advances in the Development of Carbon Nanotubes Based Flexible Sensors. *J. Electrochem. Soc.* **2020**, *167*, 047506. [CrossRef]
15. Abdullah, Z.; Othman, A.; Mwabaidur, S. Application of Carbon Nanotubes in Extraction and chromotographic analysis: A review. *Arab. J. Chem.* **2019**, *12*, 633–651.
16. Basheer, B.V.; George, J.; Siengchin, S.; Parameswaranpillai, J. Polymer grafted carbon nanotubes—Synthesis, properties, and applications: A review. *Nano-Struct. Nano-Objects* **2020**, *22*, 100429. [CrossRef]
17. Imtiaza, S.; Siddiqa, M.; Kausara, A.; Munthaa, S.J.; Ambreenb, I.B. A Review Featuring Fabrication, Properties and Appliction of Carbon Nanotubes (CNTs) Reinforced Polymer and Epoxy Nanocomposites. *Chin. J. Polym. Sci.* **2018**, *36*, 445–461. [CrossRef]
18. Gantayata, S.; Routb, D.; Swain, S.K. Carbon Nanomaterial–Reinforced Epoxy Composites: A Review. *Polym.-Plast. Technol. Eng.* **2018**, *57*, 1–16. [CrossRef]
19. Rafiquea, I.; Kausara, A.; Anwara, Z.; Muhammad, B. Exploration of Epoxy Resins, Hardening Systems, and Epoxy/Carbon Nanotube Composite Designed for High Performance Materials: A Review. *Polym.-Plast. Technol. Eng.* **2016**, *55*, 312–333. [CrossRef]
20. Kausara, A.; Rafiquea, I.; Muhammad, B. Review of Applications of Polymer/Carbon Nanotubes and Epoxy/CNT Composites. *Polym.-Plast. Technol. Eng.* **2016**, *55*, 1167–1191. [CrossRef]
21. Shahidi, S.; Moazzenchi, B. Carbon nanotube and its applications in textile industry—A review. *J. Text. Inst.* **2018**, *109*, 1653–1666. [CrossRef]
22. Rather, S.U. Preparation, characterization and hydrogen storage studies of carbon nanotubes and their composites: A review. *Int. J. Hydrogen Energy* **2020**, *45*, 3847–5110. [CrossRef]
23. Selvaraja, M.; Haia, A.; Banata, F.; Haijab, M. Application and prospects of carbon nanostructured materials in water treatment: A review. *J. Water Process Eng.* **2020**, *33*, 100996. [CrossRef]
24. Bassyouni, M.; Mansi, A.E.; Elgabry, A.; Ibrahim, B.A.; Kassem, O.A.; Alhebeshy, R. Utilization of carbon nanotubes in removal of heavy metals from wastewater: A review of the CNTs' potential and current challenges. *Appl. Phys. A* **2020**, *126*, 38. [CrossRef]
25. Ghoranneviss, M.; Elahi, A.S. Review of carbon nanotubes production by thermal chemical vapor deposition technique. *Mol. Cryst. Liq. Cryst.* **2016**, *629*, 158–164. [CrossRef]
26. Shah, K.A.; Tali, B.A. Synthesis of carbon nanotubes by catalytic chemical vapour deposition: A review on carbon sources, catalysts and substrates. *Mater. Sci. Semicond. Process.* **2016**, *41*, 67–82. [CrossRef]
27. Khanna, V.; Bakshi, B.R.; Lee, L.J. Carbon Nanofiber Production: Life cycle energy consumption and environmental impact. *J. Ind. Ecol.* **2008**, *12*, 394–410. [CrossRef]

28. Hu, L.; Song, H.; Ge, J.; Zhu, J.; Jiao, S. Capture and electrochemical conversion of CO_2 to ultrathin graphite sheets in $CaCl_2$-based melts. *J. Mat. Chem. A.* **2015**, *3*, 21211–21218. [CrossRef]
29. Liang, C.; Chen, Y.; Wu, M.; Wang, K.; Zhang, W.; Gan, Y.; Huang, H.; Chen, J.; Zhang, J.; Zheng, S.; et al. Green synthesis of graphite from CO_2 without graphitization process of amorphous carbon. *Nat. Commun.* **2021**, *12*, 119. [CrossRef]
30. Licht, S.; Wang, B.; Ghosh, S.; Ayub, H.; Jiang, D.; Ganley, J. New solar carbon capture process: STEP carbon capture. *J. Phys. Chem. Lett.* **2010**, *1*, 2363–2368. [CrossRef]
31. Ren, J.; Li, F.; Lau, J.; Gonzalez-Urbina, L.; Licht, S. One-pot synthesis of carbon nanofibers from CO_2. *Nano Lett.* **2010**, *15*, 6142–6148. [CrossRef] [PubMed]
32. Ren, J.; Licht, S. Tracking airborne CO_2 mitigation and low cost transformation into valuable carbon nanotubes. *Sci. Rep.* **2016**, *6*, 27760. [CrossRef]
33. Ren, J.; Lau, J.; Lefler, M.; Licht, S. The minimum electrolytic energy needed to convert carbon dioxide to carbon by electrolysis in carbonate melts. *J. Phys. Chem. C.* **2015**, *119*, 23342–23349. [CrossRef]
34. Licht, S.; Douglas, A.; Ren, J.; Carter, R.; Lefler, M.; Pint, C.L. Carbon nanotubes produced from ambient carbon dioxide for environmentally sustainable lithium-ion and sodium ion battery anodes. *ACS Cent. Sci.* **2016**, *2*, 162–168. [CrossRef]
35. Dey, G.; Ren, J.; El-Ghazawi, O.; Licht, S. How does an amalgamated Ni cathode affect carbon nanotube growth? A density functional calculation of nucleation and growth. *RSC Adv.* **2016**, *122*, 400–410.
36. Ren, J.; Johnson, M.; Singhal, R.; Licht, S. Transformation of the greenhouse gas CO_2 by molten electrolysis into a wide controlled selection of carbon nanotubes. *J. CO_2 Util.* **2017**, *18*, 335–344. [CrossRef]
37. Johnson, M.; Ren, J.; Lefler, M.; Licht, G.; Vicini, J.; Licht, S. Data on SEM, TEM and Raman spectra of doped, and wool carbon nanotubes made directly from CO_2 by molten electrolysis. *Data Br.* **2017**, *14*, 592–606. [CrossRef] [PubMed]
38. Johnson, M.; Ren, J.; Lefler, M.; Licht, G.; Vicini, J.; Liu, X.; Licht, S. Carbon nanotube wools made directly from CO_2 by molten electrolysis: Value driven pathways to carbon dioxide greenhouse gas mitigation. *Mater. Today Energy* **2017**, *5*, 230–236. [CrossRef]
39. Wang, X.; Liu, X.; Licht, G.; Wang, B.; Licht, S. Exploration of alkali cation variation on the synthesis of carbon nanotubes by electrolysis of CO_2 in molten carbonates. *J. CO_2 Util.* **2019**, *18*, 303–312. [CrossRef]
40. Wang, X.; Licht, G.; Licht, S. Green and scalable separation and purification of carbon materials in molten salt by efficient high-temperature press filtration. *Sep. Purif. Technol.* **2021**, *244*, 117719. [CrossRef]
41. Wang, X.; Sharif, F.; Liu, X.; Licht, G.; Lefer, M.; Licht, S. Magnetic carbon nanotubes: Carbide nucleated electrochemical growth of ferromagnetic CNTs. *J. CO_2 Util.* **2020**, *40*, 101218. [CrossRef]
42. Wang, X.; Liu, X.; Licht, G.; Licht, S. Calcium metaborate induced thin walled carbon nanotube syntheses from CO_2 by molten carbonate electrolysis. *Sci. Rep.* **2020**, *10*, 15146. [CrossRef]
43. Liu, X.; Licht, G.; Licht, S. The green synthesis of exceptional braided, helical carbon nanotubes and nanospiral platelets made directly from CO_2. *Mat. Today Chem.* **2021**, *22*, 100529. [CrossRef]
44. Liu, X.; Licht, G.; Licht, S. Data for the green synthesis of exceptional braided, helical carbon nanotubes and nano spiral platelets made directly from CO_2. *arXiv* **2021**, arXiv:2110.05398.
45. Lau, J.; Dey, G.; Licht, S. Thermodynamic assessment of CO_2 to carbon nanofiber transformation for carbon sequestration in a combined cycle gas or a coal power plant. *Energy Convers. Manag.* **2016**, *122*, 400–410. [CrossRef]
46. Licht, S. Co-production of cement and carbon nanotubes with a carbon negative footprint. *J. CO_2 Util.* **2017**, *18*, 378–389. [CrossRef]
47. Liu, X.; Ren, J.; Licht, G.; Wang, X.; Licht, S. Carbon nano-onions made directly from CO_2 by molten electrolysis for greenhouse gas mitigation. *Adv. Sustain. Syst.* **2019**, *3*, 1900056. [CrossRef]
48. Ren, J.; Yu, A.; Peng, P.; Lefler, M.; Li, F.-F.; Licht, S. Recent advances in solar thermal electrochemical process (STEP) for carbon neutral products and high value nanocarbons. *Acc. Chem. Res.* **2019**, *52*, 3177–3187. [CrossRef] [PubMed]
49. Liu, X.; Wang, X.; Licht, G.; Licht, S. Transformation of the greenhouse gas carbon dioxide to graphene. *J. CO_2 Util.* **2020**, *36*, 288–294. [CrossRef]
50. Wang, X.; Licht, G.; Liu, X.; Licht, S. One pot facile transformation of CO_2 to an unusual 3-D nan-scaffold morphology of carbon. *Sci Rep.* **2020**, *10*, 21518. [CrossRef] [PubMed]
51. Liu, X.; Licht, G.; Licht, S. Controlled Transition Metal Nucleated Growth of Unusual Carbon Allotropes by Molten Electrolysis of CO_2. *Catalysts* **2022**, *12*, 125. [CrossRef]
52. Gui, X.; Wei, J.; Wang, K.; Cao, A.; Zhu, H.; Jia, Y.; Shu, Q.; Wu, D. Carbon nanotube sponges. *Adv. Mat.* **2010**, *22*, 617–621. [CrossRef] [PubMed]
53. Yu, S.-H. Carbon nanofiber aerogels for emergent cleanup of oil spillage and chemical leakage under harsh conditions. *Sci. Rep.* **2014**, *4*, 4079.
54. Kim, K.H.; Tsui, M.N.; Islam, M.F. Graphene-Coated Carbon Nanotube Aerogels Remain Superelastic while Resisting Fatigue and Creep over −100 to +500 °C. *Chem. Mater.* **2017**, *4*, 2748–2755. [CrossRef]
55. Wan, W.; Zhang, R.; Li, W.; Liu, H.; Lin, Y.; Li, L.; Zhou, Y. Graphene–carbon nanotube aerogel as an ultra-light, compressible and recyclable highly efficient absorbent for oil and dyes. *Environ. Sci. Nano* **2016**, *3*, 107–113. [CrossRef]
56. Ozden, S.; Narayanan, T.N.; Tiwary, C.S.; Dong, P.; Hart, A.H.; Vajtai, R.; Ajayan, P.M. 3D macroporous solids from chemically cross-linked carbon. *Small* **2015**, *11*, 688–693. [CrossRef] [PubMed]
57. Schaefer, A.; Fane, A.G.; Waite, T.D. *Nanofiltration: Principles and Applications*; Elsevier: Oxford, UK, 2005.

58. Marchetti, P.; Jimenez Solomon, M.F.; Szekely, G.; Livingston, A.G. Molecular Separation with Organic Solvent Nanofiltration: A Critical Review. *Chem. Rev.* **2014**, *114*, 10735–10806. [CrossRef] [PubMed]
59. Qu, H.; Rayabharam, A.; Wu, X.; Wang, P.; Li, Y.; Fagan, J.; Aluru, N.R.; Wang, Y. Selective filling of n-hexane in a tight nanopore. *Nat Commun.* **2021**, *12*, 310. [CrossRef]
60. Hashim, D.P.; Romo-Herrera, J.M.; Muñoz Sandoval, E.; Ajayan, P.M. Covalently bonded three-dimensional carbon nanotube solids via boron induced nanojunctions. *Sci. Rep.* **2012**, *2*, 363. [CrossRef]
61. Gabay, T.; Jacobs, E. Engineered self-organization of neural networks using carbon nanotube clusters. *Physica A* **2005**, *350*, 611–621. [CrossRef]
62. Licht, S. Carbon Dioxide to carbon nanotube scale-up. *arXiv* **2017**, arXiv:1710.07246.
63. XPrize Foundation. *Turning CO_2 into Products*; XPrize Foundation: Culver City, CA, USA, 2021. Available online: http://CarbonXPrize.org (accessed on 14 December 2021).
64. XPrize Announces Winners with Each Team Creating Valuable Products from CO_2. Available online: https://www.nrg.com/about/newsroom/2021/xprize-announces-the-two-winners-of--20m-nrg-cosia-carbon-xprize.html (accessed on 14 December 2021).

Article

Controlled Growth of Unusual Nanocarbon Allotropes by Molten Electrolysis of CO_2

Xinye Liu [1], Gad Licht [2], Xirui Wang [1] and Stuart Licht [1,2,*]

[1] Department of Chemistry, George Washington University, Washington, DC 20052, USA; xyl@gwmail.gwu.edu (X.L.); xiruiwang@email.gwu.edu (X.W.)
[2] C2CNT, Carbon Corp., 1035 26 St NE, Calgary, AB T2A 6K8, Canada; glicht@wesleyan.edu
* Correspondence: slicht@gwu.edu

Abstract: This study describes a world of new carbon "fullerene" allotropes that may be synthesized by molten carbonate electrolysis using greenhouse CO_2 as the reactant. Beyond the world of conventional diamond, graphite and buckyballs, a vast array of unique nanocarbon structures exist. Until recently, CO_2 was thought to be unreactive. Here, we show that CO_2 can be transformed into distinct nano-bamboo, nano-pearl, nano-dragon, solid and hollow nano-onion, nano-tree, nano-rod, nano-belt and nano-flower morphologies of carbon. The capability to produce these allotropes at high purity by a straightforward electrolysis, analogous to aluminum production splitting of aluminum oxide, but instead nanocarbon production by splitting CO_2, opens an array of inexpensive unique materials with exciting new high strength, electrical and thermal conductivity, flexibility, charge storage, lubricant and robustness properties. Commercial production technology of nanocarbons had been chemical vapor deposition, which is ten-fold more expensive, generally requires metallo-organics reactants and has a highly carbon-positive rather than carbon-negative footprint. Different nanocarbon structures were prepared electrochemically by variation of anode and cathode composition and architecture, electrolyte composition, pre-electrolysis processing and current ramping and current density. Individual allotrope structures and initial growth mechanisms are explored by SEM, TEM, HAADF EDX, XRD and Raman spectroscopy.

Keywords: nanocarbon; carbon allotropes; carbon nanotubes; carbon nanofibers; carbon nano-onions; carbon dioxide electrolysis; molten carbonate; greenhouse gas mitigation

Citation: Liu, X.; Licht, G.; Wang, X.; Licht, S. Controlled Growth of Unusual Nanocarbon Allotropes by Molten Electrolysis of CO_2. *Catalysts* **2022**, *12*, 125. https://doi.org/10.3390/catal12020125

Academic Editor: Javier Ereña

Received: 17 December 2021
Accepted: 12 January 2022
Published: 21 January 2022

Publisher's Note: MDPI stays neutral with regard to jurisdictional claims in published maps and institutional affiliations.

Copyright: © 2022 by the authors. Licensee MDPI, Basel, Switzerland. This article is an open access article distributed under the terms and conditions of the Creative Commons Attribution (CC BY) license (https://creativecommons.org/licenses/by/4.0/).

1. Introduction

High atmospheric CO_2 levels are the largest cause of global warming. Atmospheric CO_2 concentration, which had cycled at 235 ± ~50 ppm for 400 millennia until 1850 and the advent of rising anthropogenic CO_2 emissions, is currently at 416 ppm and rising at a rapid annual rate, creating global planetary climate disruptions, habitat loss, and species extinction [1–4]. To date, international efforts to decrease CO_2 emissions have failed. The Earth is in the grips of an existential threat and a mass extinction event generally defined as loss of 75% of planetary species. This emphasizes the critical imperative of alternative pathways of CO_2 conversion into another, stable non-greenhouse material (CO_2 utilization). CO_2 is regarded as such a stable molecule that its transformation into a non-greenhouse material poses a major challenge, as summarized in our NSF workshop on Chemical Recycling and Utilization of CO_2 [5].

Graphitic carbon nanomaterials (CNMs) are high value, highly stable (with a graphite-like geologic stability) state-of-the-art materials, which have the potential to be attractive CO_2 utilization products. However, conventional methodologies of CNM production have a high CO_2 footprint. For example, chemical vapor deposition (CVD) is an energy-intensive, expensive process to produce CNM associated with an unusually massive release, as much as 600 tons of CO_2 emitted per ton of CNM product [6]. Despite the attractive properties and application of CNMs, which range from extraordinary strength as materials with the highest ultimate tensile strength to advanced electronic, thermal, electrical storage, shielding and tribological properties, its demand is limited by a high market cost. Due to high CVD production costs, the market price of CNMs is unusually high (USD 100,000 to over USD 1 million per ton). A "green", rather than high carbon footprint CNT, will increase demand. The use of CO_2 as a reactant to generate value-added CNM products will provide motivation to consume this greenhouse gas to mitigate climate change, and its transformation to CNMs can provide a stable material to store carbon removed from the environment.

In 2009 (fundamental) and 2010 (experimental), a novel solar driven methodology to split CO_2 into C and O_2 by molten carbonate electrolysis was demonstrated as a tool to mitigate climate change [7,8]. This molten carbonate process is not limited to sunlight as the electrolysis energy source. It was demonstrated that by using molten carbonate and a variety of electrolytic configurations, the product can be pure CNM, such as CNT [8–29]. For example, this novel chemistry transforms carbon dioxide to carbon nanotubes (C2CNT process), and as illustrated in Figure 1 directly captures atmospheric CO_2, or concentrated anthropogenic CO_2, such as from industrial processes. Power plant and cement exhausts have been investigated [23,24], and after a five-year-long international competition, this C2CNT process was awarded the 2021 XPrize XFactor Award for transforming CO_2 from flue gas into a valuable product using flue gas from the 860 MW Shepard natural gas power plant (Calgary, AB, Canada) [30,31]. Composites of these high-strength CNTs can be mixed with structural materials, such as CNT-cement, CNT-steel and CNT-aluminum, greatly reducing the carbon footprint of structural materials, and acting to amplify the CNT's CO_2 emission reduction [11]. As presented in Figure 1, several different CNMs can be produced by molten carbonate splitting including: magnetic CNTs [18], thin CNTs [19,29], long CNTs [15,16,29], doped, high electrical conductivity CNTs [13,15,16], high Li-ion anode storage CNTs [22], macroscopic CNT assemblies [29], a novel nanocarbon scaffold [28], graphene and carbon nanoplatelets (CNP) [27], and carbon nano-onions (CNO) [25] and helical CNTs (HCNT) [20]. This study introduces entirely distinct nanocarbon morphologies discovered by systematic variation of the electrochemical parameters of the molten carbonate splitting of CO_2. With the exception of a new methodology to form CNOs, there are no overlaps or redundancies with the previous electrolytically formed CNM morphologies that have been discovered.

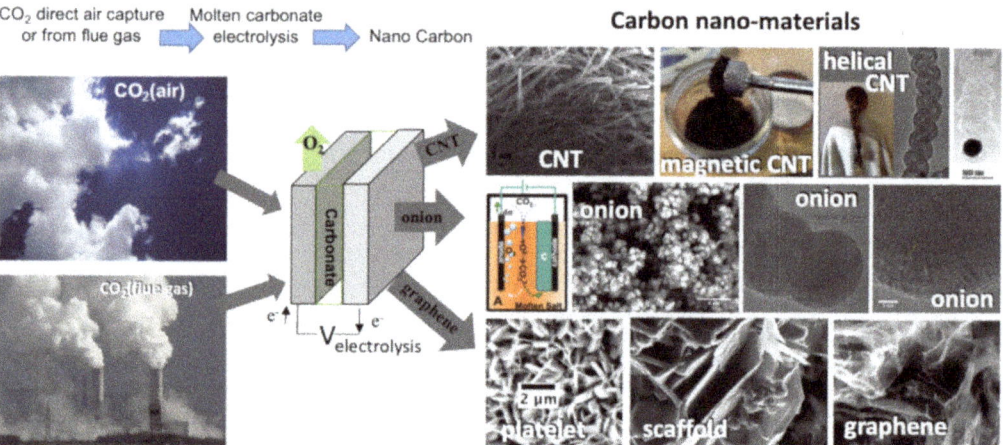

Figure 1. High-yield electrolytic synthesis of carbon nanomaterials from CO_2, either directly from the air or from smokestack CO_2, in molten carbonate.

The wide range of CNM morphologies observed shows the potential for product tuning. Different CNMs have different applications. For example, CNTs are the strongest known and most thermally conductive material along their major axis [32,33]. CNOs can make excellent lubricants and their high surface area and other characteristics make them appealing for batteries or supercapacitors, and they may have uses in refrigerants and in EMF shielding [20,25,34–39]. Platelets can contribute to the formation of lightweight and porous, but strong, nano-gels [40–42]. HCNTs have strong chiral, magnetic, piezo-electric properties, and may act as nano-springs [43–52]. Even within the same group, such as CNTs, different lengths, thicknesses, layers of graphene walls, etc. morphologies can have different properties such as differing rigidity or surface areas [15,19,47]. This would incentivize even greater CO_2 transformed to removable carbon by allowing products to have access to a wide range of markets, which collectively could increase technological progress and provide a growing demand as a buffer to remove anthropogenic CO_2.

The C2CNT process has quantified the high affinity of molten carbonates to absorb both atmospheric and flue gas CO_2 levels. It has been shown, utilizing the ^{13}C isotope of CO_2 to track the carbon from its origin (CO_2 a gas phase reactant) through its transformation to nanocarbon product, that the CO_2 originating from the gas phase serves as the renewable C building blocks in the observed CNT product [8,9]. The molten carbonate is not consumed, but renewed, catalyzing the ongoing electrolysis of CO_2. The net reaction is:

$$\text{Dissolution: } CO_2 \text{ (gas)} + Li_2O \text{ (soluble)} \rightleftharpoons Li_2CO_3 \text{ (molten)} \quad (1)$$

$$\text{Electrolysis: } Li_2CO_3 \text{ (molten)} \rightarrow C \text{ (CNT)} + Li_2O \text{ (soluble)} + O_2 \text{ (gas)} \quad (2)$$

$$\text{Net: } CO_2 \text{ (gas)} \rightarrow C \text{ (CNT)} + O_2 \text{ (gas)} \quad (3)$$

An important component of the C2CNT growth process is transition metal nucleated growth. These catalysts lead to clearly observable CNT walls with thick graphene interlayer separations. However, when these nucleation additives are excluded, rather than CNTs, instead high-yield synthesis of CNOs or graphene is accomplished [25,27].

This study systematically explores the possibility to synthesize a variety of new molten carbonate synthesized carbon "fullerene" allotropes, and opens a new world of inexpensive nanocarbons, made from CO_2, to be explored as incentivized (valuable) products for the transformation and stable removal of CO_2 and climate change mitigation. Here the conventional definition of allotrope of "different physical forms in which an element can exist" is employed, rather than the alternative "structural modifications of an element bonded together in a different manner". Specifically, this study explores which reactive pathway condition leads to the selection of new nanocarbon allotropes over another and can lead to higher purity products and better formation of a single product

2. Results and Discussion
2.1. Electrolytic Conditions Varied to Synthesize New Nanocarbon Allotropes from CO_2

This study varies conditions in which small electrolytic changes in a 770 °C molten Li_2CO_3 yield major changes to the product consisting of new, non-CNT nanocarbon allotropes. In a parallel study to this new carbon allotropes study, different systematic changes to these same electrolytic parameters have revealed a wide range of electrochemical conditions that synthesize only a high-purity, high-yield product consisting only of various carbon nanotube morphologies. That sister study focuses on the transition metal nucleation zone of CNT growth and also reveals molten electrochemical conditions which produce assemblies of CNTs [29]. Straight CNT type carbon allotropes tended to dominant when either 0.1% wt iron oxide was added or the anode contained a high amount of Fe (Nichrome C: 24%, Inconel 718 18.5%, or Inconel 718 at 18.5%). The electrochemical conditions varied here are composition and/or architecture of the cathode and anode, additives and their concentrations to the Li_2CO_3 electrolyte, and current density and time of the electrolysis. Electrolyte additives that are varied include Fe_2O_3 and nickel or chromium powder. Electrolyses are varied over a range of electrolysis current densities. Variations of the electrodes include the use of cathode metal electrodes such as Muntz brass, Monel, or Nichrome alloys. Anode variations include noble anodes such as iridium, various nickel containing anodes including nickel, Nichrome A or C, Inconel 600, 625, or 718, or specific layered combinations of these metals. Alloy composition of the metals used as electrodes is presented in Table 1. Metal variation was further refined by combining the metals in Table 1 as anodes, for example using a solid sheet of one Inconel alloy, layered with a screen or screens of another Inconel alloy, such as an anode of Inconel 625 with 3 layers of (spot welded) 100 mesh Inconel 600 screen.

Table 1. Compositions of various alloys used (weight percentage).

Alloy	Ni%	Fe%	Cu%	Zn%	Cr%	Mo%	Nb & Ta%
Nichrome C	60	24			16		
Nichrome A	80				20		
Inconel 600	52.5	18.5			19.0	3.0	3.6
Inconel 718	72% min	6–10			14–17		
Inconel 625	58	5 max			20–23	8–10	4.15–3.15
Monel	67		31.5				
Muntz Brass			60	40			

2.2. Electrochemical Conditions to Synthesize Bamboo and Pearl Nanocarbon Allotropes from CO_2

We have conducted several thousands of different electrolyses to split CO_2 in molten lithium carbonate. A fascinating, but rarely observed, product which occurred in less than 30 of those many electrolyses had nano-morphology analogous to the macro-structure of bamboo, but had been only observed as a low fraction of the total product. Table 2 summarizes the systematic optimization of electrolysis conditions in 770 °C Li_2CO_3 to optimize and maximize the electrolytic formation of this nano-bamboo. A few prior electrolyses producing nano-bamboo were associated with nickel electrodes, or started with ramping up of the current to encourage nucleation. Experiment Electrolysis I in the top row of Table 2 includes both these features including nickel as both the cathode and anode, and (not delineated in the table) an initial 10 min electrolysis at a constant 0.01 and then 0.02 A/cm^2, followed by 5 min at 0.04 and then 0.08 A/cm^2, after which the constant current electrolysis was conducted at the tabulated 0.2 A/cm^2. Nano-bamboo was evident in the product SEM, but constituted a minority (30 wt%) of the total product. As seen in Electrolysis II in Table 2, an increase in the nano-bamboo product is achieved with the direct addition of Ni and Cr powders to the electrolyte, and the anode is replaced by a noble metal (iridium) accompanied by a 5-fold decrease in current density. As noted in the table, this Electrolysis II has the first majority, 60 wt%, of the nano-bamboo product. Coulombic efficiency quantifies the measured available charge (current multiplied by the electrolysis time) to the measured number of four electrons per equivalent of C in the product. Coulombic efficiency tends to drop off with current density, and in this case the coulombic efficiency of the synthesis was 79%.

Table 2. Systematic variation of CO_2 splitting conditions in 770 °C Li_2CO_3 to optimize formation of nano-bamboo and nano-pearl carbon allotropes.

Electrolysis	Cathode	Anode	Additives (wt% Powder)	Electr Time	Current Density A/cm^2	Product Description
I	Nickel	Nickel	-	4 h	0.2	30% nano-bamboo carbon 40% regular CNT rest: graphitic carbon
II	Muntz brass	Iridium	0.4% Ni 0.4% Cr	18 h	0.08	60% nano-bamboo carbon 10% regular CNT rest: graphitic carbon
III	Muntz brass	Inconel 718 2 layers Inconel 600	0.81% Ni powder	18 h	0.08	89% 30–120 μm nano-bamboo carbon
IV	Muntz brass	Inconel 718 2 layers Inconel 600	0.81% Ni powder	18 h	0.08	94% 30–80 μm carbon nano-bamboo, 6% conical carbon nanofiber
V	Muntz brass	Inconel 718 2 layers Inconel 600	0.81% Ni powder	18 h	0.08	94% 30–80 μm carbon nano-bamboo, 6% conical carbon nanofiber
VI	Nichrome C	Nichrome C	0.4% Ni 0.4% Cr	3 h	0.4	95% nano-bamboo carbon
VII	Monel	Nichrome C	0.81% Ni	18 h	0.08	95% hollow nano-onions
VIII	Monel	Nichrome C	0.4% Ni 0.4% Cr	18 h	0.08	97% nano-pearl carbon
IX	Monel	Nichrome C	0.4% Ni 0.4% Cr	18 h	0.08	97% nano-pearl carbon

The low current ramping, pre-electrolysis conditions can have benefits and disadvantages. (1) Low current conditions may support the reduction and deposition of initial graphene layers to facilitate ongoing reduction and growth. In addition, lower current can favor transition metal deposition at the cathode and formation of nucleation sites, at low concentrations compared to carbonate (from CO_2) in the electrolyte. The analysis of bound versus free metal cations in the molten electrolyte for a reduction potential calculation has been a challenge. Without Nernst activity and temperature correction, the reduction rest potentials of Ni, Fe, Cr and Cu and CO_2 at room temperature are $CO_2(IV/0) = -1.02$, $Cr(III/0) = -0.74$, $Ni(II/0) = -0.25$, $Fe(III/0) = -0.04$, and $Cu(II/0) = 0.34$. Note, however, that the free activity of tetravalent carbon as carbonate $C(IV)O_3^{2-}$ formed by the reaction of $C(IV)O_2$ with electrolytic oxide in pure molten carbonate solutions is many orders of magnitude higher than the dissolved transition metal ion activity in the electrolysis electrolyte. This helps favor the thermodynamic and kinetic reduction of the tetravalaent carbon, over metal deposition at the cathode. The practical observation is that, for the majority of molten carbonate CO_2 electrolyses we have studied, the initial low current ramping is not observed to promote highest purity carbon deposition.

The first row of Figure 2 presents the product SEM of Electrolysis III, which continues to use a low current density, continues to exhibit a coulombic efficiency of 78%, focuses on a Ni powder addition to electrolyte, and refines the anode to Inconel 718 with two layers of Inconel 600, with an increase to 89 wt% the nano-bamboo product. Additionally, this electrolysis used an "aged" electrolyte (not delineated in the table). The freshly molten electrolyte requires time (up to 24 h) to reach a steady state equilibrium (pre-equilibration step) [16,20]. For Electrolysis III, the electrolyte was aged 24 h prior to melting and prior to immersion of the electrodes. However, it was observed that the aging is disadvantageous towards maximizing the nano-bamboo yield. A final refinement, in immediate use of the freshly melted electrolyte (elimination of the aging step), increases the nano-bamboo product to 90 wt% of the product (row 2 in Figure 2, and Electrolysis IV, and repeated as V in Table 2). Interesting, the 6% non-bamboo product in Electrolyses IV and V appears to be conical carbon nano-fiber (CNF) morphology, with its distinctive triangular-shaped voids in the morphology as seen in the second row of Figure 2. A simplified electrolysis eliminates observed CNF impurities resulting in 95% of the nano-bamboo allotorope. This Electrolysis VI is conducted without the current ramp activation at a high 0.4 A/cm^2 current density, and exhibits a 99.7% coulombic efficiency. This electrolysis was tailored to have a purposeful excess of nucleation metals accomplished both with the use of Nichrome C electrodes, which contain Ni, Fe and Cr (Table 1), and through the direct addition of Ni and Cr powders to the electrolyte.

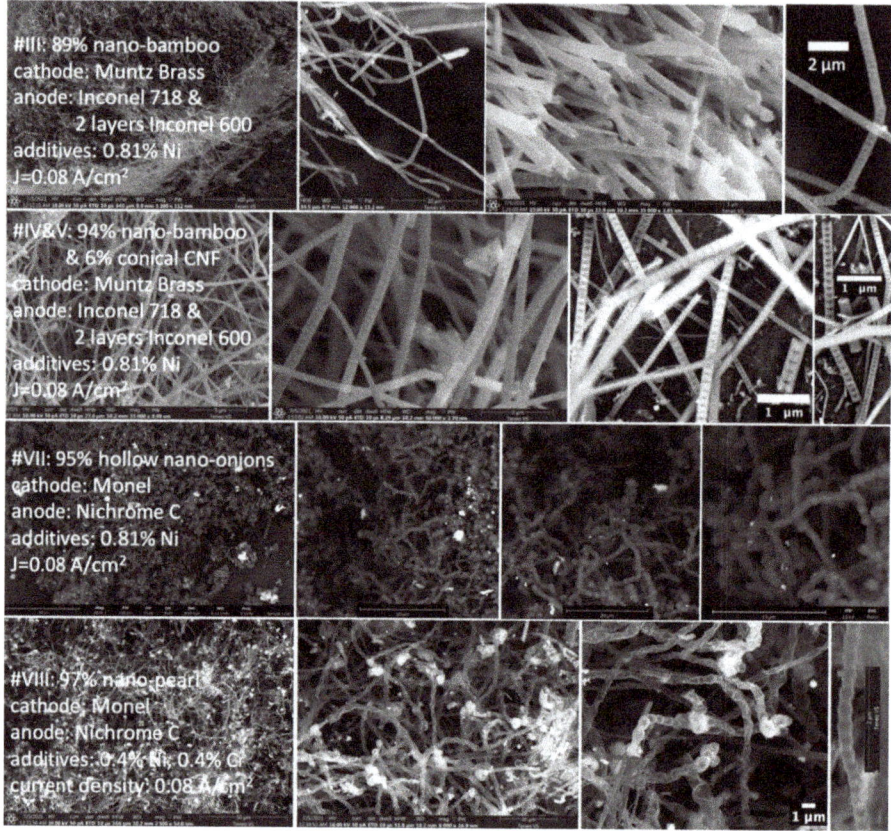

Figure 2. SEM of the synthesis product of nano-bamboo and nano-pearl allotropes of carbon by electrolytic splitting of CO_2 in 770 °C Li_2CO_3. Moving left to right in the panels, the product is analyzed by SEM with increasing magnification. Scale bars in panels (starting from left) are for panels U: 100, 10, 3 μm (different electrolysis) and 2 μm; for panels T: 5, 2, 1 and 1 μm; for panels 11: 50, 30, 20 and 15 μm; for panels X: 50 μm 10, 1 and 2 μm.

The continued use of high concentrations of added transition metal powder to the electrolyte and low current density, but a change of electrodes, yields another distinct nanocarbon allotrope termed here as "hollow nano-onions". Specifically, in Electrolysis VII in Table 2 and Figure 2, the same concentration of Ni powder that had been used as in Electrolyses VI and V was used, and again the electrolyte was not aged nor were ramped initiation currents applied. However, a Monel cathode and Nichrome C anode were used, resulting in a 95 wt% of the product having a distinctive hollow nano-onions morphology. The hollow nature of the nano-onions will be revealed by TEM, but their spheroid character is seen by SEM in the third row of Figure 2. When the pure nickel electrolyte additive was changed to half nickel and half chromium powder, as summarized in Table 2 for Electrolyses

VIII and IX, the product has a distinctive "nano-pearl" morphology with its similarity to a beaded necklace. Here, the product fraction increased to 97% of this nano-pearl carbon and is seen by SEM in the bottom row of Figure 2. Electolyses VII–IX conducted have low J = 0.0 A/cm² and exhibit a diminished coulombic efficiency of 79 to 80%.

Figure 3 compares TEM of the new nano-bamboo, nano-pearl and conical CNF nanocarbon allotropes synthesized by molten carbonate electrolysis. As seen in the top left panel of the figure, the conical carbon nanofibers (CNFs) exhibit conical voids typical of this CNF structure. Growth of the nano-bamboo is seen in the left middle of the figure and is nucleation driven, and the nucleation region appears to change shape, moving from tip to interior of the structure. We hypothesize that the lateral walls forming the bamboo "knobs" may be related to a periodic depletion of the carbon building leading walls. The walls of the nano-bamboo and nano-pearl allotropes exhibit graphene walls characterized by the typical intergraphene wall separation of 0.33 to 0.34 nm, as noted, and as measured by the observed separation between dense carbon planes in the TEM. The lower left of the figure shows the lateral multiple graphene layers separating the "knobs" of the nano-bamboo structure. The lower right of the figure shows the curved multiple graphene layers comprising the walls of the individual "beads" of the nano-pearl structure.

Figure 4 probes the elemental composition by HAADF (High Angle Annular Dark-Field TEM) and compares TEM of the new nano-bamboo and nano-pearl nanocarbon allotropes synthesized by molten carbonate electrolysis. As seen from the HAADF, the nano-bamboo product is pure carbon. That is with the exception of the presence of copper that, as shown in the lower left corner of the top left panel, is pervasively distributed at low concentration throughout, and likely originates from the grid mount of the product sample. HAADF probes two nano-pearl samples. The first exhibits a high or 100% concentration of carbon (the noise level is high) and little or no Ni, Cr or Fe. The second probes for carbon at higher resolutions and the rise and fall of carbon levels is evident as the probe moves from left to right over two separate nano-pearl structures.

The conical CNF, nano-bamboo and nano-pearl are new and unusual high-yield carbon allotropes as synthesized by molten electrolysis. Similar CVD synthesized morphologies have been synthesized by CVD. In particular, the CVD conical CNF structure has been widely characterized as shown in the upper row of Figure 5 [53–56]. In that figure, it is proposed that the morphology in CVD is due to repeated stress-induced deformation of the shape of the nucleating (Ni) metal, which causes the metal particles to jump and form the observed lateral graphene separation bridging the allotrope walls. Globular spaced nano-bamboo and nano-pearl allotropes are less common in CVD but have been observed. An example is shown in the lower left row of Figure 5, whose structures were attributed to the periodic formation of pores in the structure due to defects on the outer layers [57–59]. One specific application of bamboo CVD CNTs is as platforms for building layer by layer based biosensors [60]. Generally, carbon fibers are categorized as amorphous, or as shown on the lower right side of Figure 5, as built from graphene platelets, carbon nanotubes or conical type structures [61,62].

Figure 6 probes the TEM and the elemental composition by HAADF of the new hollow nano-onion nanocarbon allotropes synthesized by molten carbonate electrolysis. As seen, some of the nano-onion inner cores contain metal while others are void. The walls of the hollow nano-onions are composed of graphene layers as characterized by the typical intergraphene wall separation of 0.33 to 0.34 nm, as noted in Figure 6 and as measured by the observed separation between dense TEM carbon planes. As seen in the HAADF of the figure, when the core is vacant, the nano-onion is pure carbon, and when the core contains metal, the metal is either nickel or a mix of nickel and iron.

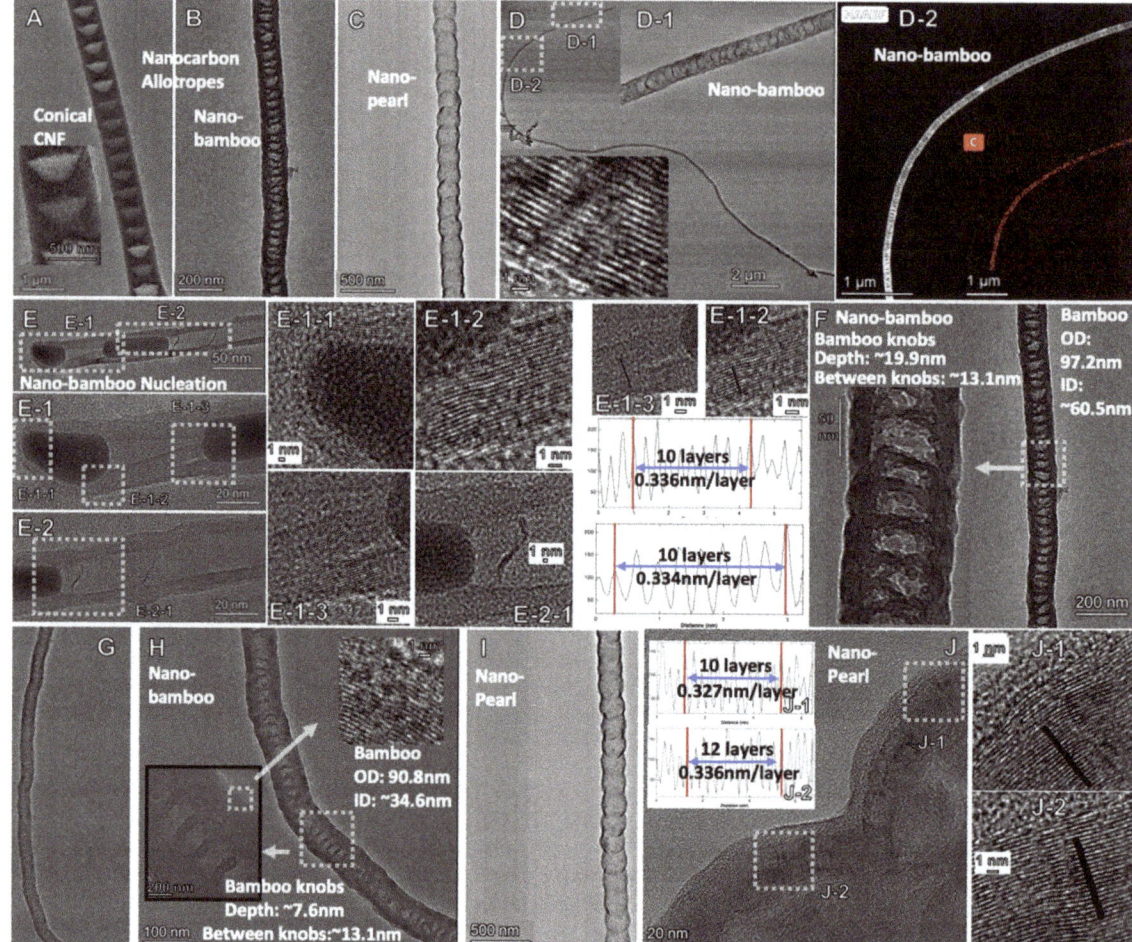

Figure 3. TEM of new molten carbonate synthesized carbon allotropes: Comparison of nano-bamboo, nano-pearl and conical CNF. (**A**): CNF; (**B**): Nano-bambpo; (**C**): Nano-pearl; (**D**), D-1, D-2 Nanobamboo; (**E**), E-1, E-2, E-1-1; E1-2 & E1-3 Nano-bamboo including measured graphene layer thickness; (**F–H**) Nano-bamboo knobs, (**I,J**) Nano-pearl; J-1, J-2 Nano-pearl with measured graphene layer thickness.

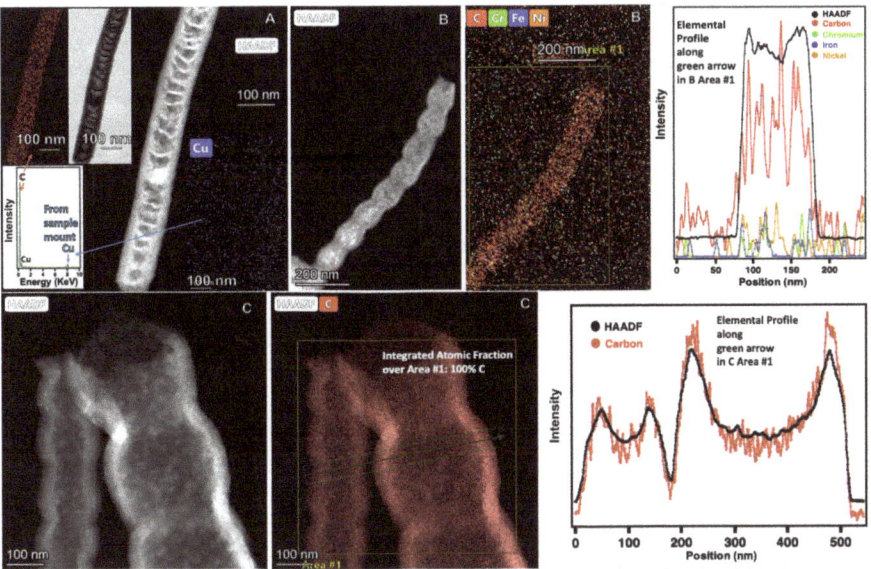

Figure 4. High angle annular dark-field TEM (HAADF) elemental analysis of nano-bamboo (panel A) and nano-pearl (panels B and C and HAADF elemental profiles) carbon allotropes synthesized by molten carbonate electrolysis. (**A**): Nano-bamboo with elemental analysis; (**B**): Nano-pearl with Elemental profile; (**C**): Nanopearl with Elemental profile.

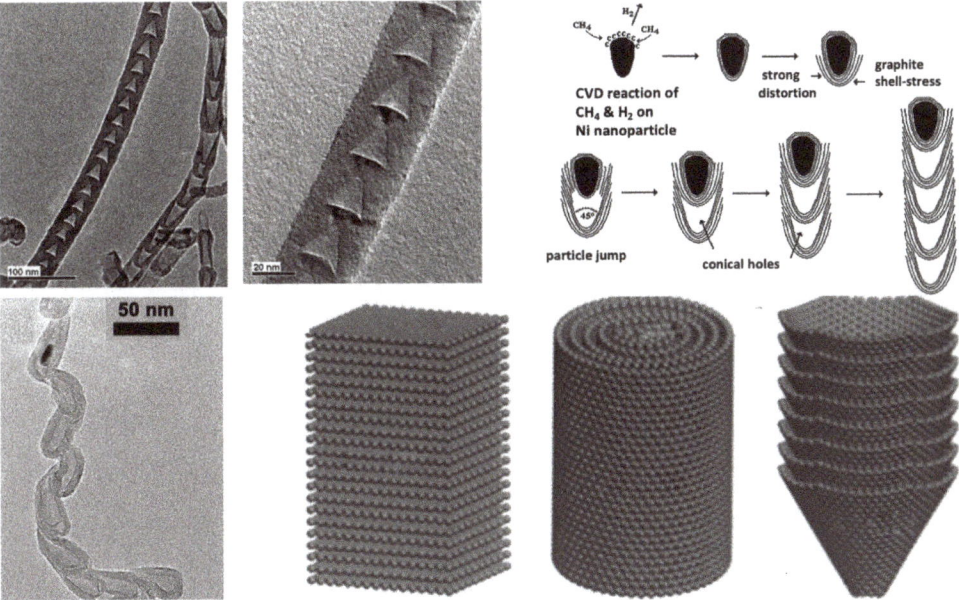

Figure 5. Top row: Conical variations of bamboo carbon nanofibers, and their proposed mechanism of growth, as formed by nickel nucleated CVD using methane and hydrogen. Reproduced open access from Reference [53] Left bottom: Knotty bamboo nanocarbon variations by CVD, and their proposed mechanism of growth. Modified from Reference [58] Right bottom: General graphene layer conformations occurring in carbon nanofibers. Modified from Reference [61].

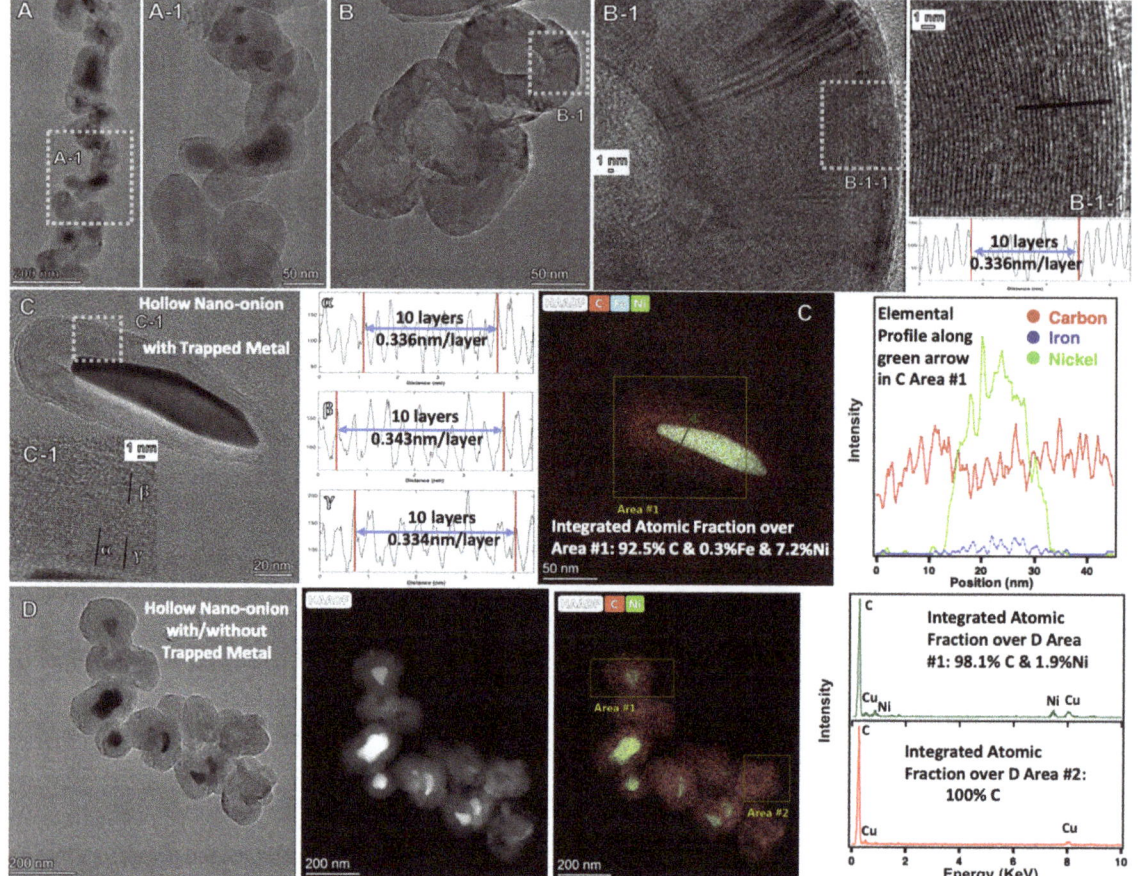

Figure 6. TEM and HAADF elemental analysis of the hollow nano-onion carbon allotrope synthesized by molten carbonate electrolysis. (**A**) and A-1: Hollow nano-onions with and without trapped metal. (**B**) and B-1: Hollow nano-onion without trapped metal. Note, measured graphene layer thickness is part of sub-figure B-1. (**C**) and C-1: Hollow nano-onion with trapped metal. Note, measured graphene layer thickncess and the elemental profile are part of sub-figure (**C**). (**D**): Hollow nano-onion with and without trapped metal. Note, measured elemental HAADF and elemental analysis are part of sub-figure (**D**).

2.3. Electrochemical Conditions to Synthesize Nickel-Coated CNTs, and Onion and Flower Nanocarbon Allotropes from CO_2

A nickel anode or an excess of added nickel leads to nickel-coated CNT. Rather than forming alternative allotropes, such as nano-bamboo or nano-pearl, the use of excess nickel, particularly (i) when employed with a stainless steel cathode, (ii) when utilized at higher electrolysis current densities, and (iii) with the activation by an initial current ramp, tends to coat the carbon nanotube with nickel. This is summarized in the top row of Table 3 as Electrolysis X, in which 0.81 wt% Ni powder is added to the Li_2CO_3 electrolyte, and Nichrome C is used as the anode. The electrolysis is conducted at 0.20 A/cm^2 and exhibits a coulombic efficiency of 98.9%. The Ni coating is further improved (appearing more uniform in the SEM) in Electrolysis XI in Table 3 and as the top row in Figure 7, when a pure nickel, rather than Nichrome C, anode is used, but no Ni powder is added to the electrolytes, and there is no current ramp employed. The electrolysis is conducted at 0.15 A/cm^2 and exhibits a coulombic efficiency of 93.4%.

Table 3. Systematic variation of CO_2 splitting conditions in 770 °C Li_2CO_3 to optimize formation of nickel-coated CNTs and onion, flower, dragon, belt and rod nanocarbon allotropes.

Electrolysis	Cathode	Anode	Additives (wt% Powder)	Electr Time	Current Density A/cm^2	Product Description
X	SST	Nichrome C	0.81% Ni	3 h	0.2	60% Ni particle coated CNT 40% 5–10 µm CNT
XI	SST	Nickel	-	4 h	0.15	89% 50–150 µm straight CNT & Ni particle coated CNT
XII	Muntz brass	Nichrome C	8% Li_3PO_4	4 h	0.2	98% nano-onions
XIII	Monel	Nichrome C	8% Li_3PO_4	18 h	0.08	97% nano-onions
XIV	Muntz brass	Nichrome C	0.81% Co	18 h	0.08	97% nano-flowers
XV	Muntz brass	Nichrome C	0.81% Co	18 h	0.08	97% nano-flowers
XVI	Monel	Inconel 718	0.1% Fe_2O_3	2 h	0.4	94% 50–100 µm nano-dragon
XVII	Muntz brass	Inconnel 718 2 layers Inconel 600	0.1% Li_2O	4 h	0.13	nano-trees: 98% 80–200 µm CNT with branches and trunk
XVIII	Muntz brass	Inconel 718	0.1% Fe_2O_3	18 h	0.08	80% nano-belt
XIX	Monel	Iridium	0.81% Ni	18 h	0.08	91% nano-rod CNT

The exclusion of transition metals from the molten electrolysis environment prevents their activity as nucleation points for carbon growth and suppresses the growth of carbon nanotubes. Suppression of the metal nucleated growth of CNTs, such as through use of a noble metal anode, was found to be an effective means to promote the growth of another nanocarbon: carbon nano-onions [25]. Here, another molten electrolysis pathway is found to ensure a high nano-onion product yield, that is through addition of lithium phosphate to the electrolyte. As summarized in Electrolyses XII and XIII in Table 3, with the addition of 8 wt% Li_3PO_4 to the Li_2CO_3 electrolyte, the product is nearly pure (97–98%) carbon nano-onions as summarized in Table 3. This nano-onion product is the observed to be the case for a wide range of electrolysynthesis current densities (0.08 to 0.20 A/cm^2), with either Muntz Brass or Monel as the cathode, and with (Electrolysis XII) or without (Electrolysis XIII) inclusion of an initial current ramp step during the electrolysis. In a future study, it will be interesting to probe whether phosphates bind or suppress specific free metal availability in molten carbonates in a manner comparable to their tendency to chelate certain metals under ambient aqueous conditions.

A variation of the low current density, Muntz brass cathode, Nichrome C anode, utilizing an aged electrolyte leads to a fascinating new high-purity molten electrolysis nanocarbon allotrope: nano-flowers. Specifically, after the 24 h aging of the electrolyte, an excess (0.081 wt%) of chromium metal powder is added to the electrolyte. The electrolysis is conducted at 0.08 A/cm^2 and exhibits a coulombic efficiency of 78%. The electrolyses are repeated (as Electrolyses XIV and XV) and yield the same results as summarized in Table 3 and shown by SEM in Figure 7. As seen in the lower right panel of Figure 7, the product does appear as hollow tubes within the flower morphology. However, the product morphology is highly unusual in several aspects. Collections of tubes seem to burst from a single point, giving the flower-like arrangement. This will require further study and could represent base, rather than tip, growth and multiple growth patterns activated from singular activation points. An alternative mechanism to be explored is tip based, in which the metal nucleation tip is sintered (decreasing in size) as growth progresses, which with continued growth would decrease the diameter of the nanocarbon product. The tubes appear as short, very straight spikes. The spikes have a diameter which diminishes towards the end of the spike. A small percentage of platelets and garnet-like material is interspersed throughout the floral arrangement. Although new as a majority molten electrolytic synthesis product, nano-flowers have been observed not only from carbon, but also from gold, platinum, and silver as well as from zinc and titanium oxides, and have been described as "a newly developed class of nanoparticles showing structure similar to flower" [62–65]. Chromium may drive both of the proposed mechanisms for nano-flower growth by making nano-metal nucleation points less fluid or bound to the electrode, promoting base growth, or, when it grows from the tip, the chromium does not keep pace with the growing CNM, causing the particle to decrease in size.

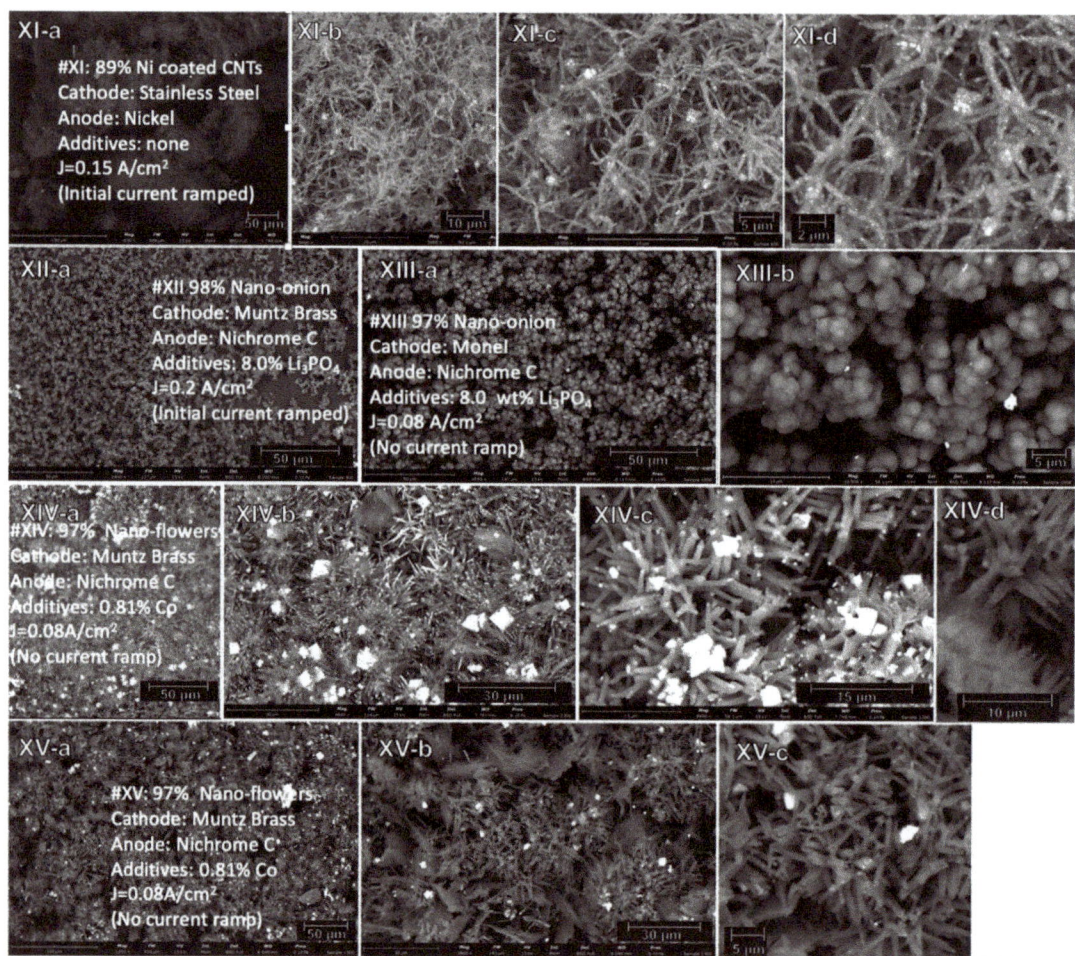

Figure 7. SEM of the synthesis product of nano-flowers, nano-broccoli, nano-onions and Ni-coated CNT allotropes of carbon by electrolytic splitting of CO_2 in 770 °C Li_2CO_3. Moving left to right in the panels, the product is analyzed by SEM with increasing magnification. Scale bars in panels (starting from left) are for panels XI: 150, 20, 15 and 2 µm; for panel XII: 50 µm; for panels XIII: 50 and 15 µm; for panels XIV: 300, 80 and 20 µm; for panels XV upper row: 100, 30, 15 and 10 µm; for panels XV lower row: 100 µm 30 and 5 µm.

2.4. Electrochemical Conditions to Synthesize Nanocarbon Dragon, Tree, Belt and Rod Allotropes from CO_2

Variation of the electrochemical conditions of CNT product formation to those of Electrolysis XVI leads to a change in allotrope from carbon nanotubes to another fascinating morphology referred to here as nano-dragons and presented in Table 3 and Figure 8. The changes from the earlier syntheses that produced CNTs under similar circumstances, include an Inconel 718 anode, rather than Nichrome C, a higher current density of 0.4, rather than 0.1, A/cm^2, (exhibiting a 100% coulombic efficiency), and that the electrolyte is not aged. Unlike the other unique electrolytically synthesized nanocarbon morphologies, carbon nano-dragons do not consist of a a simple, repeated geometric shape, but rather a complex combination of cylinders, platelets and spheres. The small "legs" observed in the Figure 8 SEM images could be smaller branched CNTs or small metal nodules of metal growth.

The addition of low levels of lithium oxide has led to high quality of CNTs [21]. With use of a specific anode (Inconel 718 with two layers of Inconel 600), the quality of the product is retained, but the morphology of the CNT changes substantially. We have previously observed larger transition metal nodule growth from the CNTs [18]. With the addition of Li_2O, branched carbon nano-trees are included as Electrolysis XVII in Table 3 and Figure 8. The electrolysis is conducted at 0.13 A/cm^2 and exhibits a coulombic efficiency of 98.7%. The nano-trees exhibit distinct growth of smaller CNT branches emanating from larger CNT trunks. The red circled area on the right panel of Electrolysis XVII in Figure 8 shows an example of y section branching. Addition of low levels of iron oxide leads to high quality CNTs. However, with 24 h aging of the electrolyte followed by subsequent addition, as in Electrolysis XVIII in Table 3 and Figure 8, an alternative flattened nanocarbon morphology is observed, which is referred to here as nano-belts. The electrolysis is conducted at 0.08 A/cm^2 and exhibits a coulombic efficiency of 79%. The nano-belt structure appears to consist of a flattened (or "deflated") carbon nanotube.

TEM and HAADF elemental analysis of the nano-dragon, nanobelt and nanotree structures are presented in Figures 9–13. In Figure 9, the nano-dragon structure is seen as a graphitic structure, albeit complex. A similar looking Pt, rather than C, structure has been previously observed and described as a bumpy surface on one-dimensional Pt nanowires [66]. The nano-tree allotrope is seen in Figure 10 to consist of CNTs, but differs from the conventional CNT structures, which generally do not contain merged CNTs. However, the nano-tree morphology includes intersecting CNTs as seen in Figure 10, whose structures merge and appear to branch off one another. A nanocarbon CVD growth branching mechanism has been suggested and is shown in Figure 11, catalyzed by fractionation of the nucleation sites leading to carbon branches [67]. In Figures 10 and 12, it can be seen that the interior of nano-trees and nano-belts can respectively contain nickel and iron, or nickel in the structure interior. As seen in Figure 12, the nano-belt product is flat and consists of graphene layers, but other than the measured presence of nickel, the mechanism of this unusual flat morphology is evident from the TEM. CVD nano-belt CNT structures have been previously synthesized with a schematic structure illustrated on the right side of Figure 11 [68].

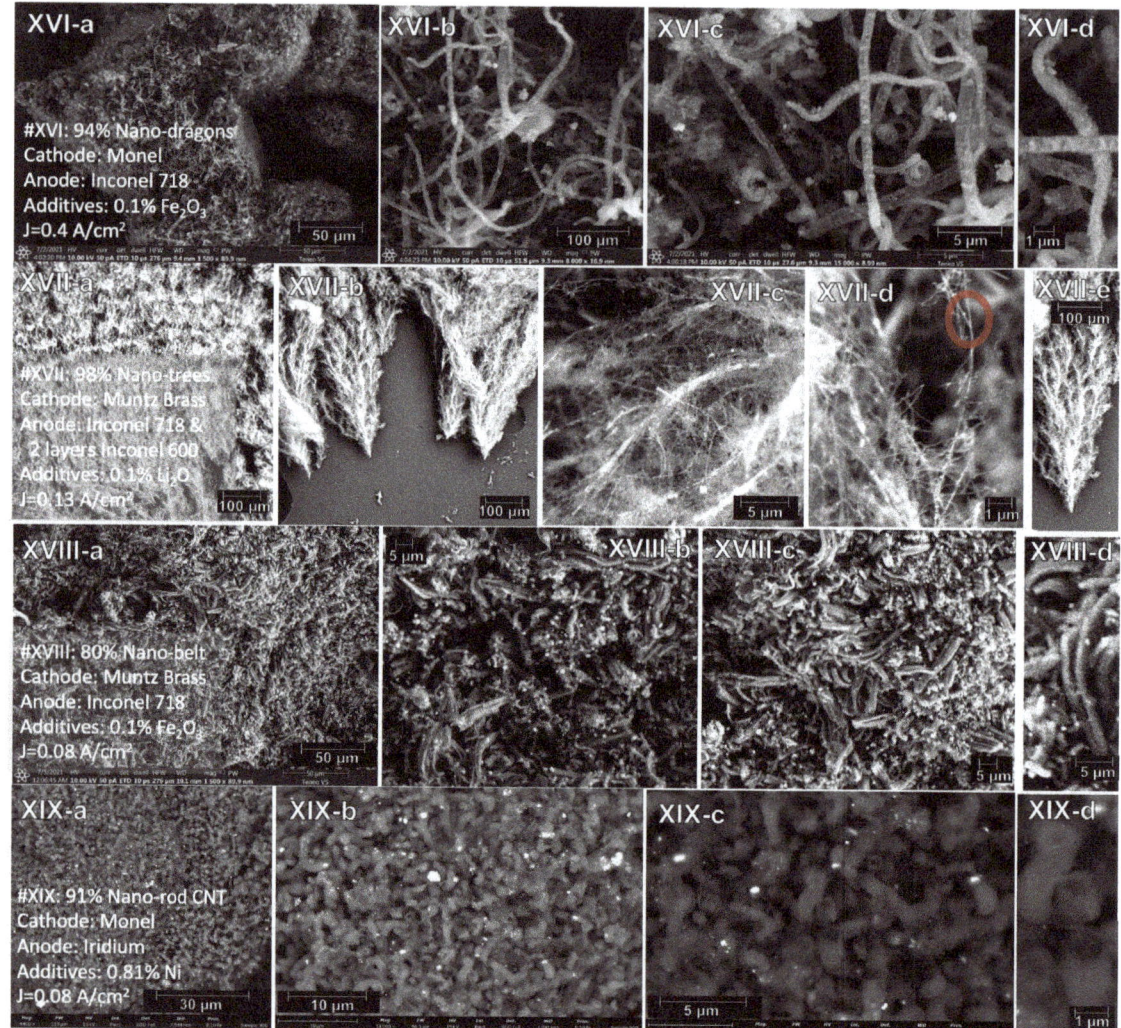

Figure 8. SEM of the synthesis product of nano-dragons, nano-trees, nano-belts and nano-rod allotropes of carbon by electrolytic splitting of CO_2 in 770 °C Li_2CO_3. Moving left to right in the panels, the product is analyzed by SEM with increasing magnification. Scale bars in panels (starting from left) are for panels k: 50, 10, 5 and 5 µm; for panels Q: 100, 100, 5, 1 and 100 µm; for panels Z: 50, 5 and 5 µm; for panels 9: 30, 10, 5 and 1 µm.

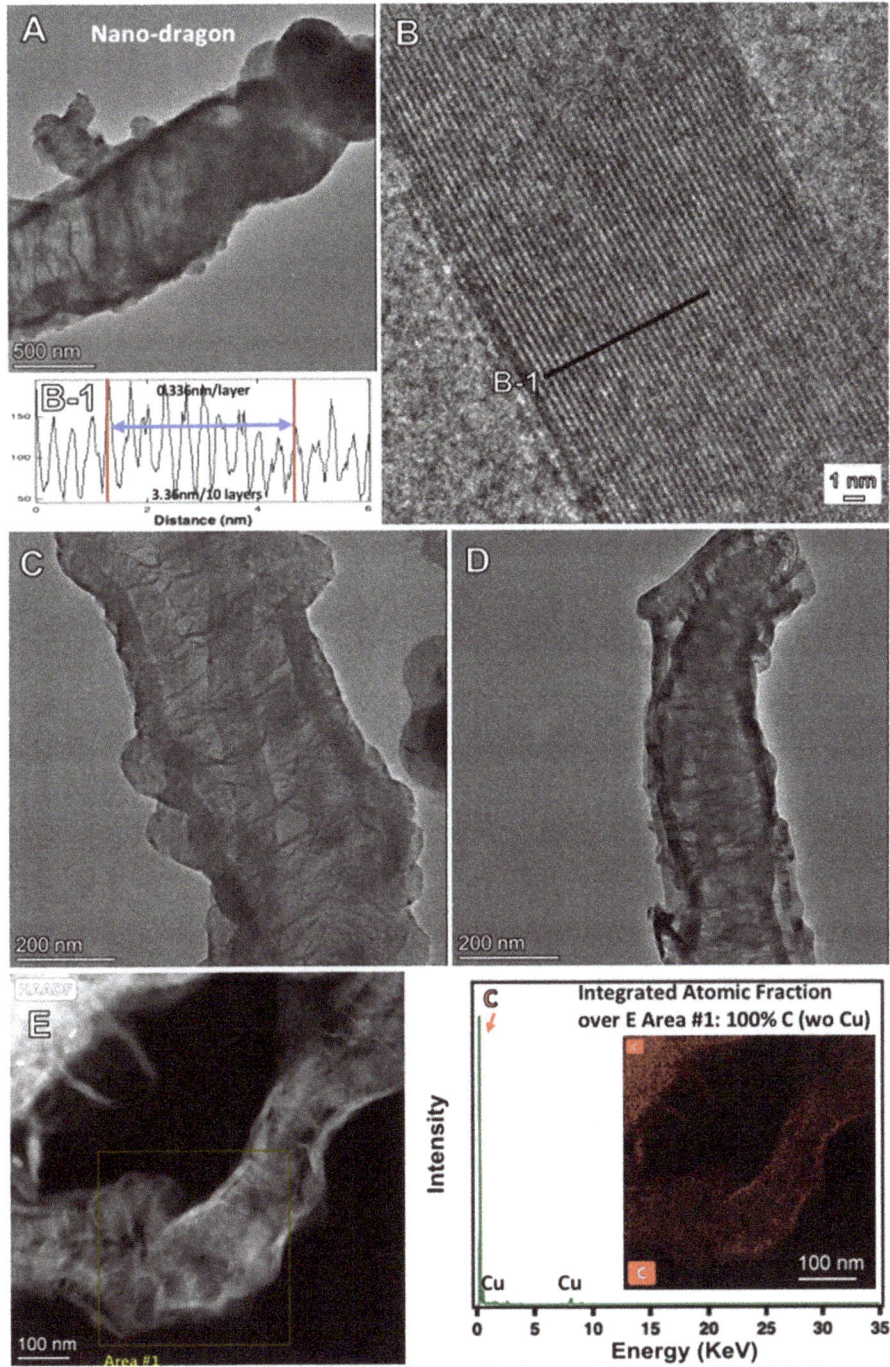

Figure 9. TEM and HAADF elemental analysis of the nano-dragons carbon allotrope synthesized by molten carbonate electrolysis. (**A,C,D**): Nano-dragon; (**B**) and B-1: Nano-dragon wall and measured graphene layer thickness (**E**); Nano-drageon and (right side) elemental analysis.

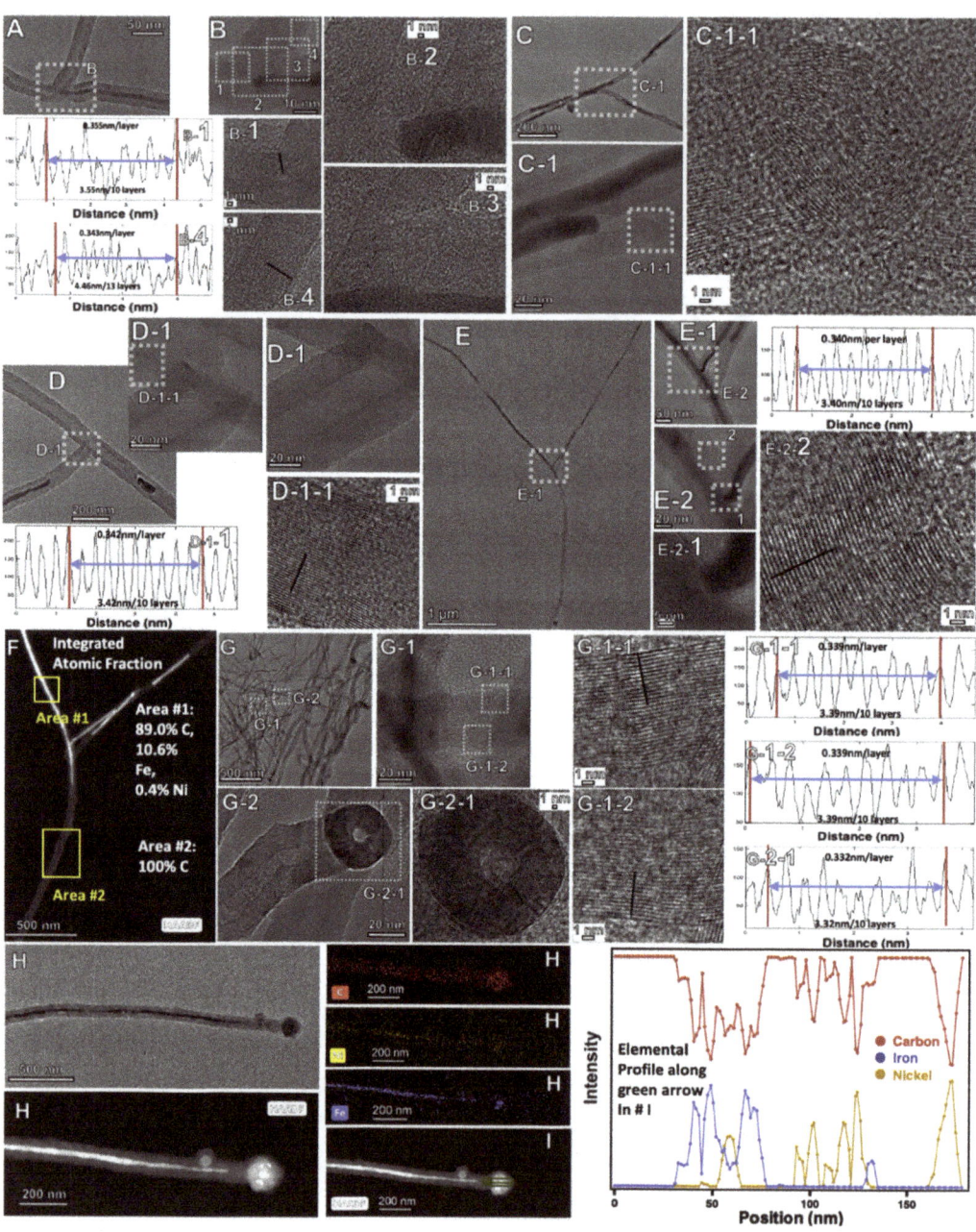

Figure 10. TEM and TEM HAADF elemental analysis of the nano-trees carbon allotrope synthesized by molten carbonate electrolysis. (**A**,**B**), B-2, B-3, (**C**), C-1, C-1-1, (**D**), D-1, (**E**), E-1, E-2-1, (**F**), F-1, (**G**), G-1. G-2, TEM of Nano-trees; B-1, B-4, D-1-1, E1, G1-1, G1-2, G2-1; TEM of Nano-trees with measured graphene layer thickness; (**H**,**I**) HAADF of Nano-tree with elemental profile (right side).

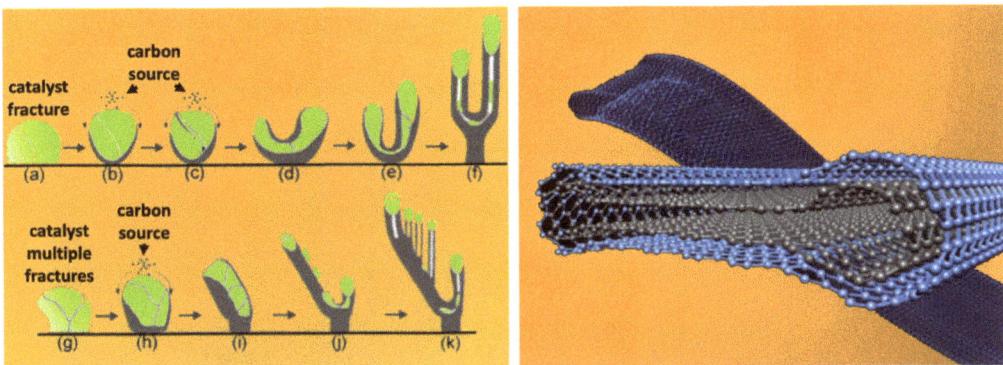

Figure 11. Left: A scheme illustrating the growth of an observed CVD synthesized amorphous branched carbon nano-tree catalyzed by iron carbide (include as the yellow domains). a–f and g–k show fractionation of the yellow iron carbide nucleation site leading to one or more purple-colored carbon branches. Modified from Reference [64]. **Right**: A scheme illustrating the structure of a CVD synthesized carbon nano-belt. Modified from Reference [65].

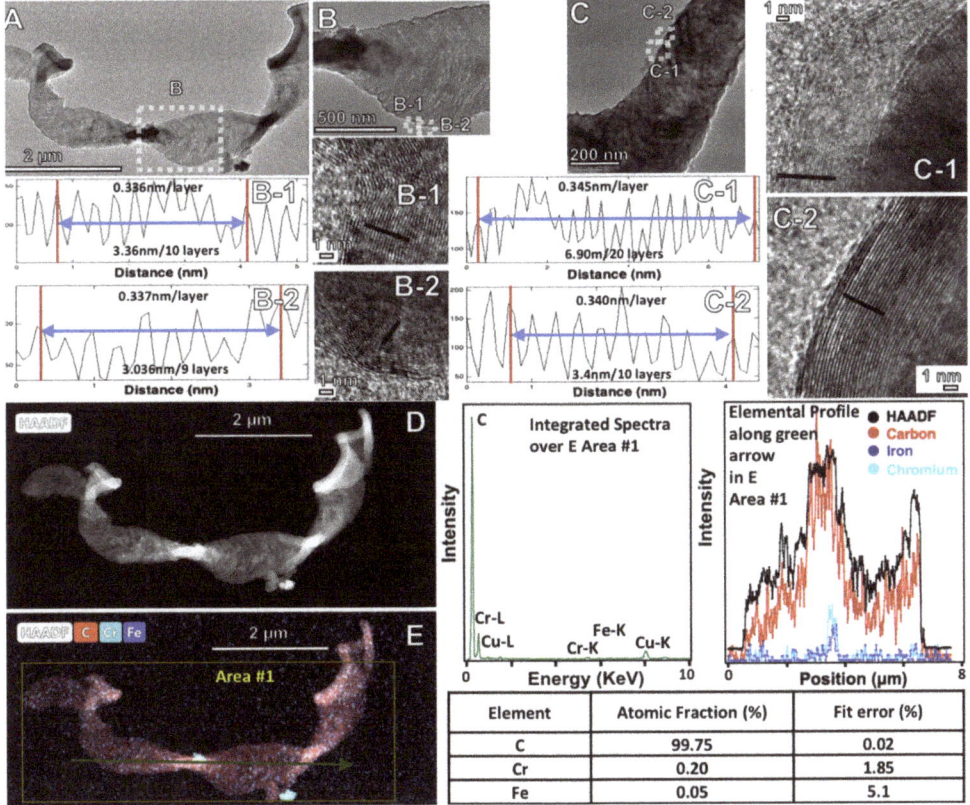

Figure 12. TEM and TEM HAADF elemental analysis of the Nano-belt carbon allotrope synthesized by molten carbonate electrolysis. (**A–C**) TEM of Nano-belts; B-1, B-2, C-1, C-2 TEM of Nano-trees with measured graphene layer thickness; (**D**) HAADF of Nano-belt with elemental analysis (**E**) and profile (middle and right side).

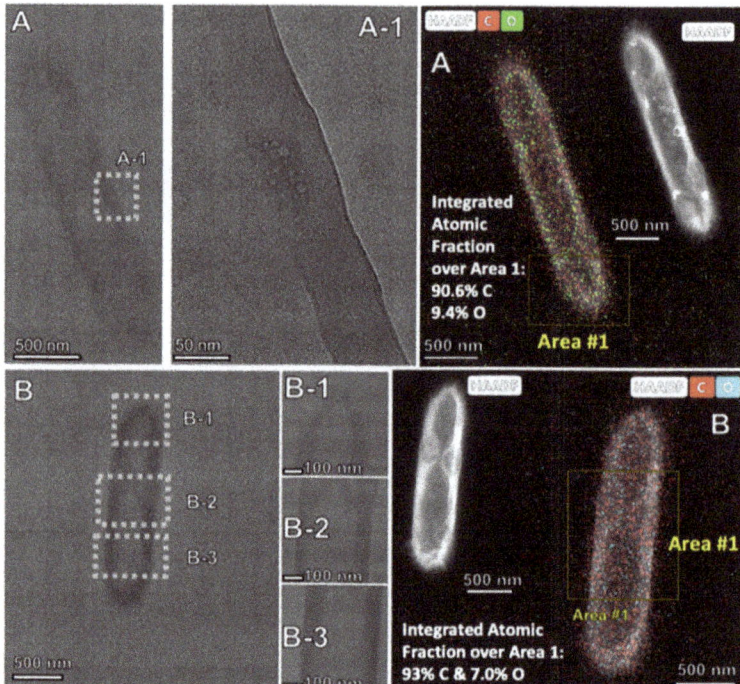

Figure 13. TEM and HAADF elemental analysis of nano-rod carbon allotrope synthesized by molten carbonate electrolysis. TEM and TEM HAADF elemental analysis of the Nano-rod carbon allotrope synthesized by molten carbonate electrolysis. (**A**,**B**) TEM with (right side) HAADF elemental analysis of Nano-rods; A-1, B-1, B-2, B-3 TEM of Nano-rods.

Without aging the electrolyte, the low current density (0.08 A/cm^2, exhibiting a coulombic efficiency of 80%), long-term growth (18 h) growth of carbon nanotubes with a Monel cathode, iridium anode, 0.81% Ni, and no ramped current activation step, leads to squat, ring-like nano-rod allotropes seen in Figure 13, and included in Table 4, as Electrolysis XIX. Of the electrolyses presented here, the product is singularly unusual from two physical chemical perspectives: (1) The TEM in Figure 13 reveals no evidence of a layered graphene structure. However, as shown in a later section, this morphology does exhibit an XRD peak and Raman spectrum typical of graphitic layered graphene structures. (2) As seen in the elemental analysis in Figure 13, the nano-rods are the only one of the new molten synthesized nanocarbon allotropes in which a significant concentration of oxygen (7.0 to 9.4%) is observed. With the bulbous rod-like morphology, rather than a growth which increases a CNT's length along with its diameter in time, this appears consistent with a long-term growth dominated by diameter, rather than length, increases.

Table 4. Raman spectra of a diverse range of carbon CNMs formed by molten electrolysis.

CO$_2$ Molten Electrolysis Product Description	ν_D (cm^{-1})	ν_G (cm^{-1})	ν_{2D} (cm^{-1})	I$_D$/I$_G$	I$_{2D}$/I$_G$
Multi-wall carbon nanotube	1342.4	1576.5	2688.7	0.30	0.60
Hollow nano-onion	1346.3	1577	2694.6	0.33	0.61
Helical carbon nanotube	1346.1	1578.2	2692.8	0.45	0.40
Nano-dragon	1346.7	1580.3	2695.0	0.67	0.62
Nano-flower	1347.9	1582.7	2692.2	0.78	0.50
Nano-tree	1343.7	1583.7	2696.4	0.82	0.47
Nano-bamboo	1352.0	1586.2	2696.9	1.04	0.72
Nano-pearl	1352.9	1588.5	2689.3	1.05	0.52
Nano-rod	1351.6	1586.0	2695.9	0.78	0.81
Carbon nanofiber	1349.3	1594.9	2696.0	1.27	0.37
Nano-belt	1348.5	1590.5	2705.1	1.30	0.41

2.5. The Diverse Range of Carbon Allotropes Formed by Molten Electrolysis

The top row and middle row of Figure 14 compares microscopy of this study's new carbon allotropes to those structures in the second row that were previously formed by molten electrolysis. The new electrolysis synthesis structures shown are conical CNF, nano-bamboo, nano-pearl, Ni coated CNT, nano-flower, nano-dragon, nano-rod, nano belt, nano-onion (also previously synthesized by alternative methodology in [25]), hollow nano-onion, and nano-tree. The previous distinct nanocarbon structures synthesized were carbon nanotubes ([11], and onward), nano-platelet [28], graphene (a two-step synthesis of CO$_2$ molten electrolysis followed by exfoliation) [27] and nano-helices [20].

Annual anthropogenic emissions of carbon amount to about 7 Gigatons. Can molten carbon splitting of CO$_2$ occur at a sufficient level to mitigate this anthropogenic carbon and mitigate global warming? Yes, but at a massive scale, as previously described [69].

Rather, than building stockpiles of CNMs to mitigate climate change, we put forth that the collective physical chemical properties of graphitic nanocarbon allotropes (including highest strength, high thermal and electrical conductivities, electronic and electrical storage properties, lubrication, medical applications, durable textiles, etc., and properties yet to be discovered) are greatly preferred over conventional materials and provide an incentive for their replacement by CNMs. Coupled with the very low cost of inorganic molten electrolysis (a low cost analogous to the industrial cost of aluminum production by splitting of aluminum oxide, but instead nanocarbon production by splitting of carbon dioxide), these new CNMs provide a value-added logical choice for replacement of these conventional materials, while eliminating CO$_2$. Collectively, replacement of today's annual 9 Gigaton (Gt) usage of these conventional materials (including, annually, production of cement = 4 Gt, steel = 1.6 Gt, aluminum = 0.058 Gt, plastics = 0.37 Gt, wood construction = 2.0 Gt, cotton and wool = 0.1 Gt, paper and cardboard = 0.4 Gt) provides the opportunity to eliminate net anthropogenic CO$_2$ and mitigate climate change.

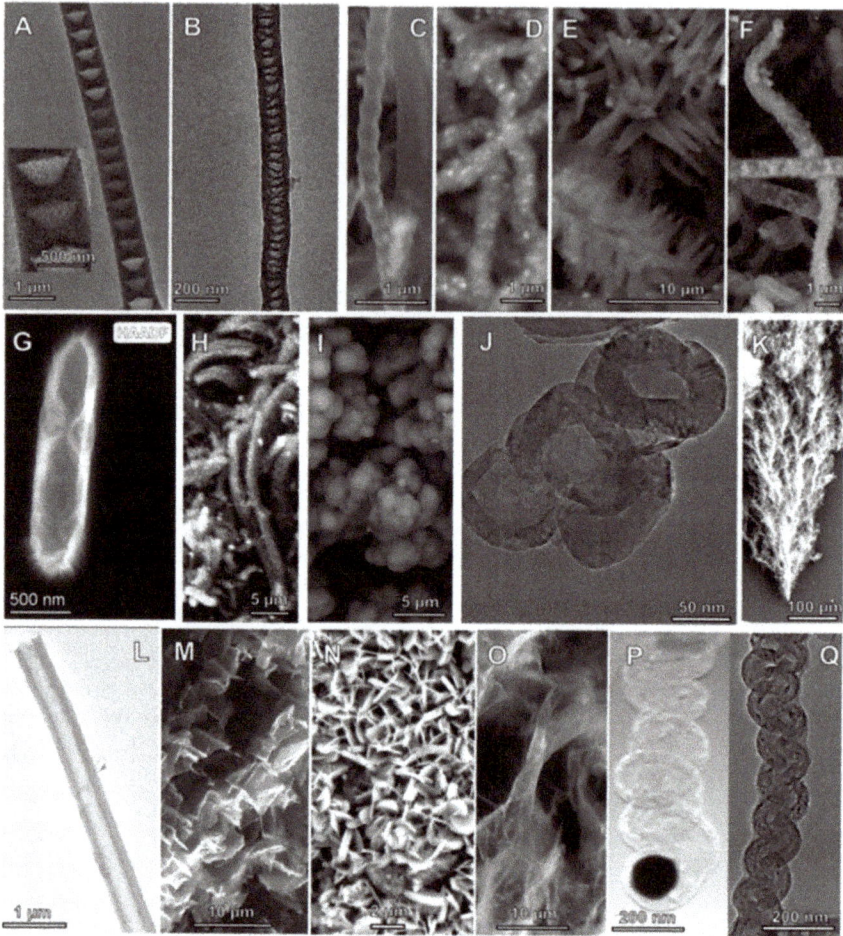

Figure 14. SEM of nanocarbon allotropes synthesized by the electrolytic splitting of CO_2 in molten carbonate. Top and middle row: nanocarbon allotropes as introduced and synthesized in this study. Bottom row: as previously synthesized. Top row (from left to right): (**A**): conical CNF, (**B**): nano-bamboo, (**C**): nano-pearl, (**D**): Ni-coated CNT, (**E**): nano-flower, (**F**): nano-dragon. Middle row: (**G**): nano-rod, (**H**): nano-belt, (**I**): nano-onion (also previously synthesized by alternative methodology in [25]), (**J**): hollow nano-onion, and (**K**): nano-tree. Bottom row (from left to right) (**L**): Carbon nanotube ([9]), (**M**): nano-scaffold ([28]), (**N**): nano-platelet ([27]), (**O**): graphene (2 step process, [27]) (**P,Q**): nano-helices ([20]).

2.6. Raman and XRD of the New Structures Formed by Molten Electrolysis

Figure 15 presents the effect of variation of the electrolysis conditions on the Raman spectra and XRD of the new carbon products of CO_2 electrolysis in 770 °C Li_2CO_3. For comparison purposes, also included are the Raman spectra of the CNTs [29,38]. The graphitic fingerprints lie in the 1880–2300 cm^{-1} and are related to different collective vibrations of sp-hybridized C-C bonds. The tangential G-band (at ~1580 cm^{-1}) is derived from the graphite-like in-plane mode of E_{2G} symmetry, and can be split into several modes, two of which are most distinct: the G_1 (1577 cm^{-1}) and G_2 (1610 cm^{-1}). The Raman spectrum exhibits two sharp peaks ~1350 and ~1580 cm^{-1}, which correspond to the disorder-induced mode (D band) and the high frequency E_{2G} first order mode (G band), respectively, and an additional peak, the 2D band, at 2700 cm^{-1}. The G′ peak at ~2300, is related to the collective stretching vibrations of sp-hybridized C–C bonds.

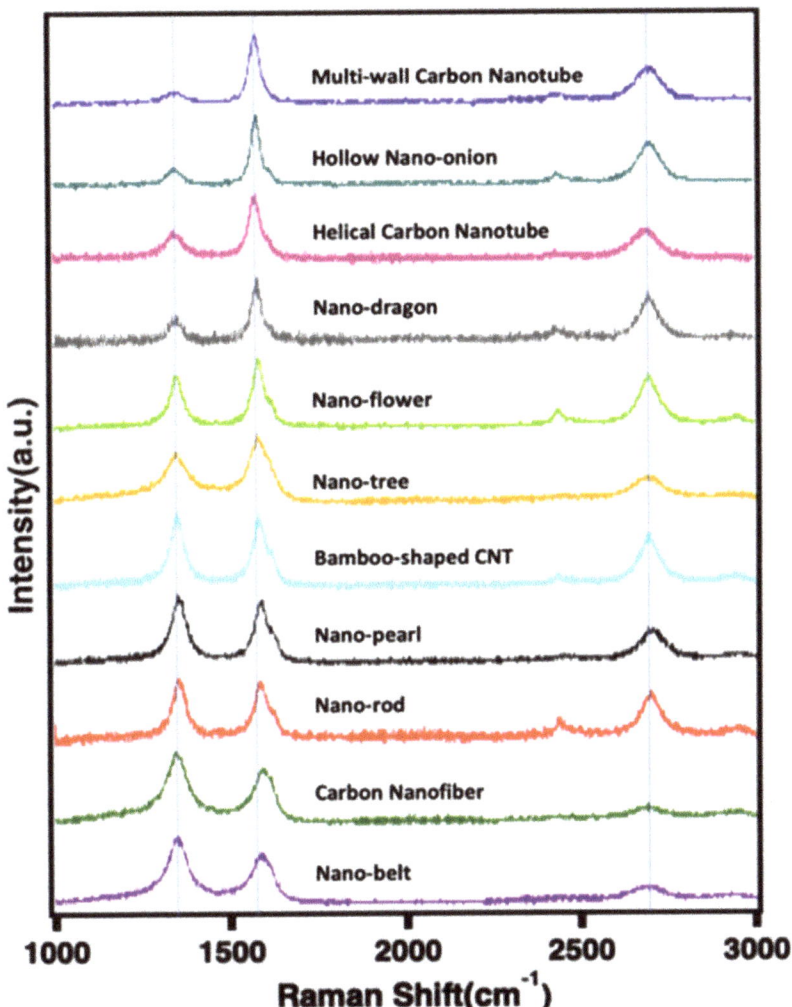

Figure 15. Raman of the synthesis product consists of various labeled CNMs and packed carbon nanotube assemblies synthesized by the electrolytic splitting of CO_2 in 770 °C Li_2CO_3 with a variety of systematically varied electrochemical conditions.

The intensity ratio between D band and G band (I_D/I_G) is a useful parameter to evaluate the relative number of defects and degree of graphitization. Table 4 summarizes Raman band peak locations and includes calculated (I_D/I_G) and (I_{2D}/I_G) peak ratios for the various carbon allotropes. A higher ratio I_D/I_G or a shift in I_G frequency[15] is a measure of increased defects in the carbon graphitic structure [70]. Defects that can occur in the graphitic structure include replacement of carbon sp^2 bonds, typical of the hexagonal carbon configuration in the graphene layers comprising the structures, with sp^3, increase in pores or missing carbon in the graphene, and enhancement of defects that cause formation of heptagonal and pentagonal, rather than the conventional hexagonal, graphene building blocks of graphene [71].

Typically, I_D/I_G for multi-walled carbon nanotubes is in the range of 0.2 to 06. Compared to these values, with the exception of the hollow nano-onions, the new carbon allotropes generally exhibit a higher than 0.6 I_D/I_G, evidence of a higher number of defects and perhaps consistent with the greater morphological complexity of these new allotropes. The nanocarbon bamboo, pearl, annular and belts each exhibit a relatively high level of defect, often associated with greater pores and twists and turns in the structure due to the higher presence of sp^3 carbons. As observed from Table 4, the order of the increasing I_D/I_G ratio is

CNT < hollow nano-onion < dragon < flower < trees < bamboo < pearl < rod < CNF < belt.

The shift to higher frequencies of the frequency, ν, of the G band generally correlates with the observed I_D/I_G variation, with variations due to near lying ratios, and with the exception of an unusually large shift for nano-bamboo.

High levels of Ni, Cr or Co added to the electrolyte (nano-bamboo, nano-pearl and nano-flower allotropes) also appear to correlate with an increase in defects, and the very high added Ni powder used in the nano-rod synthesis correlates with a very high level of defects as indicated by the shift in I_G frequency and an increase in I_D/I_G. Previously, increased concentrations of iron oxide added to the Li_2CO_3 electrolyte correlated with an increasing degree of disorder in the graphitic structure [20,21]. Interestingly, it is the synthesis with a low level of added iron oxide powder (but only added prior to the 24 h aging of the electrolyte) that results in the CNMs with the highest level of defects, the nano-belt CNM.

Lower defects are associated with applications which require high electrical conductivity and strength, while high defects are associated with applications which permit high diffusivity through the structure, such as those associated with increased intercalation and higher anodic capacity in Li-ion batteries and higher charge super capacitor.

Along with the XRD library of relevant compound spectra, XRD is presented in Figure 16 of the new nanocarbon morphologies products, prepared as summarized in Tables 2 and 3, and with SEM in Figures 2, 7 and 8. Each of the spectra exhibit the strong, sharp diffraction peak at 2θ = 27° characteristic of graphitic structures, and no indication of the broad peak indicative of amorphous carbon. In addition to graphite, the products XRD are grouped by which metal salts are present. Nano-bamboo exhibits the simplest composition with only a lithiated nickel salt present. The next most complex compositions, seen in the middle of Figure 16, are nano-dragons, nano-flowers and nano-trees, which also include the iron carbide salt Fe_3C. The next most complex composition is exhibited in the figure in the lower left hand corner, for hollow nano-onions, which exhibit each of those previous metal salts as well as a lithiated chromium(III) salt. Finally, both nano-belt and nano-pearls include an additional lithiated copper salt, and it may be noted that they were respectively synthesized with a Muntz brass and a Monel cathode which contain copper. However, the source of the copper requires further investigation. To enter the nanocarbon, the copper may need to dissolve from the cathode, which is under cathodic bias. This did not occur with the other CNM products. The nano-belt XRD spectra is distinct from the others having a dominant peak at 2θ = 43°, reflecting a higher concentration of metals than in the other products. The diminished presence of defects previously noted by the Raman spectra for the hollow nano-onion morphology, along with the XRD presence of $Li_2Ni_8O_{10}$, $LiCrO_2$ and Fe_3C provide evidence that the co-presence of Ni, Cr and Fe as nucleating agents can diminish defects in the structure compared to Ni. On the other hand, the enhanced presence of defects previously noted by the Raman spectra for the nano-belt and nano-pearl morphologies, along with the XRD presence of $LiCuO_2$, provide evidence that the copper salt increases defects in the structure compared to Ni, Fe or Cr as transition meal nucleating agents. Finally, it should be noted that the singular (amongst all the electrolyses) addition of cobalt powder to Electrolyses XIV and XV must be correlated with the subsequent observed formation of the nano-flower allotrope. However, this cobalt does not make its way into the product as analyzed by XRD in Figure 16, is observed only in trace quantities by HAADF TEM (to be delineated and probed in future studies) and presumably has another role in promoting formation of this unusual products.

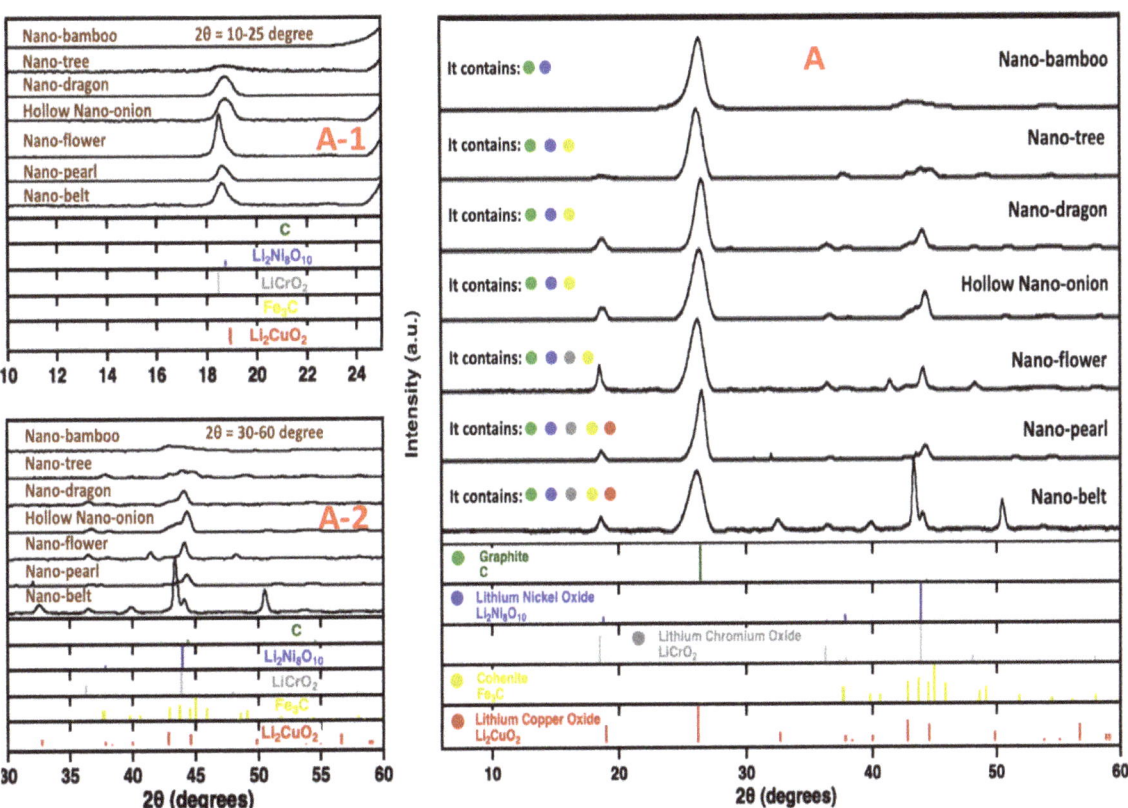

Figure 16. XRD of the synthesis product consisting of various labeled unusual nanocarbon morphologies synthesized by the electrolytic splitting of CO_2 in 770 °C Li_2CO_3 with a variety of systematically varied electrochemical conditions. (**A,A-1,A-2**): XRD over various ranges of 2θ.

3. Materials and Methods

3.1. Materials

Lithium carbonate (Li_2CO_3, 99.5%), lithium oxide (Li_2O, 99.5%), lithium phosphate Li_3PO_4 (Li_3PO_4, 99.5%), iron oxide (Fe_2O_3, 99.9%, Alfa Aesar) and boric acid (H_3BO_3, Alfa Aesar 99 + %) were used as electrolyte components in this study. For electrodes, Nichrome A (0.04-inch-thick), Nichrome C (0.04-inch-thick), Inconel 600 (0.25-inch-thick), Inconel 625 (0.25-inch-thick), Monel 400, Stainless Steel 304 (0.25-inch-thick) and Muntz Brass (0.25-inch-thick) were purchased from onlinemetals.com. Ni powder was 3–7 µm (99.9%, Alfa Aesar). Cr powder was <10 µm (99.2%, Alfa Aesar). Co powder was 1.6 µm (99.8%, Alfa Aesar). Iron oxide was 99.9% Fe_2O_3 (Alfa Aesar). Co powder was 1.6 µm (99.8%, Alfa Aesar). Inconel 600 (100 mesh) was purchased from Cleveland Cloth. The electrolysis was a conducted in a high form crucible >99.6% alumina (Advalue).

3.2. Electrolysis and Purification

Specific electrolyte compositions of each electrolyte are described in the text. The electrolyte was pre-mixed by weight in the noted ratios then metal or metal oxide additives were added if used. The cathode was mounted vertically across from the anode and immersed in the electrolyte. Generally, the electrodes are immersed subsequent to electrolyte melt. For several, as noted, electrolyses, once melted, the electrolyte was maintained at 770 °C ("aging" the electrolyte) prior to immersion of the electrolytes followed by immediate electrolysis. Generally, the electrolysis was driven with a described constant current density. As noted, for some electrolyses, the current density is ramped in several steps building to the applied electrolysis current which is then maintained at a constant current density. Instead, most of the electrolyses are initiated, and held, at a single constant current. The electrolysis temperature was 770 °C.

3.3. Product Characterization

The raw product was collected from the cathode after the experiment and cooldown, followed by an aqueous wash procedure which removes electrolyte congealed with the product as the cathode cools. The washed carbon product was separated by vacuum filtration. The washed carbon product is dried overnight in a 60 °C oven, yielding a black powder product.

The coulombic efficiency of electrolysis is the percent of applied, constant current charge that was converted to carbon determined as:

$$100\% \times C_{experimental} / C_{theoretica} \qquad (4)$$

This is measured by the mass of washed carbon product removed from the cathode, $C_{experimental}$, and calculated from the theoretical mass, $C_{theoretical} = (Q/nF) \times (12.01 \text{ g C mol}^{-1})$ which is determined from Q, the time integrated charged passed during the electrolysis, F, the Faraday (96,485 As mol^{-1} e^{-}), and the n = 4 e^{-} mol^{-1} reduction of tetravalent carbon consistent with Equation (2).

Characterization: The carbon product was washed and analyzed by PHENOM Pro Pro-X SEM (with EDX), FEI Teneo LV SEM, and by FEI Teneo Talos F200X TEM (with EDX). XRD powder diffraction analyses were conducted with a Rigaku D = Max 2200 XRD diffractometer and analyzed with the Jade software package. Raman spectra were collected with a LabRAM HR800 Raman microscope (HORIBA). This Raman spectrometer/microscope uses an incident laser light with a high resolution of 0.6 cm^{-1} at 532.14 nm wavelength. The Raman spectrometer/microscope uses an incident 0.13 mW laser light with a high resolution of 0.6 cm^{-1} at 532.14 nm wavelength with 1800 gr/mm, 800 mm focal objective, and 100 ms integration. The PHENOM Pro SEM provides over 100,000× electron optical magnification and uses up to a 15 kV acceleration voltage for imaging and analysis. Specifications of the other instruments are available online from the manufacturers.

4. Conclusions

Known conventional carbon allotropes include diamond and graphite, and more recently buckyballs, graphene, carbon nanofibers and CNTs. In this study, the range of carbon allotropes and morphologies, and in particular nanocarbon morphologies, that can be synthesized by the molten carbonate electrolysis of CO_2 has been greatly expanded. Fascinating high purity morphologies that have been obtained in this study by the systematic variation of electrolysis conditions are conical CNF, nano-bamboo, nano-pearl, Ni-coated CNT, nano-flower, nano-dragon, nano-rod, nano-belt, nano-onion (also previously synthesized by an alternative methodology in [25]), hollow nano-onions and nano-trees. Each of these CNMs have their unusual and distinctive morphologies, such as the nano-trees with their branching CNT structure, or the nano-bamboo and -pearl with their different, but repeated knob or bulb shapes. These distinctive morphologies may lead to unusual physical chemical properties with implications useful to applications, such as those utilizing the high strength, high thermal, magnetic, electronic, piezoelectronic, tribological characteristics of graphene-based materials, but which distribute these properties differently throughout the unusual geometries of this novel allotropes. For example, alternative applications such as high-capacity Li-anodes, unusual electronics, EMF shielding, improved lubricants, and new structural or polymer composites may be anticipated.

This study has explored a variety of electrochemical configurations, systematically varying electrode composition, current density and electrolysis time, current ramping initiation, and variation of electrolyte additives and their concentrations. The observed nanocarbon structures were analyzed by SEM, TEM, including with HAADF, Raman and XRD. With the exception of the nano-rod structure, each of the structures is graphitic in nature, containing graphene layers arranged in a variety of geometries. The graphene layers exhibit the characteristic inter-layer spacing of 0.33 to 0.34 nm. Except for the presence of Ni, Fe, Cr and occasionally Cu, which may serve as nucleating growth sites, each of the structures is pure carbon. Generally, intersecting graphene layers did not merge, but in the nano-tree, the graphene layers bend at intersections leading to the observed branching.

Many of the structures including nano-bamboo, nano-pearl, Ni-coated CNTs and conical CNFs exhibit walls containing concentric graphene layers. The nano-dragon and nano-belt structures include layered planer or planar-twisted graphene layers. Several of the observed structures, including nano-trees, and hollow and filled nano-onions, exhibit concentric, highly spherical graphene layers generally composed of carbon and containing a low level of internal transition metal. A new pathway to the formation of nano-onions via phosphate addition to the electrolyte is demonstrated, and the hypothesis that phosphate selectively binds transition metal ions should be pursued.

Each of the syntheses were conducted in a 770 °C Li_2CO_3 electrolyte with or without various additives, on a variety of metal or metal alloy electrodes, and with a range of current densities. In a sister paper [29], slight variations of these same synthesis parameters form high-purity carbon but only with the CNT structure. The varied anode and cathodes contained either pure Ni, or mixes also including Fe and Cr, or various mixes of an extended variety of transition metals. However, the changing conditions led to variations of the CNT morphology (length, diameter, curled or straight, added defects etc.). All syntheses in the study from Electrolysis IV onward produced a high-purity product of the stated structure, with the exception of the conical CNFs that were a minority (6%) within a majority of nano-bamboo carbon, and the moderate purity (85%) nano-belt carbon product. Coulombic efficiency of the electrolyses ranged from 79 to 80% at lower current densities of 0.08 A/cm^2, to over 99% at current densities of 0.2 A/cm^2 or higher. The high purity products each exhibited sharp XRD graphic peaks, and a moderate (0.3 to 1.3) Raman I_D/I_G ratio indicative of a moderate level of defects in the structure. In addition to a majority of pure, graphitic carbon, the XRD also exhibited different singular or mixed transition metal salts of either iron carbide, or nickel, chromium or copper lithiated oxides.

TEM HAADF of the new nanostructures shows that their inner core is generally metal-free (void, with the walls 100% carbon), but in other areas, the void is filled with transition metals: Ni, Fe and/or Cr. Except for the nano-rod product, each of the structures included distinct graphene layers with a graphene characteristic, inter-layer spacing of 0.33–0.34 nm. Depending on the nanostructure, adjacent graphene layers were organized either in a planer, cylindrical or spherical geometry. When the internal transition metal is in the tip, the layered graphene walls are observed to bend in a highly spherical fashion around the metal supporting the transition metal nucleated CNT growth mechanism. The use of a Ni-anode, or an excess of added Ni to the electrolyte, leads to Ni-coated CNTs when stainless steel is used as the cathode. Generally, intersecting graphene layers did not merge, but in the nano-tree allotrope, graphene layers bend to become part of a CNT intersection consistent with branching.

Molten carbonate electrolysis of CO_2 provides an effective path for the synthesis of a portfolio of unusual, valuable nanocarbon allotropes. Mass production of these structures from CO_2 will provide a valuable incentive to consume this greenhouse gas. Such structures are rare, or were previously non-existent, and are not generally commercially available. However, those that are in use, such as nano-onions made by pyrolysis of nano-diamonds [35], or by CVD, have high carbon footprints and associated costs at over USD 1 million/ton. CNT production by the molten carbonate electrolysis of CO_2, the C2CNT process, is an inexpensive synthesis comparable to the cost of aluminum oxide splitting in the industrial production of aluminum [14]. The scale-up of this process was awarded the 2021 Carbon XPrize XFactor award for producing the most valuable product from CO_2 [31,32]. The new synthesis conditions consist of small variations of the scaled C2CNT process with a comparable, straightforward path to scale up to contribute to consumption of CO_2 for climate change mitigation.

Author Contributions: Conceptualization, S.L. and G.L.; methodology, S.L., X.L., G.L. and X.W.; writing S.L. and G.L. All authors have read and agreed to the published version of the manuscript.

Funding: C2CNT LLC funded this research through the C2CNT LLC XPrize support funding.

Data Availability Statement: The authors confirm that the data supporting the findings of this study are available within the article.

Conflicts of Interest: The authors declare no conflict of interest.

References

1. CO_2-Earth: Daily CO_2 Values. Available online: https://www.CO2.earth/daily-CO2 (accessed on 7 December 2021).
2. NASA: Global Climate Change: The Relentless Rise of Carbon Dioxide. Available online: https://climate.nasa.gov/climate_resources/24/ (accessed on 7 December 2021).
3. Urban, M.C. Accelerating extinction risk from climate change. *Science* **2015**, *348*, 571–573. [CrossRef]
4. Pimm, S.L. Climate disruption and biodiversity. *Curr. Biol.* **2009**, *19*, R595–R601. [CrossRef]
5. Praksh, G.K.; Olah, G.A.; Licht, S.; Jackson, N.B. *Reversing Global Warming: Chemical Recycling and Utilization of CO_2*; Report of 2008 NSF Workshop; University of Southern California: Los Angeles, CA, USA, 2008. Available online: https://loker.usc.edu/ReversingGlobalWarming.pdf (accessed on 7 December 2021).
6. Khanna, V.; Bakshi, B.R.; Lee, L.J. Carbon Nanofiber Production: Life cycle energy consumption and environmental impact. *J. Ind. Ecol.* **2008**, *12*, 394–410. [CrossRef]
7. Licht, S.; Wang, B.; Ghosh, S.; Ayub, H.; Jiang, D.; Ganley, J. New solar carbon capture process: STEP carbon capture. *J. Phys. Chem. Lett.* **2010**, *1*, 2363–2368. [CrossRef]
8. Ren, J.; Li, F.; Lau, J.; Gonzalez-Urbina, L.; Licht, S. One-pot synthesis of carbon nanofibers from CO_2. *Nano Lett.* **2010**, *15*, 6142–6148. [CrossRef]
9. Ren, J.; Licht, S. Tracking airborne CO_2 mitigation and low cost transformation into valuable carbon nanotubes. *Sci. Rep.* **2016**, *6*, 27760. [CrossRef]
10. Ren, J.; Lau, J.; Lefler, M.; Licht, S. The minimum electrolytic energy needed to convert carbon dioxide to carbon by electrolysis in carbonate melts. *J. Phys. Chem. C* **2015**, *119*, 23342–23349. [CrossRef]
11. Licht, S.; Liu, X.; Licht, G.; Wang, X.; Swesi, A.; Chan, Y. Amplified CO_2 reduction of greenhouse gas emissions with C2CNT carbon nanotube composites. *Mater. Today Sustain.* **2019**, *6*, 100023. [CrossRef]
12. Dey, G.; Ren, J.; El-Ghazawi, O.; Licht, S. How does an amalgamated Ni cathode affect carbon nanotube growth? *RSC Adv.* **2016**, *122*, 400–410.
13. Ren, J.; Johnson, M.; Singhal, R.; Licht, S. Transformation of the greenhouse gas CO_2 by molten electrolysis into a wide controlled selection of carbon nanotubes. *J. CO_2 Util.* **2017**, *18*, 335–344. [CrossRef]
14. Johnson, M.; Ren, J.; Lefler, M.; Licht, G.; Vicini, J.; Licht, S. Data on SEM, TEM and Raman spectra of doped, and wool carbon nanotubes made directly from CO_2 by molten electrolysis. *Data Br.* **2017**, *14*, 592–606. [CrossRef]
15. Johnson, M.; Ren, J.; Lefler, M.; Licht, G.; Vicini, J.; Liu, X.; Licht, S. Carbon nanotube wools made directly from CO_2 by molten electrolysis: Value driven pathways to carbon dioxide greenhouse gas mitigation. *Mater. Today Energy* **2017**, *5*, 230–236. [CrossRef]
16. Wang, X.; Liu, X.; Licht, G.; Wang, B.; Licht, S. Exploration of alkali cation variation on the synthesis of carbon nanotubes by electrolysis of CO_2 in molten carbonates. *J. CO_2 Util.* **2019**, *18*, 303–312. [CrossRef]
17. Wang, X.; Licht, G.; Licht, S. Green and scalable separation and purification of carbon materials in molten salt by efficient high-temperature press filtration. *Sep. Purif. Technol.* **2021**, *244*, 117719. [CrossRef]
18. Wang, X.; Sharif, F.; Liu, X.; Licht, G.; Lefer, M.; Licht, S. Magnetic carbon nanotubes: Carbide nucleated electrochemical growth of ferromagnetic CNTs. *J. CO_2 Util.* **2020**, *40*, 101218. [CrossRef]
19. Wang, X.; Liu, X.; Licht, G.; Licht, S. Calcium metaborate induced thin walled carbon nanotube syntheses from CO_2 by molten carbonate electrolysis. *Sci. Rep.* **2020**, *10*, 15146. [CrossRef]
20. Liu, X.; Licht, G.; Licht, S. The green synthesis of exceptional braided, helical carbon nanotubes and nanospiral platelets made directly from CO_2. *Mat. Today Chem.* **2021**, *22*, 100529. [CrossRef]

21. Liu, X.; Licht, G.; Licht, S. Data for the green synthesis of exceptional braided, helical carbon nanotubes and nano spiral platelets made directly from CO_2. *arXiv* **2021**, arXiv:2110.05398.
22. Licht, S.; Douglas, A.; Carter, R.; Lefler, M.; Pint, C. Carbon Nanotubes Produced from Ambient Carbon Dioxide for Environmentally Sustainable Lithium-Ion and Sodium-Ion Battery Anodes. *ACS Cent. Sci.* **2016**, *2*, 162–168. [CrossRef] [PubMed]
23. Lau, J.; Dey, G.; Licht, S. Thermodynamic assessment of CO_2 to carbon nanofiber transformation for carbon sequestration in a combined cycle gas or a coal power plant. *Energy Conser. Manag.* **2016**, *122*, 400–410. [CrossRef]
24. Licht, S. Co-production of cement and carbon nanotubes with a carbon negative footprint. *J. CO_2 Util.* **2017**, *18*, 378–389. [CrossRef]
25. Liu, X.; Ren, J.; Licht, G.; Wang, X.; Licht, S. Carbon nano-onions made directly from CO_2 by molten electrolysis for greenhouse gas mitigation. *Adv. Sustain. Syst.* **2019**, *3*, 1900056. [CrossRef]
26. Ren, J.; Yu, A.; Peng, P.; Lefler, M.; Li, F.F.; Licht, S. Recent advances in solar thermal electrochemical process (STEP) for carbon neutral products and high value nanocarbons. *Acc. Chem. Res.* **2019**, *52*, 3177–3187. [CrossRef]
27. Liu, X.; Wang, X.; Licht, G.; Licht, S. Transformation of the greenhouse gas carbon dioxide to graphene. *J. CO_2 Util.* **2020**, *36*, 288–294. [CrossRef]
28. Wang, X.; Licht, G.; Liu, X.; Licht, S. One pot facile transformation of CO_2 to an unusual 3-D nan-scaffold morphology of carbon. *Sci. Rep.* **2020**, *10*, 21518. [CrossRef]
29. Liu, X.; Licht, G.; Licht, S. Controlled Transition Metal Nucleated Growth of Carbon Nanotubes by Molten Electrolysis of CO_2. *Catalysts* **2022**. submitted.
30. Yu, M.-F.; Lourie, O.; Dyer, M.; Moloni, K.; Kelly, T.; Ruoff, R. Strength and Breaking Mechanism of Multiwalled Carbon Nanotubes Under Tensile Load. *Science* **2000**, *287*, 637–640. [CrossRef]
31. XPrize Foundation. *Turning CO_2 into Products*; XPrize Foundation: Culver City, CA, USA, 2021. Available online: http://CarbonXPrize.org (accessed on 7 December 2021).
32. XPrize Announces Winners with Each Team Creating Valuable Products from CO_2. Available online: https://www.nrg.com/about/newsroom/2021/xprize-announces-the-two-winners-of--20m-nrg-cosia-carbon-xprize.html (accessed on 7 December 2021).
33. Chang, C.-C.; Hsu, H.-K.; Aykol, M.; Hung, W.; Chen, C.; Cronin, S. Strain-induced D band observed in carbon nanotubes. *ACS Nano* **2012**, *5*, 854–862. [CrossRef]
34. Tenne, R. Recent advances in the research of inorganic nanotubes and fullerene-like nanoparticles. *Front. Phys.* **2014**, *9*, 370–377. [CrossRef]
35. Pech, D.; Brunet, M.; Durou, H.; Huang, P.; Mochalin, V.; Gogotsi, Y.; Taberna, P.L.; Simon, P. Ultrahigh-power micrometre-sized supercapacitors based on onion-like carbon. *Nat. Nanotechnol.* **2010**, *5*, 651–654. [CrossRef] [PubMed]
36. Han, F.; Yao, B.; Bai, Y. Preparation of Carbon Nano-Onions and Their Application as Anode Materials for Rechargeable Lithium-Ion Batteries. *J. Phys. Chem. C* **2011**, *115*, 8923–8927. [CrossRef]
37. Zeiger, M.; Jackel, N.; Mochalin, V.N.; Presser, V. Review: Carbon onions for electrochemical energy storage. *J. Mater. Chem. A* **2016**, *4*, 3172. [CrossRef]
38. Sano, N.; Wang, H.; Alexandrou, I.; Chhowalla, M.; Teo, K.B.K.; Amaratunga, G.A.J.; Iimura, K. Properties of carbon onions produced by an arc discharge in water. *J. Appl. Phys.* **2002**, *92*, 2783. [CrossRef]
39. Keller, N.; Maksimova, N.I.; Roddatis, V.V.; Schur, M.; Mestl, G.; Butenko, Y.V.; Kuznetsov, V.L.; Schlogl, R. The Catalytic Use of Onion-Like Carbon Materials for Styrene Synthesis by Oxidative Dehydrogenation of Ethylbenzene. *Chem. Int.* **2002**, *41*, 1885. [CrossRef]
40. Bidsorkhi, H.B.; D'Aloia, A.G.; Tamburrano, A.; Bellis, G.; Delfini, A.; Ballirano, P.; Sarto, M.S. 3D Porous Graphene Based Aerogel for Electromagnetic Applications. *Sci. Rep.* **2019**, *9*, 15719. [CrossRef]
41. Yu, C.; Song, Y.S. Analysis of Thermoelectric Energy Harvesting with Graphene Aerogel-Supported Form-Stable Phase Change Materials. *Nanomaterials* **2021**, *11*, 2192. [CrossRef]
42. Reece, R.; Lekakou, L.; Smith, P.A.; Grilli, R.; Trapalis, C. Sulphur-linked graphitic and graphene oxide platelet-based electrodes for electrochemical double layer capacitors. *J. Alloys Compd.* **2019**, *792*, 582–593. [CrossRef]
43. Tang, N.; Wen, J.; Zhang, Y.; Liu, F.; Lin, K.; Du, Y. Helical Carbon Nanotubes: Catalytic Particle Size-Dependent Growth and Magnetic Properties. *ACS Nano* **2010**, *4*, 241. [CrossRef]
44. Gao, R.; Wang, Z.L.; Fan, S. Kinetically Controlled Growth of Helical and Zigzag Shapes of Carbon Nanotubes. *J. Phys. Chem. B* **2000**, *104*, 1227. [CrossRef]
45. Qin, Y.; Zhang, Z.; Cui, Z. Helical carbon nanofibers with a symmetric growth mode. *Carbon* **2004**, *42*, 1917. [CrossRef]
46. Suda, Y. *Chemical Vapor Deposition of Helical Carbon Nanofibers, Chemical Vapor Deposition for Nanotechnology*; IntechOpen: London, UK, 2018.
47. Bajpai, V.; Dai, L.; Ohashi, T. Large-Scale Synthesis of Perpendicularly Aligned Helical Carbon Nanotubes. *J. Am. Chem. Soc.* **2004**, *126*, 5070. [CrossRef]
48. Tak, K.; Lu, M.; Hui, D. Coiled carbon nanotubes: Synthesis and their potential applications in advanced composite structures. *Compos. Part B Eng.* **2006**, *37*, 437.
49. Zhang, M.; Li, J. Carbon nanotube in different shapes. *Mater. Today* **2009**, *12*, 12. [CrossRef]

50. Wang, W.; Yang, K.; Galliard, J.; Bandaru, P.R.; Rao, Q.M. Rational Synthesis of Helically Coiled Carbon Nanowires and Nanotubes through the Use of Tin and Indium Catalysts. *Adv. Mater.* **2008**, *20*, 179. [CrossRef]
51. Zhang, Q.; Zhao, M.; Tang, D.; Li, F.; Huang, J.; Liu, B.; Zhu, W.; Zhang, Y.; Wei, F. Carbon-Nanotube-Array Double Helices. *Angew. Chem.* **2010**, *49*, 3642. [CrossRef] [PubMed]
52. Walling, B.E.; Kuang, Z.; Hao, Y.; Estrada, D.; Wood, J.D.; Lian, F.; Miller, L.A.; Shah, A.B.; Jeffries, J.L.; Haasch, R.T. Helical Carbon Nanotubes Enhance the Early Immune Response and Inhibit Macrophage-Mediated Phagocytosis of *Pseudomonas aeruginosa*. *PLoS ONE* **2013**, *8*, e80283. [CrossRef]
53. Jia, K.; Kou, K.; Qin, M.; Wu, H.; Puleo, F.; Liotta, L.F. Controllable and Large-Scale Synthesis of Carbon Nanostructures: A review of Bamboo-Like Nanotubes. *Catalysts* **2017**, *7*, 256. [CrossRef]
54. Lobo, L.S.; Carabineiro, S.A.C. Explaining Bamboo-Like Carbon Fiber Growth Mechanism: Catalyst Shape Adjustments above Tammann Temperature. *J. Carbon Res.* **2020**, *6*, 18. [CrossRef]
55. Gonzalez, I.; De Jesus, J.; Canizales, E. Bamboo-shaped carbon nanotubes generated by methane thermal decomposition using Ni nanoparticles synthesized in water–oil emulsions. *Micron* **2011**, *42*, 819–925. [CrossRef] [PubMed]
56. Maurice, J.-L.; Pribat, D.; He, Z.; Patriarche, G.; Cojocaru, C.S. Catalyst faceting during graphene layer crystallization in the course of carbon nanofiber growth. *Carbon* **2014**, *79*, 93–102. [CrossRef]
57. Zhang, M.; Zhao, N.; Sha, J.; Liu, E.; Shi, C.; Li, J.; He, C. Synthesis of novel carbon nano-chains and their application as supercapacitors. *J. Mat. Chem. A* **2014**, *2*, 16268. [CrossRef]
58. Zhang, M.; He, C.; Liu, E.; Zhiu, S.; Shi, C.; Li, J.; Li, Q.; Zhao, N. Activated Carbon Nano-Chains with Tailored Micro-Meso Pore Structures and Their Application for Supercapacitors. *J. Phys. Chem. C* **2015**, *119*, 21810–21817. [CrossRef]
59. Wang, P.; Xiao, P.; Zhong, S.; Chen, J.; Lin, H.; Wu, X.-L. Bamboo-like carbon nanotubes derived from colloidal polymer nanoplates for efficient removal of Bisphenol A. *J. Mat. Chem. A* **2016**, *4*, 15450–15456. [CrossRef]
60. Primo, E.N.; Gutierrz, F.A.; Rubianes, M.D. Bamboo-like multiwall carbon nanotubes dispersed in double stranded calf-thymus DNA as a new analytical platform for building layer-by-layer based biosensors. *Electrochim. Acta* **2015**, *182*, 391–397. [CrossRef]
61. Yadav, D.; Amini, F.; Ehrmann, A. Recent advances in carbon nanofibers and their applications—A review. *Eur. Polym. J.* **2020**, *138*, 109963. [CrossRef]
62. Mohamed, A. Synthesis, Characterization, and Applications Carbon Nanofibers. In *Carbon-Based Nanofillers and their Rubber Nanocomposites*; Yaragaila, S., Mishra, R., Thomas, S., Kalarikkal, N., Maria, H.J., Eds.; Elsevier: Amsterdam, The Netherlands, 2019; Chapter 9.
63. Shende, P.; Kasture, P.; Gaud, R.S. Nanoflowers: The future trend of nanotechnology for multi-applications. *Artif. Cells Nanomed. Biotechnol.* **2019**, *46*, 413–422. [CrossRef]
64. Kharisov, B.I. A Review for Synthesis of Nanoflowers. *Recent Pat. Nanotechology* **2008**, *2*, 190–200. [CrossRef] [PubMed]
65. Thongtem, S.; Singjai, P.; Thongtem, T.; Preyachoti, S. Growth of carbon nanoflowers on glass slides using sparked iron as a catalyst. *Mat. Sci. Eng. A* **2006**, *423*, 209–213. [CrossRef]
66. Takai, A.; Ataee-Esfahani, H.; Doi, Y.; Fuziwara, M.; Yamauchi, Y.; Kuroda, K. Pt nanoworms: Creation of a bumpy surface on one-dimensional (1D) Pt nanowires with the assistance of surfactants embedded in mesochannels. *Chem. Comm.* **2011**, *47*, 7701–7703. [CrossRef]
67. He, Z.; Maurice, J.-L.; Lee, C.S.; Cojocaru, C.S.; Pribat, D. Growth mechanisms of carbon nanostructures with branched carbon nanofibers synthesized by plasma-enhanced chemical vapour deposition. *Cryst. Eng. Comm.* **2014**, *16*, 2990–29995. [CrossRef]
68. Lin, C.-T.; Chen, R.-H.; Chin, T.-S.; Lee, C.-Y.; Chiu, H.-T. Quasi two-dimensional carbon nanobelts synthesized using a template method. *Carbon* **2008**, *46*, 741–746. [CrossRef]
69. Licht, S. Efficient Solar-Driven Synthesis, Carbon Capture, and Desalinization, STEP: Solar Thermal Electrochemical Production of Fuels, Metals, Bleach. *Adv. Mat.* **2011**, *23*, 5592. [CrossRef] [PubMed]
70. Basheer, B.V.; George, J.; Siengchin, S.; Parameswaranpillai, J. Polymer grafted carbon nanotubes—Synthesis, properties, and applications: A review. *Nano-Struct. Nano-Objects* **2020**, *22*, 100429. [CrossRef]
71. Tian, W.; Yu, W.; Liu, X. A Review on Lattice Defects in Graphene: Types, Generation, Effects and Regulation. *Micromachines* **2017**, *8*, 163. [CrossRef]

Article

CO_2 Reduction to Valuable Chemicals on TiO_2-Carbon Photocatalysts Deposited on Silica Cloth

Antoni Waldemar Morawski [1], Katarzyna Ćmielewska [1,*], Kordian Witkowski [1], Ewelina Kusiak-Nejman [1], Iwona Pełech [1], Piotr Staciwa [1], Ewa Ekiert [1], Daniel Sibera [1,2], Agnieszka Wanag [1], Marcin Gano [1] and Urszula Narkiewicz [1]

[1] Department of Inorganic Chemical Technology and Environment Engineering, Faculty of Chemical Technology and Engineering, West Pomeranian University of Technology in Szczecin, Pulaskiego 10, 70-322 Szczecin, Poland; Antoni.Morawski@zut.edu.pl (A.W.M.); wk44867@zut.edu.pl (K.W.); Ewelina.Kusiak@zut.edu.pl (E.K.-N.); Iwona.Pelech@zut.edu.pl (I.P.); piotr.staciwa@zut.edu.pl (P.S.); Ewa.Dabrowa@zut.edu.pl (E.E.); Daniel.Sibera@zut.edu.pl (D.S.); Agnieszka.Wanag@zut.edu.pl (A.W.); Marcin.Gano@zut.edu.pl (M.G.); Urszula.Narkiewicz@zut.edu.pl (U.N.)

[2] Department of General Civil Engineering, Faculty of Civil and Environmental Engineering, West Pomeranian University of Technology in Szczecin, Piastow 50a, 70-311 Szczecin, Poland

* Correspondence: katarzyna.przywecka@zut.edu.pl

Abstract: A new photocatalyst for CO_2 reduction has been presented. The photocatalyst was prepared from a combination of a commercial P25 with a mesopore structure and carbon spheres with a microporous structure with high CO_2 adsorption capacity. Then, the obtained hybrid TiO_2-carbon sphere photocatalysts were deposited on a glass fiber fabric. The combined TiO_2-carbon spheres/silica cloth photocatalysts showed higher efficiency in the two-electron CO_2 reduction towards CO than in the eight-electron reaction to methane. The 0.5 g graphitic carbon spheres combined with 1 g of TiO_2 P25 resulted in almost 100% selectivity to CO. From a practical point of view, this is promising as it economically eliminates the need to separate CO from the gas mixture after the reaction, which also contains CH_4 and H_2.

Keywords: photocatalysis; CO_2 reduction; carbon-TiO_2

1. Introduction

Over the last 35 years, the annual CO_2 concentration in the atmosphere has increased from about 347 ppm to about 416 ppm [1]. At the same time, the average temperature increased by 0.7 °C. When the CO_2 concentration reaches 550 ppm, the estimated average temperature will increase by 2.0 °C [2]. Direct utilization of solar energy for photocatalytic reduction of carbon dioxide is still an important and elegant challenge for researchers to reduce atmospheric warming and climate change. Moreover, it can reduce the number of processes related to CO_2 disposal because it is carried out at atmospheric pressure and at room temperature. The photocatalytic conversion of CO_2 could lead to valuable products, such as hydrogen, carbon monoxide, methane, methanol, formaldehyde, formic acid, ethanol, and higher hydrocarbons. The nature of the products and the selectivity depend on the photocatalyst types and their modifications, the presence of water, and the chemical nature of their support [2–4].

Currently, many scientific publications consider the modification of TiO_2. Photocatalysis with UV-Vis radiation is one of the most intensively studied fields. According to the Scopus database, about 8400 publications in this field were published in 2021. Finding the additives that prevent the recombination of the electron-hole pair is highly desirable. They should also have the ability to shift the radiation absorption band towards the visible waves.

The modifiers are metals, metal oxides, and nonmetals, of which carbon in all its allotropes seems to be the most important [3,5]. When the reaction is considered in the gas phase, it turns out that the nature of photocatalyst support significantly alters its properties. An example is the photoreduction of carbon dioxide over Fe-, Co-, Ni-, and Cu-incorporated TiO_2 on basalt fiber films. In this case, the CO_2 reduction to methane with high selectivity was caused by a synergistic effect. This effect was induced by the promotion of the photogenerated electron-hole pair (e^-/h^+) separations and also by the enhanced CO_2 adsorption [6].

For some time, carbon and its various allotropes have been considered as effective titanium dioxide modifiers that improve the adsorption capacity of the photocatalyst, as well as its photoactivity. It prevents the recombination of the electron-hole pair and shifts the adsorption band towards visible light. These properties relate to reactions in water and gas environments.

The pioneering series of works was initiated by J.-M. Herrmann [7–10]. The introduction of commercial activated carbon in contact with TiO_2 accelerates the synergistic effect and increases the reaction rate of phenol degradation by a factor of 2.5 [7]. Different activated carbons were tested to confirm this observation [8]. When 4-chlorophenol was selected as a model aromatic pollutant, the same result was obtained [9]. Different types of hazardous wastes were studied, including herbicides [10].

In the modification of carbon, we distinguish the following types of interactions of TiO_2 with carbon: carbon-doped TiO_2, carbon-coated TiO_2, and TiO_2-loaded carbon [11]. Enhanced adsorption and interaction of carbon with oxygen vacancies were postulated to be responsible for the higher activity.

One of the directions of modification of TiO_2 for photocatalytic reduction of CO_2 is the use of carbon nanotubes [12] or graphene [3,13]. Many scientists consider the electron transfer between TiO_{2-x} and graphene through Ti-O-C bonds [5]. The enhanced photoactivity was mainly attributed to the presence of graphene, which has an excellent ability to transport and collect electrons [13].

Peng Wang et al. [14] demonstrated a new carbon-doped amorphous titanium oxide for photocatalytic CO_2 reduction, prepared by sol-gel method. The best photocatalyst was prepared after annealing at 300 °C with yields of CH_4 and CO of 4.1 and 2.5 µmol/g/h for solar light, and 0.53 and 0.63 µmol/g/h for visible light, respectively.

J. Liu et al. [15] proposed a TiO_2-graphene nanocomposite prepared using GO and TiO_2 nanoparticles mixed in a suspension with water, sonicated, and then heated in an oil bath. The photoactivity upon reduction of CO_2 to CH_4 (2.1 µmol/g/h) and CH_3OH (2.2 µmol/g/h) was attributed to the synergistic effect between TiO_2 and graphene.

Tianyu Zhang et al. [16] have shown that the modification of graphene quantum dots by functional -OH and -NH_2 electron-donating groups increases the yield of CH_4 from CO_2 electroreduction with Faradaic efficiency by 70%.

The adsorption of CO_2 on the surface or the volume absorption on the photocatalyst is an important step in the photoreduction of CO_2. It is crucial especially in the gas-phase reaction. Therefore, in this work, we have investigated the preparation of composites of TiO_2 and carbon spheres using a simple method proposed by Herrmann [7–10], which consists of the mechanical mixing of TiO_2 with carbon material. The basis for this research direction was our previous work describing the high CO_2 adsorption capacity on the microporous carbon spheres of the graphitic structures we fabricated [17]. Proper management of adsorption and photoactivity by selecting a sorbent hybridized with TiO_2 leads to an increase in CO_2 reduction efficiency [18]. Finally, the above composite material was deposited on a glass fiber cloth which formed the photocatalytic bed in the reactor.

2. Results and Discussion

2.1. Characterization of Photocatalysts

The XRD patterns of the initial carbon spheres used for the nanocomposite production (Figure 1) showed two diffraction peaks of carbon at about 23° and 43°. The first peak corresponds to the stacking carbon layer structure (002) related to the parallel and azimuthal orientation of the aromatic and carbonized structures. The high symmetry of the peak can suggest the absence of γ-bands linked to amorphous and aliphatic structures [19]. The second peak of lower intensity observed at 43° corresponds to the ordered graphitic and hexagonal carbon structures (100) [20]. The broadening of (100) peak indicates a low degree of aromatic ring condensation (low degree of graphitization) [21,22]. The (001) peak becomes more intense and sharper when carbon spheres are prepared at higher temperatures; then the degree of graphitization increases, and some graphene flakes can be observed in SEM images.

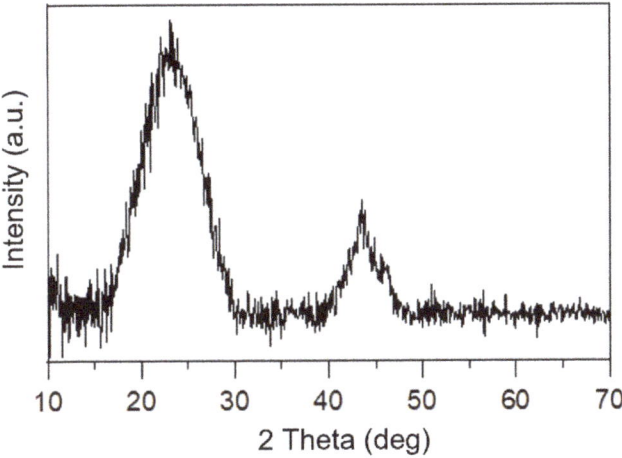

Figure 1. X-ray diffraction pattern of the starting carbon spheres.

In the case of carbon spheres used here, the temperature was lower and no graphene flakes can be observed in the SEM image shown in Figure 2. The produced spheres have a regular spherical shape and a smooth surface; neither defects nor inclusions can be observed. The size distribution is narrow, and the spheres are homogeneous and have an average diameter of about 600 nm.

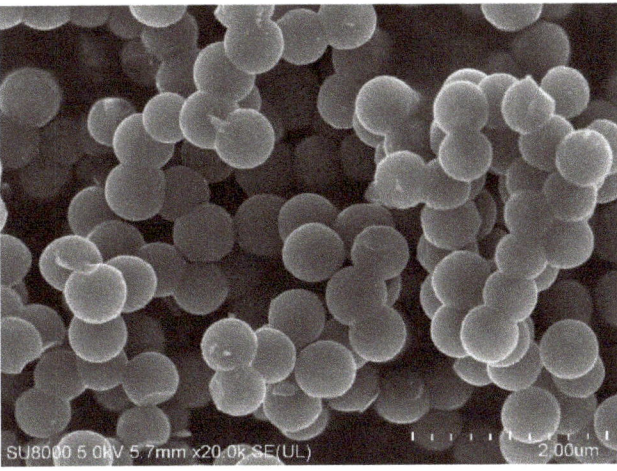

Figure 2. SEM image of the used carbon spheres.

The surface area of the spheres determined by the BET methods was 455 m^2/g, dominated by ultrapores and micropores (Table 1). These characteristics affected CO_2 adsorption to 3.25 mmol/g and 2.43 mmol/g at temperatures of 0 °C and 25 °C, respectively.

Table 1. Textural parameters and CO_2 sorption capacities of pure carbon spheres.

Material	S_{BET} (m^2/g)	TPV (cm^3/g)	V_s (<1 nm) (cm^3/g)	V_m (<2 nm) (cm^3/g)	V_{meso} (cm^3/g)	CO_2 0 °C (mmol/g)	CO_2 25 °C (mmol/g)
CS	455	0.26	0.19	0.22	0.04	3.25	2.43

S_{BET}—specific surface area; TPV—total pore volume; V_s—the volume of ultramicropores with diameters smaller than 1 nm; V_m—the volume of micropores with diameters smaller than 2 nm; V_{meso}—the volume of mesopores with diameters from 2 to 50 nm.

The diffraction pattern of pure TiO_2 P25 is shown in Figure 3. The used titanium dioxide consisted of anatase (89%) and rutile (11%) with a crystallite size of 27 nm for anatase and 43 nm for rutile, as shown in Table 2. The energy gap calculated by the Kubelka–Munk theory was estimated as E_g = 3.204 eV. The determined BET surface area was 54 m^2/g, dominated by the mesopore structure (Table 3). This is quite a difference compared to the carbon spheres. The adsorption of CO_2 was also much lower compared to the carbon spheres; 0.72 mmol/g for P25 at 30 °C [23] compared to 2.43 mmol/g at 25 °C for the carbon spheres.

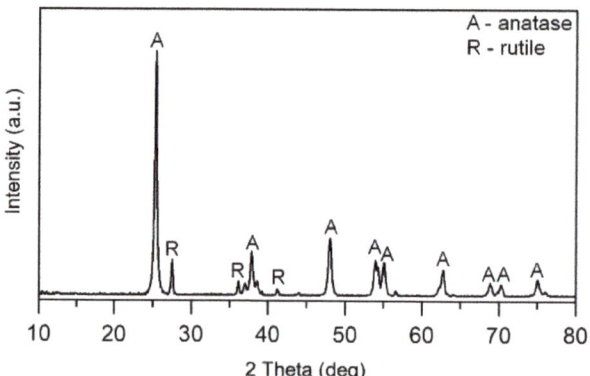

Figure 3. X-ray diffraction pattern of TiO_2 P25.

Table 2. Phase and crystallite composition of used P25.

Name	Phase Composition (%)		Crystallite Size (nm)	
	Anatase	Rutile	Anatase	Rutile
TiO_2 P25	89	11	27	43

Table 3. Textural parameters of used P25.

Material	S_{BET} (m^2/g)	$V_{total\ 0,95}$ (m^3/g)	$V_{mikro\ DR}$ (cm^3/g)	V_{mezo} (cm^3/g)	CO_2 30 °C (mmol/g)
TiO_2 P25	54	0.4	0.02	0.38	0.72

From the photos taken with the scanning microscope, it can be observed that TiO$_2$ P25 nanocrystallites form relatively large, noncircular agglomerates with dimensions in the range of about 0.5–2 µm (Figure 4).

Figure 4. SEM images for used TiO$_2$ P25.

In the SEM images of pure silica fabric (Figure 5), one can see the fibers bound with a binder that provides a matrix for the applied photocatalysts. The surfaces of the fibers are virtually clear and smooth. EDX mapping shows that only Si, O, Al, and Ca originate from the fibers and the inorganic binder.

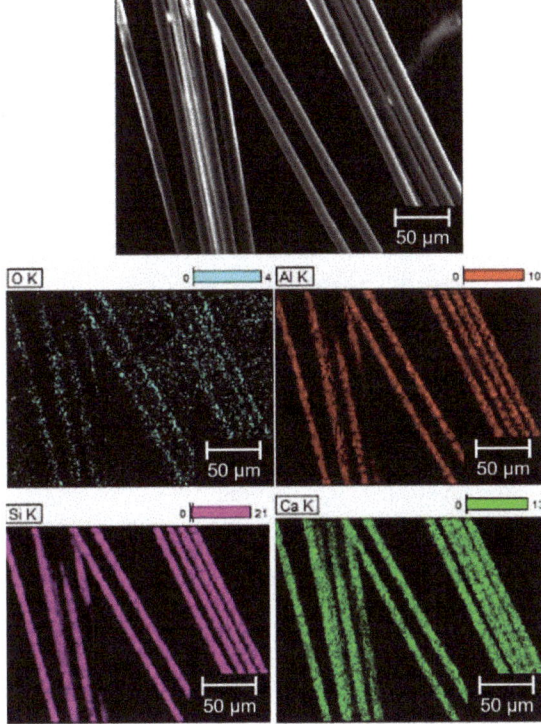

Figure 5. SEM images and EDX chemical element mappings of pure silica cloth.

Figure 6 presents photos of the prepared materials showing the dispersion of the photocatalysts on the silica fiber matrix. The dispersion of the TiO$_2$ P25 photocatalyst is quite homogenous. The EDS analysis showed that the fibers are composed of silica and alumina with a Si/Al ratio of about 5/1. The matrix also contains calcium. The active phase, Ti in the form of TiO$_2$, is present in a constant amount of about 10 wt% in each sample. The photocatalyst is dispersed on the surface of the fibers and the binder and is located in particular between the layers of the fibers.

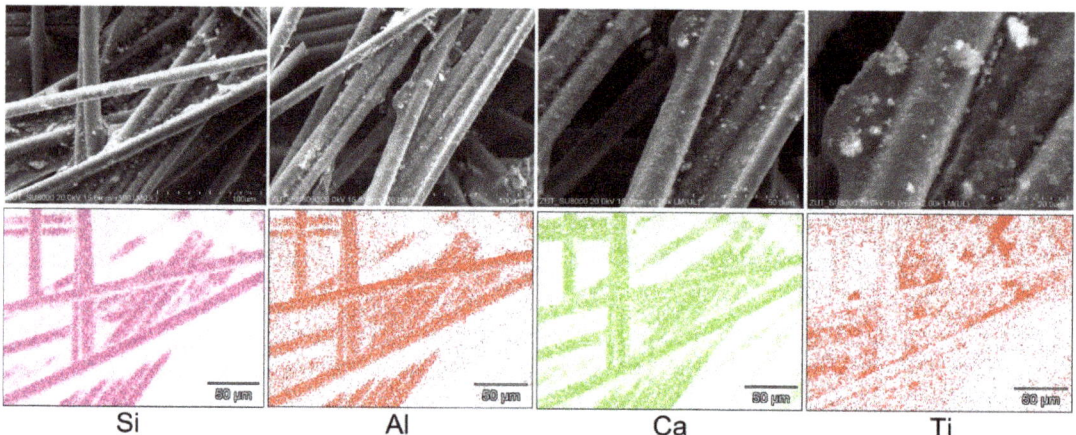

Figure 6. SEM images and EDX chemical element mappings of silica cloth coated by TiO$_2$ P25.

From the SEM photos presented in Figure 7, it can be seen that as the carbon spheres content is increased from 0.05 to 0.5, the number of agglomerates increases slightly. Of course, the shape of agglomerates characteristic of TiO$_2$ P25 dominates since the content of carbon spheres is lower in contrast to TiO$_2$.

As in the case of the silica fabric coated with pure TiO$_2$ (Figure 6), almost the same properties are observed for the samples with different carbon content (Figure 7). Therefore, Figure 8 shows element mappings for the sample P25 + C 1/0.5 as an example. The analysis of EDS has shown that the fibers are composed of silica and alumina with a ratio of Si/Al of about 5/1 and also calcium, as can be seen in Table 4. The distributions of Si, O, and Ca indicate that these elements are constituents of the fiber. The active phase of Ti in the form of TiO$_2$ is present in each sample in a constant amount of about 10 wt%. It is located on the surface of the fibers, mainly in the form of agglomerates.

2.2. Photocatalytic Activity Measurements

Figure 9a–d present the ability to photoreduce carbon dioxide with water vapor to carbon monoxide (CO), methane (CH$_4$), and hydrogen (H$_2$), respectively, with reaction times given in $\mu mol/g_{photocatalyst}/dm^3$.

The addition of carbon spheres in an amount of 0.05 g per 1 g of TiO$_2$ resulted in a slight increase in the photocatalyst activity (Figure 9b) compared to pure P25 (Figure 9a). The comparison of the photoactivity towards CO production is presented in Figure 10. The amount of carbon monoxide produced decreased slightly from 55.75 $\mu mol/g_{photocatalyst}/dm^3$ for P25 and 55.94 $\mu mol/g_{photocatalyst}/dm^3$ for P25 + C 1/0.05 to 43.62 $\mu mol/g_{photocatalyst}/dm^3$ for the sample P25 + C 1/0.1 and to 46.79 $\mu mol/g_{photocatalyst}/dm^3$ in the case of the P25 + C 1/0.5 material. At the same time, the amount of methane decreased significantly and gradually from 9.02 $\mu mol/g_{photocatalyst}/dm^3$ (pure P25) through 4.76 and 4.02 $\mu mol/g_{photocatalyst}/dm^3$ for materials P25 + C 1/0.05 and P25 + C 1/0.1, respectively; to practically 0.00 $\mu mol/g_{photocatalyst}/dm^3$ in the case of the P25 + C 1/0.5 sample.

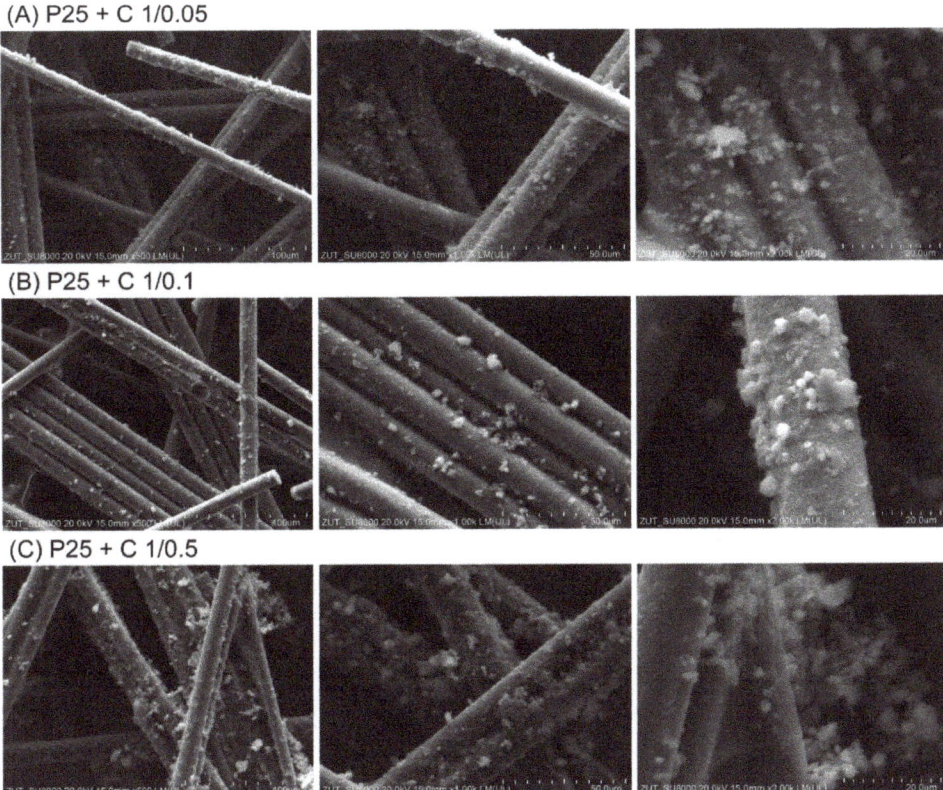

Figure 7. SEM images of silica cloth coated by TiO$_2$ P25 combined with different amount of carbon spheres; (**A**) P25 + C 1/0.05; (**B**) P25 + C 1/0.1; (**C**) P25 + C 1/0.5.

From the results obtained, it can be seen that the amount of methane captured decreased with the increasing carbon content in the material (Figure 11). The same dependence was also observed for the hydrogen production (Figure 12).

Two factors may contribute to the decrease in the photoactivity for CO, CH$_4$, and H$_2$ with the increasing carbon sphere content. The first aspect is the well-known electron interaction between amphoteric electrons from the surface of the graphite spheres and the functional groups of the TiO$_2$ surface. This prevents the recombination of the electron-hole pair. It is known that even a small interaction, specifically doping [14], is necessary to achieve a positive effect. The addition of large amounts of carbon from 0.1 g/1 g TiO$_2$ to 0.5 g/1 g TiO$_2$ causes the scattering of electrons in the structure with a large amount of carbon.

The second effect is caused by the increased CO$_2$ adsorption by the carbon spheres, as shown in Table 1. TiO$_2$ adsorbs only 0.72 mmol CO$_2$/g at 30 °C (Table 3), while carbon spheres adsorb 2.43 mmol CO$_2$/g. A large addition of carbon spheres increases the sorption and the presence of CO$_2$ on the surface of the photocatalyst, which forms CO$_3^{2-}$ ions in the presence of H$_2$O vapors, which are strong electron scavengers. The further addition of carbon spheres did not positively affect the activity. In the case of sample P25 + C 1/0.5 (Figure 9d), the only product of the process was carbon monoxide. The higher content of carbon spheres blocked the UV-Vis radiation, which can generate the TiO$_2$ electron-hole pair and, at the same time, scatter and disperse the excited electrons in the crystal lattice of the carbon spheres.

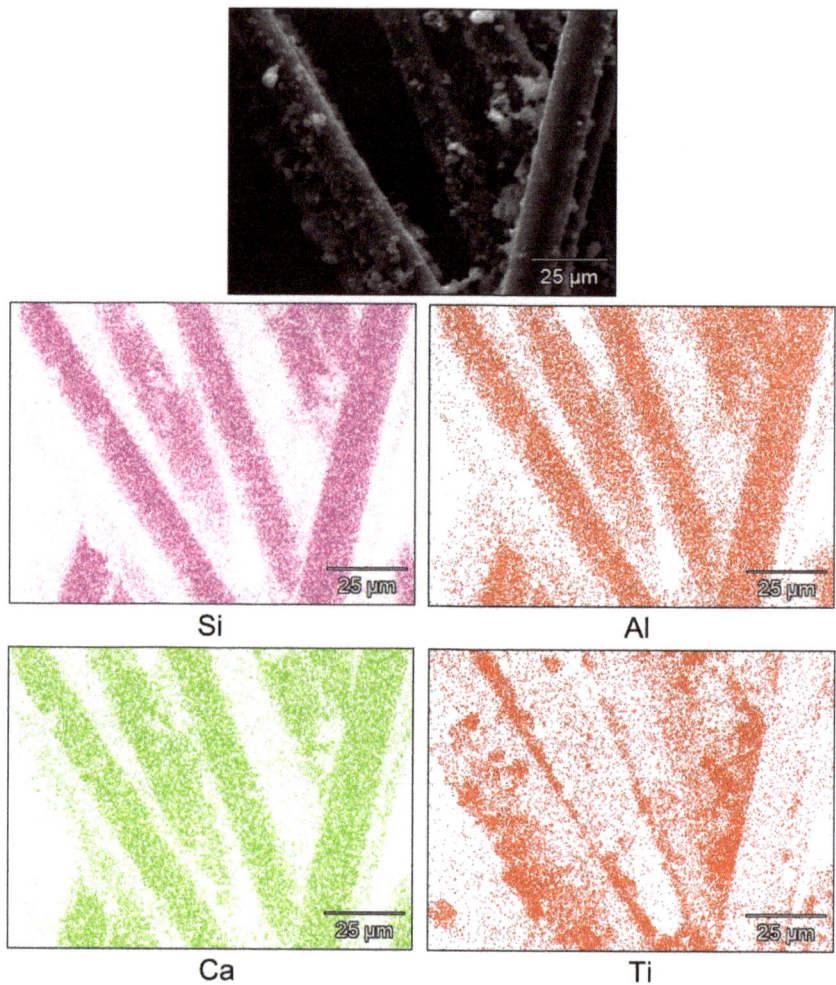

Figure 8. Example of EDX chemical element mappings of silica cloth coated by TiO$_2$ P25 combined with carbon spheres for sample P25 + C 1/0.5.

Table 4. Surface chemical element compositions from EDX of silica cloth coated by TiO$_2$ P25 combined with carbon spheres for sample P25 + C 1/0.5.

Chemical Element	Weight (wt%)	Atom (%)
C	18.69	37.17
O	9.33	13.93
Al	4.96	4.39
Si	25.42	21.62
Ca	21.77	12.97
Ti	19.55	9.75
K	0.28	0.17

It can be observed that the main products are CO, CH$_4$, and H$_2$. Carbon monoxide is also the dominant component of the postreaction gases, regardless of the amount of carbon spheres added as a modifier.

The reactivity of the photocatalysts is probably limited by the reduction of water to hydrogen, which is the first stage of the whole complex process. In our case, with the exception of sample P25 + C 1/0.5, there is some excess of hydrogen, which is a positive phenomenon.

Hydrogen is formed in the reactions at the photocatalysts, as reported by Peng Wang et al. [14] and Minoo Tasbihi et al. [24]:

$$H_2O + 2h^+ \rightarrow \frac{1}{2} O_2 + 2H^+ \tag{1}$$

$$2H^+ + 2e^- \rightarrow 2\,H_2 \tag{2}$$

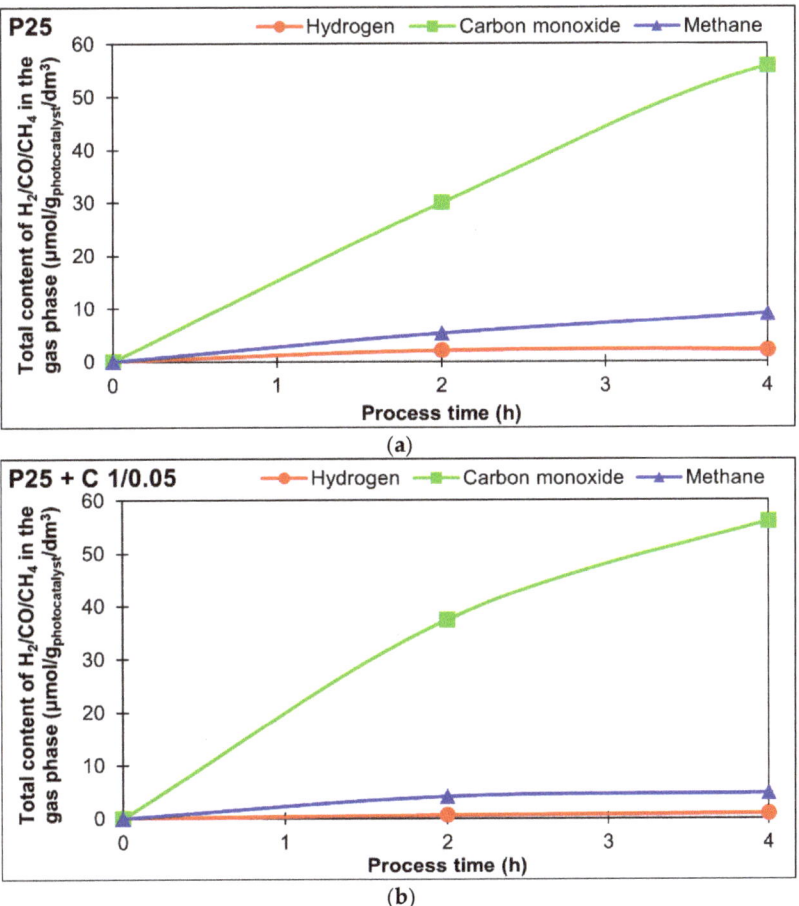

Figure 9. *Cont.*

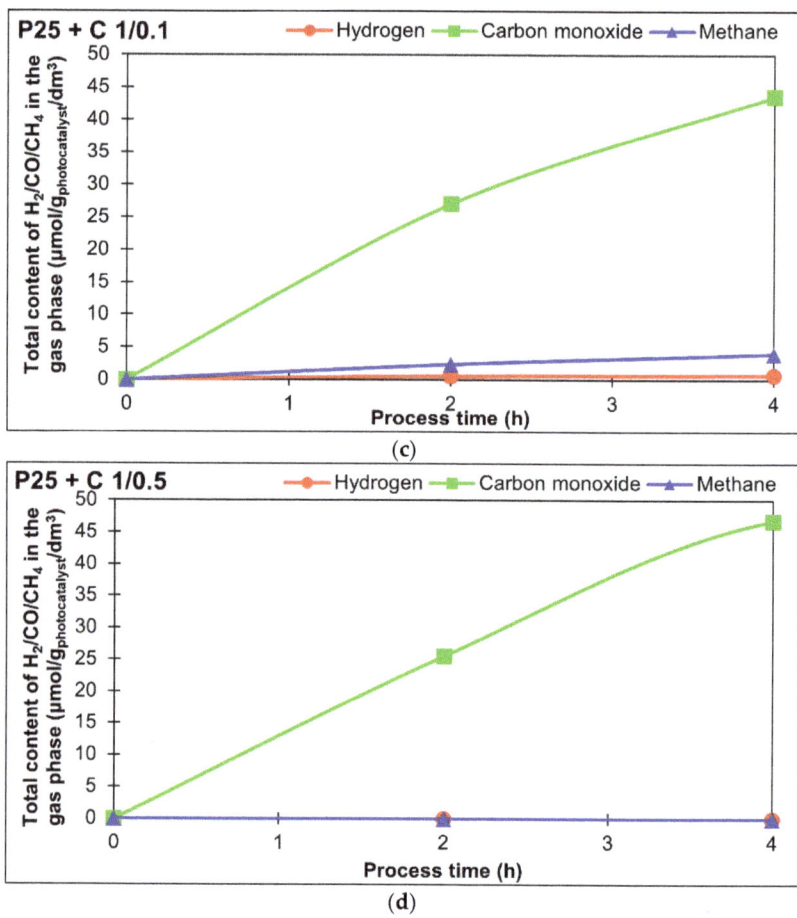

Figure 9. Total content of the products of the photoreduction of CO_2 in the process with: (**a**) P25, (**b**) P25 + C 1/0.05, (**c**) P25 + C 1/0.1, (**d**) P25 + C 1/0.5.

Carbon monoxide is the result of an easier two-electron reduction of carbon dioxide:

$$CO_2 + 2H^+ + 2e^- \rightarrow CO + H_2O \tag{3}$$

The lower efficiency of the reaction towards methane production is the result of the more difficult eight-electron CO_2 reduction reaction:

$$CO_2 + 8H^+ + 8e^- \rightarrow CH_4 + 2H_2O \tag{4}$$

On the homogeneous surface of the photocatalyst, reactions No. 1 to No. 4 are the two-electron production of hydrogen (reactions 1–2) and a two-electron to carbon monoxide (reaction No. 3), which could be considered as an intermediate step to the 8-electron reaction of No. 4 towards methane.

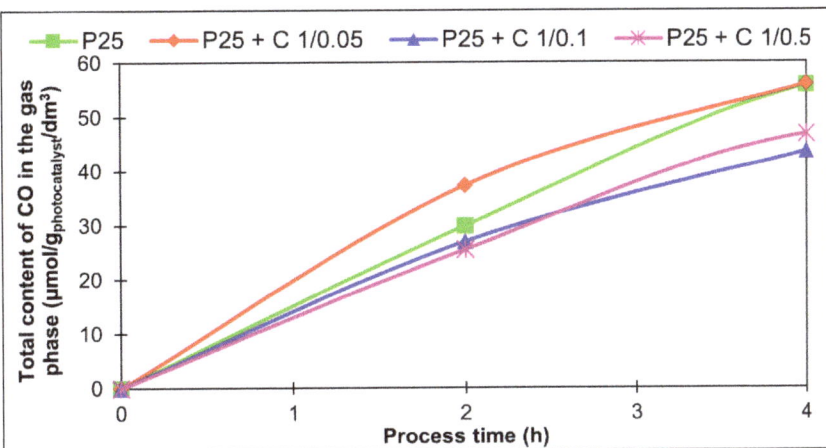

Figure 10. Total content of carbon monoxide of the photoreduction of CO_2 for the studied samples: P25; P25 + C 1/0.05; P25 + C 1/0.1, and P25 + C 1/0.5.

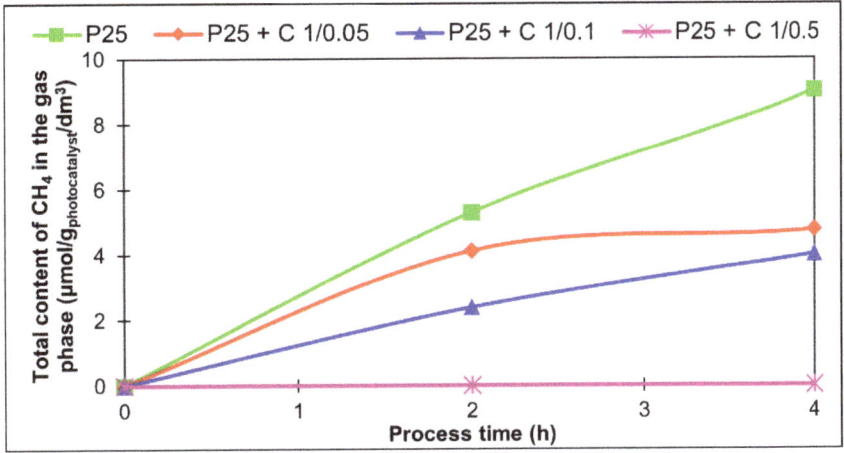

Figure 11. Total content of methane in the photoreduction of CO_2 for the following photocatalysts: P25; P25 + C 1/0.05; P25 + C 1/0.1, and P25 + C 1/0.5.

It is known that the hydrogen production reactions (reactions No. 1–2 water splitting) take place in the semiconductor valence band (on holes), and the CO_2 reduction reactions take place in the conduction band (with the participation of electrons).

For this reason, the production of hydrogen controls all CO_2 reduction reactions. In this case, only the CO formation reaction (No. 3) and the methane forming reaction (No. 4) are linearly competing with each other. Indeed, as shown in Figure 13, with the addition of carbon spheres, excess hydrogen decreases, and at the same time, methane decreases, which requires more hydrogen than the formation of carbon monoxide in reaction No. 3.

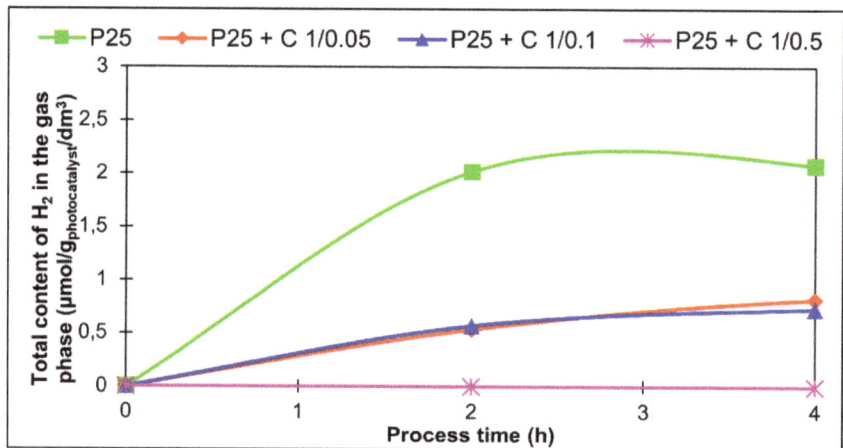

Figure 12. Total content of hydrogen in the photoreduction of CO_2 for subsequent photocatalysts: P25, P25 + C 1/0.05, P25 + C 1/0.1, and P25 + C 1/0.5.

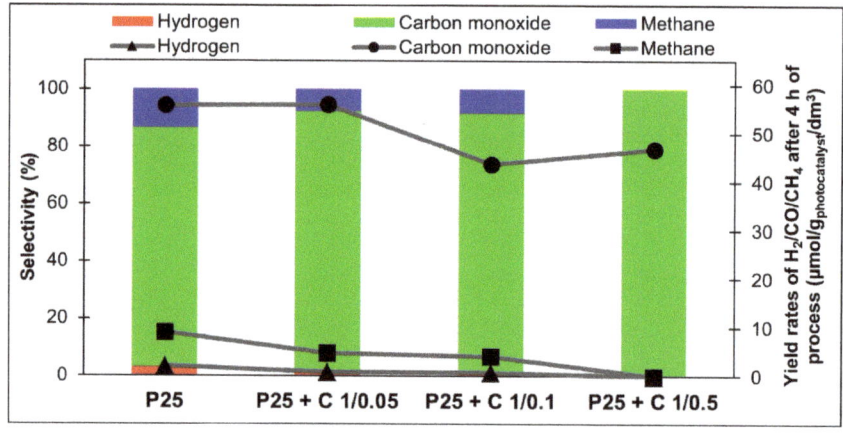

Figure 13. Selectivity and yield rates of the products after 4 h of the process.

The collected postreaction gas composition is different from that obtained on photocatalysts on mineralogical ground fibers, where only methane was obtained and no other components were identified [6]. In our case, by using silicon fibers, it is possible to obtain a mixture of CO, CH_4, and small amounts of hydrogen, with high selectivity to CO. Figure 13 shows the calculated selectivity, and the obtained amounts of each component of the gas phase produced. In all the cases tested, high selectivity to carbon monoxide is observed—from about 83.40% to 100% for the sample P25 + C 1/0.5, where the hydrogen was entirely consumed by the two-electron reaction of reduction of CO_2 to CO.

3. Experimental

3.1. Preparation of the Samples

In the experiments, photocatalysts consisting of TiO_2 P25, manufactured by Degussa (Evonik Industries AG, Germany), and carbon material were tested. The carbon material was prepared with the use of resorcinol and formaldehyde. The description of the original method of producing carbon spheres used in this work is presented elsewhere [25]. Three powders with different mass ratios of titanium dioxide to carbon were prepared by grinding in a mortar. The mass ratios were: 1:0.05; 1:0.1, and 1:0.5. Aqueous suspensions of the prepared materials were applied to the fiberglass cloth. Glass fiber fabric with an area weight of 40 g/m^2 was supplied by Fiberglass Fabrics (Opole, Poland). Then the fibers with the photocatalysts were dried at 110 °C for 1 h.

3.2. Photoreduction Process

Experiments were performed in a cylindrical quartz reactor with a working volume of 392 cm^3 (Figure 14). Four Actinic BL TL-E Philips lamps were used with a total power of 88 W, emitting UV-A radiation in the wavelength range of 350–400 nm. The lamps were located outside the reactor and formed a ring. The reactor was sealed in a thermostatic chamber to exclude other light sources and ensure a stable process temperature. An amount of 1 cm^3 of distilled water was poured into the reactor. A photocatalyst previously applied to glass fibers was added to the reactor. Then the whole system was purged with pure CO_2 (Messer, Poland) for 30 min. After this time, the system was tightly sealed and the lamps were turned on. Both during purging and during the process, the gas was stirred using a pump with a flow rate of 1.6 dm^3/h. The process was carried out for 4 h. Gas samples were collected every 2 h for analysis.

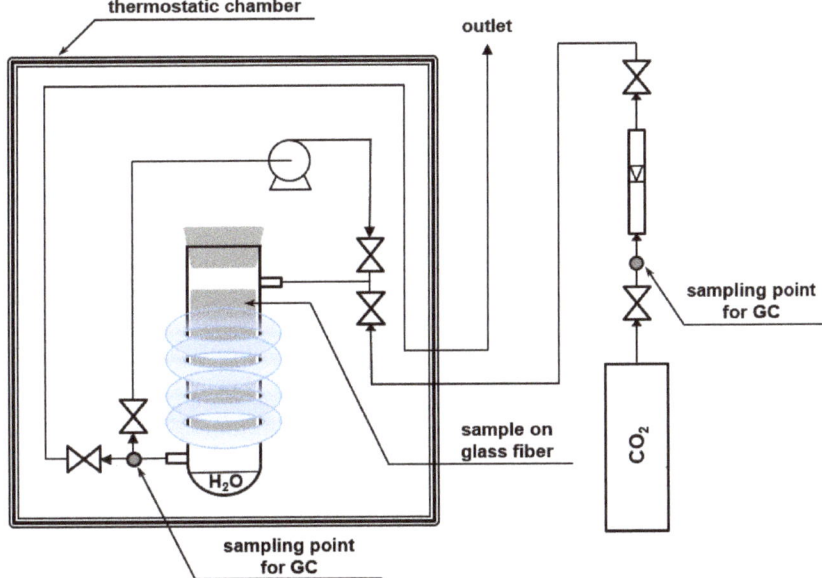

Figure 14. The scheme of the reactor for photocatalytic reduction of CO_2 in the gas phase.

3.3. The Analysis of the Gas Phase

The gas phase composition was analyzed using SRI 310C gas chromatograph (SRI Instruments, Torrance, CA, USA), equipped with a 5Å molecular sieve column and an HID detector (Helium Ionization Detector). The carrier gas was helium. The analyses were performed under isothermal conditions at 60 °C. The gas flow through the column was 60 cm^3/min, while the volume of the gas sample was 1 cm^3. The content of each component in the gas phase was calculated in the subsequent measurements based on the calibration curve.

3.4. XRD Analysis

The crystalline structure of used titanium dioxide and carbon spheres was studied with X-ray powder diffraction (CuKα radiation, Malvern PANalytical B.V., The Netherlands). The mean crystallites sizes were calculated based on Scherrer's equation:

$$D = \frac{K\lambda}{\beta \cos\theta} \quad (5)$$

where:

D—mean crystallite size (nm),
λ—wavelength of Cu Kα radiation (nm),
θ—Bragg's angle (°),
β—calibrated width of a diffraction peak at half maximum intensity (rad).

The percentage of anatase in the crystalline phase (%A) was calculated according to the equation:

$$\%A = \frac{I_A}{I_A + I_R} \cdot 100\% \quad (6)$$

where I_A and I_R are the diffraction intensities of the anatase peak at 25.4° and rutile peak at 27.5°, respectively.

3.5. SEM/EDS Measurements

The surface morphology of the samples was examined using a scanning electron microscope (SEM Hitachi SU 8020, Japan). The SEM and EDS analysis parameters were: acceleration voltage of 20 kV and current of 10 μA. The samples for investigation using SEM were firstly vapor-deposited with a 5 nm thin chromium layer to protect the samples from the electrical charge.

3.6. Textural Parameters and CO_2 Sorption Capacity Analysis

The specific surface areas (S_{BET}) were calculated from the Brunauer–Emmett–Teller equation, based on the nitrogen adsorption isotherms measured at 77 K using QUADRA-SORB evo™ Gas Sorption analyzer (Anton Paar GmbH, Austria) in the relative pressure range of 0.05–0.2. Before the measurements, the materials were degassed for 12 h at 105 °C under the high vacuum. The total pore volume (V_{total}) was determined on the basis of the maximum adsorption of nitrogen vapor at $p/p_0 = 0.95$. Micropore volume (V_{micro}) was calculated using the Dubinin-Radushkevich method, while the mesopore volume (V_{meso}) was obtained from the difference between V_{total} and V_{micro}.

Adsorption of the carbon dioxide was performed at 0 °C and 25 °C using the same Quadrasorb™ automatic system (Quantachrome Instruments) in the pressure range between 0.01 and 0.98 bar.

Adsorption of the carbon dioxide was also measured at 30 °C with the use of the TGA method (Netzsch STA 449 C, Netzsch GmbH, Germany). The flow rate of carbon dioxide during the analysis was 30 cm^3/min. The process of adsorption of CO_2 by the tested material lasted 120 min.

4. Conclusions

A relatively simple method for the preparation of a photocatalyst was presented in which TiO_2 was combined with carbon spheres, which were then deposited on a silica fiber cloth to provide a good bed for the photocatalytic reduction of CO_2 in the gas phase. The obtained TiO_2-carbon spheres/silica fiber photocatalysts showed high selectivity towards carbon monoxide. This is very promising from a practical point of view since the step of separating the carbon monoxide from the postreaction mixture can be avoided with only a slight reduction in the activity. For TiO_2 combined with a higher amount of graphitic carbon spheres, the two-electron reduction of CO_2 to CO is more privileged than an eight-electron reaction to methane.

Author Contributions: Conceptualization, A.W.M. and U.N.; methodology, A.W.M., E.K.-N., A.W. and K.Ć.; software, M.G.; validation, A.W.M., U.N., E.K.-N. and I.P.; formal analysis, K.Ć., K.W., E.E. and M.G.; investigation, A.W.M., K.Ć. and K.W.; resources, U.N.; data curation, A.W.M. and U.N.; writing—original draft preparation, A.W.M. and K.Ć.; writing—review and editing, K.Ć., A.W.M., U.N., P.S. and D.S.; visualization, K.Ć.; supervision, A.W.M. and U.N.; project administration, U.N.; funding acquisition, U.N. All authors have read and agreed to the published version of the manuscript.

Funding: The research leading to these results received funding from the Norway Grants 2014–2021 via the National Centre for Research and Development under the grant number NOR/POLNORCCS/PhotoRed/0007/2019-00.

Data Availability Statement: Not applicable.

Conflicts of Interest: The authors declare no conflict of interest.

References

1. Daily CO_2. Available online: https://www.co2.earth/daily-co2 (accessed on 14 December 2021).
2. Al Jitan, S.; Palmisano, G.; Garlisi, C. Synthesis and Surface Modification of TiO_2-Based Photocatalysts for the Conversion of CO_2. *Catalysts* **2020**, *10*, 227. [CrossRef]
3. Fu, Z.; Yang, Q.; Liu, Z.; Chen, F.; Yao, F.; Xie, T.; Zhong, Y.; Wang, D.; Li, J.; Li, X.; et al. Photocatalytic Conversion of Carbon Dioxide: From Products to Design the Catalysts. *J. CO_2 Util.* **2019**, *34*, 63–73. [CrossRef]
4. Fu, J.; Jiang, K.; Qiu, X.; Yu, J.; Liu, M. Product Selectivity of Photocatalytic CO_2 Reduction Reactions. *Mater. Today* **2020**, *32*, 222–243. [CrossRef]
5. Zhang, J.; Tian, B.; Wang, L.; Xing, M.; Lei, J. *Photocatalysis. Fundamentals, Materials and Application*; Springer: Singapore, 2018.
6. Do, J.Y.; Kwak, B.S.; Park, S.M.; Kang, M. Effective Carbon Dioxide Photoreduction over Metals (Fe-, Co-, Ni-, and Cu-) Incorporated TiO_2/Basalt Fiber Films. *Int. J. Photoenergy* **2016**, *2016*, 5195138. [CrossRef]
7. Matos, J.; Laine, J.; Herrmann, J.-M. Synergy Effect in the Photocatalytic Degradation of Phenol on a Suspended Mixture of Titania and Activated Carbon. *Appl. Catal. B Environ.* **1998**, *18*, 281–291. [CrossRef]
8. Matos, J.; Laine, J.; Herrmann, J.-M. Association of Activated Carbons of Different Origins with Titania in the Photocatalytic Purification of Water. *Carbon N. Y.* **1999**, *37*, 1870–1872. [CrossRef]
9. Herrmann, J.-M.; Matos, J.; Disdier, J.; Guillard, C.; Laine, J.; Malato, S.; Blanco, J. Solar Photocatalytic Degradation of 4-Chlorophenol Using the Synergistic Effect between Titania and Activated Carbon in Aqueous Suspension. *Catal. Today* **1999**, *54*, 255–265. [CrossRef]
10. Matos, J.; Laine, J.; Herrmann, J.-M. Effect of the Type of Activated Carbons on the Photocatalytic Degradation of Aqueous Organic Pollutants by UV-Irradiated Titania. *J. Catal.* **2001**, *200*, 10–20. [CrossRef]
11. Morawski, A.; Janus, M.; Tryba, B.; Toyoda, M.; Tsumura, T.; Inagaki, M. Carbon Modified TiO_2 Photocatalysts for Water Purification. *Pol. J. Chem. Technol.* **2009**, *11*, 46–50. [CrossRef]
12. Rodríguez, V.; Camarillo, R.; Martínez, F.; Jiménez, C.; Rincón, J. CO_2 Photocatalytic Reduction with CNT/TiO_2 Based Nanocomposites Prepared by High-Pressure Technology. *J. Supercrit. Fluids* **2020**, *163*, 104876. [CrossRef]
13. Morawski, A.W.; Kusiak-Nejman, E.; Wanag, A.; Narkiewicz, U.; Edelmannová, M.; Reli, M.; Kočí, K. Influence of the Calcination of TiO_2-Reduced Graphite Hybrid for the Photocatalytic Reduction of Carbon Dioxide. *Catal. Today* **2021**, *380*, 32–40. [CrossRef]
14. Wang, P.; Yin, G.; Bi, Q.; Huang, X.; Du, X.; Zhao, W.; Huang, F. Efficient Photocatalytic Reduction of CO_2 Using Carbon-Doped Amorphous Titanium Oxide. *ChemCatChem* **2018**, *10*, 3854–3861. [CrossRef]
15. Liu, J.; Niu, Y.; He, X.; Qi, J.; Li, X. Photocatalytic Reduction of CO_2 Using TiO_2-Graphene Nanocomposites. *J. Nanomater.* **2016**, *2016*, 6012896. [CrossRef]
16. Zhang, T.; Li, W.; Huang, K.; Guo, H.; Li, Z.; Fang, Y.; Yadav, R.M.; Shanov, V.; Ajayan, P.M.; Wang, L.; et al. Regulation of Functional Groups on Graphene Quantum Dots Directs Selective CO_2 to CH_4 Conversion. *Nat. Commun.* **2021**, *12*, 5265. [CrossRef] [PubMed]
17. Morawski, A.W.; Staciwa, P.; Sibera, D.; Moszyński, D.; Zgrzebnicki, M.; Narkiewicz, U. Nanocomposite Titania–Carbon Spheres as CO_2 and CH_4 Sorbents. *ACS Omega* **2020**, *5*, 1966–1973. [CrossRef]
18. Liu, L.; Zhao, C.; Xu, J.; Li, Y. Integrated CO_2 Capture and Photocatalytic Conversion by a Hybrid Adsorbent/Photocatalyst Material. *Appl. Catal. B Environ.* **2015**, *179*, 489–499. [CrossRef]
19. Lu, L.; Kong, C.; Sahajwalla, V.; Harris, D. Char structural ordering during pyrolysis and combustion and its influence on char reactivity. *Fuel* **2002**, *81*, 1215–1225. [CrossRef]
20. Liu, X.; Song, P.; Hou, J.; Wang, B.; Xu, F.; Zhang, X. Revealing the Dynamic Formation Process and Mechanism of Hollow Carbon Spheres: From Bowl to Sphere. *ACS Sustain. Chem. Eng.* **2018**, *6*, 2797–2805. [CrossRef]
21. Kukułka, W.; Wenelska, K.; Baca, M.; Chen, X.; Mijowska, E. From Hollow to Solid Carbon Spheres: Time-Dependent Facile Synthesis. *Nanomaterials* **2018**, *8*, 861. [CrossRef]

22. Juhl, A.C.; Schneider, A.; Ufer, B.; Brezesinski, T.; Janek, J.; Fröba, M. Mesoporous hollow carbon spheres for lithium-sulfur batteries: Distribution of sulfur and electrochemical performance. *Beilstein J. Nanotechnol.* **2016**, *7*, 1229–1240. [CrossRef]
23. Kapica-Kozar, J.; Piróg, E.; Wrobel, R.J.; Mozia, S.; Kusiak-Nejman, E.; Morawski, A.W.; Narkiewicz, U.; Michalkiewicz, B. TiO_2/Titanate Composite Nanorod Obtained from Various Alkali Solutions as CO_2 Sorbents from Exhaust Gases. *Microporous Mesoporous Mater.* **2016**, *231*, 117–127. [CrossRef]
24. Tasbihi, M.; Kočí, K.; Troppová, I.; Edelmannová, M.; Reli, M.; Čapek, L.; Schomäcker, R. Photocatalytic Reduction of Carbon Dioxide over Cu/TiO_2 Photocatalysts. *Environ. Sci. Pollut. Res.* **2018**, *25*, 34903–34911. [CrossRef] [PubMed]
25. Pełech, I.; Sibera, D.; Staciwa, P.; Kusiak-Nejman, E.; Kapica-Kozar, J.; Wanag, A.; Narkiewicz, U.; Morawski, A.W.; Pl, A.W.M. ZnO/Carbon Spheres with Excellent Regenerability for Post-Combustion CO_2 Capture. *Materials* **2021**, *14*, 6478. [CrossRef] [PubMed]

Article

The Effect of Si on CO$_2$ Methanation over Ni-xSi/ZrO$_2$ Catalysts at Low Temperature

Li Li [1], Ye Wang [2], Qing Zhao [3] and Changwei Hu [1,2,*]

[1] Key Laboratory of Green Chemistry and Technology, Ministry of Education, College of Chemistry, Sichuan University, Chengdu 610064, China; lili2209362583@163.com
[2] College of Chemical Engineering, Sichuan University, Chengdu 610064, China; xihazhezhe@163.com
[3] Analytical & Testing Center, Sichuan University, Chengdu 610064, China; zhaoqing8026@gmail.com
* Correspondence: changweihu@scu.edu.cn; Tel.: +86-028-8541-1105

Abstract: A series of Ni-xSi/ZrO$_2$ (x = 0, 0.1, 0.5, 1 wt%, the controlled contents of Si) catalysts with a controlled nickel content of 10 wt% were prepared by the co-impregnation method with ZrO$_2$ as support and Si as a promoter. The effect of different amounts of Si on the catalytic performance was investigated for CO$_2$ methanation with the stoichiometric H$_2$/CO$_2$ molar ratio (4/1). The catalysts were characterized by BET, XRF, H$_2$-TPR, H$_2$-TPD, H$_2$-chemisorption, CO$_2$-TPD, XRD, TEM, XPS, and TG-DSC. It was found that adding the appropriate amount of Si could improve the catalytic performance of Ni/ZrO$_2$ catalyst at a low reaction temperature (250 °C). Among all the catalysts studied, the Ni-0.1Si/ZrO$_2$ catalyst showed the highest catalytic activity, with H$_2$ and CO$_2$ conversion of 73.4% and 72.5%, respectively and the yield of CH$_4$ was 72.2%. Meanwhile, the catalyst showed high stability and no deactivation within a 10 h test. Adding the appropriate amount of Si could enhance the interaction between Ni and ZrO$_2$, and increase the Ni dispersion, the amounts of active sites including surface Ni^0, oxygen vacancies, and strong basic sites on the catalyst surface. These might be the reasons for the high activity and selectivity of the Ni-0.1Si/ZrO$_2$ catalyst.

Keywords: CO$_2$ methanation; Ni-xSi/ZrO$_2$; Si promotion; oxygen vacancies

Citation: Li, L.; Wang, Y.; Zhao, Q.; Hu, C. The Effect of Si on CO$_2$ Methanation over Ni-xSi/ZrO$_2$ Catalysts at Low Temperature. *Catalysts* **2021**, *11*, 67. https://doi.org/10.3390/catal11010067

Received: 19 November 2020
Accepted: 3 January 2021
Published: 5 January 2021

Publisher's Note: MDPI stays neutral with regard to jurisdictional claims in published maps and institutional affiliations.

Copyright: © 2021 by the authors. Licensee MDPI, Basel, Switzerland. This article is an open access article distributed under the terms and conditions of the Creative Commons Attribution (CC BY) license (https://creativecommons.org/licenses/by/4.0/).

1. Introduction

With the release of a large amount of CO$_2$ into the atmosphere, the greenhouse effect has increased in recent years [1,2]. Facing this great challenge, there are three main ways to control CO$_2$ emissions: 1. Reducing CO$_2$ emissions, 2. capture and storage of CO$_2$, and 3. chemical conversion and utilization of CO$_2$ [3–5]. Converting CO$_2$ into value-added chemicals is by far the most cost-effective method [6–8]. Clean and renewable energy resources (wind, solar, and tidal energy) produce discontinuous electricity, which is not capable of being connected to the grid. Whereas, hydrogen can be generated by electrolysis using this discontinuous electricity [9–11]. With this H$_2$ supply, the CO$_2$ methanation reaction is attracting more and more interest due to the use of both carbon dioxide and hydrogen obtained from renewable energy. Compared with hydrogen, methane has many advantages, which could be easily liquefied, stored, transported, and used by the natural gas infrastructure [12,13]. In 1902, Sabatier et al. came up with the CO$_2$ methanation reaction (also called the Sabatier reaction) firstly, CO$_2$ + 4 H$_2$ → CH$_4$ + 2 H$_2$O, $\Delta G = -130.8$ KJ·mol^{-1}, $\Delta H = -165$ KJ·mol^{-1} [14]. It can be seen that the CO$_2$ methanation reaction is highly exothermic, which is thermodynamically favorable but kinetically constrained [15]. Meanwhile, the heat released during the reaction will lead to the sintering of metal particles and deactivation of the catalysts [9,16]. Therefore, designing a high active and stable catalyst operating at low temperature is important [17].

Many studies have shown that VIII metals exhibit a catalytic performance for CO$_2$ methanation, especially Ru [18], Rh [19], Pd [20], and Ni [21]. Although noble metal catalysts exhibit a higher catalytic activity and CH$_4$ selectivity at low temperature, their large-scale industrial applications are limited by the high cost [21,22]. Therefore, Nickel as the most practical active metal has been widely investigated in the CO_2 methanation reaction due to its abundance, low cost, and high activity [23–25]. Moreover, the properties of support play an important role in catalytic performance including the surface properties, the ability to disperse the active phase, and metal-support interaction. It is crucial to choose an appropriate support in preparing effective catalysts [26,27]. Many different metal oxides, such as CeO_2 [28], MgO [29], Al_2O_3 [30], TiO_2 [31,32], SiO_2 [9,33], ZrO_2 [34–38], and Y_2O_3 [39], have been used as support to promote CO_2 methanation. ZrO_2 becomes the most promising support and is getting increased attention since it has a higher concentration of oxygen defects [5]. Xu et al. [35] studied the CO_2 methanation mechanism on the Ni/ZrO_2 catalyst by in-situ FTIR and DFT methods, and found that c-ZrO_2 could improve the electron mobility, the reducibility of Ni, and thus increase the catalytic activity of the catalysts. Martínez et al. [36] found that Ni/ZrO_2 had a higher catalytic activity due to the strong interaction between Ni and ZrO_2 by comparing the catalytic performance of Ni/ZrO_2, Ni/SiO_2, and $Ni/MgAl_2O_4$ catalysts. Hu et al. [37] also used ZrO_2 as support, due to its good synergetic function, to explore the effect of La on the CO_2 methanation reaction. Jia et al. [38] found that the structure of the catalyst had a significant influence on the catalytic performance. In addition, 71.9% CO_2 conversion and 69.5% CH_4 yield at 300 °C could be obtained on the Ni/ZrO_2 catalyst prepared by the DBD plasma decomposition of nickel nitrate, while the CO_2 conversion and CH_4 yield were only 32.9% and 30.3% on the catalyst prepared thermally. Many researchers also reported that $Ni/CeZrO_2$ catalysts exhibited an excellent low temperature catalytic performance for CO_2 methanation due to the oxygen vacancies and high oxygen storage capacity of $CeZrO_2$ [40–42].

In addition, promoters could further improve the catalytic activity. Therefore, the addition of a second element into nickel-based catalysts is considered as an effective method to improve the catalytic activity and stability at low temperature [14,21,43]. There are many types of promoters, including alkaline earth metals (Mg, Ca) [44,45], noble metals (Pt, Pd, Rh) [43], rare earth metals (La, Ce, Sm) [46], etc. The activation of CO_2 can be enhanced by changing the surface basicity of the catalyst and the metal-support interaction [30,44,46]. Guilera et al. [30] studied the metal-oxide promoted Ni/Al_2O_3 catalyst, which exhibited a higher catalytic performance compared with the Ni/Al_2O_3 catalyst, due to the increase in basic sites and nickel dispersion. Xu et al. [46] found that the surface basicity and the intensity of CO_2 chemisorption on Ni-based catalysts promoted by rare earth metals greatly increased, which could enhance the low-temperature catalytic activity of the catalysts. Wang et al. [47] found that the $Ni-Si/ZrO_2$ catalyst (Si as promoter) had a better catalytic performance on the DRM reaction than the $Ni-Zr/SiO_2$ catalyst (Zr as promoter). The Si on the $Ni-Si/ZrO_2$ catalyst could promote the dispersion of Ni and improve the stability of metal Ni in the DRM reaction process. The highly dispersed Ni species on $Ni-Si/ZrO_2$ increased the activation of CH_4 and CO_2, thus increasing the catalytic activity.

Since CO_2 adsorption and activation were the key steps in both the DRM reaction and CO_2 methanation, and CO_2, CH_4, and H_2 co-existed in both the two systems. This study tried to apply the $Ni-Si/ZrO_2$ catalyst to CO_2 methanation to improve the activation of CO_2 and promote CO_2 methanation. Therefore, $Ni-xSi/ZrO_2$ (Ni is 10 wt%, x = 0, 0.1, 0.5, 1 wt%) catalysts with different contents of Si were prepared by the co-impregnation method. The catalysts were characterized by different methods, including BET, H_2-TPR, H_2-TPD, H_2-chemisorption, CO_2-TPD, XRD, TEM, XPS, and TG-DSC, to explore the influence of Si.

2. Results and Discussion

2.1. The Catalytic Performance of Ni-xSi/ZrO$_2$ Catalysts

The catalytic performance of $Ni-xSi/ZrO_2$ catalysts was investigated over the temperature range from 200 to 400 °C under atmospheric pressure. The results of the catalytic test were presented in Figure 1, including the conversion of H_2 and CO_2, and the selectivity and yield of CH_4. The catalytic activity increased as the temperature increased until it was thermodynamically limited by the equilibrium. Among all catalysts, the Ni-0.1Si/ZrO$_2$ catalyst was the most active over the whole temperature range, followed by the Ni/ZrO$_2$ catalyst, which exhibited an excellent catalytic activity at 250 °C. On the unpromoted Ni/ZrO$_2$ catalyst, the CO$_2$ conversion and CH$_4$ yield were 66.6% and 66.2%, respectively with the 99.4% CH$_4$ selectivity. Si-promoted Ni/ZrO$_2$ catalysts exhibited a different catalytic activity, which changed with the amount of Si. The Ni-0.1Si/ZrO$_2$ catalyst exhibited the highest CO$_2$ conversion of 72.5% and H$_2$ conversion of 73.4% among the studied catalysts. Simultaneously, it showed the highest CH$_4$ selectivity of 99.6% and the highest CH$_4$ yield of 72.2%, whose catalytic activity was about 6% higher than the Ni/ZrO$_2$ catalyst. However, the Ni-0.5Si/ZrO$_2$ catalyst showed a rather low activity, with 10% H$_2$ conversion and 9.8% CO$_2$ conversion. The lowest catalytic activity was obtained on the Ni-1Si/ZrO$_2$ catalyst, with only 1% CH$_4$ yield. In general, adding the appropriate Si was beneficial to CO$_2$ methanation.

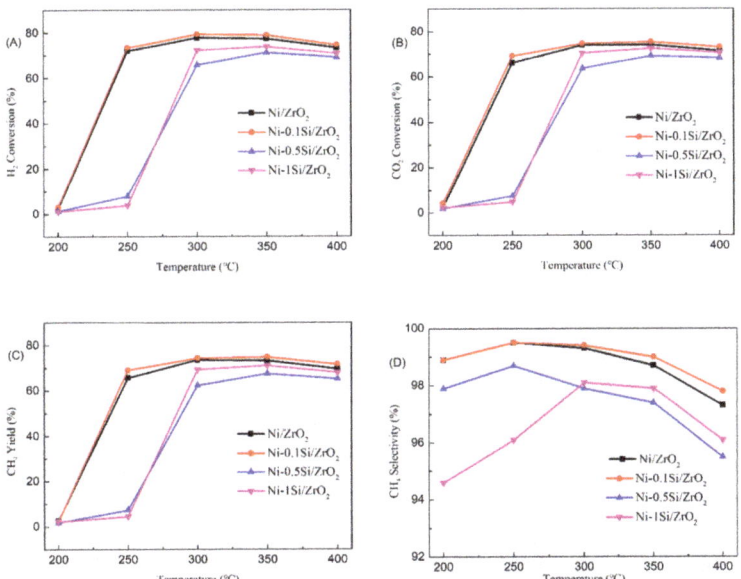

Figure 1. The conversion of H$_2$ (**A**) and CO$_2$ (**B**), the selectivity of CH$_4$ (**C**), and the yield of CH$_4$ (**D**) on Ni-xSi/ZrO$_2$ catalysts over the temperature range from 200 to 400 °C.

Then, the catalytic stability of Ni-xSi/ZrO$_2$ catalysts was investigated at 250 °C where the catalyst exhibited a high activity. The results of the catalytic test were presented in Figure 2. All the catalysts exhibited the high stability with 10 h on stream. TG-DSC analysis results showed that the weight of all spent catalysts did not decrease during the heating process (Figure S1), which illustrated that there was no carbon deposition on the catalysts after reaction.

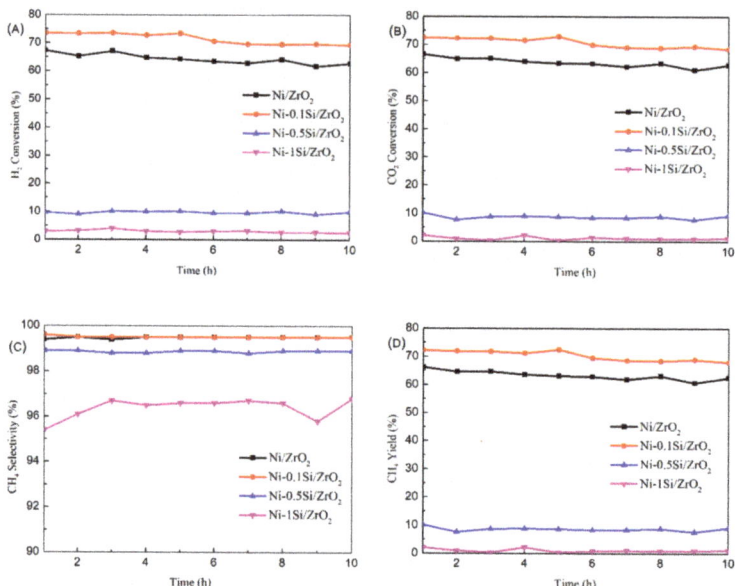

Figure 2. The conversion of H_2 (**A**) and CO_2 (**B**), the selectivity of CH_4 (**C**), and the yield of CH_4 (**D**) on Ni-xSi/ZrO_2 catalysts at 250 °C with a time on stream (H_2/CO_2 = 4/1, F = 150 mL/min).

2.2. The Textural Properties of Ni-xSi/ZrO$_2$ Catalysts

The physical properties of Ni-xSi/ZrO_2 catalysts were characterized by N_2 adsorption-desorption experiments. Figure 3 showed the N_2 adsorption-desorption isotherms, pore volume, and size distribution of the catalysts. All isotherms of the catalysts were assigned to the type IV isotherm and the P/P_0 for the hysteresis loop was 0.7~0.9, which indicated that all the catalysts were with the mesoporous structure, and that also could be proved by the pore size distribution in Figure 2B [48–50]. The textural properties of catalysts were summarized in Table 1. The specific surface area was calculated by the Brunauer-Emmett-Teller (BET) method and the pore size and volume were calculated by the Barrett-Joyner-Halenda (BJH) method. Obviously, it could be seen that the BET results of catalysts with a different Si content had no significant differences, which suggested that the different loading of Si did not damage the pore structure of the Ni/ZrO_2 catalyst.

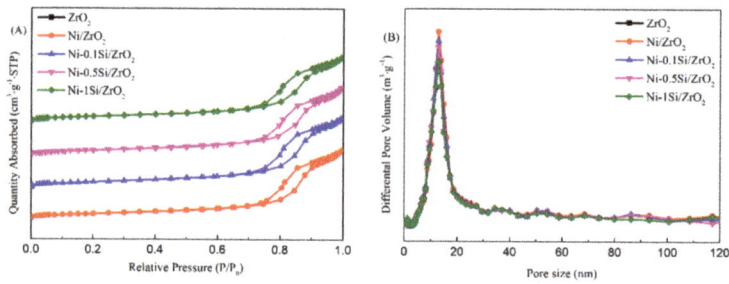

Figure 3. N_2 adsorption-desorption isotherms (**A**) and pore volume and size distribution (**B**) of Ni-xSi/ZrO_2 catalysts.

Table 1. Textural properties and element contents of Ni-xSi/ZrO$_2$ catalysts.

Catalyst	S$_{BET}$ [a] (m^2·g^{-1})	V$_{BJH}$ [b] (m^3·g^{-1})	Dp [c] (nm)	Ni (%) [d]	Si (%) [d]
Ni/ZrO$_2$	42.5	0.15	13.5	6.74	-
Ni-0.1Si/ZrO$_2$	46.0	0.16	12.8	6.66	0.11
Ni-0.5Si/ZrO$_2$	44.6	0.15	13.2	6.51	0.34
Ni-1Si/ZrO$_2$	44.4	0.15	13.2	5.79	0.58

[a]: Surface area (S$_{BET}$) determined by the BET method; [b]: BJH adsorption cumulative volume of pores; [c]: BJH adsorption average pore diameter; [d]: Obtained by XRF.

The actual loadings of Ni and Si were determined by the X-ray fluorescence (XRF) test. The content of Ni on Ni/ZrO$_2$, Ni-0.1Si/ZrO$_2$, Ni-0.5Si/ZrO$_2$, and Ni-1Si/ZrO$_2$ were 6.74%, 6.66%, 6.51%, and 5.79%, respectively. The actual contents of Ni on these catalysts were very close, indicating the successful loading of Ni onto ZrO$_2$. The content of Si on Ni-0.1Si/ZrO$_2$, Ni-0.5Si/ZrO$_2$, and Ni-1Si/ZrO$_2$ were 0.11%, 0.34%, and 0.58%.

2.3. The Reducibility of Ni-xSi/ZrO$_2$ Catalysts

The TPR profiles of calcined Ni-xSi/ZrO$_2$ catalysts and support ZrO$_2$ were presented in Figure 4. There was a broad peak at around 500 °C on the ZrO$_2$. Three main different reduction peaks (α, β, and γ) could be observed on the Ni-xSi/ZrO$_2$ catalysts, which suggested that there were three kinds of nickel oxide species. The α peak (317–327 °C) was attributed to the reduction of the bulk NiO species on the surface of the catalysts. The β peak (340–366 °C) and γ peak (419–428 °C) were assigned to the reduction of NiO species that interacted weakly and strongly with support, respectively. The amount of each NiO species was summarized in Table 2. The minimum amount of bulk NiO species could be observed, indicating that a little bulk NiO species existed on all catalysts. The amount of β peak was over 60% on both Ni-0.5Si/ZrO$_2$ and Ni-1Si/ZrO$_2$ catalysts. Associated with the catalytic activity, the NiO species that interacted with ZrO$_2$ weakly was not good for the CO$_2$ methanation reaction after reduction, resulting in a lower catalytic activity. While Ni-0.1Si/ZrO$_2$ and Ni/ZrO$_2$ catalysts showed the higher amount of the γ peak contributing to a higher catalytic activity, suggesting that the strong interaction between the nickel and the support could promote the dispersion of Ni on the support [51]. Furthermore, the β and γ peaks of promoted Ni/ZrO$_2$ catalysts shifted to a lower temperature compared with that of the Ni/ZrO$_2$ catalyst, suggesting that the appropriate Si (0.1 wt%) promoted the dispersion of Ni [50,51]. The peaks of the Ni-0.1Si/ZrO$_2$ catalyst shifted to the lowest temperature (the β at 340 °C and the γ at 419 °C), indicating the highest Ni dispersion, which could increase the catalytic activity of catalysts. Therefore, the strong nickel-support interaction promoted the dispersion of Ni on the support, which was beneficial to the catalytic performance [1,50].

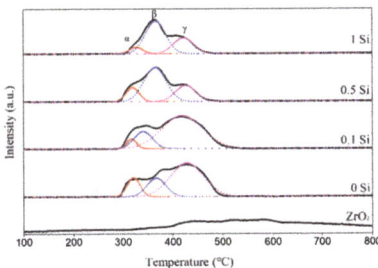

Figure 4. H$_2$-TPR profiles of Ni-xSi/ZrO$_2$ catalysts and ZrO$_2$.

Table 2. The amount of each nickel oxide species on the Ni-xSi/ZrO$_2$ catalysts.

Catalyst	α		β		γ	
	Position (°C)	Content (%)	Position (°C)	Content (%)	Position (°C)	Content (%)
Ni/ZrO$_2$	321	13.9	366	20.5	428	65.6
Ni-0.1Si/ZrO$_2$	317	5.0	340	17.5	419	77.5
Ni-0.5Si/ZrO$_2$	319	14.2	365	60.1	426	25.7
Ni-1Si/ZrO$_2$	327	7.8	366	62.1	422	30.1

2.4. The H$_2$-TPD and H$_2$-Chemisorption of Ni-xSi/ZrO$_2$ Catalysts

The amounts of active sites of reduced Ni-xSi/ZrO$_2$ catalysts were measured by H$_2$-TPD and H$_2$-chemisorption. Two peaks could be clearly observed in the patterns of reduced catalysts depicted in Figure 5. The first peak appeared at around 92–111 °C, which was assigned to the weak active site, and a broad peak around at 500 °C could be found over the catalysts, corresponding to the strong active site with the strong H$_2$ adsorption [51,52]. The temperature of two H$_2$ desorption peaks was different with different contents of Si. The highest temperature with 111 and 517 °C of H$_2$ desorption peaks on the Ni-0.1Si/ZrO$_2$ catalyst suggested the strong H$_2$ adsorption to the active sites, which could be assigned to the high dispersion of Ni and the increasing number of active sites [51–53]. Considering the possible spill over in TPD over the catalysts, the pulse chemisorption of H$_2$ was also carried out. The dispersion of Ni obtained by H$_2$-chemisorption was shown in Table 3. The Ni dispersion of Ni/ZrO$_2$, Ni-0.1Si/ZrO$_2$, Ni-0.5Si/ZrO$_2$, and Ni-1Si/ZrO$_2$ were 0.65%, 0.67%, 0.34%, and 0.28%, respectively. The Ni-0.1Si/ZrO$_2$ catalyst showed the highest dispersion of Ni, which was slightly higher than the Ni/ZrO$_2$ catalyst. The amounts of adsorbed H$_2$ decreased significantly on Ni-0.5Si/ZrO$_2$ and Ni-1Si/ZrO$_2$ catalysts, suggesting that the amounts of Ni decreased, which corresponded to their lower catalytic activity.

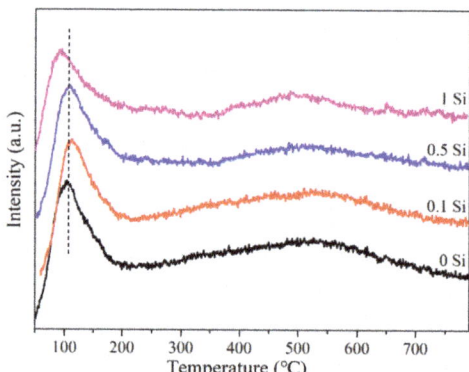

Figure 5. H$_2$-TPD patterns of Ni-xSi/ZrO$_2$ catalysts.

Table 3. The H_2 uptake and Ni dispersion of Ni-xSi/ZrO_2 catalysts.

Catalyst	Peak 1		Peak 2		Total H_2 Uptake (μmol/g) [a]	Ni Dispersion (%) [b]
	T (°C)	H_2 Uptake (μmol/g)	T (°C)	H_2 Uptake (μmol/g)		
Ni/ZrO_2	102	4.64	512	5.02	9.66	0.65
Ni-0.1Si/ZrO_2	111	4.64	517	4.95	9.59	0.67
Ni-0.5Si/ZrO_2	103	4.28	505	2.59	6.87	0.34
Ni-1Si/ZrO_2	92	4.06	503	2.51	6.57	0.28

[a]: Obtained by H_2-TPD; [b]: Obtained by H_2 chemisorption.

2.5. The Results of CO_2-TPD on Ni-xSi/ZrO_2 Catalysts

The CO_2 desorption capability was explored by the CO_2-TPD measurement and the profiles were shown in Figure 6, which was used to describe the basicity of catalysts usually. Two main CO_2 desorption peaks could be observed over each catalyst, which were classified as the adsorbed CO_2 at weak basic sites (361–410 °C) and strong basic sites (577–604 °C), respectively. Based on the pioneer studies, the CO_2 adsorbed on the weak basic sites could be desorbed at a low temperature and that absorbed on the strong basic sites could be desorbed at a high temperature [44]. The contents of weak and strong basic sites were calculated by the area of the peaks (Table 4). It could be seen that the content of strong basic sites on Ni-0.1Si/ZrO_2 was the highest (71%) among all the catalysts studied, which could inhabit carbon formation effectively [54,55]. Le et al. [56] found that the stronger CO_2 desorption peak around 500 °C on the Ni-CeO_2 catalyst suggested the strong CO_2 adsorption ability, which had a positive effect on the catalytic performance. In addition, the CO_2 adsorption peak of strong basic sites slightly shifted to a higher temperature with the increasing of Si, suggesting that the addition of Si promoted the adsorption of CO_2. The intensified surface basicity of the Ni-0.1Si/ZrO_2 catalyst could promote the adsorption of CO_2 thus promoting the CO_2 methanation reaction [42,44,46].

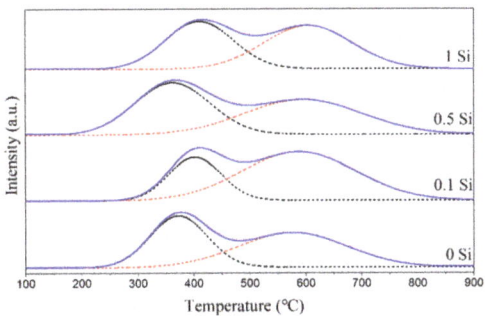

Figure 6. CO_2-TPD profiles of Ni-xSi/ZrO_2 catalysts.

Table 4. Peak position and basic sites of Ni-xSi/ZrO_2 catalysts obtained by CO_2-TPD.

Catalyst	Peak 1		Peak 2		Total Basicity (μmol/g)
	Position (°C)	Content (%)	Position (°C)	Content (%)	
Ni/ZrO_2	372	44	577	56	123
Ni-0.1Si/ZrO_2	402	29	588	71	124
Ni-0.5Si/ZrO_2	361	49	595	51	127
Ni-1Si/ZrO_2	410	46	604	54	124

2.6. Crystallite Structure of Ni-xSi/ZrO₂ Catalysts

The XRD patterns of catalysts could be seen in Figure 7. Before reduction, the diffraction peaks at 37.2, 43.3, and 62.9° were attributed to NiO (PDF no. 47-1049), as shown in Figure 7a [1,26]. The diffraction peaks at 44.5, 51.8, and 76.4° corresponded to the Ni metal (PDF no. 04-0850) over reduced catalysts, as shown in Figure 7b, and there was no diffraction peak of NiO [57]. In addition, no obvious diffraction peak of Si species was detected on Ni-xSi/ZrO₂ catalysts, suggesting that Si was in a high dispersion or amorphous state [51]. The crystallite sizes of NiO and Ni metal were calculated by the Scherrer equation at 2 θ = 43.3° and 2 θ = 44.5°, respectively, and the results were listed in Table 5. It could be seen that the particle sizes of NiO had no significant changes and the crystallite sizes of NiO were all around 25 nm on Ni-xSi/ZrO₂ catalysts before reduction. The grain sizes of the Ni metal were around 22 nm on reduced Ni-xSi/ZrO₂ catalysts, which also showed no obvious variation.

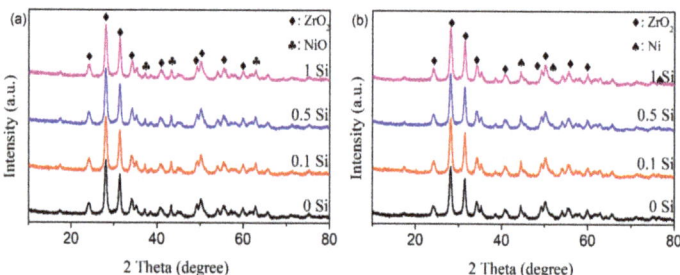

Figure 7. XRD patterns of Ni-xSi/ZrO₂ catalysts (**a**) before reduction, (**b**) after reduction.

Table 5. The crystallite sizes of NiO and Ni metal on Ni-xSi/ZrO₂ catalysts.

Catalyst	Before Reduction	After Reduction
	NiO	Ni Metal
Ni/ZrO₂	24 nm	22 nm
Ni-0.1Si/ZrO₂	26 nm	23 nm
Ni-0.5Si/ZrO₂	26 nm	23 nm
Ni-1Si/ZrO₂	25 nm	21 nm

2.7. The TEM Images of Ni-xSi/ZrO₂ Catalysts

The TEM images of reduced Ni-xSi/ZrO₂ catalysts were presented in Figure 8. It can be seen that all the catalysts exhibited a similar morphology, and the particle sizes were distributed between 10 and 30 nm. From Figure 8B, Ni was more evenly distributed with particle sizes of 18–20 nm on the Ni-0.1Si/ZrO₂ catalyst. However, on the Ni-1Si/ZrO₂ catalyst, the size distribution of nickel particles was uneven, and there were big nickel particles. From Figure 8D, bulk particles were found on the Ni-1Si/ZrO₂ catalyst, which might be bulk Ni particles or the mixture of Ni and Si.

2.8. Chemical State of the Elements on Ni-xSi/ZrO₂ Catalysts

The surface element composition and chemical state of reduced Ni-xSi/ZrO₂ catalysts were obtained by the XPS experiment, which were shown in Figure 9. There were four main peaks in the Ni 2p spectra. The peak at around 852 eV was assigned to the characteristic peak of Ni⁰ [7]. There were two characteristic peaks of Ni²⁺, P1 (about 854 eV) was the low energy peak and belonged to the peak of NiO, while P2 (about 856 eV) corresponded to Ni (OH)₂ [25]. The peak at around 860 eV was the companioning peak of Ni²⁺, produced by the orbital spin splitting [50]. The percentages of different elements on the surface of catalysts were summarized in Table 6. The surface content of Ni⁰ was the highest on the

Ni-0.1Si/ZrO$_2$ catalyst, which was 2.15%. More Ni^0 could provide more active sites and facilitate CO$_2$ methanation [58]. It could be found that the surface content of Si was higher than its actual loading, suggesting that Si enriched on the catalysts surface and part of Ni might be covered by Si, especially on Ni-0.5Si/ZrO$_2$ and Ni-1Si/ZrO$_2$ catalysts. The O 1s spectrum exhibited two types of oxygen species, as shown in Figure 9B. The peak at 530–530.46 eV was attributed to the lattice oxygen (O$_\alpha$) and the peak at 528.9–529.1 eV belonged to the surface oxygen (O$_\beta$) [28,59]. Based on the areas of O$_\alpha$ and O$_\beta$, the ratios of the oxygen vacancies could be obtained by calculating the ratio of O$_\beta$ to O$_T$ (O$_T$ = O$_\alpha$ + O$_\beta$) [5,22,38]. In Table 6, the reduced Ni-0.1Si/ZrO$_2$ catalyst exhibited the highest ratio (0.589) of O$_\beta$ to O$_T$ compared with other catalysts, suggesting the highest amount of oxygen vacancies. The presence of oxygen vacancies was beneficial to the adsorption of CO$_2$, which could promote CO$_2$ activation [23,58]. Jiang et al. [17] found that the content of surface oxygen on the Mn promoted Ni/bentonite catalyst was 83.55%, which was higher than that of the unpromoted Ni/bentonite catalyst (74.85%), representing the higher amount of oxygen vacancies. The increased oxygen vacancies were helpful to the adsorption and dissociation of CO$_2$ on the catalyst. To sum up, the Ni-0.1Si/ZrO$_2$ catalyst exhibited the highest catalytic activity and stability due to the highest amount of Ni^0 and oxygen vacancies on the surface.

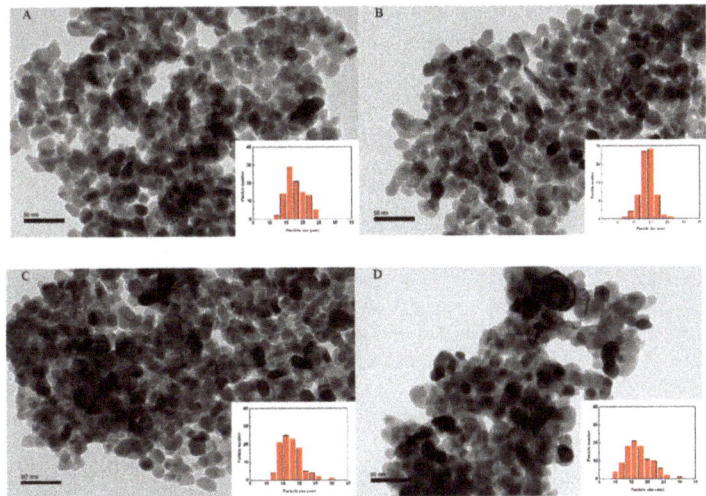

Figure 8. TEM images of Ni-xSi/ZrO$_2$ catalysts (**A**) Ni/ZrO$_2$ catalyst, (**B**) Ni-0.1Si/ZrO$_2$ catalyst, (**C**) Ni-0.5Si/ZrO$_2$ catalyst, and (**D**) Ni-1Si/ZrO$_2$ catalyst.

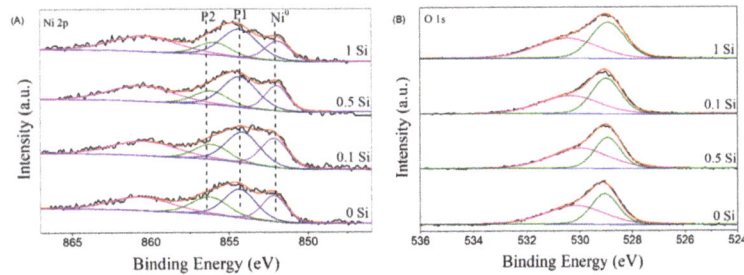

Figure 9. XPS spectra of reduced Ni-xSi/ZrO$_2$ catalysts (**A**) Ni 2p and (**B**) O 1s.

Table 6. Surface contents on reduced Ni-xSi/ZrO$_2$ catalysts.

Catalyst	Ni^0 (%)	Si (%)	O$_\beta$/O$_T$
Ni/ZrO$_2$	1.88	0	0.56
Ni-0.1Si/ZrO$_2$	2.15	0.45	0.59
Ni-0.5Si/ZrO$_2$	1.78	1.91	0.51
Ni-1Si/ZrO$_2$	1.27	3.74	0.50

3. Materials and Methods

3.1. Catalysts Preparation

The ZrO$_2$ support was prepared by the precipitation method. A certain amount of Zr(NO$_3$)$_4$·5H$_2$O (ChengduKelong, China) was dissolved in deionized water with continuous stirring until dissolved completely. Then, NH$_3$·H$_2$O was added to the above solution to achieve a pH value of 9. After stirring for 2 h, the mixture was aged for 24 h at room temperature. After that, the mixture was filtered and washed three times with deionized water. The obtained sample was dried at 110 °C for 4 h and then calcined at 500 °C for 5 h (with the rate of 2 °C/min) to obtain the ZrO$_2$ support.

The impregnation method was used to synthesize the Ni-xSi/ZrO$_2$ catalysts (ZrO$_2$ was used as the support and Si was used as the promoter with the load of 0, 0.1, 0.5, and 1 wt%). The designed amount of Ni(NO$_3$)$_2$·6H$_2$O (ChengduKelong, Chengdu, China) and different amounts of (C$_2$H$_5$O)$_4$Si (Alfa Aesar Chemicals, Shanghai, China) were dissolved in a certain amount of absolute ethanol (Chengdu Chron Chemical, Chengdu, China). The ZrO$_2$ support was impregnated with the aforesaid ethanol solution for 24 h at room temperature. Then, these samples were dried at 80 °C for 2 h and 110 °C for 4 h. Finally, the above mixtures were heated to 500 °C (with the rate of 2 °C/min) and calcined at 500 °C for 5 h under air flow, then, the Ni-xSi/ZrO$_2$ catalysts were obtained (x = 0, 0.1, 0.5, 1 wt%).

3.2. Catalytic Activity Test

The catalytic activity test was carried out in a fixed-bed continuous flow micro-quartz-tube reactor with 10 mm in diameter at atmospheric pressure. There was a thermocouple near the reactor close to the fixed-bed, which was used to follow the temperature of the catalysts during the test. Before the activity test, 0.50 g of the catalyst was heated up to 450 °C (10 °C/min) and then reduced at a constant temperature of 450 °C for 1 h in H$_2$/Ar (F(H$_2$) = F(Ar) = 30 mL/min) mixture gas. After that, the catalyst was cooled down to the reaction temperature before the introduction of the mixture of reactants (H$_2$/CO$_2$ = 4, F = 150 mL/min) for the CO$_2$ methanation reaction. The effluent from the reactor passed through a condensing device and was analyzed online by a gas chromatograph (plot-C2000 capillary column) per hour.

$$X_{CO_2} = \frac{n_{CO_2,in} - n_{CO_2,out}}{n_{CO_2,in}} \times 100\%, \qquad (1)$$

$$X_{H_2} = \frac{n_{H_2,in} - n_{H_2,out}}{n_{H_2,in}} \times 100\%, \qquad (2)$$

$$S_{CH_4} = \frac{n_{CH_4,our}}{n_{CH_4,our} + n_{CO,our}} \times 100\%, \qquad (3)$$

$$Y_{CH_4} = X_{H_2} \times S_{CH_4} \qquad (4)$$

where X_{CO_2} and X_{H_2} were the conversion of CO$_2$ and H$_2$, S_{CH_4} was the selectivity of CH$_4$, and Y_{CH_4} was the yield of CH$_4$.

3.3. Catalysts Characterization

The physical property test was conducted in the Micromeritics Tristar II 3020 instrument by using the N$_2$ adsorption-desorption method. Before the measurements, about 0.1 g samples were outgassed at 150 °C for 2 h and then at 300 °C for 2 h under a vacuum.

The actual Ni and Si loadings of the fresh catalysts were determined by the X-ray fluorescence (XRF) test. Ni and Si were detected by the Ni Kα line and Si Kα line, respectively.

The hydrogen temperature-programmed reduction (H_2-TPR) measurement was carried out with a Micromeritics AutoChem II Chemisorption Analyzer. At first, about 100 mg of the catalyst was pretreated with Ar flow at 150 °C for 30 min. Then, the TPR experiment was performed from 50 to 800 °C in the H_2/Ar (10/90 vol%) flow with the heating rate of 8 °C/min. The TCD detector was used to monitor the H_2 consumption.

The temperature-programmed desorption of H_2 (H_2-TPD) was performed on the same equipment for H_2-TPR. About 100 mg of the reduced catalyst was pretreated at 500 °C for 1 h in Ar flow. Next, H_2 was absorbed at 50 °C for 1 h in 10% H_2/Ar. After cleaning the excess unabsorbed H_2, the catalyst was heated to 800 °C with a heating rate of 10 °C/min under Ar flow. The results were detected by a TCD detector. The H_2 pulse chemisorption was also processed on the equipment. About 100 mg of the reduced catalyst was pretreated at 450 °C for 1 h in Ar flow and cooled down to 50 °C at the same atmosphere for beginning the H_2 adsorption. A gas mixture of 10% H_2 balance Ar was pulsed over the catalyst for chemisorption measurements.

Before the temperature-programmed desorption of CO_2 (CO_2-TPD), about 100 mg of the reduced catalyst (reduced at 450 °C) was pretreated at 500 °C in He flow for 1 h to remove surface impurities. Then, CO_2 was absorbed at 50 °C for 1 h in 10% CO_2/He. After cleaning the excess unabsorbed CO_2, the catalyst was heated to 900 °C with a heating rate of 10 °C/min in He flow. The observed curves were fitted into two Gaussian peaks.

The X-ray diffraction (XRD) was performed on a DX-1000 CSC diffractometer instrument, operating at 40 kV and 25 mA with a Cu Kα radiation source for the calcined and reduced catalysts. The data was recorded over the scattering angle range of 2θ from 10 to 80°, with a scan step with of 0.03°.

Transmission electron microscopy (TEM) was used to characterize the reduced catalysts on the Tecnai G2 F20 machine. The twin instrument with the 0.20 nm resolution was used, and the acceleration voltage was 200 Kv.

The analysis of X-ray photoelectron spectroscopy (XPS) was carried out on a KRATOS spectrometer with an AXIS Ultra DLD. The Al Kα monochromatized line was operated at the accelerating power of 25 W. In addition, the binding energy was calibrated with C 1 s 284.6 eV.

Thermogravimetric (TG) and differential scanning calorimetry analysis (DSC) was used to characterize the deposited carbon of the spent Ni-xSi/ZrO_2 catalysts, using the NETZSCH TG209F1 instrument. Before the test, the sample was placed until a better gas equilibrium. Then, the temperature was increased from 30 to 800 °C with a 5 °C·min^{-1} heating rate in air flow with a rate of 60 mL·min^{-1}.

4. Conclusions

Adding the appropriate amount of Si could increase the catalytic activity of Ni/ZrO_2 catalyst, and the Ni-0.1Si/ZrO_2 catalyst showed the highest catalytic activity and stability. The strong interaction between Ni and ZrO_2 could promote the dispersion of Ni on the support, and the strong basic sites on the catalyst were beneficial to the absorption of CO_2, thus to the CO_2 methanation reaction on the Ni-0.1Si/ZrO_2 catalyst. In addition, the higher amount of surface Ni^0 could provide more active sites, and the more oxygen vacancies were beneficial to the absorption and activation of CO_2 on the 0.1Si/ZrO_2 catalyst.

Supplementary Materials: The following are available online at https://www.mdpi.com/2073-4344/11/1/67/s1. Figure S1. The TG-DSC profile of the spent Ni-xSi/ZrO_2 catalysts (A) Ni/ZrO_2 catalyst, (B) Ni-0.1Si/ZrO_2 catalyst, (C)Ni-0.5Si/ZrO_2 catalyst, and (D) Ni-1Si/ZrO_2 catalyst.

Author Contributions: Conceptualization, methodology, L.L., Y.W. and C.H.; software, validation, formal analysis, investigation, resources, data curation, writing—original draft preparation, L.L. and Q.Z.; writing—review and editing, visualization, supervision, Y.W. and C.H.; project administration, funding acquisition, C.H. All authors have read and agreed to the published version of the manuscript.

Funding: This research was supported by the National Key R and D Program of China (2018YFB1501404), the 111 program (B17030), and Fundamental Research Funds for the Central Universities.

Institutional Review Board Statement: Not applicable.

Informed Consent Statement: Not applicable.

Data Availability Statement: Not applicable.

Acknowledgments: We would like to thank the Analytical and Testing center of Sichuan University for the characterization and we are grateful to Yunfei Tian for his help in the XPS experiments.

Conflicts of Interest: The authors declare no conflict of interest.

References

1. Iglesias, I.; Quindimil, A.; Mariño, F.; De-La-Torre, U.; González-Velasco, J.R. Zr promotion effect in CO_2 methanation over ceria supported nickel catalysts. *Int. J. Hydrogen Energy* **2019**, *44*, 1710–1719. [CrossRef]
2. Liu, Q.; Bian, B.; Fan, J.; Yang, J. Cobalt doped Ni based ordered mesoporous catalysts for CO_2 methanation with enhanced catalytic performance. *Int. J. Hydrogen Energy* **2018**, *43*, 4893–4901. [CrossRef]
3. Stangeland, K.; Kalai, D.; Li, H.; Yu, Z. CO_2 Methanation: The Effect of Catalysts and Reaction Conditions. *Energy Procedia* **2017**, *105*, 2022–2027. [CrossRef]
4. Anwar, M.N.; Fayyaz, A.; Sohail, N.F.; Khokhar, M.F.; Baqar, M.; Yasar, A.; Rasool, K.; Nazir, A.; Raja, M.U.F.; Rehan, M.; et al. CO_2 utilization: Turning greenhouse gas into fuels and valuable products. *J. Environ. Manag.* **2020**, *260*, 110059–110072. [CrossRef] [PubMed]
5. Li, W.; Nie, X.; Jiang, X.; Zhang, A.; Ding, F.; Liu, M.; Liu, Z.; Guo, X.; Song, C. ZrO_2 support imparts superior activity and stability of Co catalysts for CO_2 methanation. *Appl. Catal. B Environ.* **2018**, *220*, 397–408. [CrossRef]
6. Do, J.Y.; Park, N.-K.; Seo, M.W.; Lee, D.; Ryu, H.-J.; Kang, M. Effective thermocatalytic carbon dioxide methanation on Ca-inserted $NiTiO_3$ perovskite. *Fuel* **2020**, *271*, 117624–117641. [CrossRef]
7. Liu, Q.; Wang, S.; Zhao, G.; Yang, H.; Yuan, M.; An, X.; Zhou, H.; Qiao, Y.; Tian, Y. CO_2 methanation over ordered mesoporous NiRu-doped $CaO-Al_2O_3$ nanocomposites with enhanced catalytic performance. *Int. J. Hydrogen Energy* **2018**, *43*, 239–250. [CrossRef]
8. Wang, Y.; Yao, L.; Wang, S.; Mao, D.; Hu, C. Low-temperature catalytic CO_2 dry reforming of methane on Ni-based catalysts: A review. *Fuel Process. Technol.* **2018**, *169*, 199–206. [CrossRef]
9. Li, S.; Guo, S.; Gong, D.; Kang, N.; Fang, K.-G.; Liu, Y. Nano composite composed of $MoO_x-La_2O_3$-Ni on SiO_2 for storing hydrogen into CH_4 via CO_2 methanation. *Int. J. Hydrogen Energy* **2019**, *44*, 1597–1609. [CrossRef]
10. Alarcón, A.; Guilera, J.; Díaz, J.A.; Andreu, T. Optimization of nickel and ceria catalyst content for synthetic natural gas production through CO_2 methanation. *Fuel Process. Technol.* **2019**, *193*, 114–122. [CrossRef]
11. Lin, J.; Ma, C.; Wang, Q.; Xu, Y.; Ma, G.; Wang, J.; Wang, H.; Dong, C.; Zhang, C.; Ding, M. Enhanced low-temperature performance of CO_2 methanation over mesoporous $Ni/Al_2O_3-ZrO_2$ catalysts. *Appl. Catal. B Environ.* **2019**, *243*, 262–272. [CrossRef]
12. Yan, X.; Sun, W.; Fan, L.; Duchesne, P.N.; Wang, W.; Kübel, C.; Wang, D.; Kumar, S.G.H.; Li, Y.F.; Tavasoli, A.; et al. Nickel@Siloxene catalytic nanosheets for high-performance CO_2 methanation. *Nat. Commun.* **2019**, *10*, 2608–2618. [CrossRef] [PubMed]
13. Bukhari, S.N.; Chong, C.C.; Teh, L.P.; Vo, D.-V.N.; Ainirazali, N.; Triwahyono, S.; Jalil, A.A.; Setiabudi, H.D. Promising hydrothermal technique for efficient CO_2 methanation over Ni/SBA-15. *Int. J. Hydrogen Energy* **2019**, *44*, 20792–20804. [CrossRef]
14. Xu, L.; Lian, X.; Chen, M.; Cui, Y.; Wang, F.; Li, W.; Huang, B. CO_2 methanation over Co-Ni bimetal-doped ordered mesoporous Al_2O_3 catalysts with enhanced low-temperature activities. *Int. J. Hydrogen Energy* **2018**, *43*, 17172–17184. [CrossRef]
15. Vita, A.; Italiano, C.; Pino, L.; Laganà, M.; Ferraro, M.; Antonucci, V. High-temperature CO_2 methanation over structured Ni/GDC catalysts: Performance and scale-up for Power-to-Gas application. *Fuel Process. Technol.* **2020**, *202*, 106365–106375. [CrossRef]
16. Atzori, L.; Cutrufello, M.G.; Meloni, D.; Monaci, R.; Cannas, C.; Gazzoli, D.; Sini, M.F.; Deiana, P.; Rombi, E. CO_2 methanation on hard-templated $NiO-CeO_2$ mixed oxides. *Int. J. Hydrogen Energy* **2017**, *42*, 20689–20702. [CrossRef]
17. Jiang, Y.; Huang, T.; Dong, L.; Su, T.; Li, B.; Luo, X.; Xie, X.; Qin, Z.; Xu, C.; Ji, H. Mn Modified Ni/Bsentonite for CO_2 Methanation. *Catalysts* **2018**, *8*, 646. [CrossRef]
18. Xu, J.; Lin, Q.; Su, X.; Duan, H.; Geng, H.; Huang, Y. CO_2 methanation over $TiO_2-Al_2O_3$ binary oxides supported Ru catalysts. *Chin. J. Chem. Eng.* **2016**, *24*, 140–145. [CrossRef]
19. Younas, M.; Sethupathi, S.; Kong, L.L.; Mohamed, A.R. CO_2 methanation over Ni and Rh based catalysts: Process optimization at moderate temperature. *Int. J. Hydrogen Energy* **2018**, *42*, 3289–3302. [CrossRef]

20. Martin, N.M.; Velin, P.; Skoglundh, M.; Bauer, M.; Carlsson, P.-A. Catalytic hydrogenation of CO_2 to methane over supported Pd, Rh and Ni catalysts. *Catal. Sci. Technol.* **2017**, *7*, 1086–1094. [CrossRef]
21. Zhang, L.; Bian, L.; Zhu, Z.; Li, Z. La-promoted Ni/Mg-Al catalysts with highly enhanced low-temperature CO_2 methanation performance. *Int. J. Hydrogen Energy* **2018**, *43*, 2197–2206. [CrossRef]
22. Yu, Y.; Chan, Y.M.; Bian, Z.; Song, F.; Wang, J.; Zhong, Q.; Kawi, S. Enhanced performance and selectivity of CO_2 methanation over g-C_3N_4 assisted synthesis of Ni CeO_2 catalyst: Kinetics and DRIFTS studies. *Int. J. Hydrogen Energy* **2018**, *43*, 15191–15204. [CrossRef]
23. Zhou, G.; Liu, H.; Cui, K.; Jia, A.; Hu, G.; Jiao, Z.; Liu, Y.; Zhang, X. Role of surface Ni and Ce species of Ni/CeO_2 catalyst in CO_2 methanation. *Appl. Surf. Sci.* **2016**, *383*, 248–252. [CrossRef]
24. Lin, J.; Ma, C.; Luo, J.; Kong, X.; Xu, Y.; Ma, G.; Wang, J.; Zhang, C.; Li, Z.; Ding, M. Preparation of Ni based mesoporous Al_2O_3 catalyst with enhanced CO_2 methanation performance. *RSC Adv.* **2019**, *9*, 8684–8694. [CrossRef]
25. Garbarino, G.; Wang, C.; Cavattoni, T.; Finocchio, E.; Riani, P.; Flytzani-Stephanopoulos, M.; Busca, G. A study of Ni/La-Al_2O_3 catalysts: A competitive system for CO_2 methanation. *Appl. Catal. B Environ.* **2019**, *248*, 286–297. [CrossRef]
26. Gnanakumar, E.S.; Chandran, N.; Kozhevnikov, I.V.; Grau-Atienza, A.; Ramos Fernández, E.V.; Sepulveda-Escribano, A.; Shiju, N.R. Highly efficient nickel-niobia composite catalysts for hydrogenation of CO_2 to methane. *Chem. Eng. Sci.* **2019**, *194*, 2–9. [CrossRef]
27. Czuma, N.; Zarębska, K.; Motak, M.; Gálvez, M.E.; Da Costa, P. Ni/zeolite X derived from fly ash as catalysts for CO_2 methanation. *Fuel* **2020**, *267*, 117139–117147. [CrossRef]
28. Ye, R.-P.; Li, Q.; Gong, W.; Wang, T.; Razink, J.J.; Lin, L.; Qin, Y.-Y.; Zhou, Z.; Adidharma, H.; Tang, J.; et al. High-performance of nanostructured Ni/CeO_2 catalyst on CO_2 methanation. *Appl. Catal. B Environ.* **2020**, *268*, 118474–118484. [CrossRef]
29. Loder, A.; Siebenhofer, M.; Lux, S. The reaction kinetics of CO_2 methanation on a bifunctional Ni/MgO catalyst. *J. Ind. Eng. Chem.* **2020**, *85*, 196–207. [CrossRef]
30. Guilera, J.; Del Valle, J.; Alarcón, A.; Díaz, J.A.; Andreu, T. Metal-oxide promoted Ni/Al_2O_3 as CO_2 methanation micro-size catalysts. *J. CO_2 Util.* **2019**, *30*, 11–17. [CrossRef]
31. Yang, Y.; Liu, J.; Liu, F.; Wu, D. Reaction mechanism of CO_2 methanation over Rh/TiO_2 catalyst. *Fuel* **2020**, *276*, 118093–118103. [CrossRef]
32. Zhou, R.; Rui, N.; Fan, Z.; Liu, C.-J. Effect of the structure of Ni/TiO_2 catalyst on CO_2 methanation. *Int. J. Hydrogen Energy* **2016**, *41*, 22017–22025. [CrossRef]
33. Moghaddam, S.V.; Rezaei, M.; Meshkani, F.; Daroughegi, R. Synthesis of nanocrystalline mesoporous Ni/Al_2O_3-SiO_2 catalysts for CO_2 methanation reaction. *Int. J. Hydrogen Energy* **2018**, *43*, 19038–19046. [CrossRef]
34. Zeng, L.; Wang, Y.; Li, Z.; Song, Y.; Zhang, J.; Wang, J.; He, X.; Wang, C.; Lin, W. Highly Dispersed Ni Catalyst on Metal–Organic Framework-Derived Porous Hydrous Zirconia for CO_2 Methanation. *ACS Appl. Mater. Interfaces* **2020**, *12*, 17436–17442. [CrossRef] [PubMed]
35. Xu, X.; Tong, Y.; Huang, J.; Zhu, J.; Fang, X.; Xu, J.; Wang, X. Insights into CO_2 methanation mechanism on cubic ZrO_2 supported Ni catalyst via a combination of experiments and DFT calculations. *Fuel* **2021**, *283*, 118867–118876. [CrossRef]
36. Martínez, J.; Hernández, E.; Alfaro, S.; López Medina, R.; Valverde Aguilar, G.; Albiter, E.; Valenzuela, M. High Selectivity and Stability of Nickel Catalysts for CO_2 Methanation: Support Effects. *Catalysts* **2018**, *9*, 24. [CrossRef]
37. Hu, L.; Urakawa, A. Continuous CO_2 capture and reduction in one process: CO_2 methanation over unpromoted and promoted Ni/ZrO_2. *J. CO_2 Util.* **2018**, *25*, 323–329. [CrossRef]
38. Jia, X.; Zhang, X.; Rui, N.; Hu, X.; Liu, C.-J. Structural effect of Ni/ZrO_2 catalyst on CO_2 methanation with enhanced activity. *Appl. Catal. B Environ.* **2019**, *244*, 159–169. [CrossRef]
39. Yan, Y.; Dai, Y.; Yang, Y.; Lapkin, A.A. Improved stability of Y_2O_3 supported Ni catalysts for CO_2 methanation by precursor-determined metal-support interaction. *Appl. Catal. B Environ.* **2018**, *237*, 504–512. [CrossRef]
40. Shang, X.; Deng, D.; Wang, X.; Xuan, W.; Zou, X.; Ding, W.; Lu, X. Enhanced low-temperature activity for CO_2 methanation over Ru doped the Ni/$Ce_xZr_{(1-x)}O_2$ catalysts prepared by one-pot hydrolysis method. *Int. J. Hydrogen Energy* **2018**, *43*, 7179–7189. [CrossRef]
41. Ashok, J.; Ang, M.L.; Kawi, S. Enhanced activity of CO_2 methanation over Ni/CeO_2-ZrO_2 catalysts: Influence of preparation methods. *Catal. Today* **2017**, *281*, 304–311. [CrossRef]
42. Pan, Q.; Peng, J.; Sun, T.; Gao, D.; Wang, S.; Wang, S. CO_2 methanation on Ni/$Ce_{0.5}Zr_{0.5}O_2$ catalysts for the production of synthetic natural gas. *Fuel Process. Technol.* **2014**, *123*, 166–171. [CrossRef]
43. Mihet, M.; Lazar, M.D. Methanation of CO_2 on Ni/γ-Al_2O_3: Influence of Pt, Pd or Rh promotion. *Catal. Today* **2018**, *306*, 294–299. [CrossRef]
44. Xu, L.; Yang, H.; Chen, M.; Wang, F.; Nie, D.; Qi, L.; Lian, X.; Chen, H.; Wu, M. CO_2 methanation over Ca doped ordered mesoporous Ni-Al composite oxide catalysts: The promoting effect of basic modifier. *J. CO_2 Util.* **2017**, *21*, 200–210. [CrossRef]
45. Baysal, Z.; Kureti, S. CO_2 methanation on Mg-promoted Fe catalysts. *Appl. Catal. B Environ.* **2020**, *262*, 118300–118310. [CrossRef]
46. Xu, L.; Wang, F.; Chen, M.; Nie, D.; Lian, X.; Lu, Z.; Chen, H.; Zhang, K.; Ge, P. CO_2 methanation over rare earth doped Ni based mesoporous catalysts with intensified low-temperature activity. *Int. J. Hydrogen Energy* **2017**, *42*, 15523–15539. [CrossRef]
47. Wang, Y.; Yao, L.; Wang, Y.; Wang, S.; Zhao, Q.; Mao, D.; Hu, C. Low-Temperature Catalytic CO_2 Dry Reforming of Methane on Ni-Si/ZrO_2 Catalyst. *ACS Catal.* **2018**, *8*, 6495–6506. [CrossRef]

48. Sing, K.S.W.; Everett, D.H.; Haul, R.A.W.; Moscou, L.; Pierotti, R.A.; Rouquerol, J.; Siemieniewska, T. Reporting physisorption data for gas/solid systems with special reference to the determination of surface area and porosity. *Pure Appl. Chem.* **1985**, *57*, 603–619. [CrossRef]
49. Thommes, M.; Kaneko, K.; Neimark, A.V.; Olivier, J.P.; Rodriguez-Reinoso, F.; Rouquerol, J.; Sing, K.S.W. Physisorption of gases, with special reference to the evaluation of surface area and pore size distribution (IUPAC Technical Report). *Pure Appl. Chem.* **2015**, *87*, 1051–1069. [CrossRef]
50. Jiang, Y.; Huang, T.; Dong, L.; Qin, Z.; Ji, H. Ni/bentonite catalysts prepared by solution combustion method for CO_2 methanation. *Chin. J. Chem. Eng.* **2018**, *26*, 2361–2367. [CrossRef]
51. Wang, X.; Zhu, L.; Liu, Y.; Wang, S. CO_2 methanation on the catalyst of Ni/MCM-41 promoted with CeO_2. *Sci. Total Environ.* **2018**, *625*, 686–695. [CrossRef] [PubMed]
52. Liu, Y.; Zhu, L.; Wang, X.; Yin, S.; Leng, F.; Zhang, F.; Lin, H.; Wang, S. Catalytic methanation of syngas over Ni-based catalysts with different supports. *Chin. J. Chem. Eng.* **2017**, *25*, 602–608. [CrossRef]
53. Ray, K.; Deo, G. A potential descriptor for the CO_2 hydrogenation to CH_4 over Al_2O_3 supported Ni and Ni-based alloy catalysts. *Appl. Catal. B Environ.* **2017**, *218*, 525–537. [CrossRef]
54. Kim, H.-M.; Jang, W.-J.; Yoo, S.-Y.; Shim, J.-O.; Jeon, K.-W.; Na, H.-S.; Lee, Y.-L.; Jeon, B.-H.; Bae, J.W.; Roh, H.-S. Low temperature steam reforming of methane using metal oxide promoted Ni-$Ce_{0.8}Zr_{0.2}O_2$ catalysts in a compact reformer. *Int. J. Hydrogen Energy* **2018**, *43*, 262–270. [CrossRef]
55. Roh, H.-S.; Jun, K.-W.; Dong, W.-S.; Chang, J.-S.; Park, S.-E.; Joe, Y.U.-I. Highly active and stable Ni/Ce–ZrO_2 catalyst for H_2 production from methane. *J. Mol. Catal. A Chem.* **2002**, *181*, 137–142. [CrossRef]
56. Le, T.A.; Kim, M.S.; Lee, S.H.; Kim, T.W.; Park, E.D. CO and CO_2 methanation over supported Ni catalysts. *Catal. Today* **2017**, *293–294*, 89–96.
57. Wang, Y.; Li, L.; Wang, Y.; Costa, P.D.; Hu, C. Highly Carbon-Resistant Y Doped NiO–ZrO_m Catalysts for Dry Reforming of Methane. *Catalysts* **2019**, *9*, 1055. [CrossRef]
58. Rui, N.; Zhang, X.; Zhang, F.; Liu, Z.; Cao, X.; Xie, Z.; Zou, R.; Senanayake, S.D.; Yang, Y.; Rodriguez, J.A.; et al. Highly active Ni/CeO_2 catalyst for CO_2 methanation: Preparation and characterization. *Appl. Catal. B Environ.* **2021**, *282*, 119581–119592. [CrossRef]
59. Wang, Y.; Zhao, Q.; Wang, Y.; Hu, C.; Da Costa, P. One-Step Synthesis of Highly Active and Stable Ni–ZrO_x for Dry Reforming of Methane. *Ind. Eng. Chem. Res.* **2020**, *59*, 11441–11452. [CrossRef]

Article

Enhanced Carbon Dioxide Decomposition Using Activated SrFeO$_{3-\delta}$

Jaeyong Sim [1,2], Sang-Hyeok Kim [1,2], Jin-Yong Kim [1,2], Ki Bong Lee [2,*], Sung-Chan Nam [1] and Chan Young Park [1,*]

[1] Greenhouse Gas Research Laboratory, Korea Institute of Energy Research, 152 Gajeong-ro, Yuseong, Daejeon 34129, Korea; resumemat@naver.com (J.S.); ksh0630ksh@kier.re.kr (S.-H.K.); tyui15678@kier.re.kr (J.-Y.K.); scnam@kier.re.kr (S.-C.N.)
[2] Department of Chemical and Biological Engineering, Korea University, 145 Anam-ro, Seongbuk-gu, Seoul 02841, Korea
* Correspondence: kibonglee@korea.ac.kr (K.B.L.); cpark@kier.re.kr (C.Y.P.); Tel.: +82-2-3290-4851 (K.B.L.); +82-42-860-3069 (C.Y.P.)

Received: 28 September 2020; Accepted: 30 October 2020; Published: 3 November 2020

Abstract: Today, climate change caused by global warming has become a worldwide problem with increasing greenhouse gas (GHG) emissions. Carbon capture and storage technologies have been developed to capture carbon dioxide (CO_2); however, CO_2 storage and utilization technologies are relatively less developed. In this light, we have reported efficient CO_2 decomposition results using a nonperovskite metal oxide, $SrFeCo_{0.5}O_x$, in a continuous-flow system. In this study, we report enhanced efficiency, reliability under isothermal conditions, and catalytic reproducibility through cyclic tests using $SrFeO_{3-\delta}$. This ferrite needs an activation process, and 3.5 vol% H_2/N_2 was used in this experiment. Activated oxygen-deficient $SrFeO_{3-\delta}$ can decompose CO_2 into carbon monoxide (CO) and carbon (C). Although $SrFeO_{3-\delta}$ is a well-known material in different fields, no studies have reported its use in CO_2 decomposition applications. The efficiency of CO_2 decomposition using $SrFeO_{3-\delta}$ reached ≥90%, and decomposition (≥80%) lasted for approximately 170 min. We also describe isothermal and cyclic experimental data for realizing commercial applications. We expect that these results will contribute to the mitigation of GHG emissions.

Keywords: greenhouse gas; climate change; CO_2 decomposition; CO_2 utilization; $SrFeO_{3-x}$

1. Introduction

Climate change caused by global warming has emerged as a problem worldwide owing to the increase in greenhouse gas (GHG) emissions. Carbon dioxide (CO_2) emissions account for more than 90% of global GHG emissions [1,2]. According to the Intergovernmental Panel on Climate Change report in 2018, CO_2 emissions were estimated to be at 32 and 40 billion tons in 2010 and 2020, respectively [3]. Carbon capture and storage technologies have been developed to capture CO_2 emissions, especially those from power plants [4–6]. However, CO_2 storage is vulnerable to earthquakes and can cause pollution and is, therefore, recognized as only a temporary method [7]. Therefore, there is an urgent need to develop and implement techniques for utilizing captured CO_2. For example, CO_2 reforming of methane has been proposed; however, it has high cost and energy requirements [8,9].

One of the treatment methods for captured CO_2 is catalytic decomposition using oxygen-deficient metal oxides. Tamauara et al. reported that oxygen-deficient magnetite ($Fe_{3+\delta}O_4$, $\delta = 0.127$) decomposed up to 100% of CO_2 and H_2O at 290 °C [10]. Subsequently, CO_2 decomposition using ferrites with divalent metals such as Ni^{2+} and Cu^{2+} was investigated [11]. Mn^{2+}- and Zn^{2+}-based ferrites were also reported as having CO_2 decomposition efficiencies of 66% and 90%, respectively [12,13]. Even trivalent,

Ni-Cu ferrites were tested for an identical purpose [14]. It was reported that nickel and copper substitutions at the A-site of ferrites were the most beneficial for reduction and oxidation reactions and demonstrated meaningful results. However, all these approaches are difficult to apply in practical applications because the experimental data were obtained from a small stagnant batch-type reactor.

The first attempt to go beyond the batch system was made in 2001. Shin et al. reported CO_2 decomposition data obtained through thermogravimetric analysis using activated $CuFe_2O_4$ [11]. Furthermore, Kim et al. investigated CO_2 decomposition using activated $Ni_{0.5}Zn_{0.5}Fe_2O_{4-\delta}$ in a continuous flow of 10% CO_2-balanced N_2 [15,16]. They reported that trivalent ferrites (i.e., $(Ni_xZn_{1-x})Fe_2O_4$, x = 0.3, 0.5, 0.7, and 1) showed a higher CO_2 decomposition efficiency than divalent $NiFe_2O_4$ ferrite. They ascertained that the ferrites could completely decompose 10% CO_2 for 5 to 7 min. They also asserted that Ni/Zn-ferrite synthesized by the hydrothermal method displayed better CO_2 decomposition performance than that synthesized by the coprecipitation method. However, they did not perform a blank test and quantitative analysis. Although their results were elementary and had some weaknesses, their trials were invaluable in that they can be applied in practical applications. Therefore, the accumulated data of CO_2 decomposition in a continuous system should be obtained for realizing economically efficient CO_2 treatment.

In our previous work [17], we demonstrated the possibility of continuous CO_2 decomposition by using oxygen-deficient metal oxides and suggested its reaction mechanism. Compared to Ni-ferrites, a nonperovskite-type metal oxide (i.e., $SrFeCo_{0.5}O_x$) was much more effective for CO_2 decomposition: Ni-ferrites decomposed only up to 20% of CO_2, whereas $SrFeCo_{0.5}O_x$ displayed a CO_2 decomposition efficiency of up to 90%. These results were obtained based on our suggested mechanism that high electrical and ionic conductivities affect CO_2 decomposition. Currently, the obtainment of suitable isothermal and regeneration data will be more helpful for practical applications. In our ongoing research project, we have found that another material, $SrFeO_{3-\delta}$, shows greater promise for this purpose.

Originally, $SrFeO_{3-\delta}$ was used as an oxygen transport material [18–23] and as a catalyst for methane combustion and chemical looping processes [24,25]. Perovskite-type $SrFeO_{3-\delta}$ ($0 \leq \delta \leq 0.5$) is a nonstoichiometric metal oxide containing Fe ions in a mixed valence, such as Fe^{4+} and Fe^{3+} [26]. Under reducing conditions, $SrFeO_{3-\delta}$ produces oxygen vacancies; the number of oxygen vacancies depends on the temperature and the oxygen partial pressure [27]. Recently, Marek et al. reported the stable use of $SrFeO_{3-\delta}$ in chemical looping systems; the material reduced above $\delta = 0.5$ could be reoxidized with either CO_2 or air, resulting in $SrFeO_{3-\delta}$ ($0 \leq \delta \leq 0.5$) [25]. Therefore, we consider it a promising material for CO_2 decomposition. Several studies have reported on the use of $SrFeO_{3-\delta}$ in various fields. However, no study has reported the use of $SrFeO_{3-\delta}$ for CO_2 decomposition in a continuous-flow system. In this report, we describe the reduction behavior and redox reaction of $SrFeO_{3-\delta}$. Furthermore, through cyclic experiments, we demonstrate that it exhibits consistently high CO_2 decomposition performance under isothermal conditions. We also demonstrate its structural stability as a catalytic material for practical applications.

2. Results

2.1. Characterization

The crystal structure of $SrFeO_{3-\delta}$ was analyzed by X-ray powder diffraction (XRD) at 40 kV and 200 mA. The XRD powder patterns of the samples were obtained in 0.02° steps over the range of $20° \leq 2\theta \leq 80°$. It has been reported that the structure of $SrFeO_{3-\delta}$ could be changed susceptibly by δ values, namely, cubic at δ = 0–0.12, tetragonal at δ = 0.16–0.24, and orthorhombic at δ = 0.25 [28,29]. The lattice constant of $SrFeO_{3-\delta}$ obtained from XRD data was a = 5.479(5) Å, b = 7.729(8) Å, c = 5.521(2) Å, and V = 233.8(5) Å^3. It was determined to be an orthorhombic perovskite, which is in good agreement with the reported value (PDF# 01-077-9154). Figure 1a shows XRD powder patterns of as-synthesized $SrFeO_{3-\delta}$. We also obtained secondary electron images. These are discussed with those measured after CO_2 decomposition tests at the end of this section. The chemical composition was reasonably acceptable, and the surface area of $SrFeO_{3-\delta}$ was determined to be 3.19 m^2/g.

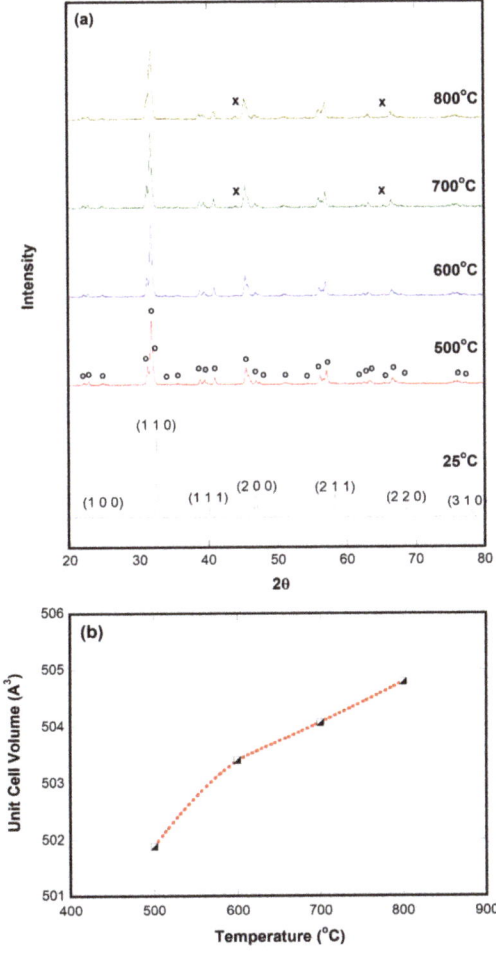

Figure 1. In-situ XRD results of $SrFeO_{3-\delta}$: (**a**) In-situ XRD powder pattern and (**b**) unit cell volume at $500 \leq T \leq 800$ °C. The symbols indicate Fe metal (x) and brownmillerite (o).

2.2. Oxygen-Deficient $SrFeO_{3-\delta}$

As discussed in our previous report [17], sample activation is an essential step to decompose CO_2. An understanding of the reduction behavior is also needed because CO_2 decomposition could mainly be affected by the number of oxygen vacancies and their mobility. CO_2 decomposition is induced by the incorporation of O^{2-} into oxygen vacancies. In-situ XRD was performed during high-temperature reduction with 3.5 vol% H_2/Ar to identify structural changes occurring in the metal oxide. Figure 1 includes the in-situ X-ray powder patterns of $SrFeO_{3-\delta}$ obtained at $500 \leq T \leq 800$ °C.

$SrFeO_{3-\delta}$ was observed in the perovskite phase at room temperature, and the pattern was very similar to that of $SrFeO_{2.75}$ (PDF# 01-077-9154). As the temperature increased, the perovskite would lose more oxygen and change to a brownmillerite phase. Phase changes from $SrFeO_{3-\delta}$ to $SrFeO_{2.5}$ at 500 °C are attributed to the partial reduction of Fe^{4+} to Fe^{3+} [30]. An almost pure brownmillerite phase

(SrFeO$_{2.5}$, PDF# 01-070-0836) was observed at 500–600 °C. The XRD pattern at 500 °C was completely indexed with an orthorhombic unit cell with lattice parameter a = 5.69(9) Å, b = 15.80(2) Å, c = 5.57(2) Å, and V = 501.8(0) Å^3. This indicated that the cell volume increased by more than twice via the expansion of one side in the orthorhombic unit cell, especially the b-axis. Sr$_{n+1}$Fe$_n$O$_{3n+1}$ and Fe^0 peaks have been reported to appear when SrFeO$_{2.5}$ was reduced further by increasing temperature and reaction time [25]. SrO and Fe^0 peaks are considered the final products of the reduction. In our patterns, only a trace of Fe metal (PDF# 01-080-3817) peaks appeared at 2θ ≈ 44.1° and 65.4° at ≥700 °C. Typical Fe^0 peaks could be observed at 2θ ≈ 44.0° and 65.3°. The Sr$_3$Fe$_2$O$_{6.14}$ phase might be possible; however, it overlaps with brownmillerite peaks. In addition, based on our thermogravimetric result (not shown), the change in the nonstoichiometric value (δ) was determined to be ~0.8 at 25 °C ≤ T ≤ 800 °C. Figure 1b shows the calculated unit cell volumes as a function of temperature. They demonstrate linearity over 600 °C [31]. It has been noted that SrFeO$_{3-\delta}$ should be activated without complete structural collapse. If these phase changes are reversible, it would be beneficial for catalyst redox reactions or in a chemical looping system.

2.3. Effect of Conductivity on CO$_2$ Decomposition

As we suggested a mechanism in our previous report, CO$_2$ decomposition is considered to be affected significantly by the conductivity of the sample [17]. To decompose CO$_2$ effectively, the metals in the ferrite should easily provide electrons for CO$_2$; therefore, the electronic conductivity plays an important role at first. As the oxygen in neutral CO$_2$ has sufficient electrons, electron transfer from metals on the surface of the ferrite to CO$_2$ is not likely to occur naturally. Therefore, the activation process should be performed before exposure to CO$_2$, and the produced oxygen vacancies become the driving force of the redox reaction. Based on such reasoning, the amount of oxygen vacancies could be the most important factor in this reaction. In addition, the activation process would be easier if the metal oxide contained metals with variable oxidation number.

Once the oxygen ions (i.e., O^{2-}) fill the oxygen vacant sites on the sample surface, the ability to decompose CO$_2$ to CO or C is lost owing to the saturation (or deactivation) of the sample surface. However, if oxygen ions migrate well inside the lattice and oxygen vacancies are reformed on the sample surface, CO$_2$ decomposition could be continued until all vacancies are filled. This is why oxygen ionic conductivity should also be considered. It would be interesting to determine how much of a role oxygen ion conductivity plays in the decomposition of carbon dioxide; however, this will be reported in a separate paper. After all, the oxygen ions accepted through the electronic conductivity effect could move to inside defects via oxygen ionic conducting properties. Therefore, samples with good electrical and ionic conductivity would decompose CO$_2$ more effectively. The total conductivity of SrFeO$_{3-\delta}$ is 31.6 S cm^{-1} at 800 °C [32], which is higher than that of SrFeCo$_{0.5}$O$_x$ (17 S cm^{-1}) [33]. The total conductivity of SrFeO$_{3-\delta}$ shows good agreement with the reference value, and it was determined to be 33.9 S cm^{-1} at 800 °C from our own measurement. This feature was the reason that SrFeO$_{3-\delta}$ was selected for the CO$_2$ decomposition experiment in this paper.

2.4. CO$_2$ Decomposition

To the best of our knowledge, only two studies have reported CO$_2$ decomposition in a continuous gas-flow reactor before our previous report [17]. One [11] provided TGA measurement results, and the other [15] provided extremely limited information. We have used several metal oxides for CO$_2$ decomposition experiments in a continuous-flow reactor. In our previous report, we reported CO$_2$ decomposition with SrFeCo$_{0.5}$O$_x$ using data obtained under nonisothermal conditions. Even if nonisothermal data are insufficient to cover all practical applications, they can serve as a cornerstone to determine the most economically efficient temperature region for CO$_2$ decomposition. Furthermore, temperature fluctuates during both activation and decomposition processes. Considering these applications, we performed nonisothermal tests, isothermal tests, and cyclic experiments for CO$_2$

decomposition with a noncobalt metal oxide, $SrFeO_{3-\delta}$. A comparison of nonisothermal data for $SrFeCo_{0.5}O_x$ and $SrFeO_{3-\delta}$ is presented below.

Nonisothermal CO_2 decomposition: Figure 2 shows a comparison of the results of nonisothermal CO_2 decomposition using $SrFeO_{3-\delta}$ and $SrFeCo_{0.5}O_x$ for temperatures ranging between 25 and 800 °C. Data for $SrFeCo_{0.5}O_x$ were extracted from our previous report [17], and the same experimental conditions were applied. Initially, we started CO_2 decomposition with $NiFe_2O_4$ as an oxygen-deficient ferrite; however, it decomposed only up to 20% of CO_2 in the continuous gas-flow system. We obtained a ~90% efficiency of CO_2 decomposition using $SrFeCo_{0.5}O_x$ selected based on our proposed mechanism. Here, we demonstrated several enhanced CO_2 decomposition results obtained by using $SrFeO_{3-\delta}$. First, $SrFeO_{3-\delta}$ could be activated at a much lower temperature and for shorter duration. $SrFeO_{3-\delta}$ was primarily activated at $280 \leq T \leq 600$ °C, as shown in Figure 2a. The H_2 concentration during reduction decreased rapidly up to ≈460 °C, which indicated the phase changes from perovskite to brownmillerite. This behavior can be confirmed from the in-situ XRD data shown in Figure 1. It is believed that oxygen vacancies are created the most at these temperatures. Second, the amount of CO_2 decomposed using $SrFeO_{3-\delta}$ is approximately 2.2 times higher than that decomposed using $SrFeCo_{0.5}O_x$ based on the calculated result of ≥50% CO_2 decomposition, as shown in Figure 2b. As the temperature increased, the CO_2 decomposition efficiencies of both metal oxides increased by up to ~90%. After reaching 800 °C, the CO_2 decomposition efficiency of $SrFeCo_{0.5}O_x$ decreased, whereas the high decomposition efficiency of $SrFeO_{3-\delta}$ was maintained over 100 min. This indicates that $SrFeO_{3-\delta}$ might be a more appropriate material for CO_2 decomposition than $SrFeCo_{0.5}O_x$. The amount of CO produced using $SrFeO_{3-\delta}$ was also slightly higher than that produced using $SrFeCo_{0.5}O_x$.

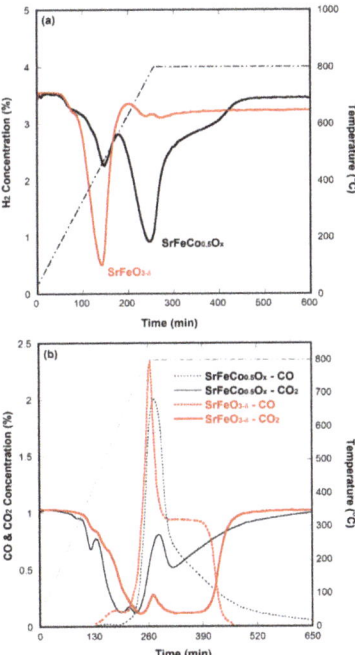

Figure 2. CO_2 decomposition results: (a) Consumed H_2 concentration and (b) CO_2 and CO concentrations during sample activation and CO_2 decomposition tests, respectively. The black lines indicate the results of $SrFeCo_{0.5}O_x$ extracted from our previous work [17], and the straight dotted lines indicate temperature profiles (i.e., 3 °C/min).

The CO_2 decomposition ability could be expressed in units of millimoles of decomposed CO_2 and generated CO per gram of sample loaded (i.e., mmol g^{-1}). This calculation was made using several assumptions. For example, the final decomposition was determined to be the point at which CO_2 decomposition ceased, resulting in the revelation of the initial CO_2 concentration. The final CO_2 decomposition time was determined during the point at which $SrFeO_{3-\delta}$ started to decompose at 200 °C and continued for over 4 h, even reaching 800 °C. This calculation was made using the decomposition time limit that ranged between 54 and 500 min (see Figure 2b). Secondly, in spite of nonisothermal CO_2 decomposition, the ideal gas law was used to calculate the decomposed amount (i.e., mmol g^{-1}) of input CO_2. The exact same conditions were applied for the $NiFe_2O_{3-\delta}$ and $SrFeCo_{0.5}O_x$ samples. The results are summarized in Table 1. Both CO_2 decomposition and CO generation using a perovskite ($SrFeO_{3-\delta}$) demonstrated enhanced performance compared to those using a spinel ($NiFe_2O_{3-\delta}$) or a nonperovskite ($SrFeCo_{0.5}O_x$). In addition, $SrFeO_{3-\delta}$ is a cobalt-free compound that is economical and environmentally friendly. Generally, cobalt-containing metal oxides display good catalytic behavior but have several shortcomings, such as structural instability even at intermediate operating temperatures (500 to 800 °C) in a long-term test [34]. In the case of $NiFe_2O_{3-\delta}$, other shortcomings were observed, such as too long an oxidation time.

Table 1. The rate of decomposed CO_2 and generated CO in nonisothermal experiments.

Sample	Decomposed CO_2 (mmol/g)	Produced CO (mmol/g)	Reference
$NiFe_2O_{4-\delta}$	2.25	2.89	[17]
$SrFeCo_{0.5}O_x$	2.35	2.65	[17]
$SrFeO_{3-\delta}$	3.30	2.95	This work

Isothermal CO_2 decomposition: For practical applications, data for CO_2 decomposition using $SrFeO_{3-\delta}$ at constant temperature should be determined. Figure 3 shows the isothermal results. The measurements were performed at 500, 600, 625, 650, 700, and 800 °C. The temperatures for sample activation and decomposition were identically controlled, and a blank test was also performed in the same reactor. GC was used to determine data points every ~4 min after switching the gas with 1 vol% CO_2/He. Fresh powder samples were used in each measurement. As the temperature increased, the amounts of CO_2 decomposition and CO production increased (see Figure 3a,b). This is probably attributable to the amount and high mobility of oxygen vacancies at higher temperatures. The ionic conductivities are proportional to the mobility of perovskite metal oxides (i.e., $\sigma = n \times e \times \mu$, where σ is the specific conductivity; n, the number of charge carriers of a species; e, its charge; and μ, its mobility) and generally increases with the temperature [35].

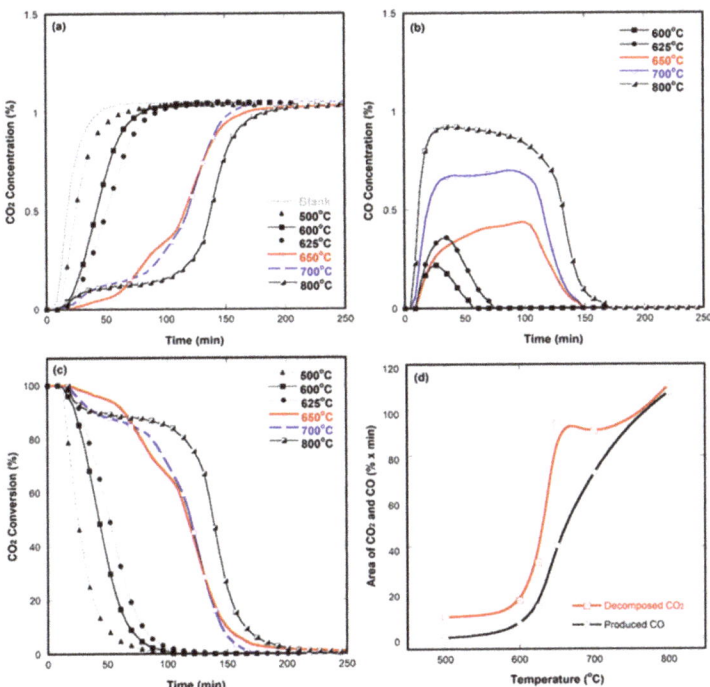

Figure 3. Isothermal CO_2 decomposition results obtained using $SrFeO_{3-\delta}$ at various temperatures: (**a**) Decomposed CO_2 concentration, (**b**) produced CO concentration, (**c**) CO_2 conversion as a function of time, and (**d**) decomposed CO_2 and produced CO area extracted from (**a**) and (**b**).

The CO_2 decomposition results between 600 and 700 °C were noteworthy. As the operating temperature increased from 625 to 650 °C, the amount of CO_2 decomposition doubled. We performed in-situ XRD and TGA experiments to analyze this unusual behavior in this temperature region (not shown). However, no special structural phase or weight changes were seen in the sample activation process. The amounts of consumed hydrogen for sample activation and cell parameters also demonstrated no considerable difference. The reason for the sudden increase in CO_2 decomposition upon increasing temperature by only 25 °C remains unclear. We presume that the thermal energy at 650 °C might boost CO_2 decomposition and the reverse Boudouard reaction (i.e., $C(s) + CO_2 \rightarrow 2CO$). The mobility increase caused by the thermal energy might be an important factor because other factors, such as the unit cell volume, oxygen ion vacancy concentration, and weight change from TGA, did not change abruptly. These issues are discussed further by comparing sample characteristics before and after performing measurements in Section 2.5.

Based on the obtained data, CO_2 conversion rates were calculated using Equation (1).

$$CO_2 \text{ Conversion } (\%) = \frac{CO_2 \text{ In} - CO_2 \text{ Out}}{CO_2 \text{ In}} \times 100 \qquad (1)$$

Although Figure 3c illustrates the same data in the same format as Figure 3a, the conversion degree is easier to distinguish from the CO_2 conversion plot. Furthermore, ≥90% of CO_2 conversion lasted for ≈65 min at 650 °C. This drastic change was much more evident from the area plots of the decomposed CO_2 and produced CO shown in Figure 3d. We calculated these areas by subtracting those obtained in the isothermal blank tests. It should be noted that the area for CO_2 (i.e., amount of CO_2 decomposition) was unusually high at 650 °C. It was even slightly higher than that at 700 °C. Further, CO production increased rapidly until the temperature was increased up to 800 °C. The shape of the isothermal CO_2 decomposition curve at 650 °C also slightly differed from those of the others. We plan to analyze these behaviors using temperature-programed reduction and temperature-programed oxidation in a separate paper.

2.5. Stability Tests

Irrespective of how high the efficiency is, samples should have reproducibility and long-term stability, especially in severe redox reactions. This is the most important criterion for practical applications. In this report, we performed CO_2 decomposition in five reproducibility tests. We also performed cyclic CO_2 decomposition tests with partially activated $SrFeO_{3-\delta}$; however, we did not include the data, owing to their overlapped content. After these stability tests, the changes in the structural behavior and surface morphology of the used $SrFeO_{3-\delta}$ were investigated.

Figure 4 shows the results of the five cyclic reproducibility tests for CO_2 decomposition using $SrFeO_{3-\delta}$ at 700 °C. The sample was activated for 800 min in each cycle. The CO_2 decomposition of each cycle was subtracted from the amount of the blank test. The measurement data of the last cycle (i.e., the fifth cycle) did not perfectly match those of the first. However, they were reasonably close, and the difference could be attributed to the annealing effect of lasting temperatures. Average amounts of decomposed CO_2 and generated CO were determined to be 1.38 and 1.02 mmol per gram of catalyst, respectively. This result indicated a certain level of coke generation or possible adsorption of part of the carbon dioxide instead of decomposition. Although $SrFeO_{3-\delta}$ demonstrated good reproducibility even after repeating the redox experiment several times, it still took too long to activate the sample. In these days, we are using coke oven gas from the steel industry for sample activation. As it contains 55% to 60% hydrogen, the activation time will be much faster. In addition, $SrFeO_{3-\delta}$ impregnated with a small amount of precious metals such as Ru and Rh is tested for low-temperature CO_2 decomposition. Table 2 summarizes the test results.

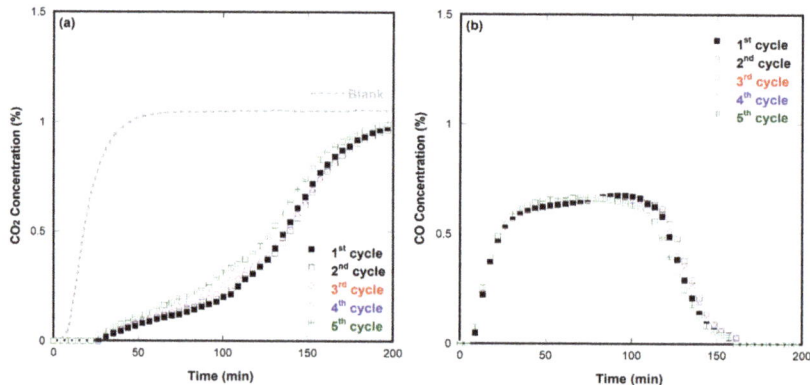

Figure 4. Five cyclic reproducibility tests for CO_2 decomposition using $SrFeO_{3-\delta}$ at 700 °C: (a) Decomposed CO_2 concentration and (b) produced CO concentration.

Table 2. Results of isothermal cyclic experiments with $SrFeO_{3-\delta}$.

Temperature (°C)	Cycle (Number)	Decomposed CO_2 (mmol/g)	Produced CO (mmol/g)	Cell Parameters (Å)
700	1	1.40	1.00	
	2	1.40	1.03	a = 5.66(8)
	3	1.39	1.03	b = 15.59(2)
	4	1.42	1.04	c = 5.53(4)
	5	1.29	0.98	V = 489.1 Å^3
	Average	1.38	1.02	

To investigate the structural changes, XRD measurements were performed with the tested SrFeO$_{3-\delta}$ powders. Figure 5a shows the sample used for the nonisothermal CO$_2$ decomposition experiment at up to 800 °C. Figure 5b,d illustrate the powder patterns of SrFeO$_{3-\delta}$ that underwent cyclic tests at 700 and 650 °C, respectively. The brownmillerite phase remained, and it was hard to find impurities, including even traces of Fe-metal peaks. This indicates that the redox reaction is reversible and that SrFeO$_{3-\delta}$ can presumably serve as an excellent reactant. For comparison, the in-situ XRD result extracted from Figure 1 was added in Figure 5c. Three oxidized XRD patterns (i.e., Figure 5a,b,d) shifted to a higher 2θ angle; Figure 5c shows the pattern of the reduced sample. When the sample reduced, the oxygen vacancy concentration increased, resulting in its increased volume. The unit cell volume of reduced SrFeO$_{3-\delta}$ at 700 °C was 504.1 Å^3, and that of the oxidized sample decreased to 489.1 Å^3. Table 3 lists the fully indexed unit cell parameters for SrFeO$_{3-\delta}$ tested at 700 °C.

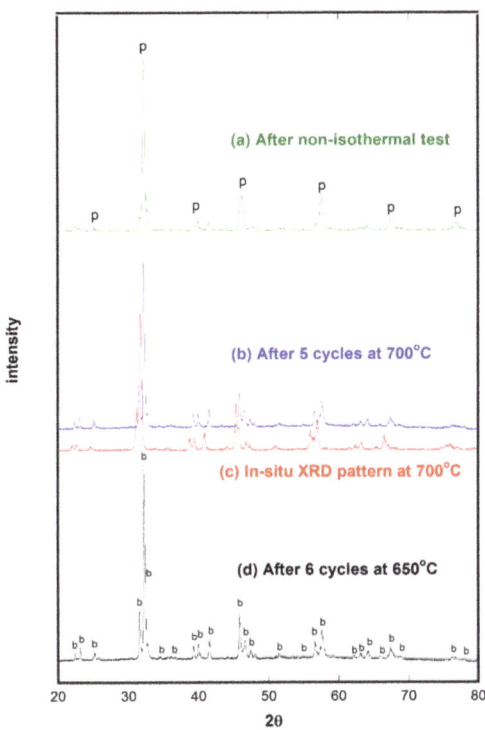

Figure 5. XRD powder patterns of SrFeO$_{3-\delta}$ tested for CO$_2$ decomposition measurements: (**a**) After nonisothermal test, (**b**) after five cycles at 700 °C, (**c**) in-situ XRD at 700 °C, and (**d**) after six cycles at 650 °C. The symbols indicate perovskite (p) and brownmillerite (b).

Table 3. Experimental conditions of nonisothermal and isothermal tests with SrFeO$_{3-\delta}$.

Type	Temp. (°C)	Source	Conditions
Nonisothermal	25–800	Figure 2	
Isothermal	500	Figure 3	Activation: 3.5 vol% H$_2$/N$_2$
	600		CO$_2$ decomposition: 1 vol% CO$_2$/He
	625		Flow rate: 50 mL/min
	650		Ramp rate: 3 °C/min
	700		
	800		

It should be noted that the powder pattern of the sample tested at 800 °C shows a mixed phase, that is, perovskite and brownmillerite, whereas the sample tested at ≤700 °C primarily shows a brownmillerite phase. This result is relevant to the increased amounts of decomposed CO_2 at 800 °C (see Figure 3a). For a perovskite phase to exist at 800 °C, more oxygen vacancies need to be filled. This condition is induced by CO_2 decomposition. It should also be noted that three oxidized samples were slightly reduced with N_2 because 1 vol% CO_2/He gas was switched with N_2 after decomposition tests when cooling down to room temperature.

The microstructure of $SrFeO_{3-\delta}$ was examined using SEM before and after CO_2 decomposition experiments. Figure 6a shows a secondary electron image of pristine $SrFeO_{3-\delta}$, indicating that many small-sized (≤100 nm) particles are attached and dispersed on bigger ones. These small particles grew twice as large after the redox tests at 650 °C, and some large particles appeared to aggregate to each other (see Figure 6b). Even larger agglomerates developed and were observed in the sample tested at 700 °C, as shown in Figure 6c. Furthermore, the agglomerated particles (even ≥1 µm) displayed distinct grain boundaries. As the activity of samples is generally believed to depend on their surface area, a detailed microstructural investigation will be conducted in a separate study.

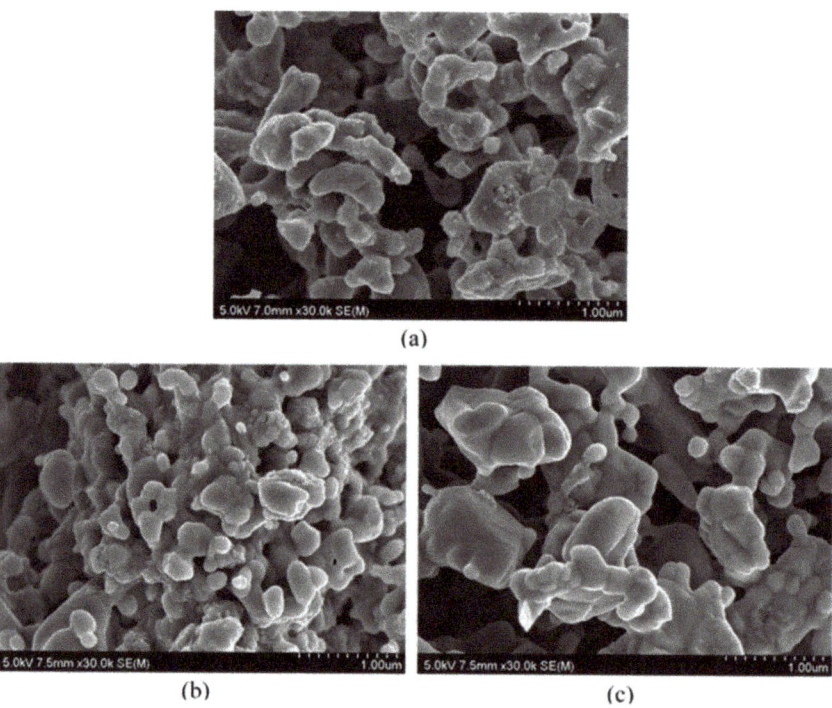

Figure 6. SEM images of $SrFeO_{3-\delta}$: (**a**) Before the redox test, (**b**) after six cycles of redox measurements at 650 °C, and (**c**) after five cycles of reproducibility tests at 700 °C.

3. Materials and Methods

3.1. $SrFeO_{3-\delta}$ Preparation and Characterization

$SrFeO_{3-\delta}$ perovskite-type metal oxide was prepared using solid-state synthesis for use as a CO_2 decomposition material. The metal oxide was purchased from K-ceracell Co. Ltd. (Taejeon, Korea), and $SrCO_3$ (Alfa Aesar, >99%, Ward Hill, MA, USA) and Fe_2O_3 (Alfa Aesar, >99.9%, Ward Hill, MA, USA) were used as starting materials. After weighing the appropriate amount of materials, the powders were mixed in ethanol (Samchun Chemicals, >99.9%, Seoul, Korea). The mixed powder was ball-milled with zirconia balls (φ3–5 mm) for 48 h and then dried to remove the solvent. $SrFeO_{3-\delta}$ powder was heated at 1000 °C for 3 h in ambient air. The final powders were obtained by ball-milling with ethanol for 24 h and then drying in an oven at 80 °C for 24 h.

The synthesized powders were characterized using X-ray powder diffraction (XRD, Rigaku D/MAX-2500, Tokyo, Japan), scanning electron microscopy (SEM, S-4800 Hitachi, Hitachi, Japan), and energy-dispersive X-ray spectroscopy (EDS, Thermo Scientific, Waltham, MA, USA) to analyze the phase structure and purity, surface morphology, and chemical composition, respectively. In-situ XRD (Rigaku D/MAX-2500, Tokyo, Japan) and thermogravimetric analysis (TGA, TA Instruments, Milford, MA, USA) were used to investigate the reduction behaviors of the synthesized powder. In-situ XRD and the change in weight were measured with $\approx$100 mL min^{-1} of 3.5 vol% H_2/Ar.

3.2. CO_2 Decomposition Experiments

We used a continuous-flow reactor to investigate the products of the CO_2 decomposition reaction in the tests. Sample activation and CO_2 decomposition were performed with 3.5 vol% H_2/N_2 and 1 vol% CO_2/He, respectively. Approximately 1.5 g of sample powder with zirconia balls (φ2–3 mm, 10 g) was placed at the center of a quartz tube (I.D.: 12 mm, O.D.: 16 mm, height: 600 mm). The flow rate for the sample activation and CO_2 decomposition was 50 mL min^{-1}. The CO_2 decomposition experiments were performed under increasing temperature at $25 \leq T \leq 800$ °C (nonisothermal). The ramp rate was 3 °C min^{-1}. The same measurements were performed at certain fixed temperatures between 500 and 800 °C (isothermal). The residence time was calculated to be 3.393 s. CO_2 decomposition cycle tests were also performed to check the regeneration, reproducibility, and stability of the sample. The temperature was selected with the results based on the isothermal CO_2 decomposition results. Five cyclic tests under a fixed condition were performed at 700 °C. The blank tests were carried out under identical conditions by using only zirconia balls instead of the sample powder. Table 3 summarizes the experimental conditions for both activation and decomposition. The produced gas concentrations were measured with an Agilent 8890 gas chromatograph (GC, ShinCarbon ST 100/120 micropacked column, Bellefonte, PA, USA). The details of this experiment have been described in a previous report [27].

4. Conclusions

We demonstrated that $SrFeO_{3-\delta}$ could serve as a promising material for effective CO_2 decomposition in a continuous gas-flow system. In this study, we performed three categorized CO_2 treatment experiments using $SrFeO_{3-\delta}$: nonisothermal, isothermal, and stability tests. In nonisothermal experiments, the maximum CO_2 decomposition rate reached ~90% and the decomposition rate ($\geq$80%) lasted for around 170 min. Both CO_2 decomposition and CO generation by $SrFeO_{3-\delta}$ exhibited better performance than those of $NiFe_2O_{4-\delta}$ or $SrFeCo_{0.5}O_x$. The results of isothermal tests indicated that the optimized temperature for CO_2 decomposition should range between 650 and 700 °C. As the operating temperature increased from 625 to 650 °C, the amount of CO_2 decomposition increased unusually, even though no special structural phase or weight changes occurred. In addition, $SrFeO_{3-\delta}$ maintained the CO_2 decomposition efficiency during isothermal cyclic and reproducibility experiments. Therefore, $SrFeO_{3-\delta}$ is expected to have the capacity to contribute to the mitigation of CO_2, a greenhouse gas. Nonetheless, studies of CO_2 utilization should be continued because many unsolved problems remain in this field.

Author Contributions: Conceptualization, C.Y.P. and S.-C.N.; methodology, J.S.; investigation, S.-H.K. and J.-Y.K.; data curation, C.Y.P. and J.S.; writing—original draft preparation, J.S.; writing—review and editing, K.B.L. and C.Y.P.; supervision, C.Y.P.; project administration, S.-C.N.; funding acquisition, S.-C.N. All authors have read and agreed to the published version of the manuscript.

Funding: The authors would like to acknowledge the financial support of the National Research Foundation under the "Next Generation Carbon Upcycling Project" (Project No. 2017M1A2A2043109) of the Ministry of Science and ICT, Republic of Korea.

Acknowledgments: The analytical support from the Platform Technology Laboratory at Korea Institute of Energy Research is much appreciated, especially the invaluable assistance provided for in-situ XRD and surface analysis.

Conflicts of Interest: The authors declare no conflict of interest.

References

1. Meinshausen, M.; Meinshausen, N.; Hare, W.; Raper, S.C.B.; Frieler, K.; Knutti, R.; Frame, D.J.; Allen, M.R. Greenhouse-gas emission targets for limiting global warming to 2 °C. *Nat. Cell Biol.* **2009**, *458*, 1158–1162. [CrossRef] [PubMed]
2. Edwards, J.H. Potential sources of CO_2 and the options for its large-scale utilization now and in the future. *Catal. Today* **1995**, *23*, 59–66. [CrossRef]
3. Masson-Delmotte, V.; Zhai, P.; Pörtner, H.O.; Roberts, D.; Skea, J.; Shukla, P.R.; Pirani, A.; Moufouma-Okia, W.; Péan, C.; Pidcock, R.; et al. Summary for policymakers: Global warming of 1.5 °C. *IPCC* **2018**, *1*, 1–32.
4. Bergman, P.D.; Winter, E.M. Disposal of carbon dioxide in aquifers in the U.S. *Energy Convers. Manag.* **1995**, *36*, 523–526. [CrossRef]
5. Rubin, E.S.; Rao, A.B. A Technical, Economic and Environmental Assessment of Amine-Based CO_2 Capture Technology for Power Plant Greenhouse Gas Control. In *A Technical, Economic and Environmental Assessment of Amine-Based CO_2 Capture Technology for Power Plant Greenhouse Gas Control*; Office of Scientific and Technical Information (OSTI): Oak Ridge, TN, USA, 2002; Volume 36, pp. 4467–4475.
6. Desideri, U.; Paolucci, A. Performance modelling of a carbon dioxide removal system for power plants. *Energy Convers. Manag.* **1999**, *40*, 1899–1915. [CrossRef]
7. Leung, D.Y.; Caramanna, G.; Maroto-Valer, M.M. An overview of current status of carbon dioxide capture and storage technologies. *Renew. Sustain. Energy Rev.* **2014**, *39*, 426–443. [CrossRef]
8. Frontera, P.; Macario, A.; Ferraro, M.; Antonucci, P. Supported catalysts for CO_2 methanation: A review. *Catalysts* **2017**, *7*, 59. [CrossRef]
9. Mark, M.F.; Maier, W.F. CO_2-Reforming of methane on supported Rh and Ir catalysts. *J. Catal.* **1996**, *164*, 122–130. [CrossRef]
10. Tamaura, Y.; Tahata, M. Complete reduction of carbon dioxide to carbon using cation-excess magnetite. *Nat. Cell Biol.* **1990**, *346*, 255–256. [CrossRef]
11. Shin, H.C.; Choi, S.C.; Jung, K.D.; Han, S.H. Mechanism of M ferrites (M = Cu and Ni) in the CO_2 Decomposition reaction. *Chem. Mater.* **2001**, *13*, 1238–1242. [CrossRef]
12. Tabata, M.; Nishida, Y.; Kodama, T.; Mimori, K.; Yoshida, T.; Tamaura, Y. CO_2 decomposition with oxygen-deficient Mn(II) ferrite. *J. Mater. Sci.* **1993**, *28*, 971–974. [CrossRef]
13. Kodama, T.; Tabata, M.; Tominaga, K.; Yoshida, T.; Tamaura, Y. Decomposition of CO_2 and CO into carbon with active wustite prepared form Zn(II)-bearing ferrite. *J. Mater. Sci.* **1993**, *28*, 547–552. [CrossRef]
14. Shin, H.C.; Oh, J.H.; Lee, J.C.; Han, S.H.; Choi, S.C. The carbon dioxide decomposition reaction with $(Ni_xCu_{1-x})Fe_2O_4$ solid solution. *Phys. Stat. Sol. A* **2002**, *189*, 741–745. [CrossRef]
15. Kim, J.S.; Ahn, J.R. Characterization of wet processed (Ni, Zn)-ferrites for CO_2 decomposition. *J. Mater. Sci.* **2001**, *36*, 4813–4816. [CrossRef]
16. Kim, J.S.; Ahn, J.R.; Lee, C.W.; Murakami, Y.; Shindo, D. Morphological properties of ultra-fine (Ni,Zn)-ferrites and their ability to decompose CO_2. *J. Mater. Chem.* **2001**, *11*, 3373–3376. [CrossRef]
17. Kim, S.H.; Jang, J.T.; Sim, J.; Lee, J.H.; Nam, S.C.; Park, C.Y. Carbon dioxide decomposition using $SrFeCo_{0.5}O_x$, a nonperovskite-type metal oxide. *J. CO2 Util.* **2019**, *34*, 709–715. [CrossRef]
18. Yang, J.; Zhao, H.; Liu, X.; Shen, Y.; Xu, L. Bismuth doping effects on the structure, electrical conductivity and oxygen permeability of $Ba_{0.6}Sr_{0.4}Co_{0.7}Fe_{0.3}O_{3-\delta}$ ceramic membranes. *Int. J. Hydrogen Energy* **2012**, *37*, 12694–12699. [CrossRef]
19. Li, X.; Kerstiens, T.; Markus, T. Oxygen permeability and phase stability of $Ba_{0.5}Sr_{0.5}Co_{0.8}Fe_{0.2}O_{3-\delta}$ perovskite at intermediate temperatures. *J. Membr. Sci.* **2013**, *438*, 83–89. [CrossRef]
20. Kovalevsky, A. Processing and oxygen permeability of asymmetric ferrite-based ceramic membranes. *Solid State Ionics* **2008**, *179*, 61–65. [CrossRef]

21. Leo, A.; Liu, S.; Da Costa, J.C.D. Development of mixed conducting membranes for clean coal energy delivery. *Int. J. Greenh. Gas Control.* **2009**, *3*, 357–367. [CrossRef]
22. Shao, Z.; Haile, S.M. A high-performance cathode for the next generation of solid-oxide fuel cells. *Chemin* **2004**, *35*. [CrossRef]
23. Fuks, D.; Mastrikov, Y.A.; Kotomin, E.A.; Maier, J. Ab initio thermodynamic study of (Ba,Sr)(Co,Fe)O_3 perovskite solid solutions for fuel cell applications. *J. Mater. Chem. A* **2013**, *1*, 14320. [CrossRef]
24. Falcón, H.; Barbero, J.A.; Alonso, J.A.; Martínez-Lope, M.J.; Fierro, J.L.G. SrFeO$_{3-\delta}$ Perovskite oxides: Chemical features and performance for methane combustion. *Chem. Mater.* **2002**, *14*, 2325–2333. [CrossRef]
25. Marek, E.; Hu, W.; Gaultois, M.; Grey, C.P.; Scott, S.A. The use of strontium ferrite in chemical looping systems. *Appl. Energy* **2018**, *223*, 369–382. [CrossRef]
26. Takeda, Y.; Kanno, K.; Takada, T.; Yamamoto, O.; Takano, M.; Nakayama, N.; Bando, Y. Phase relation in the oxygen nonstoichiometric system, SrFeO$_x$ (2.5 ≤ x ≤ 3.0). *J. Solid State Chem.* **1986**, *63*, 237–249. [CrossRef]
27. Xiao, G.; Liu, Q.; Wang, S.; Komvokis, V.G.; Amiridis, M.D.; Heyden, A.; Ma, S.; Chen, F. Synthesis and characterization of Mo-doped SrFeO$_{3-\delta}$ as cathode materials for solid oxide fuel cells. *J. Power Sources* **2012**, *202*, 63–69. [CrossRef]
28. Ji, K.; Dai, H.; Deng, J.; Zhang, L.; Wang, F.; Jiang, H.; Au, C.T. Three-dimensionally ordered macroporous SrFeO$_{3-\delta}$ with high surface area: Active catalysts for the complete oxidation of toluene. *Appl. Catal. A Gen.* **2012**, *425*, 153–160. [CrossRef]
29. Hombo, J.; Matsumoto, Y.; Kawano, T. Electrical conductivities of SrFeO$_{3-\delta}$ and BaFeO$_{3-\delta}$ perovskites. *J. Solid State Chem.* **1990**, *84*, 138–143. [CrossRef]
30. Hodges, J.; Short, S.; Jorgensen, J.; Xiong, X.; Dabrowski, B.; Mini, S.; Kimball, C. Evolution of oxygen-vacancy ordered crystal structures in the perovskite series Sr$_n$Fe$_n$O$_{3n-1}$ (n = 2, 4, 8, and ∞), and the relationship to electronic and magnetic properties. *J. Solid State Chem.* **2000**, *151*, 190–209. [CrossRef]
31. Khare, A.; Lee, J.; Park, J.; Kim, G.Y.; Choi, S.Y.; Katase, T.; Roh, S.; Yoo, T.S.; Hwang, J.; Ohta, H.; et al. Directing oxygen vacancy channels in SrFeO$_{2.5}$ epitaxial thin films. *ACS Appl. Mater. Interfaces* **2018**, *10*, 4831–4837. [CrossRef]
32. Vashuk, V.V.; Kokhanovskii, L.V.; Yushkevich, I.I. Electrical conductivity and oxygen stoichiometry of SrFeO$_{3-\delta}$. *Inorg. Mater.* **2000**, *36*, 79–83. [CrossRef]
33. Ma, B.; Victory, N.I.; Balachandran, U.; Mitchell, B.J.; Richardson, J.W. Study of the mixed-conducting SrFeCo$_{0.5}$O$_y$ system. *J. Am. Ceram. Soc.* **2004**, *85*, 2641–2645. [CrossRef]
34. Park, C.; Lee, T.; Dorris, S.; Park, J.H.; Balachandran, U. Ethanol reforming using Ba$_{0.5}$Sr$_{0.5}$Cu$_{0.2}$Fe$_{0.8}$O$_{3-\delta}$/Ag composites as oxygen transport membranes. *J. Power Sources* **2012**, *214*, 337–343. [CrossRef]
35. West, A.R. *Solid State Chemistry and Its Application*; John Wiley & Sons Ltd.: Hoboken, NJ, USA, 1984.

Publisher's Note: MDPI stays neutral with regard to jurisdictional claims in published maps and institutional affiliations.

© 2020 by the authors. Licensee MDPI, Basel, Switzerland. This article is an open access article distributed under the terms and conditions of the Creative Commons Attribution (CC BY) license (http://creativecommons.org/licenses/by/4.0/).

Article

Efficient Electrochemical Reduction of CO_2 to CO in Ionic Liquid/Propylene Carbonate Electrolyte on Ag Electrode

Fengyang Ju [1], Jinjin Zhang [2] and Weiwei Lu [2],*

[1] School of Food and Drug, Luoyang Normal University, Luoyang 471934, China; jufengyang@lynu.edu.cn
[2] School of Chemical Engineering and Pharmaceutics, Henan University of Science and Technology, Luoyang 471003, China; zhangjinjin_2020@126.com
* Correspondence: luweiwei@haust.edu.cn

Received: 30 July 2020; Accepted: 23 September 2020; Published: 24 September 2020

Abstract: The electrochemical reduction of CO_2 is a promising way to recycle it to produce value-added chemicals and fuels. However, the requirement of high overpotential and the low solubility of CO_2 in water severely limit their efficient conversion. To overcome these problems, in this work, a new type of electrolyte solution constituted by ionic liquids and propylene carbonate was used as the cathodic solution, to study the conversion of CO_2 on an Ag electrode. The linear sweep voltammetry (LSV), Tafel characterization and electrochemical impedance spectroscopy (EIS) were used to study the catalytic effect and the mechanism of ionic liquids in electrochemical reduction of CO_2. The LSV and Tafel characterization indicated that the chain length of 1-alkyl-3-methyl imidazolium cation had strong influences on the catalytic performance for CO_2 conversion. The EIS analysis showed that the imidazolium cation that absorbed on the Ag electrode surface could stabilize the anion radical ($CO_2^{\bullet-}$), leading to the enhanced efficiency of CO_2 conversion. At last, the catalytic performance was also evaluated, and the results showed that Faradaic efficiency for CO as high as 98.5% and current density of 8.2 mA/cm^2 could be achieved at -1.9 V (vs. Fc/Fc$^+$).

Keywords: electrochemical reduction of CO_2; ionic liquids; propylene carbonate; imidazolium cation

1. Introduction

The unprecedented increase of CO_2 concentration in the atmosphere has led to many concerns about global warming, and even predictable environmental disasters, which make us feel urged to limit the CO_2 emission and effectively utilize them [1–4]. The catalytic transformation of CO_2 into value-added chemicals and fuels has been regarded as one of the most promising ways to realize the valorization of CO_2 [5]. In this context, several strategic options, including electrochemical [6–12], thermochemical [13–15], photocatalytic [16–18] and photoelectrochemical [19–22] approaches, have been developed to undertake the CO_2 conversion. Among them, the electrochemical reduction of CO_2 is regarded as the most prospective way, because it allows one to combine with carbon capture and storage technology, and to utilize renewable energy (such as solar energy and wind energy), as inputting energy and water as a reductant to reduce CO_2 into various carbon-based fuels and chemicals (e.g., CO, HCOOH, CH_4, C_2H_4, and CH_3OH) in a modular electrochemical reactor under ambient temperature and pressure [7–11]. However, the linear CO_2 molecule is thermodynamically stable and kinetically inert to be reduced, due to its low electron affinity and large energy gap between its highest occupied molecular orbital (HOMO) and lowest unoccupied molecular orbital (LUMO) [23]. It has been reported that the main hurdle in CO_2 electroreduction lay in the first-step one-electron reduction of CO_2 to form an anion radical ($CO_2^{\bullet-}$), because this activation step requires a much high reduction potential of -1.9 V (vs. NHE) [24,25]. Therefore, to accelerate this process, considerable efforts have been devoted to study the electrocatalysts, because the structure of electrocatalysts provides active sites to activate the reactant of CO_2, and great progress has been witnessed in recent years [26].

While the electrocatalysts are important in research efforts, the electrolyte on the other side plays an equally pivotal role in catalysis, by interacting with the reactant and intermediate species, ultimately influencing the overall reduction reaction [8,27]. In this aspect, ionic liquids (ILs), which are composed of relatively large organic cations and small inorganic anions, have shown their advantages in the reduction of CO_2, and have been extensively studied in recent years [6,7,28–34]. The first consideration is that ILs have a high capacity for CO_2 capture [35,36] and unique electrochemical properties [37–40], such as wide electrochemical windows, high conductivity, and high stability. More importantly, ILs could interact with CO_2 or the reaction intermediate species, and eventually improve the catalytic activity and influence the product selectivity [7,34,41–47]. For example, by using the 18 mol% 1-ethyl-3-methylimidazolium tetrafluoroborate [Emim][BF$_4$] aqueous solution as the cathode electrolyte, Rosen et al. [45] reported that Faradaic efficiency (FE) greater than 96% for CO from CO_2 could be achieved at very low overpotential. In the subsequent work [48], they proposed that the formation of an adsorbed CO_2–[Emim]$^+$ complex provided a low-energy pathway for CO_2 conversion to CO, accounting for the enhanced performance at the presence of ILs. In another ILs assistant CO_2 conversion work, Sun et al. [49] reported that the cation of 1-ethyl-3-methylimidazolium bis[(trifluoromethyl)sulfonyl]imide [Emim][NTf$_2$] played the role of stabilizing effect to prevent the close approach and dimerization between two $CO_2^{\bullet-}$ to form oxalate, and thus switched the C-derived product to CO. Furthermore, they proposed that the formation of an imidazolium carboxylate through the coordination of [Emim]$^+$ with $CO_2^{\bullet-}$ appeared as a feasible pathway for the CO_2 conversion to CO [49]. However, due to the vast number of ILs through different combinations of various cations and anions, an understanding of the interactions between ILs electrolyte and CO_2 molecule or intermediate species, as well as the behind decisive role of ILs is necessary for properly selecting the ILs to enhance the efficiency for CO_2 electroreduction.

Because of the high cost and high viscosity, ILs are often mixed with different molecular solvents of water or organic solvent, and then used as supporting electrolytes, as well as active co-catalysts [50–54]. For water solution, the hurdles for efficient electrochemical conversion of CO_2 stem from the low solubility of CO_2 in aqueous solution, the complicated CO_2 species in water, and the competitive H_2 evolution from H_2O reduction [55]. Compared with water, organic solvents have high solubility of CO_2 and have been alternatively investigated as solvents of ILs for the conversion of CO_2 [10,50,51]. However, these widely investigated organic solvents, such as acetonitrile (AN), N,N-dimethylformamide (DMF), and dimethyl sulfoxide (DMSO), have some severe shortages in practical use. AN is unsuitable for practical use because of its high toxicity and volatility. DMF is also not an appropriate candidate, since it is prone to hydrolysis in water. As for DMSO, owing to its low melting point (18.4 °C), its utilization is highly restricted at ambient conditions. Therefore, the seeking of suitable liquid solvents that can be utilized in CO_2 conversion is still a challenging and urgent task. In recent years, propylene carbonate (PC), as a polar aprotic liquid, has attracted much attention and has been regarded as a promising and "green" sustainable alternative solvent in various chemical and electrochemical transformations fields [56–58]. This is mainly due to its wide electrochemical window and unique physicochemical properties, such as the low toxicity and vapor pressure, as well as the non-corrosive and biodegradable nature of PC. More importantly, PC also has a high capacity of CO_2, which makes PC a promising alternative solvent to overcome the afore-mentioned problem faced in CO_2 conversion. Despite these advantages, PC has only been seldom used as a solvent in the electrocatalytic conversion of CO_2 [59,60].

Inspired by these signs of progress, in this study, the electrocatalytic conversion of CO_2 was performed in imidazolium-based ILs/PC solution, in which the ILs act as active component and electrolyte, and PC as the solvent. First, the onset potential and the main kinetic parameters of CO_2 reduction in PC solution of 1-alkyl-3-methyl imidazolium tetrafluoroborate with different chain length were measured using linear voltammetry (LSV) and Tafel characterization, respectively. Then, the electrochemical impedance spectroscopy (EIS) and equivalent circuit analysis were carried out to investigate the catalytic role of ILs in the course of CO_2 reduction. At last, the catalytic performance in the presence of ILs and tetrabutylammonium tetrafluoroborate salt was evaluated and compared.

2. Results and Discussion

2.1. Reference Electrode Calibration in Non-Aqueous Solution

In this work, we used $Pt/(I^-/I_3^-)$ as the reference electrode. For a facile comparison of electrochemical potentials measured in the different non-aqueous electrolyte solution, the electrode potentials versus I^-/I_3^- were first calibrated using ferrocene/ferrocenium (Fc/Fc^+) redox couple as an internal standard [61], as recommended by International Union of Pure and Applied Chemistry. This calibration method based on the measurement of cyclic voltammograms (CV) at the presence of Fc/Fc^+ redox couple in the corresponding catholyte (0.1 M ILs in PC in this work). The obtained CV curves are shown in Figure 1.

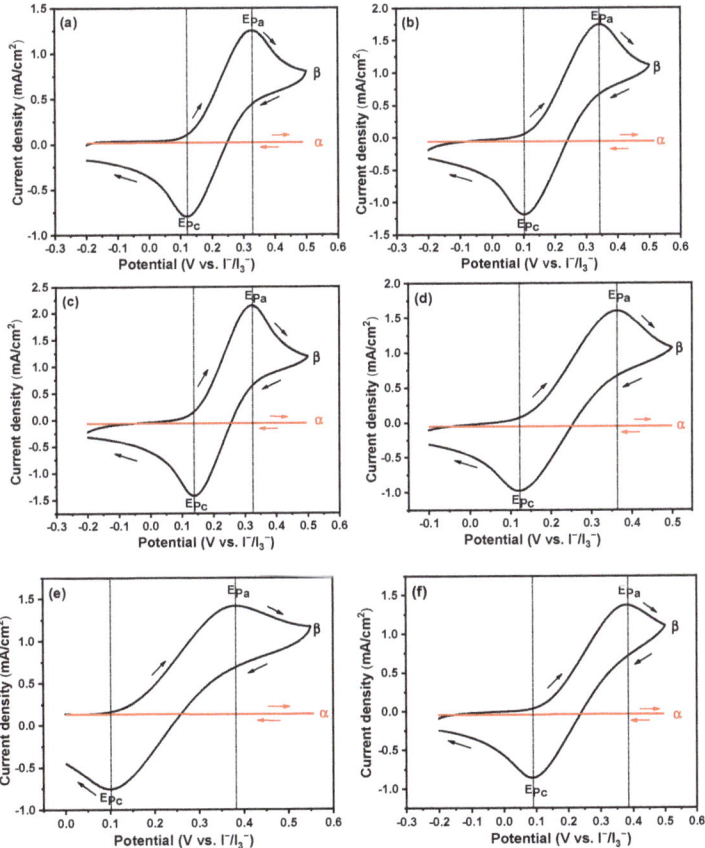

Figure 1. Cyclic voltammetry measured in [Emim]BF_4/PC (**a**), [Bmim]BF_4/PC (**b**) [Hmim]BF_4/PC (**c**), [Omim]BF_4/PC (**d**), [Dmim]BF_4/PC (**e**), [Bu$_4$N]BF_4/PC (**f**); α: blank, β: 10 mM ferrocene; [Emim]BF_4: 1-ethyl-3-methylimidazolium tetrafluoroborate; [Bmim]BF_4: 1-butyl-3-methylimidazolium tetrafluoroborate; [Hmim]BF_4: 1-hexyl-3-methylimidazolium tetrafluoroborate; [Omim]BF_4: 1-octyl-3-methylimidazolium tetrafluoroborate; [Dmim]BF_4: 1-decyl-3-methylimidazolium tetrafluoroborate; and [Bu$_4$N]BF_4: tetrabutylammonium tetrafluoroborate.

For comparison, the CV curve in the commonly used tetrabutylammonium salt [Bu$_4$N]BF$_4$ catholyte solution is also included. As shown in Figure 1, in all cases, the redox peaks are not observed across the CV curves (the red ones) in the absence of ferrocene. Otherwise, typical redox peaks (the black curves) are observed at the presence of 10 mM ferrocene. The obtained half-wave potentials ($E_{1/2} = (E_{Pc} + E_{Pa})/2$) are 0.242 V, 0.231 V, 0.223V, 0.222 V, 0.241 V and 0.242 V in 0.1 M [Emim]BF$_4$/PC, [Bmim]BF$_4$/PC, [Hmim]BF$_4$/PC, [Omim]BF$_4$/PC, [Dmim]BF$_4$/PC and [Bu$_4$N]BF$_4$/PC, respectively. Then, all the potentials in different catholyte solutions are calibrated versus the internal standard if Fc/Fc$^+$ according to the equation: E (vs. Fc/Fc$^+$) = E (vs. I$^-$/I$_3^-$) − $E_{1/2}$ [62].

2.2. Linear Sweep Voltammetry Measurements

Linear sweep voltammograms (LSV) curves are recorded to study the effect of the alkyl chain length of the imidazolium cations on the electroreduction of CO_2, and the results are shown in Figure 2. For comparison, the LSV curve of [Bu$_4$N]BF$_4$ with the same anion BF$_4^-$ as catholyte is also recorded. In agreement with previous reports [49], the onset potential for CO_2 reduction was defined as the potential that results in a current density of 0.6 mA/cm^2. As shown in Figure 2, though all CV curves show similar profiles, the onset potentials shift anodically when ILs replace [Bu$_4$N]BF$_4$ as supporting electrolytes, indicating that ILs play a promoting role in the process of CO_2 electrochemical reduction. Furthermore, it can be observed that the onset potentials change with the alkyl chain length of imidazolium cations. Among these five studied ILs, the [Bmim]BF$_4$ gives the most positive onset potential of −1.702 V (vs. Fc/Fc$^+$), which shifted positively by 133 mV compared with that in [Bu$_4$N]BF$_4$/PC solution (−1.835 V (vs. Fc/Fc$^+$)).

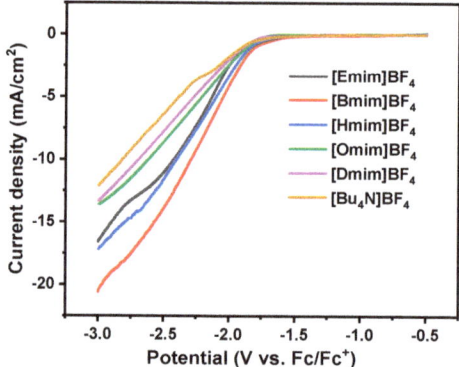

Figure 2. Linear sweep voltammograms (LSV) curves in CO_2-saturated propylene carbonate (PC) solution of ionic liquids (ILs) and [Bu$_4$N]BF$_4$.

2.3. Tafel Analysis

Tafel plots of CO_2 reduction in PC solution with different ILs and [Bu$_4$N]BF$_4$ as electrolytes are further conducted, and the results are shown in Figure 3. It can be determined from Figure 3 that the equilibrium potential of CO_2 reduction in [Emim]BF$_4$/PC, [Bmim]BF$_4$/PC, [Hmim]BF$_4$/PC, [Omim]BF$_4$/PC, [Dmim]BF$_4$/PC and [Bu$_4$N]BF$_4$/PC solution is −1.512 V, −1.275 V, −1.561 V, −1.633 V, −1.672 V, −1.751 V (vs. Fc/Fc$^+$), respectively. In consistency with the results of LSV characterization, the [Bmim]BF$_4$ IL has the higher catalytic promotion effect than other ILs and the [Bu$_4$N]BF$_4$ salt. The equilibrium potential of CO_2 in [Bmim]BF$_4$/PC solution shifted positively 476 mV by comparison with that in [Bu$_4$N]BF$_4$/PC solution. The kinetic parameters of CO_2 reduction in different electrolytes were further calculated according to Equations (1) and (2) [63], and the results are listed in Table 1.

$$\eta = a + b\log|i|, \tag{1}$$

$$a = -2.303RT\log i_0/(n\gamma F), b = 2.303RT/(n\gamma F), \qquad (2)$$

where η is the overpotential, a is the Tafel constant, b is the Tafel slope, γ is the charge transfer coefficient to indicate the symmetry of the energy barrier, i_0 is the exchange current density, n is the number of electrons transferred, F is the Faradaic constant, T is the absolute temperature, and R is the gas constant.

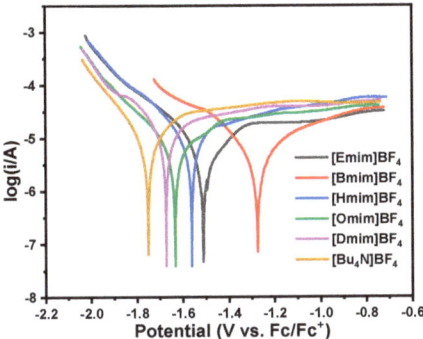

Figure 3. Tafel plots of CO_2 reduction in CO_2-saturated PC solution of ILs and $[Bu_4N]BF_4$.

Table 1. Electrochemical kinetic parameters of CO_2 reduction in $[Emim]BF_4$/PC, $[Bmim]BF_4$/PC, $[Hmim]BF_4$/PC, $[Omim]BF_4$/PC, $[Dmim]BF_4$/PC, and $[Bu_4N]BF_4$/PC electrolyte solution.

Catholyte	a/(V)	b/(V)	γ	i_0/(A/cm^2)
$[Emim]BF_4$/PC	0.899	0.167	0.177	4.168×10^{-6}
$[Bmim]BF_4$/PC	0.791	0.160	0.185	1.14×10^{-5}
$[Hmim]BF_4$/PC	0.872	0.165	0.179	5.19×10^{-6}
$[Omim]BF_4$/PC	0.941	0.176	0.168	4.49×10^{-6}
$[Dmim]BF_4$/PC	0.976	0.181	0.163	4.06×10^{-6}
$[Bu_4N]BF_4$/PC	1.01	0.186	0.159	3.175×10^{-6}

It can be seen from Table 1 that, for Tafel constant a and Tafel slope b, these two parameters in all ILs/PC solution are smaller than that in $[Bu_4N]BF_4$/PC solution. From Equation (2), the smaller value of a and b means that the activation over potential η will also be lower at the same current density. Moreover, the charge transfer coefficient γ and the exchange current density i_0 in all ILs/PC solution are higher than that in $[Bu_4N]BF_4$/PC solution. Therefore, these kinetic parameters derived from Tafel characterization confirm that the electrolytes of imidazolium-based ILs can lower the activation energy, and then is expected to enhance the electrochemical conversion efficiency of CO_2. Furthermore, $[Bmim]BF_4$ electrolyte gives the smallest value of a (0.791 V) and b (0.160 V/dec), and on the other side, has the largest value of γ (0.185) and i_0 (1.14×10^{-5} A/cm^2), which means that $[Bmim]BF_4$ would have the highest catalytic activity among all the studied ILs.

2.4. Electrochemical Impedance Analysis

Based on LSV and Tafel characterization, it has been confirmed that ILs, especially $[Bmim]BF_4$, showed an enhanced ability for activating CO_2 in the electrochemical conversion process. To further study the underlying role of ILs in CO_2 reduction on the Ag electrode, we further performed electrochemical impedance spectroscopy (EIS) analysis. The EIS results in different ILs/PC and $[Bu_4N]BF_4$/PC solutions are shown in Figure 4. As shown in Figure 4a, Nyquist plot in $[Bu_4N]BF_4$/PC electrolyte shows a semicircle in the region of high frequency, which can be ascribed to the Faradaic electron transfer process, and a nearly straight line can be observed in the low-frequency region, which resulted from the diffusion-control process. The deviation of the line from the typical 45° may be due to the rough surface of the Ag electrode. However, different from that in the [Bu$_4$N]BF$_4$/PC electrolyte, the Nyquist plots in all ILs/PC electrolytes (Figure 4b) show an additional big semicircle. The apparent differences of the Nyquist plot between ILs/PC systems and [Bu$_4$N]BF$_4$/PC electrolyte mean that the reduction mechanism of CO$_2$ is changed when IL is used as the electrolyte.

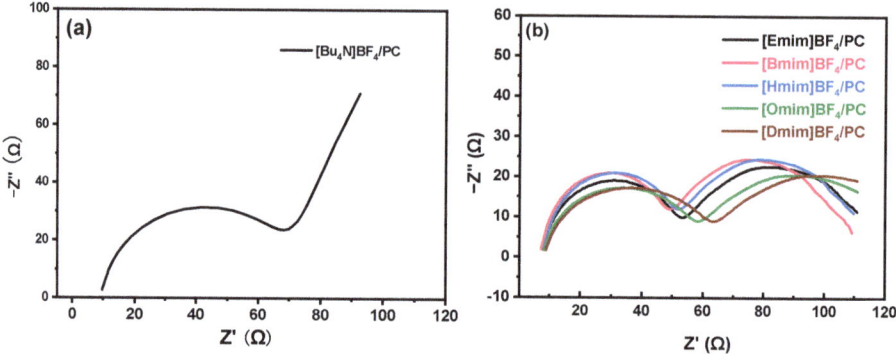

Figure 4. Nyquist plots of electrochemical impedance spectroscopy (EIS) in 0.1 M CO$_2$-saturated [Bu$_4$N]BF$_4$/PC (**a**) and different ILs/PC (**b**) electrolyte solutions.

The equivalent circuits of EIS results were further derived and were shown in Figure 5. Figure 5a is a typical Randles circuit that can well feature the Nyquist plot in [Bu$_4$N]BF$_4$/PC electrolyte (see Figure 4a), in which R_s is the solution resistance, R_{ct} is the charge transfer resistance, C is the double-layer capacitance, and W is the Warburg impedance. The diagram in Figure 5b represented the equivalent circuit of the Nyquist plot in ILs/PC solution. The presence of two semicircles in the Nyquist plot (see Figure 4b) means that there exist two different time constants. Since each time constant is related to an RC component [64], an additional RC component should be added in the equivalent circuit. When ILs/PC are used as the electrolyte in CO$_2$ reduction, CO$_2$ molecules have to pass through the absorbed ILs film layer to reach electrode surface, and then anticipate into an electroreduction reaction. Therefore, the additional RC circuit should be connected in parallel with another RC circuit [64]. So, in the equivalent circuit diagram of Figure 5b, the R_1 represents the migration resistance of CO$_2$, through the adsorbed layer of IL. In addition, the second semicircle in Figure 4b is an arc and deviated obviously from the semicircular trajectory. This deviation is due to a dispersion behavior of the real capacitive component. Therefore, a constant phase element (CPE) has been commonly introduced and defined as in Equation (3) [65,66]:

$$Z_{(CPE)} = (j\omega)^{-n}/Y_0 \tag{3}$$

where Y_0 is the constant phase coefficient, n is the dispersion coefficient (if $n = 0$, it is equivalent to a resistance; if $n = 1$, it is equivalent to a capacitance), j is the imaginary unit, and ω is angular frequency.

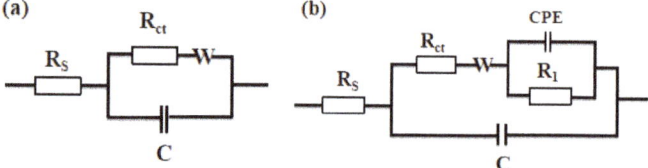

Figure 5. The equivalent circuit diagram of CO$_2$ in [Bu$_4$N]BF$_4$/PC (**a**), [Bmim]BF$_4$/PC (**b**) electrolyte. R_s is the solution resistance, R_{ct} is the charge transfer resistance, C is the double layer capacitance, W is the Warburg impedance, R_1 is the migration resistance of CO$_2$ through the ionic liquid adsorption layer, and constant phase element (CPE) is the constant phase element.

According to the equivalent circuit shown in Figure 5, the impedance parameters are further obtained, and the results are shown in Table 2. For example, in [Bmim]BF$_4$/PC solution, the value of the dispersion coefficient n of the CPE is 0.8325. This means its deviation from a pure capacitance, indicating that [Bmim]BF$_4$ formed a film layer on the surface of the Ag electrode and then caused a dispersion behavior [67].

Table 2. Parameter values of equivalent circuit components.

Catholyte	$R_s/(\Omega)$	$R_{ct}/(\Omega)$	n	$R_1/(\Omega)$
[Bu$_4$N]BF$_4$/PC	10.02	59.13	-	-
[Emim]BF$_4$/PC	7.839	49.71	0.8130	58.87
[Bmim]BF$_4$/PC	6.069	43.07	0.8325	49.46
[Hmim]BF$_4$/PC	7.950	47.32	0.8362	57.62
[Omim]BF$_4$/PC	8.323	53.38	0.7635	89.50
[Dmim]BF$_4$/PC	8.905	55.09	0.7825	94.26

It can be seen from Table 2 that the R_{ct} values of all studied ILs/PC systems are also smaller than that of [Bu$_4$N]BF$_4$/PC solution, confirming the promotion effect of ILs on the electrochemical reduction of CO_2. This promotion effect of ILs/PC system is expected because of the catalytic role of the absorbed IL for the activation of CO_2 through the complexation and stabilization of the $CO_2^{\bullet-}$ radical anion [7,31,47]. In addition, the R_{ct} values of ILs/PC systems have the following trend: [Bmim]BF$_4$/PC (43.07 Ω) < [Hmim]BF$_4$/PC (47.32 Ω) < [Emim]BF$_4$/PC (49.07 Ω) < [Omim]BF$_4$/PC (53.38 Ω) < [Dmim]BF$_4$/PC (55.09 Ω). Among all the studied ILs with different chain lengths, [Bmim]BF$_4$ has the smallest R_{ct} value, indicating its highest catalytic effect for CO_2 reduction. This result of EIS characterization is consistent with that obtained by LSV and Tafel analyses.

Furthermore, from the above electrochemical characterizations, it was found that, although the exact promotion effect of ILs on the electrochemical reduction of CO_2 is not very clear, previous reports have generally indicated that the ILs adsorbed on the electrode can complex with $CO_2^{\bullet-}$ reactive intermediate [45,46,54,68–71], resulting in the reduction of the overpotential and the ultimate facilitation of the CO_2 conversion. From this point of view, the adsorption behavior of ILs on the electrode can influence the interaction between the imidazolium cation and $CO_2^{\bullet-}$, and plays a decisive role in the course of CO_2 electrochemical conversion [7,29,47]. However, the adsorption behavior (i.e., the adsorption strength, the spatial structure, and the density) is influenced by many factors, such as the material and the potential of electrode, the solvent, the concentration of ILs, and the type of anion. In our study, when decreasing the chain length at N1-position of imidazolium cation from octyl to butyl, the catalytic activities of ILs increase, which can be ascribed to the higher adsorbed quantity with a lower steric hindrance of shorter chain length. However, further deceasing the chain length from butyl to ethyl, the adsorbed [Emim]$^+$ may further increase, and cause the film layer too dense to let CO_2 molecules diffuse across, which, on the contrary, is detrimental to CO_2 conversion.

2.5. The Catalytic Performance and the Catalytic Mechanism

After the electrochemical characterizations, we then evaluated the CO_2 conversion performances in [Bu$_4$N]BF$_4$/PC and [Bmim]BF$_4$/PC electrolyte solution, and the results of Faradaic efficiency (FE) and current density are shown in Figure 6. It can be seen from Figure 6a that the current density in both [Bmim]BF$_4$/PC and [Bu$_4$N]BF$_4$/PC electrolytes increase as the applied potential decrease, and at each potential, the current density enhanced when ILs of [Bmim]BF$_4$ replace [Bu$_4$N]BF$_4$ as the electrolyte. For example, at -1.90 V (vs. Fc/Fc$^+$), the current density in [Bmim]BF$_4$/PC (8.2 mA/cm^2) was about three times that in the traditional [Bu$_4$N]BF$_4$/PC (2.7 mA/cm^2). The FE results in Figure 6b show that the FE of CO in [Bmim]BF$_4$/PC is much higher than that in [Bu$_4$N]BF$_4$/PC at each applied potential. Correspondingly, the FE of byproduct H_2 is reduced when the [Bmim]BF$_4$/PC is alternatively used as the electrolyte solution. More importantly, it can be seen that the FE as high as 98.5% can be obtained in [Bmim]BF$_4$/PC solution when the applied potential is at −1.90 V (vs. Fc/Fc$^+$).

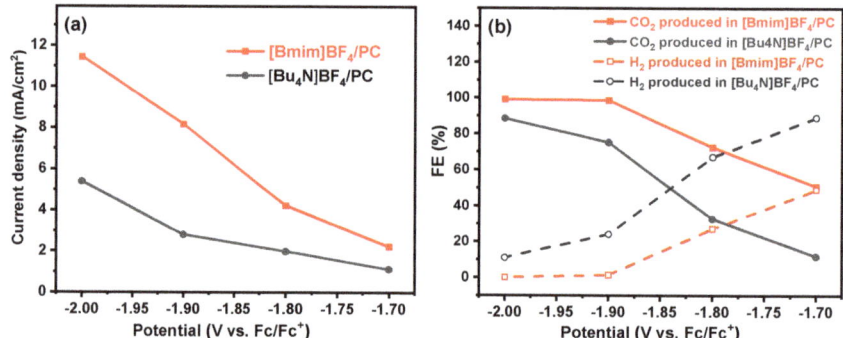

Figure 6. The current density (**a**) and Faradaic efficiency (FE) of CO and H$_2$ (**b**) for electrocatalytic CO$_2$ conversion in [Bu$_4$N]BF$_4$/PC and [Bmim]BF$_4$/PC electrolyte solution.

Based on the above electrochemical characterization, the performance results and previous reports, the reduction mechanism of CO$_2$ at the presence of [Bmim]BF$_4$ is proposed, and the corresponding schematic diagram is shown in Figure 7. Firstly, the imidazolium cation [Bmim]$^+$ adsorbs on the surface of the Ag electrode and forms a film layer of ILs [43]. Subsequently, the CO$_2$ molecules diffuse through the film layer of ILs and reach the Ag electrode. Then, the CO$_2$ molecule obtains one single electron, resulting in the formation of radical CO$_2$$^{\bullet-}$, and the generated CO$_2$$^{\bullet-}$ interacts with [Bmim]$^+$ and forms a [Bmim-CO$_2$]$_{ad}$ complex intermediate [45,46,68–70], in which the cation of [Bmim]$^+$ plays the role of stabilizing the radical CO$_2$$^{\bullet-}$, and in consequence, reduces the required activation energy for the overall reduction of CO$_2$ [7,31,47]. The formed [Bmim-CO$_2$]$_{ad}$ intermediate is further combined with another electron and two protons H$^+$ to ultimately produce CO. It should be noted here that the H$^+$ was supplied from the anodic electrolyte (sulfuric acid) and passed through the Nafion N-117 membrane to reach the cathode cell, to participate into the CO$_2$ reduction reaction. At last, the generated CO diffuses into the solution and overflows the liquid surface.

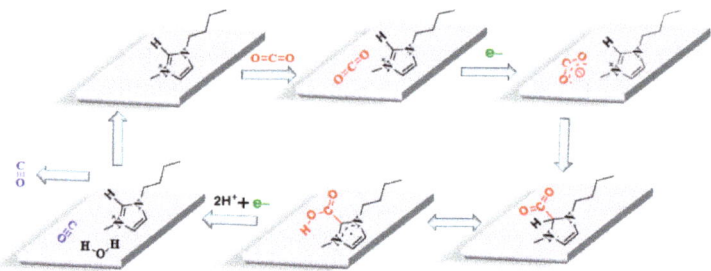

Figure 7. Schematic diagram of electroreduction of CO$_2$ in [Bmim]BF$_4$/PC solution with Ag as working electrode.

3. Materials and Methods

3.1. Materials

Notably, 1-ethyl-3-methylimidazolium tetrafluoroborate ([Emim]BF$_4$, ≥98%), 1-butyl-3-methylimidazolium tetrafluoroborate ([Bmim]BF$_4$, ≥99%), 1-hexyl-3-methylimidazolium tetrafluoroborate

([Hmim]BF$_4$, ≥98%), 1-octyl-3-methylimidazolium tetrafluoroborate ([Omim]BF$_4$, ≥98%), 1-decyl-3-methylimidazolium tetrafluoroborate ([Dmim]BF$_4$, ≥98%), and tetrabutylammonium tetrafluoroborate ([Bu$_4$N]BF$_4$, ≥99%) were all supplied by Chengjie Chem. Co., Ltd. (Shanghai, China). Propylene carbonate (PC, ≥99.9%), iodine (I$_2$, ≥99.5%), and tetrabutylammonium iodide (TBAI, ≥99.9%) were purchased from Shanghai Aladdin Biochemical Technology Co., Ltd. (Shanghai, China). Ferrocene (Fc, ≥99.9%) was purchased from Tianjin Deen Chemical Reagent Co., Ltd. (Tianjin, China). Nafion N-117 membrane (0.180 mm thick, ≥0.90 meg/g exchange capacity) was purchased from Alfa Aesar China Co., Ltd. (Tianjin, China). Ag electrodes (10mm × 10mm, 1mm in thickness, >99.99%) and graphite rod (5 mm diameter, length 15 cm, ≥99.99%) were acquired from Tianjin Aida Hengsheng Technology Development Co., Ltd. (Tianjin, China).

3.2. The Construction of Pt/(I$^-$/I$_3^-$) Reference Electrode

In this work, Pt/(I$^-$/I$_3^-$) was used as the reference electrode in the ILs/PC non-aqueous system. The Pt/(I$^-$/I$_3^-$) reference electrode was prepared by inserting Pt wire into a mixture solution of I$_2$ (0.05 M) and TBAI (0.1 M). In addition, in order to avoid the junction potential and the pollution of the cathodic electrolyte solution from the electrolyte in Pt/(I$^-$/I$_3^-$) reference electrode, a salt bridge filled with the same cathodic electrolyte solution is also used. Both bottoms of the Pt/(I$^-$/I$_3^-$) reference electrode and the salt bridge were sealed with porous polytetrafluoroethylene.

3.3. Electrochemical Characterization

All electrochemical experiments were carried out on the CHI 660E electrochemical workstation (Shanghai Chenhua, Shanghai, China). The scan rate of linear sweeping voltammetry (LSV) is 20 mV/s and sweeping region from −0.5~−3.0 V (vs. Fc/Fc$^+$). The Tafel scan rate is 10 mV/s. The electrochemical impedance (ESI) is tested at a constant potential, with an amplitude of 10 mV and frequency ranging from 10 kHz to 1 Hz.

3.4. The CO$_2$ Conversion Performance Test

The electrochemical reduction of CO$_2$ was performed in an H-type dual-chamber reactor. The Nafion N-117 ion-exchange membrane was used to separate the cathode chamber and the anode chamber. A silver plate (1cm × 1cm × 0.3mm) and a graphite rod were used as the working electrode and counter electrode, respectively. ILs/PC mixture solution (0.1 M) and H$_2$SO$_4$ aqueous solution (0.1 M) were used as the cathodic and anodic electrolytes, respectively.

In a typical procedure, the reactor was connected to a gas circulation and online sampling system (Labsolar-III AG, Beijing Perfectlight Technology Co., Ltd. (Beijing, China), and the details are stated in our previous work [46,72]). Subsequently, the air solubilized in the electrolyte solution, and the air in the circulation channel was evacuated for 30 min. Then, CO$_2$ was introduced into the electrolyte solution from the bottom of the cathodic cell for one hour to reach the solubility equilibrium of CO$_2$. Then two hours of the reaction was carried out to make sure that the concentration of products was high enough to exceed the detection limit of the detectors for reducing the analytic errors.

3.5. The Products Analysis and Calculation of Faradaic Efficiency

The products of H$_2$ and CO were online sampled and in-situ quantified by gas chromatography (GC 9790II, Zhejiang Fu Li Analytical Instrument Co. Ltd., Zhejiang, China). The separation of H$_2$, CO, and CO$_2$ feed gas was realized by a TDX-01 column and the separated gas was subsequently brought into two paths. In one path, the products of H$_2$ and CO were guided into a Molsieve5 A column, and the H$_2$ was then measured using a thermal conductivity detector (TCD), while the gas CO was passed through a mechanizer to be transformed into methane by the nickel catalyst at 380 °C, and was ultimately detected by a flame ionization detector (FID). In another path, CO$_2$ and other possible hydrocarbons that have longer retention time than H$_2$ and CO in the TDX-01 column were introduced into a Porapak N column. Finally, the CO_2 gas was vented out, and the gas of hydrocarbons were measured by FID.

The FE is calculated from the product analysis by the equation:

$$FE = \frac{n_i z_i}{n_e}$$

where n_i is the mole number of a specific product (mol), z_i is the number of electrons transferred for this product i ($z_i = 2$ for CO and H_2), and n_e is the total mole number of electrons passed through the circuit (mol). Furthermore, the n_e (mol) can be determined by the equation:

$$n_e = \frac{Q}{F}$$

where Q is the passed charge (C) that can be obtained from the integration of the recorded chronoamperometric (i-t) curve, and F is the Faradaic constant (96500 C/mol).

4. Conclusions

In summary, the ILs/PC mixture solution was constructed and investigated as the electrolyte for electrochemical reduction of CO_2 on the Ag plate electrode. The investigation on the alkyl length of imidazole-based ILs indicates that the IL of [Bmim]BF_4 gives the lower onset potential and Tafel slope, as well as higher exchange current density, compared to other studied ILs and the traditional [Bu_4N]BF_4 salt. The EIS characterization further indicated that the imidazolium cation could absorb on the electrode surface and reduce the overpotential through complexing with anion radical ($CO_2^{\bullet-}$) and stabilizing them. The [Bmim]BF_4 IL gives the highest performance for the conversion of CO_2. The performance tests show that in [Bmim]BF_4/PC electrolyte solution, FE for CO as high as 98.5% and current density of 8.2 mA/cm^2 can be achieved at −1.9 V (vs. Fc/Fc$^+$).

Author Contributions: Conceptualization, F.J. and W.L.; Funding acquisition, W.L.; Synthesis, characterization, performance test and writing—original draft preparation, F.J. and J.Z.; Formal analysis and writing—review and editing, F.J., W.L. and J.Z. All authors have read and agreed to the published version of the manuscript.

Funding: This research was funded by the National Natural Science Foundation of China, grant number 21673067.

Conflicts of Interest: The authors declare no conflict of interest.

References

1. Davis, S.J.; Caldeira, K.; Matthews, H.D. Future CO_2 emissions and climate change from existing energy infrastructure. *Science* **2010**, *329*, 1330–1333. [CrossRef]
2. Lewis, N.S. Research opportunities to advance solar energy utilization. *Science* **2016**, *351*, aad1920. [CrossRef] [PubMed]
3. Bui, M.; Adjiman, C.S.; Bardow, A.; Anthony, E.J.; Boston, A.; Brown, S.; Fennell, P.S.; Fuss, S.; Galindo, A.; Hackett, L.A.; et al. Carbon capture and storage (CCS): The way forward. *Energy Environ. Sci.* **2018**, *11*, 1062–1176. [CrossRef]
4. Davis, S.J.; Lewis, N.S.; Shaner, M.; Aggarwal, S.; Arent, D.; Azevedo, I.L.; Benson, S.M.; Bradley, T.; Brouwer, J.; Chiang, Y.M.; et al. Net-zero emissions energy systems. *Science* **2018**, *360*, eaas9793. [CrossRef] [PubMed]
5. Olah, G.A.; Prakash, G.K.S.; Goeppert, A. Anthropogenic Chemical Carbon Cycle for a Sustainable Future. *J. Am. Chem. Soc.* **2011**, *133*, 12881–12898. [CrossRef] [PubMed]
6. Alvarez-Guerra, M.; Albo, J.; Alvarez-Guerra, E.; Irabien, A. Ionic liquids in the electrochemical valorisation of CO_2. *Energy Environ. Sci.* **2015**, *8*, 2574–2599. [CrossRef]
7. Faggion, D.; Goncalves, W.D.G.; Dupont, J. CO_2 Electroreduction in Ionic Liquids. *Front. Chem.* **2019**, *7*, 102. [CrossRef]
8. Gao, D.; Aran-Ais, R.M.; Jeon, H.S.; Cuenya, B.R. Rational catalyst and electrolyte design for CO_2 electroreduction towards multicarbon products. *Nat. Catal.* **2019**, *2*, 198–210. [CrossRef]

9. Sánchez, O.G.; Birdja, Y.Y.; Bulut, M.; Vaes, J.; Breugelmans, T.; Pant, D. Recent advances in industrial CO_2 electroreduction. *Curr. Opin. Green Sust. Chem.* **2019**, *16*, 47–56. [CrossRef]
10. Yang, D.; Zhu, Q.; Chen, C.; Liu, H.; Liu, Z.; Zhao, Z.; Zhang, X.; Liu, S.; Han, B. Selective electroreduction of carbon dioxide to methanol on copper selenide nanocatalysts. *Nat. Chem.* **2019**, *10*, 677. [CrossRef]
11. Zheng, Y.; Vasileff, A.; Zhou, X.; Jiao, Y.; Jaroniec, M.; Qiao, S.-Z. Understanding the Roadmap for Electrochemical Reduction of CO_2 to Multi-Carbon Oxygenates and Hydrocarbons on Copper-Based Catalysts. *J. Am. Chem. Soc.* **2019**, *141*, 7646–7659. [CrossRef] [PubMed]
12. Lu, W.; Zhang, Y.; Zhang, J.; Xu, P. Reduction of Gas CO_2 to CO with High Selectivity by Ag Nanocube-Based Membrane Cathodes in a Photoelectrochemical System. *Ind. Eng. Chem. Res.* **2020**, *59*, 5536–5545. [CrossRef]
13. Álvarez, A.; Bansode, A.; Urakawa, A.; Bavykina, A.V.; Wezendonk, T.A.; Makkee, M.; Gascon, J.; Kapteijn, F. Challenges in the Greener Production of Formates/Formic Acid, Methanol, and DME by Heterogeneously Catalyzed CO_2 Hydrogenation Processes. *Chem. Rev.* **2017**, *117*, 9804–9838. [CrossRef] [PubMed]
14. Prieto, G. Carbon Dioxide Hydrogenation into Higher Hydrocarbons and Oxygenates: Thermodynamic and Kinetic Bounds and Progress with Heterogeneous and Homogeneous Catalysis. *ChemSusChem* **2017**, *10*, 1056–1070. [CrossRef] [PubMed]
15. Saeidi, S.; Najari, S.; Fazlollahi, F.; Nikoo, M.K.; Sefidkon, F.; Klemes, J.J.; Baxter, L.L. Mechanisms and kinetics of CO_2 hydrogenation to value-added products: A detailed review on current status and future trends. *Renew. Sustain. Energy Rev.* **2017**, *80*, 1292–1311. [CrossRef]
16. Guo, L.-j.; Wang, Y.-j.; He, T. Photocatalytic Reduction of CO_2 over Heterostructure Semiconductors into Value-Added Chemicals. *Chem. Rec.* **2016**, *16*, 1918–1933. [CrossRef]
17. Qureshi, M.; Takanabe, K. Insights on Measuring and Reporting Heterogeneous Photocatalysis: Efficiency Definitions and Setup Examples. *Chem. Mater.* **2017**, *29*, 158–167. [CrossRef]
18. Xie, H.; Wang, J.; Ithisuphalap, K.; Wu, G.; Li, Q. Recent advances in Cu-based nanocomposite photocatalysts for CO_2 conversion to solar fuels. *J. Energy Chem.* **2017**, *26*, 1039–1049. [CrossRef]
19. Zhang, L.; Zhao, Z.-J.; Wang, T.; Gong, J. Nano-designed semiconductors for electro-and photoelectro-catalytic conversion of carbon dioxide. *Chem. Soc. Rev.* **2018**, *47*, 5423–5443. [CrossRef]
20. Zhang, N.; Long, R.; Gao, C.; Xiong, Y. Recent progress on advanced design for photoelectrochemical reduction of CO_2 to fuels. *Sci. China Mater.* **2018**, *61*, 771–805. [CrossRef]
21. Chang, X.; Wang, T.; Yang, P.; Zhang, G.; Gong, J. The Development of Cocatalysts for Photoelectrochemical CO_2 Reduction. *Adv. Mater.* **2019**, *31*, 1804710. [CrossRef] [PubMed]
22. Lu, W.; Ju, F.; Yao, K.; Wei, X. Photoelectrocatalytic Reduction of CO_2 for Efficient Methanol Production: Au Nanoparticles as Electrocatalysts and Light Supports. *Ind. Eng. Chem. Res.* **2020**, *59*, 4348–4357. [CrossRef]
23. Yu, S.; Wilson, A.J.; Kumari, G.; Zhang, X.; Jain, P.K. Opportunities and Challenges of Solar-Energy-Driven Carbon Dioxide to Fuel Conversion with Plasmonic Catalysts. *ACS Energy Lett.* **2017**, *2*, 2058–2070. [CrossRef]
24. Koppenol, W.; Rush, J. Reduction potential of the carbon dioxide/carbon dioxide radical anion: A comparison with other C1 radicals. *J. Phys. Chem.* **1987**, *91*, 4429–4430. [CrossRef]
25. Paik, W.; Andersen, T.N.; Eyring, H. Kinetic studies of electrolytic reduction of carbon dioxide on mercury electrode. *Electrochim. Acta* **1969**, *14*, 1217–1232. [CrossRef]
26. Chen, T.Y.; Shi, J.; Shen, F.X.; Zhen, J.Z.; Li, Y.F.; Shi, F.; Yang, B.; Jia, Y.J.; Dai, Y.N.; Hu, Y.Q. Selection of Low-Cost Ionic Liquid Electrocatalyst for CO_2 Reduction in Propylene Carbonate/Tetrabutylammonium Perchlorate. *ChemElectroChem* **2018**, *5*, 2295–2300. [CrossRef]
27. Sharma, P.P.; Zhou, X.D. Electrocatalytic conversion of carbon dioxide to fuels: A review on the interaction between CO_2 and the liquid electrolyte. *WIRES Energy Environ.* **2017**, *6*, e239. [CrossRef]
28. Lu, Q.; Rosen, J.; Jiao, F. Nanostructured Metallic Electrocatalysts for Carbon Dioxide Reduction. *ChemCatChem* **2015**, *7*, 38–47. [CrossRef]
29. Shukia, S.K.; Khokarale, S.G.; Bui, T.Q.; Mikkola, J.P.T. Ionic Liquids: Potential Materials for Carbon Dioxide Capture and Utilization. *Front. Mater.* **2019**, *6*, 42. [CrossRef]
30. Chen, Y.; Mu, T.C. Conversion of CO_2 to value-added products mediated by ionic liquids. *Green Chem.* **2019**, *21*, 2544–2574. [CrossRef]
31. Lim, H.K.; Kim, H. The Mechanism of Room-Temperature Ionic-Liquid-Based Electrochemical CO_2 Reduction: A Review. *Molecules* **2017**, *22*, 536. [CrossRef] [PubMed]

32. Wang, S.; Wang, X. Imidazolium Ionic Liquids, Imidazolylidene Heterocyclic Carbenes, and Zeolitic Imidazolate Frameworks for CO_2 Capture and Photochemical Reduction. *Angew. Chem. Int. Ed.* **2016**, *55*, 2308–2320. [CrossRef] [PubMed]

33. Rees, N.V.; Compton, R.G. Electrochemical CO_2 sequestration in ionic liquids: A perspective. *Energy Environ. Sci.* **2011**, *4*, 403–408. [CrossRef]

34. Asadi, M.; Kim, K.; Liu, C.; Addepalli, A.V.; Abbasi, P.; Yasaei, P.; Phillips, P.; Behranginia, A.; Cerrato, J.M.; Haasch, R.; et al. Nanostructured transition metal dichalcogenide electrocatalysts for CO_2 reduction in ionic liquid. *Science* **2016**, *353*, 467–470. [CrossRef] [PubMed]

35. Zhang, S.J.; Sun, J.; Zhang, X.C.; Xin, J.Y.; Miao, Q.Q.; Wang, J.J. Ionic liquid-based green processes for energy production. *Chem. Soc. Rev.* **2014**, *43*, 7838–7869. [CrossRef] [PubMed]

36. Zeng, S.; Zhang, X.; Bai, L.; Zhang, X.; Wang, H.; Wang, J.; Bao, D.; Li, M.; Liu, X.; Zhang, S. Ionic-Liquid-Based CO_2 Capture Systems: Structure, Interaction and Process. *Chem. Rev.* **2017**, *117*, 9625–9673. [CrossRef] [PubMed]

37. Hapiot, P.; Lagrost, C. Electrochemical reactivity in room-temperature ionic liquids. *Chem. Rev.* **2008**, *108*, 2238–2264. [CrossRef]

38. Liu, H.; Liu, Y.; Li, J. Ionic liquids in surface electrochemistry. *Phys. Chem. Chem. Phys.* **2010**, *12*, 1685–1697. [CrossRef]

39. MacFarlane, D.R.; Pringle, J.M.; Howlett, P.C.; Forsyth, M. Ionic liquids and reactions at the electrochemical interface. *Phys. Chem. Chem. Phys.* **2010**, *12*, 1659–1669. [CrossRef]

40. Zhang, G.R.; Etzold, B.J.M. Ionic liquids in electrocatalysis. *J. Energy Chem.* **2016**, *25*, 199–207. [CrossRef]

41. Rudnev, A.V.; Kiran, K.; Lopez, A.C.; Dutta, A.; Gjuroski, I.; Furrer, J.; Broekmann, P. Enhanced electrocatalytic CO formation from CO_2 on nanostructured silver foam electrodes in ionic liquid/water mixtures. *Electrochim. Acta* **2019**, *306*, 245–253. [CrossRef]

42. Wu, H.R.; Song, J.L.; Xie, C.; Hu, Y.; Han, B.X. Highly efficient electrochemical reduction of CO_2 into formic acid over lead dioxide in an ionic liquid-catholyte mixture. *Green Chem.* **2018**, *20*, 1765–1769. [CrossRef]

43. Lim, H.K.; Kwon, Y.; Kim, H.S.; Jeon, J.; Kim, Y.H.; Lim, J.A.; Kim, B.S.; Choi, J.; Kim, H. Insight into the Microenvironments of the Metal-Ionic Liquid Interface during Electrochemical CO_2 Reduction. *ACS Catal.* **2018**, *8*, 2420–2427. [CrossRef]

44. Medina-Ramos, J.; Lee, S.S.; Fister, T.T.; Hubaud, A.A.; Sacci, R.L.; Mullins, D.R.; DiMeglio, J.L.; Pupillo, R.C.; Velardo, S.M.; Lutterman, D.A.; et al. Structural Dynamics and Evolution of Bismuth Electrodes during Electrochemical Reduction of CO_2 in Imidazolium-Based Ionic Liquid Solutions. *ACS Catal.* **2017**, *7*, 7285–7295. [CrossRef]

45. Rosen, B.A.; Salehi-Khojin, A.; Thorson, M.R.; Zhu, W.; Whipple, D.T.; Kenis, P.J.A.; Masel, R.I. Ionic Liquid-Mediated Selective Conversion of CO_2 to CO at Low Overpotentials. *Science* **2011**, *334*, 643–644. [CrossRef]

46. Lu, W.; Jia, B.; Cui, B.; Zhang, Y.; Yao, K.; Zhao, Y.; Wang, J. Efficient photoelectrochemical reduction of carbon dioxide to formic acid: A functionalized ionic liquid as an absorbent and electrolyte. *Angew. Chem. Int. Edit.* **2017**, *56*, 11851–11854. [CrossRef]

47. Feng, J.; Zeng, S.; Feng, J.; Dong, H.; Zhang, X. CO_2 Electroreduction in Ionic Liquids: A Review. *Chin. J. Chem.* **2018**, *36*, 961–970. [CrossRef]

48. Rosen, J.; Hutchings, G.S.; Lu, Q.; Rivera, S.; Zhou, Y.; Vlachos, D.G.; Jiao, F. Mechanistic Insights into the Electrochemical Reduction of CO_2 to CO on Nanostructured Ag Surfaces. *ACS Catal.* **2015**, *5*, 4293–4299. [CrossRef]

49. Sun, L.Y.; Ramesha, G.K.; Kamat, P.V.; Brennecke, J.F. Switching the Reaction Course of Electrochemical CO_2 Reduction with Ionic Liquids. *Langmuir* **2014**, *30*, 6302–6308. [CrossRef]

50. Zhu, Q.; Ma, J.; Kang, X.; Sun, X.; Liu, H.; Hu, J.; Liu, Z.; Han, B. Efficient Reduction of CO_2 into Formic Acid on a Lead or Tin Electrode using an Ionic Liquid Catholyte Mixture. *Angew. Chem. Int. Ed.* **2016**, *55*, 9012–9016. [CrossRef]

51. Sun, X.; Zhu, Q.; Kang, X.; Liu, H.; Qian, Q.; Zhang, Z.; Han, B. Molybdenum-Bismuth Bimetallic Chalcogenide Nanosheets for Highly Efficient Electrocatalytic Reduction of Carbon Dioxide to Methanol. *Angew. Chem. Int. Ed.* **2016**, *55*, 6770–6774. [CrossRef] [PubMed]

52. Huan, T.N.; Simon, P.; Rousse, G.; Genois, I.; Artero, V.; Fontecave, M. Porous dendritic copper: An electrocatalyst for highly selective CO_2 reduction to formate in water/ionic liquid electrolyte. *Chem. Sci.* **2017**, *8*, 742–747. [CrossRef] [PubMed]
53. Rudnev, A.V.; Fu, Y.-C.; Gjuroski, I.; Stricker, F.; Furrer, J.; Kovacs, N.; Vesztergom, S.; Broekmann, P. Transport Matters: Boosting CO_2 Electroreduction in Mixtures of BMIm BF_4/Water by Enhanced Diffusion. *ChemPhysChem* **2017**, *18*, 3153–3162. [CrossRef] [PubMed]
54. Lau, G.P.S.; Schreier, M.; Vasilyev, D.; Scopelliti, R.; Gratzel, M.; Dyson, P.J. New Insights Into the Role of Imidazolium-Based Promoters for the Electroreduction of CO_2 on a Silver Electrode. *J. Am. Chem. Soc.* **2016**, *138*, 7820–7823. [CrossRef] [PubMed]
55. Singh, M.R.; Clark, E.L.; Bell, A.T. Effects of electrolyte, catalyst, and membrane composition and operating conditions on the performance of solar-driven electrochemical reduction of carbon dioxide. *Phys. Chem. Chem. Phys.* **2015**, *17*, 18924–18936. [CrossRef] [PubMed]
56. Forero, J.S.B.; Munoz, J.A.H.; Jones, J.; da Silva, F.M. Propylene Carbonate in Organic Synthesis: Exploring its Potential as a Green Solvent. *Curr. Org. Synth.* **2016**, *13*, 834–846. [CrossRef]
57. Parker, H.L.; Sherwood, J.; Hunt, A.J.; Clark, J.H. Cyclic Carbonates as Green Alternative Solvents for the Heck Reaction. *ACS Sustain. Chem. Eng.* **2014**, *2*, 1739–1742. [CrossRef]
58. Hfaiedh, A.; Yuan, K.D.; Ben Ammar, H.; Ben Hassine, B.; Soule, J.F.; Doucet, H. Eco-Friendly Solvents for Palladium-Catalyzed Desulfitative C-H Bond Arylation of Heteroarenes. *ChemSusChem* **2015**, *8*, 1794–1804. [CrossRef]
59. Shen, F.X.; Shi, J.; Chen, T.Y.; Shi, F.; Li, Q.Y.; Zhen, J.Z.; Li, Y.F.; Dai, Y.N.; Yang, B.; Qu, T. Electrochemical reduction of CO_2 to CO over Zn in propylene carbonate/tetrabutylammonium perchlorate. *J. Power Sources* **2018**, *378*, 555–561. [CrossRef]
60. Shi, J.; Shen, F.-x.; Shi, F.; Song, N.; Jia, Y.-j.; Hu, Y.-Q.; Li, Q.-Y.; Liu, J.-x.; Chen, T.-Y.; Dai, Y.-N. Electrochemical reduction of CO_2 into CO in tetrabutylammonium perchlorate/propylene carbonate: Water effects and mechanism. *Electrochim. Acta* **2017**, *240*, 114–121. [CrossRef]
61. Gagne, R.R.; Koval, C.A.; Lisensky, G.C. Ferrocene as an internal standard for electrochemical measurements. *Inorg. Chem.* **1980**, *19*, 2854–2855. [CrossRef]
62. Shi, J.; Li, Q.Y.; Shi, F.; Song, N.; Jia, Y.J.; Hu, Y.Q.; Shen, F.X.; Yang, D.W.; Dai, Y.N. Design of a Two-Compartment Electrolysis Cell for the Reduction of CO_2 to CO in Tetrabutylammonium Perchlorate/Propylene Carbonate for Renewable Electrical Energy Storage. *J. Electrochem. Soc.* **2016**, *163*, G82–G87. [CrossRef]
63. Fang, Y.-H.; Liu, Z.-P. Tafel Kinetics of Electrocatalytic Reactions: From Experiment to First-Principles. *ACS Catal.* **2014**, *4*, 4364–4376. [CrossRef]
64. Lates, V.; Falch, A.; Jordaan, A.; Peach, R.; Kriek, R.J. An electrochemical study of carbon dioxide electroreduction on gold-based nanoparticle catalysts. *Electrochim. Acta* **2014**, *128*, 75–84. [CrossRef]
65. González, J.E.G.; Santana, A.F.J.H.; Mirza-rosca, J.C. Effect of bacterial biofilm on 316 SS corrosion in natural seawater by eis. *Corros. Sci.* **1998**, *40*, 2141–2154. [CrossRef]
66. Torresi, R.M.; Lodovico, L.; Benedetti, T.M.; Alcântara, M.R.; Debiemme-Chouvy, C.; Deslouis, C. Convective mass transport in ionic liquids studied by electrochemical and electrohydrodynamic impedance spectroscopy. *Electrochim. Acta* **2013**, *93*, 32–43. [CrossRef]
67. Pajkossy, T. Impedance spectroscopy at interfaces of metals and aqueous solutions—Surface roughness, CPE and related issues. *Solid State Ion.* **2005**, *176*, 1997–2003. [CrossRef]
68. Rosen, B.A.; Haan, J.L.; Mukherjee, P.; Braunschweig, B.; Zhu, W.; Salehi-Khojin, A.; Dlott, D.D.; Masel, R.I. In Situ Spectroscopic Examination of a Low Overpotential Pathway for Carbon Dioxide Conversion to Carbon Monoxide. *J. Phys. Chem. C* **2012**, *116*, 15307–15312. [CrossRef]
69. Braunschweig, B.; Mukherjee, P.; Haan, J.L.; Dlott, D.D. Vibrational sum-frequency generation study of the CO_2 electrochemical reduction at Pt/EMIM-BF_4 solid/liquid interfaces. *J. Electroanal. Chem.* **2017**, *800*, 144–150. [CrossRef]
70. Wang, Y.Q.; Hatakeyama, M.; Ogata, K.; Wakabayashi, M.; Jin, F.M.; Nakamura, S. Activation of CO_2 by ionic liquid EMIM-BF_4 in the electrochemical system: A theoretical study. *Phys. Chem. Chem. Phys.* **2015**, *17*, 23521–23531. [CrossRef] [PubMed]

71. Asadi, M.; Kumar, B.; Behranginia, A.; Rosen, B.A.; Baskin, A.; Repnin, N.; Pisasale, D.; Phillips, P.; Zhu, W.; Haasch, R.; et al. Robust carbon dioxide reduction on molybdenum disulphide edges. *Nat. Commun.* **2014**, *5*, 4470. [CrossRef] [PubMed]
72. Xu, P.; Lu, W.; Zhang, J.; Zhang, L. Efficient Hydrolysis of Ammonia Borane for Hydrogen Evolution Catalyzed by Plasmonic Ag@Pd Core-Shell Nanocubes. *ACS Sustain. Chem. Eng.* **2020**, *8*, 12366–12377. [CrossRef]

© 2020 by the authors. Licensee MDPI, Basel, Switzerland. This article is an open access article distributed under the terms and conditions of the Creative Commons Attribution (CC BY) license (http://creativecommons.org/licenses/by/4.0/).

Article

Synthesis and Characterization of *p-n* Junction Ternary Mixed Oxides for Photocatalytic Coprocessing of CO_2 and H_2O

Davide M. S. Marcolongo [1], Francesco Nocito [1], Nicoletta Ditaranto [1], Michele Aresta [2] and Angela Dibenedetto [1,3,*]

1. Dipartimento di Chimica, Università degli Studi di Bari, Via Orabona, 4, 70125 Bari, Italy; davide.marcolongo@uniba.it (D.M.S.M.); francesco.nocito@uniba.it (F.N.); nicoletta.ditaranto@uniba.it (N.D.)
2. IC2R srl, Tecnopolis, Via Casamassima km 3, Valenzano, 70010 Bari, Italy; michele.aresta@ic2r.com
3. CIRCC-Interuniversity Consortium on Chemical reactivity and Catalysis, Via Celso Ulpiani, 27, 70126 Bari, Italy
* Correspondence: angela.dibenedetto@uniba.it; Tel.: +39-080-544-3606

Received: 5 August 2020; Accepted: 18 August 2020; Published: 31 August 2020

Abstract: In the present paper, we report the synthesis and characterization of both binary (Cu_2O, Fe_2O_3, and In_2O_3) and ternary (Cu_2O-Fe_2O_3 and Cu_2O-In_2O_3) transition metal mixed-oxides that may find application as photocatalysts for solar driven CO_2 conversion into energy rich species. Two different preparation techniques (High Energy Milling (HEM) and Co-Precipitation (CP)) are compared and materials properties are studied by means of a variety of characterization and analytical techniques UV-Visible Diffuse Reflectance Spectroscopy (UV-VIS DRS), X-ray Photoelectron Spectroscopy (XPS), X-Ray Diffraction (XRD), Transmission Electron Microscopy (TEM), and Energy Dispersive X-Ray spectrometry (EDX). Appropriate data elaboration methods are used to extract materials bandgap for $Cu_2O@Fe_2O_3$ and $Cu_2O@In_2O_3$ prepared by HEM and CP, and foresee whether the newly prepared semiconductor mixed oxides pairs are useful for application in CO_2-H_2O coprocessing. The experimental results show that the synthetic technique influences the photoactivity of the materials that can correctly be foreseen on the basis of bandgap experimentally derived. Of the mixed oxides prepared and described in this work, only $Cu_2O@In_2O_3$ shows positive results in CO_2-H_2O photo-co-processing. Preliminary results show that the composition and synthetic methodologies of mixed-oxides, the reactor geometry, the way of dispersing the photocatalyst sample, play a key role in the light driven reaction of CO_2–H_2O. This work is a rare case of full characterization of photo-materials, using UV-Visible DRS, XPS, XRD, TEM, EDX for the surface and bulk analytical characterization. Surface composition may not be the same of the bulk composition and plays a key role in photocatalysts behavior. We show that a full material knowledge is necessary for the correct forecast of their photocatalytic behavior, inferred from experimentally determined bandgaps.

Keywords: CO_2–H_2O photo-co-processing; VIS-light driven reactions; CO_2 reduction; photocatalysts properties

1. Introduction

Combustion of fossil fuels (fossil-C) is actually the main source (80.2% as for 2018) to fulfil human hunger for energy, but natural resources are not infinite and are expected to get exhausted in 160 y or so. Moreover, the use of fossil-C is responsible for the emission of 37 Gt/y of CO_2 and other green-house gases considered to be the origin of climate change. According to Earth System Research Laboratory's—ESRL's Global Monitoring Laboratory, in January 2020, atmospheric concentration of

CO_2 reached 412 ppm [1]. However, avoiding massive and continuous CO_2 emission and utilization of alternative sustainable primary energy sources is necessary [2]. On the other hand, CO_2 represents a readily available building block for chemicals and source of carbon for fuels, which can be produced through a conversion driven by C-free energy sources [3]. Solar radiations carry a quantity of energy to Earth surface, sufficient to be considered to perform CO_2 conversion [4]. This can be realized by means of photocatalysis [5], in a semiconductor-assisted light-driven process during which light is absorbed and converted into chemical energy, such as CO, CH_3OH, CH_4, Cn-species, or even H_2, produced in water-splitting [6,7].

Unfortunately, for the moment, the photocatalytic processes still suffer low efficiency and are not ready for an industrial commercially viable application [5], despite the research started with work by T. Inoue [8] and J.M. Lehn [9] that dates back to 40 years ago. Searching for the best photocatalysts [10], a variety of semiconductors have been studied, ranging from those based onto Group 4 elements to more classical Group 3–5 semiconductors [11,12], Group 6 chalcogenides [13] and more "exotic" semiconductors [14]. Among the latter materials, Earth-abundant transition and post-transition metal oxides such as TiO_2, Fe_2O_3, Cu_2O, CuO, ZnO, NiO, Ta_2O_5, Ga_2O_3, In_2O_3, and WO_3 revealed very promising to act both as photo-catalysts [10–12] or co-catalysts [15]. They are usually cheap and yet widely used as chemo-catalysts in industrial applications, easily fabricated at micro- and nano-size and can absorb light in the UV-Visible region [16,17]. Further, they can be efficiently coupled to enhance properties such as visible light absorption [18], band edge levels' position, and photogenerated charge transfer, and separation processes [15]. Copper oxides, both CuO and Cu_2O, were recently demonstrated able to convert CO_2 or water into solar fuels under VIS-light irradiation [19]. These *p-type* semiconductors, that can efficiently be used either bare [20] or as co-catalysts [21–24], are affected by high recombination rate and photodegradation. The formation of a heterojunction by coupling with a suitable *n-type* semiconductor, is a widely adopted strategy to inhibit charge recombination, enhance stability and provide alternative energy levels to carry out photocatalytic reactions [11,12]. The formation of such junctions even at very small particle size is one of the keys for leading to fabrication of active photocatalysts. The semiconductor coupling strategy can also be adopted for copper oxides, in which properties as photostability and charge separation are found to be significantly affected by coupling with *n-type* metal oxides [19]. At the same time, the addition of copper oxides is useful to shift the absorption spectrum of semiconductor partner towards visible range [12,18].

In the present work, two different *n*-type metal oxides were selected as potential partners for Cu_2O: indium oxide, In_2O_3, and iron oxide in the form of hematite, α-Fe_2O_3. Thanks to band energy levels position, affinity towards CO_2 and electronic properties, both these oxides are recognized in literature as potential photocatalysts or cocatalysts for solar fuels production [12,25,26]. In particular, indium oxide has been recently experimented for CH_4 and H_2 photocatalytic production, coupled with other semiconductors or cocatalysts [27,28], or even with a thermal input during the reaction [29,30]. Instead, iron oxide is widely experimented for photocatalytic dye degradation [31] and water cleaning [26], but recently water splitting application is under study [32].

To the best of our knowledge, semiconductor pairs composed of Cu_2O with In_2O_3 or α-Fe_2O_3 are very scarcely characterized and tested for solar fuels production. Thin films are generally preferred over powders and particles. For the latter, 2D semiconductors as graphene or carbon nitride have been used as platforms for achieving enhanced charge separation and better lattice matching [10–12]. This is especially true for the Cu_2O-In_2O_3 pair, where lattice parameters mismatch can hinder the formation of heterojunctions, but samples prepared by hydrothermal/co-precipitation methods are possibly active in photocatalytic degradation and hydrogen evolution [33,34]. In the literature, the Cu_2O-Fe_2O_3 pair is reported to a bigger extent than that with Cu_2O-In_2O_3 and samples prepared by solvothermal, co-precipitation or electrodeposition methods were tested for photocatalytic degradation [35], hydrogen production [36,37] and CO_2 reduction to carbon monoxide, methanol and methane [38,39]. A point to mention is that the Cu/In,Fe-mixed-oxides were not always fully characterized and often they were added with other compounds (especially noble metals) and hole scavengers, adding complexity to the already not clearly defined system and making difficult to understand the role of each partner.

In this work, mixed-oxide nano-particles were prepared through two different techniques: High Energy Milling—HEM and Co-Precipitation—CP. Starting oxides were either commercial samples or synthesized in our laboratory. Neat nano-sized powder samples were analyzed by Energy Dispersive X-Ray spectrometry (EDX), characterized by UV-Visible Diffuse Reflectance Spectroscopy (UV-VIS DRS), and X-ray Photoelectron Spectroscopy (XPS) and then tested in gas phase CO_2–H_2O co-processing under VIS-light irradiation, at room temperature, without addition of noble metal co-catalysts or hole scavenger species to evaluate the properties of the single mixed oxides. Evaluation and tailoring of properties of cited materials, with particular attention to electronic band structure and optical absorption, was the final goal of this work, targeting a correlation among properties of the materials and their photocatalytic activity in co-processing H_2O and CO_2 under VIS-light in different reactor geometries.

2. Results and Discussion

2.1. Synthesis, Composition, and Size of Ternary Oxides

Binary oxides powders used in this work were either commercial samples (C-Oxide) or synthesized in the laboratory (S-Oxide). Ternary oxides powders were fabricated through HEM and CP techniques, as reported in the Materials and Methods section. Such mixed oxides are labelled in Table 1 according to the preparation technique employed and numbered according to increasing Cu/In or Cu/Fe molar ratio, as measured by EDX spectrometry and calculated considering all copper, indium, and iron present in the form of Cu_2O, In_2O_3, and Fe_2O_3, respectively.

Table 1. Composition of Mixed-Oxide samples prepared by High Energy Milling (HEM) or Co-Precipitation (CP), listed by molar ratio.

Sample Name	Cu/In Ratio	Sample Name	Cu/Fe Ratio
HEM-Cu/In-1	0.60	HEM-Cu/Fe-1	0.59
HEM-Cu/In-2	1.08	HEM-Cu/Fe-2	0.99
HEM-Cu/In-3	2.20	HEM-Cu/Fe-3	1.92
CP-Cu/In-1	0.25	CP-Cu/Fe-1	0.23
CP-Cu/In-2	1.58	CP-Cu/Fe-2	0.66
CP-Cu/In-3	2.96	CP-Cu/Fe-3	1.21

Preliminary Transmission Electron Microscopy (TEM) measurements were carried out on ternary mixed oxides containing Cu_2O and In_2O_3. TEM micrographs show that both techniques (HEM and CP) were able to produce single particles whit linear size below 100 nm, which aggregate into sub-micrometric clusters, which show some differences. While in HEM prepared samples, clusters are produced by association of particles upon collision, with relevant presence of amorphous phase and no precise morphology, in CP prepared samples clusters are produced by stacking of well-formed particles of cubic morphology [24], which is common to crystal lattice of both component binary oxides [25,40], and particles clearly show a core-shell coverage.

Powder X-Ray Diffraction (XRD) measurements, used to study crystallinity and crystal phase of selected ternary oxides samples, allowed determine that, while samples prepared by HEM miss any crystallinity, samples prepared by CP show small size crystalline domains, with Cu_2O is in its usual cubic phase and Fe_2O_3 in its rhombohedral α phase (hematite).

2.2. XPS Analysis

XPS measurements were carried out onto all single metal oxides and even a set of selected mixed oxides, in order to investigate their surface elemental composition, measured as atomic percentage, with oxidation state speciation. Such a study is fundamental in the present work, for evaluating how the different preparation techniques influence the properties of the materials. In binary metal oxide samples, regions relative to specific core levels and Auger transitions are here the object for detailed analysis of high resolution XPS spectra.

Copper oxidation state and speciation are commonly studied looking at high resolution spectra for Cu $2p_{3/2}$ core level in the Binding Energy (BE) range 925–950 eV and for Cu $L_3M_{4,5}M_{4,5}$ (Cu LMM) Auger transition (BE 555–600 eV). Considerations about peak shape, curve fitting, shake-up peaks and the value of the proper modified Auger parameter are needed [41,42]. The latter equals 1849.0 eV in both commercial (C) and synthesized (S) samples. This value and the inspection of the Auger peak shape exclude any occurrence of Cu metal. Moreover, the lack of shake-up peaks in S-Cu_2O confirms the almost exclusive presence of the Cu(I) oxidation state for copper atoms, while shake-up peaks observed in C-Cu_2O, support the presence of CuO [41,42]. These considerations and curve fitting of O 1s XPS peak are used to determine the ratio Cu(I)/Cu(II), that equals 95/5 in S-Cu_2O and 70/30 in C-Cu_2O, showing that the surface of commercial Cu_2O sample is significantly oxidized.

Similar arguments are used for speciation of Indium atoms: analysis involves the In 3d core level doublet (BE 440–460 eV) and the In $M_5N_{4,5}N_{4,5}$ (In MNN) Auger transition (BE 1060–1090 eV) [43]. Energy positions of these are used to calculate a proper modified Auger parameter, which equals 850.8 eV in S-In_2O_3 and 851.1 eV in C-In_2O_3. The core level peaks resulted broad and symmetric, and no plasmon loss feature at higher binding energies were observed, as it would be expected if metallic Indium was present [43,44]. It can be thus concluded that In_2O_3 (both C and S) samples are formed exclusively of In(III)-oxide.

In samples containing iron atoms, the study about the Fe-oxidation state focused onto Fe 2p core level multiplet, occurring in the 705–730 eV region [45,46]. It is recorded as the envelope of different signals coming from electrostatic and spin interactions, from crystal field interactions and from spin-orbit coupling between the 2p and 3d states. This signal requires careful curve fitting operation able to identify the $2p_{3/2}$ main peak center of gravity and satellite peaks structures, due to shake-up and charge transfer processes, whose binding energy separation is used as parameter [45]. The recorded signal indicates that only the Fe^{3+} oxidation state is present in the prevalent α-Fe_2O_3 form in both C- and S-samples, which, nevertheless, show a different multiplet splitting and a different energy separation. The latter equals 7.4 eV in the S-Fe_2O_3 and 8.4 eV in the C-Fe_2O_3, with the higher value being indicative for the presence of $Fe(OH)_3$ [45]. From the data above, one can conclude that S-samples are more reliable than commercial ones.

In mixed metal oxides samples, the surface composition (atomic percentage) was obtained through quantification from the single high resolution XPS spectra for specific peaks: C1s, O1s, and $Cu2p_{3/2}$ elements were detected in all samples. Table 2 accounts for atomic percentage concentration data, which are specific for surface composition and can thus differ from those measured by EDX, which are instead relative to bulk concentration. By comparison with data in Table 1, a higher Cu atoms occurrence at the surface is observed for all samples, except for the CP-Cu/Fe-2 sample.

Table 2. Surface composition in mixed oxide samples determined by XPS. Atomic percentages are reported as mean values ± 1S (values averaged out of at least three replicates).

Sample	Surface Content (at. %)					Metal Content Ratio	
	C	O	Cu	In	Fe	Cu/In	Cu/Fe
HEM-Cu/In-2	42 ± 7	39.1 ± 0.7	13 ± 5	7 ± 2	-	1.95 ± 0.20	-
CP-Cu/In-2	26 ± 3	42.8 ± 1.0	17.4 ± 0.9	13.5 ± 1.2	-	1.29 ± 0.06	-
HEM-Cu/Fe-2	40.0 ± 0.5	36.4 ± 0.5	13.9 ± 0.2	-	9.7 ± 0.2	-	1.47 ± 0.12
CP-Cu/Fe-2	31.3 ± 1.0	45.8 ± 1.5	9.2 ± 0.2	-	14 ± 2	-	0.68 ± 0.13

High resolution spectra for Cu speciation in mixed metal oxides samples are shown in Figure 1. By comparing samples prepared by HEM (Figure 1a,b) or CP (Figure 1c,d), a nearly perfect trace overlapping is observed in both Cu/In and Cu/Fe pairs, showing no difference for the preparation techniques.

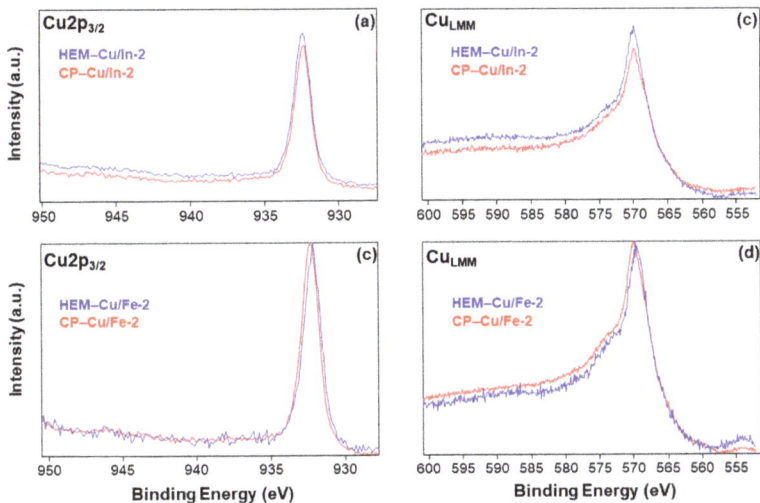

Figure 1. X-ray Photoelectron spectra for Cu $2p_{3/2}$ and Cu $L_3M_{4,5}M_{4,5}$ (Cu LMM) Auger transition in (**a**,**b**) Cu/In and in (**c**,**d**) Cu/Fe pairs.

Moreover, Cu atoms on the surface are exclusively encountered as Cu(I), with a CuO component that is observed only in traces and not quantifiable [42]. This is true also for samples prepared using commercial-Cu_2O, which contains 30% CuO (*vide infra*), showing that, where present, Cu(II) is in the bulk more than on the surface.

The spectra for In-3d core level doublet were recorded in HEM and CP mixed oxides samples and they are shown in Figure 2a: the only chemical state observed was In_2O_3 [43,44]. Figure 2b shows the Fe 2p core level spectra as recorded in mixed oxides samples. Curve fitting results and peak analysis show the presence of the solely α-Fe_2O_3 form of iron species [45,46] in both measured samples.

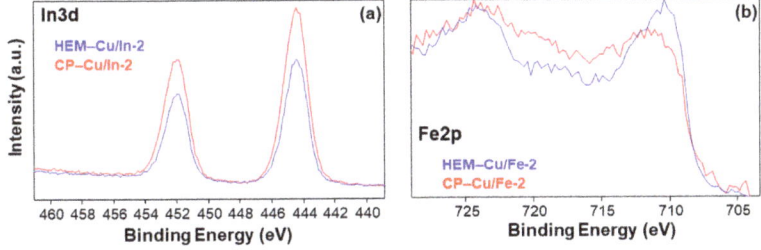

Figure 2. XP spectra for (**a**) In 3d and (**b**) Fe 2p core levels in mixed metal oxides samples.

In conclusion, samples prepared by HEM and CP do not show differences for what concerns the surface composition, and more interestingly C- and S-samples of ternary oxides show very similar surface composition and properties, even if C-Cu_2O contains 30% of Cu(II) which remains confined in the bulk.

Valence Band Maximum Evaluation by XPS

XPS analysis can be used to measure Valence Band Maximum (VBM) energy level, which is fundamental for subsequent band structure evaluation. This task can be accomplished adopting the procedure developed by Kraut and co-workers [47,48], then correctly extended to not covalent and oxide-based semiconductors by Chambers and co-workers [49]. The procedure requires acquisition of the XP spectrum in the low binding energy region, near the zero-value which corresponds to the Fermi level. Here, the XP spectrum reflects the electron density in the low energy states [47–49]. The VBM energy is determined as the intersection of two straight lines, obtained from least square fitting: the first line fits the spectrum baseline and background over the Fermi level, the second line fits the leading edge of the spectrum towards increasing binding energy [49]. In this work, while the band edge fitting operation was optimized through the maximization of R^2 correlation coefficient, the background fitting operation was performed by inclusion of all data points measured at negative binding energies coordinates. Figure 3 shows XP spectra in the Valence Band region, between 11 and -3 eV, with comparison of Commercial and Synthesized binary metal oxide samples. In Cu_2O samples, a 0.74±0.20 eV difference in extracted band edges is observed, though spectra are similar in shape and sharp steep band edges do appear. Spectra are shown in Figure 3a with values equal 1.65±0.20 and 0.86±0.14 eV in C-Cu_2O and S-Cu_2O samples, respectively, the difference being ascribed to the presence of CuO in the C-sample, for which a more positive VB edge is commonly attested [19,40,50]. Results agree with both theoretical [40,51] and experimental common literature references [19,20,50,52].

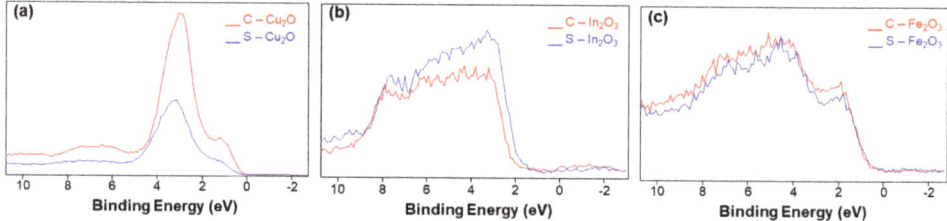

Figure 3. XP Valence Band Spectra in (a) Cu_2O, (b) In_2O_3, and (c) Fe_2O_3 binary oxide samples.

XP spectra for valence band region in In_2O_3 samples are shown in Figure 3b, where a certain difference is observed: the synthesized sample shows a higher intensity peak than the commercial one. This feature is indicative of a higher number of localized electrons within the VB energy levels, and its origin has been identified in the occurrence of random O-vacancies, whose levels build up the VB ones by orbital mixing [53], and in the crystalline domains size. This last could result reduced enough to hinder efficient electron transfer to the CB and so decrease the naturally occurring n-type character of this semiconductor material [25,54]: in fact, VBM results are closer to the Fermi level. Extracted values equal 2.20±0.11 and 1.80±0.15 eV for Commercial and Synthesized samples respectively, and both result less positive than literature values [25,30,34,52], making the material closer to hydrogen evolution potential.

Figure 3c reports XP spectra for valence band region in Fe_2O_3 samples, and differences between samples are visible, similar to those observed in In_2O_3 samples. In this case, the Commercial sample is characterized by a higher number of electrons in the VB, supposed to derive from localized states within the energy gap, due to defective particles, especially oxygen vacancies [55], and agree with the presence of iron hydroxide traces shown by XPS speciation measurements.

A less intense signal is recorded in the synthesized sample, and it is thought to be caused by defective particles and small crystalline domains. XPS data for Fe-containing systems are typically difficult to fit/convolve. Extracted VBM position are 1.53 ± 0.22 and 1.67 ± 0.23 eV for C-Fe_2O_3 and S-Fe_2O_3 samples, respectively, with a small difference between them and at lower energy than some literature theoretical [55,56] and experimental data [36,39,57]. Table 3 lists VBM extracted values for all binary metal oxide samples described above.

Table 3. Valence Band Maxima extracted from XPS measurements. The error values were determined from the regression method extrapolation.

Sample	VBM (eV)	Sample	VBM (eV)
C-Cu$_2$O	1.65 ± 0.20	S-Cu$_2$O	0.86 ± 0.14
C-In$_2$O$_3$	2.20 ± 0.11	S-In$_2$O$_3$	1.80 ± 0.15
C-Fe$_2$O$_3$	1.53 ± 0.22	S-Fe$_2$O$_3$	1.67 ± 0.23

2.3. UV-Visible Spectroscopy Characterization

The first step for photocatalytic processes is light absorption, thus UV-Visible Diffuse Reflectance Spectroscopy (UV-VIS DRS) was used to measure the optical properties for all samples in 200–800 nm range. The UV-VIS DRS properties directly depend on band gap and electronic energy structure and they affect the photocatalytic activity too [10,11,18]. Changes of light absorption properties with composition were observed in mixed oxide samples in present work. In samples containing the Cu$_2$O/In$_2$O$_3$ pair, absorption spectra (Figure 4) clearly change with composition. Lines corresponding to binary metal oxides are plotted for comparison. Whether HEM or CP preparation technique is adopted, the increase in Cu/In ratio induces a general redshift in absorption spectra and a corresponding significant absorption at wavelengths above 500 nm, where fundamental transition of Cu$_2$O is [19]. While the absorption tail observed in CP prepared samples (Figure 4a) can be attributed mainly to scattering phenomena, due to aggregation of small particle size, the reduction in size for Cu$_2$O component is at the origin of the blueshift observed in spectra of HEM prepared samples (Figure 4b). Such a shift and the marked absorption tails observed at long wavelengths for samples prepared using C-Cu$_2$O are instead attributed to the not negligible CuO presence in commercial Cu$_2$O, because CuO fundamental absorption occurs at lower energy [19].

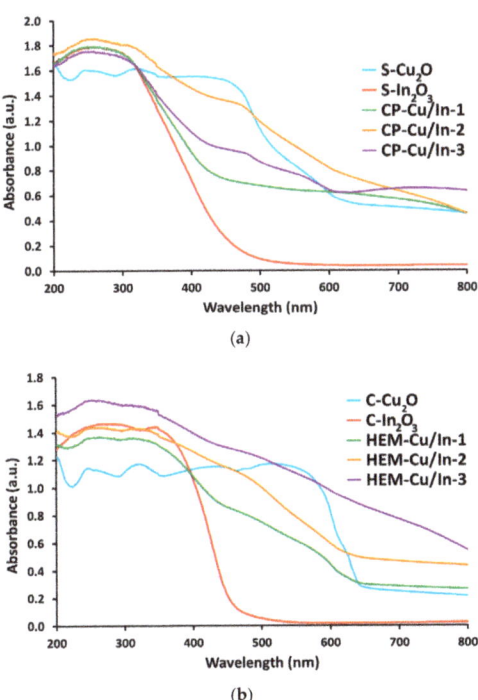

Figure 4. UV-Visible Diffuse Reflectance Spectroscopy (UV-VIS DRS) for Cu/In mixed oxides pairs prepared by (**a**) CP and by (**b**) HEM.

Absorption spectra recorded for samples containing the Cu_2O/Fe_2O_3 pair are shown in Figure 5, together with lines of binary metal oxides for comparison. Here, the increase in Cu/Fe is found to change absorption spectra for both preparation techniques, but in a different way accounting for absorption features of the two components at wavelengths longer than 500 nm [19,26]. In CP prepared samples (Figure 5a) the component oxides present fundamental absorption region different enough to observe a linear trend with increasing Cu/Fe ratio and thus a slight blueshift and increasing similarity towards the Cu_2O spectrum. This feature has been attributed to typical α-Fe_2O_3 spectra characteristics which appear over 550 nm [26,55], well over than S-Cu_2O fundamental absorption [19], thus, the addition of this specific component did not result in a redshift. On the contrary, in samples prepared by HEM using C-Cu_2O (Figure 5b), spectra of precursor components (C-Cu_2O and C-Fe_2O_3) show a similar fundamental absorption region and differences only in absorption intensity. Hence, an increase in Cu/Fe ratio corresponds to a higher absorption at longer wavelengths, which has been mainly attributed to CuO impurities in C-Cu_2O precursor.

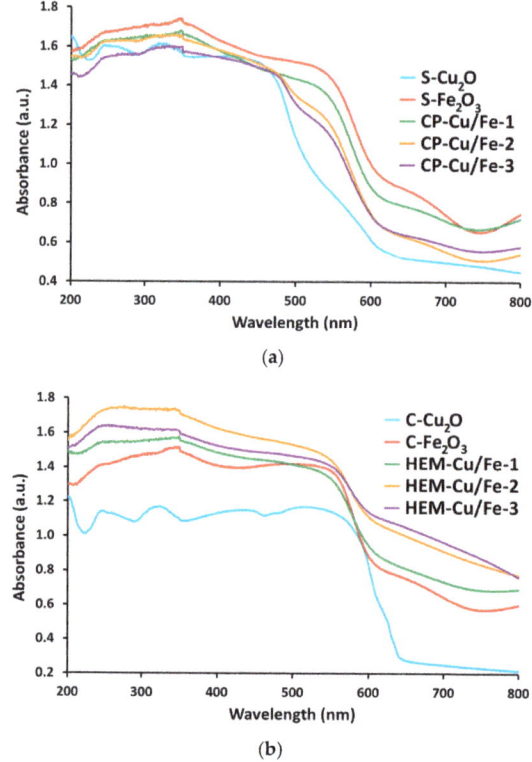

Figure 5. UV-Visible DRS for Cu/Fe mixed oxides pairs prepared by (**a**) CP and by (**b**) HEM.

However, DRS absorption measurements demonstrated that prepared nanocomposites are photoactive in almost the whole UV-Visible range in function of their composition. In particular, the introduction of Cu_2O has given In_2O_3 a better response in the VIS-region, while the interaction between Cu_2O and α-Fe_2O_3 has induced less predictable effects because of similar fundamental absorption. Studying these features is crucial because they are involved in enhancement of solar light harvesting properties for application in photocatalysis. Anyway, in both pairs, properties are very sensitive to nanocomposites preparation procedures, and thermal treatments.

Band-Gap Evaluation by UV-Visible Spectroscopy

The optical energy gap ($E_{g,opt}$) is a fundamental property in semiconductors and it equals the minimum energy required to excite an electron from VB to CB by means of light absorption. This energy gap can be directly measured through UV-Visible spectroscopy, if a single fundamental absorption is clearly distinguished. In case this is not possible or solid-state samples are studied, as in the present work, a simple and widely adopted data elaboration method can be used, which is described in detail elsewhere [58,59]. Briefly, the Kubelka–Munk function (K–M), F(R), is calculated starting from the experimental reflectance spectrum and is related to linear absorption coefficient α and to $E_{g,opt}$ through a power law (1) describing the optical absorption strength in function of photon energy.

$$F(R) \cdot (h\nu) = A \cdot (h\nu - E_g)^n \tag{1}$$

The exponent n assumes different values depending on the type of electronic transition. Provided there is some knowledge about the occurring electronic transition, the plot of product (2) versus radiation energy, $(h\nu)$, shows up a linear trend in the region corresponding to fundamental absorption and energy gap [58,59].

$$(F(R) \cdot (h\nu))^{1/n} \tag{2}$$

The linear least square fitting in this region allows for the extraction of the $E_{g,opt}$ value as the intersection of straight line with the energy axis, according to the Tauc Plot extrapolation procedure. It can be applied to pure or lightly doped semiconductors, but it does not produce reliable results if fundamental absorption edges are not separable or if a simple combination of individual optical gaps cannot be assumed, as in highly doped semiconductors or in nanocomposites. In the present work, mixed oxide samples fall in the second case, therefore $E_{g,opt}$ value was measured only for all single metal oxide samples. The linear regression operations have been optimized by merging two criteria: (a) maximization of correlation coefficient R^2, ensuring a better description by linear model and (b) consideration of a calculation range containing a minimum of 20 data points, to give procedure a statistical validation. Matching those two criteria allows for a good extrapolation result. Tauc Plots for different single metal oxide samples are shown in Figure 6.

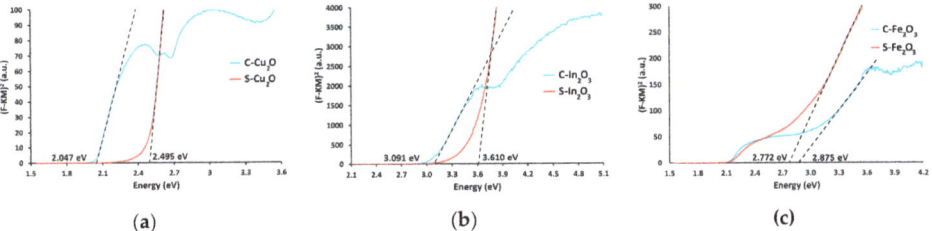

Figure 6. Tauc Plots for (**a**) Cu_2O, (**b**) In_2O_3, and (**c**) Fe_2O_3 binary oxide samples.

In the case of Cu_2O, absorption spectra are reported as light blue traces in both Figures 4 and 5 and large difference do appear between C-Cu_2O and S-Cu_2O samples. Absorption in the commercial sample extends up to 600 nm, while the S-Cu_2O sample shows a significant absorption in the UV region, decreasing significantly over 500 nm, where some absorption is guaranteed by a pronounced tail extending at longer wavelengths. The extended presence of CuO (30%) in the commercial precursor and particle size cause the differences. Figure 6a shows Tauc Plots for Cu-oxide samples. These show a steeper rise for S-Cu_2O sample and two very different values for estimated $E_{g,opt}$ energies are observed, and both have been calculated assuming a direct allowed electronic transition occurs between VBM and CBM [19,40], and n coefficient is, thus, set equal to 0.5. Values within 2.0 and 2.2 eV are generally attested in literature for Cu_2O [40], though quantum size effect is widely recognized able to markedly influence its energy gap. In this work, $E_{g,opt}$ values equal to 2.047 and 2.495 eV for C-Cu_2O and

S-Cu$_2$O sample, respectively, are found in accordance with literature [20,36,39]. This is also true for the synthesized sample, where a UV-shifted value is justified by quantum size confinement and where a pronounced tail, recorded before gap and reflected in absorption spectrum, is attributed to crystalline disorder and broad size dispersion.

Absorption spectra for In$_2$O$_3$ samples are shown in Figure 4 as red traces which do not show large differences at a first examination: both commercial and synthesized samples absorb radiation mainly in the near UV range (below 450 nm). Tauc Plots (Figure 6b) enhance features previously not evident, such as a large tail in the S-In$_2$O$_3$ trace or its steeper rise with respect to the C-In$_2$O$_3$. Extracted $E_{g,opt}$ values are equal to 3.091 eV in C-In$_2$O$_3$ and to 3.610 eV in the S-In$_2$O$_3$ sample. The determination was performed considering that a direct allowed electronic transition occurs, as for recent studies [25,54]. Though band gap nature and electronic structure for In$_2$O$_3$ are somewhat controversial, a fundamental band gap ranging from 2.6 to 2.9 eV is now commonly accepted [25], and it slightly differs from the optical gap, attested within 2.3 and 3.8 eV [30,34], therefore values measured in the present work result in accordance with literature. The 0.6 eV difference observed has been deemed coming from the preparation technique. In this case, its contribution affects band structure mainly through the introduction of lattice defects, especially oxygen vacancies, and creation of mid-gap states, a phenomenon which is commonly found in *n*-type wide gap semiconductor oxides [54], the effect appears more pronounced in the S-In$_2$O$_3$ sample, where the trace shows a large tail extending towards low energies.

Both the Fe$_2$O$_3$ samples exhibit good light harvesting properties in the whole UV-Visible region, with significant absorption up to 600 nm, higher in S-Fe$_2$O$_3$ sample, as shown by analysis of red traces in Figure 5. Four different regions are commonly identified in absorption spectra [31] and the fundamental band gap is recognized at 2.2 eV [26,55], with discussion whether its direct or indirect nature and thus its coincidence with the optical band gap. Furthermore, complication can arise in $E_{g,opt}$ determination because it usually merges with an exciton absorption, which is produced by an indirect transition between 3d–3d orbitals and dominates spectra over 550 nm, giving hematite its typical red color [55]. The exciton is clearly observed in Tauc Plots as the low intensity shoulder with onset at 2.1 eV (Figure 6c). Because of this indirect nature, absorption below 2.1–2.2 eV is considered not able to produce useful separated electron-hole pairs, suffering from very fast recombination, a widely recognized drawback in α-Fe$_2$O$_3$ [26,55]. Therefore, to study photocatalysis-useful optical absorption, a direct allowed transition has been assumed for Tauc Plots. Extracted $E_{g,opt}$ equal 2.772 and 2.875 eV for S-Fe$_2$O$_3$ and C-Fe$_2$O$_3$ sample, respectively. Although these values result unusually larger than the commonly accepted and measured ones [31,32,36,39], such a discrepancy can be explained by peculiar interacting electronic levels in iron atoms and presence of lattice defective particles. All these elements can lead to mixing between the standard direct gap and higher energy features (Ligand to Metal Charge Transfer (LMCT) processes occurring at 2.9–3.1 eV) [55]. Aggregation of very small particles towards polycrystalline clusters formation was cited able to enhance this mixing and can be also identified as the source for the large tailing trend in S-Fe$_2$O$_3$ trace.

Table 4 lists single metal oxide samples, with indication of optical energy gap extracted values and corresponding absorption wavelength. Except for iron oxide, S-samples show higher optical energy gaps than C-ones. Moreover, S-samples reveal a tailing trend more pronounced than in C-samples, indicative of broader size dispersion and lower average size (10–50 nm).

Table 4. Optical energy gap and corresponding absorption wavelength in binary oxide samples.

Sample	Optical E_g (eV)	Absorption Wavelength [1] (nm)
C-Cu$_2$O	2.047	605.7
S-Cu$_2$O	2.495	497.0
C-In$_2$O$_3$	3.091	401.1
S-In$_2$O$_3$	3.610	337.9
C-Fe$_2$O$_3$	2.875	431.3
S-Fe$_2$O$_3$	2.772	447.3

[1] Calculated by the relation $\lambda = 1240/E_{g,opt}$ [18].

2.4. Band Structure Evaluation and Discussion

Possible photocatalytic activity performances in solar fuels production by utilization of ternary metal oxides prepared in this work has been outlined on the basis of electronic band structure of binary and ternary mixed metal oxides samples. In fact, the energy position of band edges corresponds to redox potential of electrons and holes in semiconductor materials, thus it determines the redox behavior of photogenerated charge carriers and their possible transfer to adsorbed chemical species, as needed for photocatalytic reaction to occur [6,7,10,11].

An estimation for VBM and CBM is thus fundamental in characterization of semiconductors, and this task is usually accomplished by combination of theoretical considerations and of experimental data obtained by use of different techniques [10–12]. For energy band structure evaluation in the present work, experimental data for VB edges, as determined by XPS, and for $E_{g,opt}$, as determined by UV-Visible DRS, have been used in combination with theoretical arguments, found in literature, about energy band structure of specific semiconductor metal oxide considered. Band structures plots are shown in Figure 7 for ternary metal oxides, and they are reported as composed by those of binary metal oxides and compared to redox potentials involved in solar fuels production [2,11].

In all band structure schemes (Figure 7), the left side refers to Cu$_2$O energy bands, both for C- and S-samples. These were determined by considering only VBM and $E_{g,opt}$, where this last equals the fundamental E_g [19]. A big discrepancy observed in energy gap values and band position has been ascribed to CuO impurities, which were detected in C-samples used as precursor for HEM, and to quantum size confinement effect. Both of these arguments concur to the whole S-Cu$_2$O sample structure being moved upward in energy with respect to the C-Cu$_2$O sample, for which a different plot for Cu$_2$O and CuO is not possible. Moreover, it is worth noting that both CuO and Cu$_2$O are *p-type* semiconductors materials [19,50], but the unpredictable formation of junctions between Cu$_2$O and CuO can affect the Fermi level position in C-Cu$_2$O [50], thus hiding the *p-type* character expected in the schemes of Figure 7a,c, which is on the contrary easily detectable in schemes of Figure 7b,d, referred to S-Cu$_2$O.

Band structures schemes for C-In$_2$O$_3$ and S-In$_2$O$_3$ samples are reported in Figure 7a,b, respectively, where theoretical arguments found in literature about somewhat controversial and long debated electronic structure were considered [25]. In this frame, the direct optical gap measured involves a level which lays 0.81 eV below the VBM determined by XPS. Thus, the fundamental E_g for In$_2$O$_3$ equals 2.28 and 2.80 eV for C-In$_2$O$_3$ and S-In$_2$O$_3$ samples, respectively, in accordance with a maximum possible 2.9 eV value [25]. Also, schemes report correct *n-type* semiconductor behavior in both the samples [25,54].

In the case of Fe$_2$O$_3$ samples, band structure schemes are plotted in Figure 7c,d for C-Fe$_2$O$_3$ and S-Fe$_2$O$_3$ samples, respectively. While optical energy gap is considered to occur between the VBM determined by XPS and a level above the CBM, the fundamental energy gap involving the real CBM equals the energy required for exciton formation [55]. This energy has been measured by studying the corresponding optical absorption with Tauc Plot procedure, with assumption of an indirect transition (n equals 2) which incorporates exciton absorption and thus reflects the interested energy difference [31]. Fundamental E_g equals 2.01 and 1.97 eV for C-Fe$_2$O$_3$ and S-Fe$_2$O$_3$ samples, respectively, values slightly lower than common 2.1 eV expected for exciton in hematite [55]. As for the In_2O_3 samples, expected *n-type* semiconductor character results evident in all α-Fe_2O_3 samples [26,55].

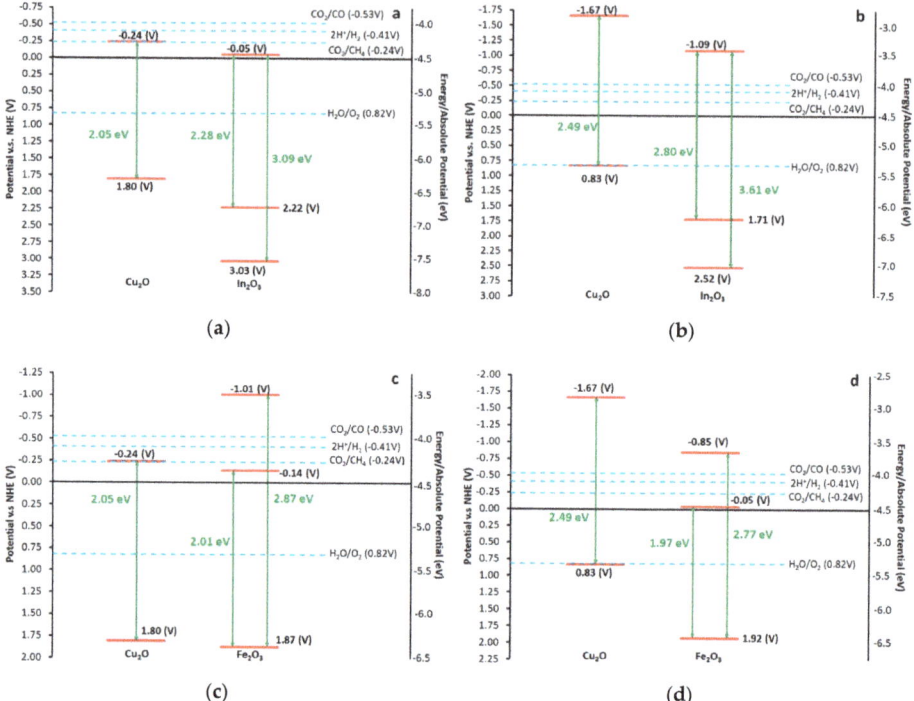

Figure 7. Band structures schemes for Cu_2O/In_2O_3 pair, prepared by (**a**) HEM and (**b**) CP, and for Cu_2O/Fe_2O_3 pair, prepared by (**c**) HEM and (**d**) CP.

Results obtained by this method are shown in Table 5, where band edges and energy gaps, both optical and fundamental are pointed out.

Table 5. Resume of energy gaps and band edges levels in semiconductor metal oxides.

Sample	E_g (eV)		VBM (eV)	CBM (eV)
	Fundamental	Optical		
C-Cu_2O	2.05	2.05	1.80	−0.24
S-Cu_2O	2.49	2.49	0.83	−1.67
C-In_2O_3	2.28	3.09	2.22	−0.05
S-In_2O_3	2.80	3.61	1.71	−1.09
C-Fe_2O_3	2.01	2.87	1.87	−0.14
S-Fe_2O_3	1.97	2.77	1.92	−0.05

From data in Table 5, some photocatalytic activity result in solar fuels production can be expected only for the CP prepared Cu_2O/In_2O_3 samples, whose band scheme structure is shown in Figure 7b. This pair could reveal useful in H_2 generation coming from photocatalytic water splitting, a reaction in which it is able to participate thank to proper band levels alignment of two metal oxide components. In particular, electrons excited in Cu_2O can move into In_2O_3 CB and leave behind holes in VB, a level adequate to oxidize water molecules. At the same time, the electrons photogenerated in In_2O_3 are in a level adequate to reduce H^+ and produce H_2, and this level also collects electrons coming from Cu_2O

CB, while the photogenerated holes move into Cu_2O VB, where they can oxidize water molecules. On the other hand, large charge carrier recombination occurring in Cu_2O could be responsible for widely recognized poor activity in general.

At the same time, CO_2 reduction can show different trend whether a 1e⁻ or multielectron-transfer is considered. In fact, the CB potential of the two components, results not negative enough for one electron transfer to CO_2 molecule (−1.90 V vs. NHE is needed) [2]. As a matter of fact, this is not the reaction we are considering: the Proton Coupled Electron Transfer (PCET) is the process that should operate in the present case.

On the basis of similar considerations, poor activity for solar fuels production can be foreseen for other oxides pairs studied in the present work which do not provide the potential for one-electron transfer to CO_2 molecule and do not show a proper level alignment for water splitting. Additionally, the presence of CuO impurities can be at the origin of enhanced charge carrier recombination, a phenomenon recognized in this material, up to withdrawn all photogenerated charge carriers from reaction or charge carrier separation processes. More, in both Cu_2O/Fe_2O_3 pairs (Figure 7c,d), the not proper level alignment adds to their relative positions, which results not adequate for photogenerated charge carrier separation, as in HEM prepared pair (Figure 7c), or even in enhanced interband recombination, as in CP prepared pair (Figure 7d).

2.5. Photocatalytic Activity

Therefore, among all materials prepared in this work, only CP-Cu_2O/In_2O_3 was tested in CO_2–H_2O coprocessing, according to procedure described in Methods section, as reputed the only one able to carry out the redox process. Three different reactors were used for running the reaction, all under Xe-lamp irradiation (Figure 8).

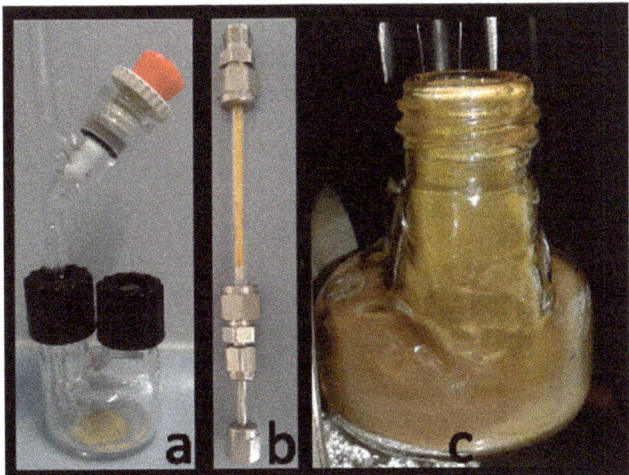

Figure 8. Reactors used for running the photochemical reaction. (**a**) Bulk nano-sized catalyst in a batch-reactor. (**b**) Flow-reactor loaded with bulk nano-catalyst. (**c**) Nano-sized catalyst finely dispersed on the wall of the reactor, simulating a nano-film.

When reactor (a) or (c) was used, the reaction gas was sampled with a gas-syringe, and with reactor (b) the gas was directly injected into the GC-column. Hydrogen and reduced species of CO_2 were monitored at regular intervals of time up to 6 h. (see Figure S2 in Supplementary Materials) Table 6 gives the results for the reactors (a)–(c). Using bulk nano-catalysts, after 3 h of irradiation of a H_2O-saturated CO_2 stream at 298 K no reduction products were observed. Conversely, when the nano-catalyst was finely dispersed on the surface of the reactor the same mixture gave positive formation of H_2 and traces of reduced CO_2 species.

Table 6. Photocatalytic reaction for three different reactors.

Reaction System	Gaseous Reaction Products		
	H_2	CO	CH_4
(a)	–	–	–
(b)	–	–	–
(c)	+++	+	+

Results in Table 6 shift attention towards the fundamental role played by reactor design and reaction setup in carrying out a process difficult as the solar fuel production. It is clearly visible how the catalyst film distribution (reactor c), allowing for a better and more homogeneous light penetration to the photo-active centers with respect to the massive-powdered-material (reactors a,b), allows the reaction to go. In practice, a film-like distribution of the photomaterial is preferred to its bulk packing as the larger surface of the photo-catalyst allows more photons to be active. Most likely, if the illumination system is changed with respect to the one we have used (Xe-lamp), the reactor (b) will be working too. The lesson learned with such synthesis of photo-materials and photocatalytic experiments, prompts us to a more focused approach to photocatalyst development coupled to reactor design and engineering for a more active conversion of CO_2 and water into energy products.

3. Materials and Methods

The following reagents were used as received and without any further treatment: Cu_2O powder (Fluka AG, in Sigma-Aldrich, Steinheim, Germany), In_2O_3 powder (Sigma-Aldrich, Steinheim, Germany), Fe_2O_3 powder (Sigma-Aldrich, Steinheim, Germany), anhydrous $CuSO_4$ (BDH Chemicals ltd, Poole, England), $In(NO_3)_3 \cdot xH_2O$ (Sigma-Aldrich, Steinheim, Germany), $FeCl_3$ (Carlo Erba Reagents srl, Milano, Italy), NaOH pellets (Fluka AG in Sigma Aldrich, Steinheim, Germany), and L-Ascorbic Acid (Alfa Aesar GmbH, Karlsruhe, Germany). Deionized water was used for all syntheses in which a solvent was required.

3.1. Preparation of Oxides

Single metal oxide samples were prepared by precipitation method, procedures adopted differ upon specific metal oxide. If not differently indicated, all operations were performed at ambient temperature and under air condition.

3.1.1. Cu_2O

Copper (I) oxide samples were prepared by adding 20 mL of a 0.5 M NaOH aqueous solution, dropwise and under stirring, to 10 mL of a 0.5 M $CuSO_4$ aqueous solution, thus producing copper (II) hydroxide. An adequate volume of a 0.1 M L-ascorbic acid aqueous solution was added dropwise, in order to reduce the Cu(II)-hydroxide into copper(I) oxide, Cu_2O. Particles were separated by centrifugation (6000 rpm, 15 min), washed with deionized water, and dried at 80 °C overnight.

3.1.2. In_2O_3

To 15 mL of a 1.1 M $In(NO_3)_3 \cdot H_2O$ aqueous solution, 50 mL of a 1.2 M NaOH aqueous solution were added dropwise, under stirring. The resulting indium(III) hydroxide suspension was kept under constant stirring for one further hour, then the solid was separated by centrifugation (6000 rpm, 15 min), washed with deionized water, and dried at 80 °C overnight. The dried solid was grinded in a ceramic mortar and the powder was heated in air at 300 °C for 2 h and then at 400 °C for 1 h, to convert hydroxide into indium(III) oxide, In_2O_3.

3.1.3. Fe_2O_3

Iron (III) oxide was prepared adding drop by drop 50 mL of a 1.2 M NaOH aqueous solution to the same volume of a 0.4 M $FeCl_3$ aqueous solution, under stirring. The so prepared suspension contains mixed oxide-hydroxide iron particles, which were separated by centrifugation (6000 rpm, 15 min), washed with deionized water and dried at 80 °C overnight. It was then grinded in a ceramic mortar and heated in air at 450 °C for 1 h, for complete conversion into iron(III) oxide, Fe_2O_3.

3.1.4. Preparation of Mixed Oxides by Coprecipitation Using Synthesized Binary Oxides

General Procedure: Cu_2O was deposited on the nanoparticles of the second oxide with formation of core-shell In_2O_3 or Fe_2O_3 particles covered with Cu_2O. We have attempted to produce Cu_2O core shell covered with In_2O_3 or Fe_2O_3 but so far substantial oxidation of Cu(I) to Cu(II) was observed during the dehydration of In- or-Fe-hydroxides. In 20 mL of a 0.5 M $CuSO_4$ aqueous solution, weighted (see below) amounts of the nano-sized powder for the *n-type* partner oxide were dispersed. 40 mL of a 0.5 M NaOH aqueous solution were dropwise added to such solution under stirring. When the addition was completed, an adequate amount of a 0.1 M L-ascorbic acid aqueous solution was added dropwise as a reductant (glucose has also been used, but with lower yield) to produce Cu_2O that deposited on the *n-type* semiconductor. Particles were separated by centrifugation (6000 rpm, 15 min), washed with deionized water, and dried at 80 °C overnight.

(a) 3.5929, 1.5698, and 0.8121 g of S-In_2O_3 nano-powder were dispersed in $CuSO_4$ aqueous solution for the ratios Cu/In-1, 2, and 3, respectively, and treated as reported above.
(b) 2.1781, 0.8799, and 0.7287 g of S-Fe_2O_3 powder were dispersed in $CuSO_4$ aqueous solution for ratios Cu/Fe-1, 2, and 3, respectively and reacted as reported in the general procedure.

3.1.5. Preparation of Mixed Oxides by High Energy Milling-HEM Using Commercial Samples

General Procedure: Mixed oxides were prepared by weighting defined quantities of commercial metal oxides powders and pouring them in cylindrical agate jars (46 mL), together with 3 agate spheres, 1 cm in diameter. Jars were sealed and placed in a planetary mill (Pulverisette 7-Fritsch, GmbH, Idar-Oberstein, Germany) and subjected to two cycles at 800 rpm, each with a duration of 90 min.

(a) for Cu/In-1, 0.5469 g of C-Cu_2O and 1.9863 g of C-In_2O_3 were mixed.
(b) for Cu/In-2, 0.8579 g of C-Cu_2O and 1.6354 g of C-In_2O_3 were mixed.
(c) for Cu/In-3, 1.2705 g of C-Cu_2O and 1.2304 g of C-In_2O_3 were mixed.
(d) For Cu/Fe-1, 0.7740 g of C-Cu_2O and 1.7332 g of C-Fe_2O_3 were mixed.
(e) For Cu/Fe-2, 1.4252 g of C-Cu_2O and 1.5915 g of C-Fe_2O_3 were mixed.
(f) For Cu/Fe-3, 1.6086 g of C-Cu_2O and 0.8969 g of C-Fe_2O_3 were mixed.

3.2. Characterization

Powder materials composition was determined by EDX measurements using an EDX-720 Shimadzu Spectrometer (Shimadzu Europe GmbH, Duisburg, Germany).

UV-Visible DRS spectra were recorded in the 200–800 nm region with a Cary-5000 spectrophotometer (Agilent Technologies, Santa Clara, CA, United States), equipped with an integration sphere covered by polymer internal coating, with a standard sample of the same material and with sample-holder for powder material.

X-ray Photoelectron Spectroscopy (XPS) analyses were run on a PHI 5000 Versa Probe II Scanning XPS Microprobe spectrometer (ULVAC-PHI Inc., Kanagawa, Japan). The measurements were done with a monochromatised Al Kα source (X-ray spot 200 µm), at a power of 50.3 W. Wide scans and detailed spectra were acquired in Fixed Analyzer Transmission (FAT) mode with a pass energy of 117.40 eV and 29.35 eV, respectively. An electron gun was used for charge compensation (1.0 V 20.0 µA). Data processing was performed by using the MultiPak software v. 9.9.0.8.

Gas mixtures were analyzed using a GC (Thermo Scientific Focus GC), equipped with a SUPELCO Carboxen™ 1010 PLOT (30 m × 0.32 mm) and with a TCD detector.

A 150 W Xe XBO lamp (Osram) was used as source of light for irradiation.

3.3. Photocatalytic Activity

Materials activity in solar fuels production was evaluated by using three different reactor systems (Figure 8) fed with a reagent gas mixture composed of CO_2 bubbled in deionized water to saturation. Such gas stream was flown for some minutes through the reactors during their loading operation in order to blow air away. Samples of gas were analyzed at regular intervals of time (30 min) over a period of 3–6 h.

In case (a), 0.1 g of nano-sized photocatalyst was placed in a 40 mL glass reactor, provided with two openings closed by hollow plastic caps. One cap holds a glass tube with a three-way valve for reagent gas mixture loading, the other is provided with a rubber septum for gas sampling using a 250 µL gas-tight syringe. The reactor was loaded with 0.15 MPa of CO_2 saturated with H_2O and stirred under illumination for 3 h. A gas-sample was withdrawn and analyzed using a GC.

In case (b), CO_2 saturated with water at 298 K was flown through a 2 mm i.d. tube (equipped with valves for feeding, sampling and reactor closing and filled with the same nanosized-catalyst) at a rate of 0.1 mL/min under illumination and recycled using a micropump. After three hours cycling the gas was injected into a GC. Alternatively, the tube was filled with CO_2–H_2O and the closed system illuminated for 3–6 h. The gas was then injected into the GC column.

In case (c), 0.1 g of the nanosized photocatalyst were dispersed on the internal surface of the same vessel used in (a) using a suspension of the catalyst in a volatile liquid (acetone or pentane) that was then slowly evaporated by flowing N_2. A thin, homogeneous solid film remained adherent to the surface of the reactor that was charged with CO_2–H_2O and illuminated. Illumination was applied for 3–6 h and a sample of gas was withdrawn with the gas-syringe at fixed intervals of time and analyzed by GC.

4. Conclusions

The experiments carried out in this work have confirmed that the properties of photocatalysts based on mixed oxides strongly depend on the starting materials and the way the binary- and mixed-oxides are prepared. Even their real composition depends on the technique used for their preparation. Careful analyses are necessary to confirm their composition that can or cannot match the stoichiometric ratio used in their preparation. The surface composition can be different from bulk and this can influence the reactions, should they occur on the surface or into channels in the bulk.

The electronic properties of the photo-materials change with their composition and mode of synthesis. Using two different techniques (HEM and CP), the fundamental properties of the photocatalysts have been measured, including the band-gap and electrochemical potential. Some of the catalysts prepared, based on their band gap and value of E, have been tested in the gas-phase photoreduction of CO_2 + H_2O. The experimental results show that the synthetic technique influences the photoactivity of the materials that can correctly be foreseen on the basis of bandgap experimentally derived. Of the mixed oxides prepared and described in this work, only $Cu_2O@In_2O_3$ prepared by co-precipitation from synthesized binary oxides have shown positive results in CO_2–H_2O photo-co-processing. Preliminary results show that the composition and synthetic methodologies of mixed-oxides, the reactor geometry, the way of dispersing the photocatalyst sample, play a key role in the light driven reaction of CO_2–H_2O. Hydrogen plus reduced species of CO_2 (in lower amount) have been observed, depending on the geometry of the reactor used and the photocatalyst used. In order to observe the formation of reduction products it is necessary that the catalyst is finely dispersed (thin film) and well illuminated. Massive amounts of photocatalyst are not active, at least under the illumination technique used in this work, most likely because the number of photons that reach the photoactive centers is quite low.

This work is a rare case of full characterization of photo-materials, using UV-Visible DRS, XPS, XRD, TEM, and EDX for the surface and bulk analytical characterization. We show that surface composition may not be the same of the bulk composition and plays a key role in photocatalysts behavior and a full material knowledge is necessary for the correct forecast of their photocatalytic behavior, inferred from experimentally determined bandgaps. Coupling UV-Vis DRS and XPS with EDX is necessary for getting the correct information about the composition of the materials and their surface-bulk characterization. Further studies are planned in order to discover the most active species and the best performing reactor geometry under best illumination conditions, using the systems which gave positive results so far. All of the systems described above are even under evaluation for discovering how their properties are changed with addition of partners such as noble metals or hole scavengers and attribute the correct role to each component of the photomaterial.

Supplementary Materials: The followings are available online at http://www.mdpi.com/2073-4344/10/9/980/s1, Figure S1: Comparison of CP-Fe_2O_3 XRD patterns with HEM-Cu/Fe-1 (a) and CP-Cu/Fe-1 (b) along with peak positions for reference diffraction patterns. Figure S2: High resolution XP spectra of all the samples. (a) C-Cu_2O sample: (a1) Cu $2p_{3/2}$ spectral region, (a2) Cu LMM Auger transition and (a3) O1s spectral region. (b) S-Cu_2O sample: (b1) Cu$2p_{3/2}$ spectral region, (b2) Cu-LMM Auger transition and (b3) O1s spectral region. (c) C-In_2O_3 sample: (c1) In3d spectral region, (c2) In MNN Auger transition and (c3) O1s spectral region. (d) S-In_2O_3 sample: (d1) In3d spectral region, (d2) In MNN Auger transition and (d3) O1s spectral region. Fe2p spectral region for (e1) C-Fe_2O_3 and (e2) S-Fe_2O_3. Figure S3: H_2 evolution with time by using CP-Cu/In mixed oxides under VIS light irradiation.

Author Contributions: Conceptualization, M.A. and A.D.; synthesis methodology and GC analytical techniques, F.N.; synthesis and spectroscopic characterization of photomaterials, D.M.S.M.; XPS analysis, N.D.; writing—original draft preparation, D.M.S.M.; writing—review and editing, M.A. and A.D.; supervision, A.D.; project administration, A.D.; funding acquisition, A.D. All authors have read and agreed to the published version of the manuscript.

Funding: This research received MIUR-IT funding. It was executed within the frame of a preliminary work of an international collaboration that has brought to a common EU-Project application.

Acknowledgments: The authors thank Roberto Comparelli (Istituto per i Processi Chimico-Fisici, Consiglio Nazionale delle Ricerche, c/o Dipartimento di Chimica, Università di Bari) and Teresa Sibillano (Istituto di Cristallografia, Consiglio Nazionale delle Ricerche, Bari) for performing and interpreting preliminary TEM and powder XRD measurements, respectively.

Conflicts of Interest: The authors declare no conflict of interest.

References

1. Global Monitoring Laboratory-Global Greenhouse Gas Reference Network. Available online: https://www.esrl.noaa.gov/gmd/ccgg/trends/global.html (accessed on 8 April 2020).
2. Aresta, M.; DiBenedetto, A.; Quaranta, E. State of the art and perspectives in catalytic processes for CO2 conversion into chemicals and fuels: The distinctive contribution of chemical catalysis and biotechnology. *J. Catal.* **2016**, *343*, 2–45. [CrossRef]
3. Aresta, M.; DiBenedetto, A.; Angelini, A. The changing paradigm in CO_2 utilization. *J. CO_2 Util.* **2013**, *3*, 65–73. [CrossRef]
4. Pesnell, D.; Addison, K. SDO | Solar Dynamics Observatory. Available online: https://sdo.gsfc.nasa.gov/ (accessed on 13 April 2020).
5. Ali, S.; Flores, M.C.; Razzaq, A.; Sorcar, S.; Hiragond, C.B.; Kim, H.R.; Park, Y.H.; Hwang, Y.; Kim, H.S.; Kim, H.; et al. Gas phase photocatalytic CO_2 reduction, "A brief overview for benchmarking". *Catalysts* **2019**, *9*, 727. [CrossRef]
6. Clarizia, L.; Russo, D.; Di Somma, I.; Andreozzi, R.; Marotta, R. Hydrogen generation through solar photocatalytic processes: A review of the configuration and the properties of effective metal-based semiconductor nanomaterials. *Energies* **2017**, *10*, 1624. [CrossRef]

7. Babu, V.J.; Vempati, S.; Uyar, T.; Ramakrishna, S. Review of one-dimensional and two-dimensional nanostructured materials for hydrogen generation. *Phys. Chem. Chem. Phys.* **2015**, *17*, 2960–2986. [CrossRef]
8. Inoue, T.; Fujishima, A.; Konishi, S.; Honda, K. Photoelectrocatalytic reduction of carbon dioxide in aqueous suspensions of semiconductor powders. *Nature* **1979**, *277*, 637–638. [CrossRef]
9. Lehn, J.-M.; Ziessel, R. Photochemical generation of carbon monoxide and hydrogen by reduction of carbon dioxide and water under visible light irradiation. *Proc. Natl. Acad. Sci. USA* **1982**, *79*, 701–704. [CrossRef]
10. Stolarczyk, J.K.; Bhattacharyya, S.; Polavarapu, L.; Feldmann, J. Challenges and prospects in solar water splitting and CO_2 reduction with inorganic and hybrid nanostructures. *ACS Catal.* **2018**, *8*, 3602–3635. [CrossRef]
11. Li, K.; Peng, B.; Peng, T. Recent advances in heterogeneous photocatalytic CO_2 conversion to solar fuels. *ACS Catal.* **2016**, *6*, 7485–7527. [CrossRef]
12. Nikokavoura, A.; Trapalis, C. Alternative photocatalysts to TiO_2 for the photocatalytic reduction of CO_2. *Appl. Surf. Sci.* **2017**, *391*, 149–174. [CrossRef]
13. Baran, T.; Wojtyła, S.; DiBenedetto, A.; Aresta, M.; Macyk, W. Zinc sulfide functionalized with ruthenium nanoparticles for photocatalytic reduction of CO_2. *Appl. Catal. B Environ.* **2015**, *178*, 170–176. [CrossRef]
14. Baran, T.; Wojtyła, S.; DiBenedetto, A.; Aresta, M.; Macyk, W. Photocatalytic carbon dioxide reduction at p-Type Copper(I) Iodide. *ChemSusChem* **2016**, *9*, 2933–2938. [CrossRef] [PubMed]
15. Li, X.; Yu, J.; Jaroniec, M.; Chen, X.-B. Cocatalysts for selective photoreduction of CO_2 into solar fuels. *Chem. Rev.* **2019**, *119*, 3962–4179. [CrossRef]
16. Védrine, J.C. Heterogeneous catalysis on metal oxides. *Catalysts* **2017**, *7*, 341. [CrossRef]
17. Roy, D.; Samu, G.F.; Hossain, M.K.; Janáky, C.; Rajeshwar, K. On the measured optical bandgap values of inorganic oxide semiconductors for solar fuels generation. *Catal. Today* **2018**, *300*, 136–144. [CrossRef]
18. Wang, Z.; Liu, Y.; Huang, B.; Dai, Y.; Lou, Z.; Wang, G.; Zhang, X.; Qin, X. Progress on extending the light absorption spectra of photocatalysts. *Phys. Chem. Chem. Phys.* **2014**, *16*, 2758. [CrossRef] [PubMed]
19. Christoforidis, K.C.; Fornasiero, P. Photocatalysis for hydrogen production and Co_2 reduction: The case of copper-catalysts. *ChemCatChem* **2018**, *11*, 368–382. [CrossRef]
20. Luévano-Hipólito, E.; Torres-Martínez, L.; Martínez, D.S.; Cruz, M.A. Cu_2O precipitation-assisted with ultrasound and microwave radiation for photocatalytic hydrogen production. *Int. J. Hydrog. Energy* **2017**, *42*, 12997–13010. [CrossRef]
21. Aguirre, M.E.; Zhou, R.; Eugene, A.J.; Guzman, M.I.; Grela, M.A. Cu_2O/TiO_2 heterostructures for CO_2 reduction through a direct Z-scheme: Protecting Cu_2O from photocorrosion. *Appl. Catal. B Environ.* **2017**, *217*, 485–493. [CrossRef]
22. An, X.; Li, K.; Tang, J. Cu_2O/reduced graphene oxide composites for the photocatalytic conversion of CO_2. *ChemSusChem* **2014**, *7*, 1086–1093. [CrossRef]
23. Xiong, Z.; Lei, Z.; Kuang, C.-C.; Chen, X.; Gong, B.; Zhao, Y.; Zhang, J.; Zheng, C.; Wu, J.C. Selective photocatalytic reduction of CO_2 into CH_4 over Pt-Cu_2O TiO_2 nanocrystals: The interaction between Pt and Cu_2O cocatalysts. *Appl. Catal. B Environ.* **2017**, *202*, 695–703. [CrossRef]
24. Handoko, A.D.; Tang, J. Controllable proton and CO_2 photoreduction over Cu_2O with various morphologies. *Int. J. Hydrog. Energy* **2013**, *38*, 13017–13022. [CrossRef]
25. Bierwagen, O. Indium oxide—A transparent, wide-band gap semiconductor for (opto)electronic applications. *Semicond. Sci. Technol.* **2015**, *30*, 024001. [CrossRef]
26. Mishra, M.; Chun, D.-M. α-Fe_2O_3 as a photocatalytic material: A review. *Appl. Catal. A Gen.* **2015**, *498*, 126–141. [CrossRef]
27. Cao, S.-W.; Liu, X.-F.; Yuan, Y.-P.; Zhang, Z.; Liao, Y.; Fang, J.; Loo, S.C.J.; Sum, T.C.; Xue, C. Solar-to-fuels conversion over In_2O_3/g-C_3N_4 hybrid photocatalysts. *Appl. Catal. B Environ.* **2014**, *147*, 940–946. [CrossRef]
28. Tahir, M.; Tahir, B.; Amin, N.A.S.; Muhammad, A. Photocatalytic CO_2 methanation over NiO/In_2O_3 promoted TiO_2 nanocatalysts using H_2O and/or H_2 reductants. *Energy Convers Manag.* **2016**, *119*, 368–378. [CrossRef]
29. Hoch, L.B.; He, L.; Qiao, Q.; Liao, K.; Reyes, L.M.; Zhu, Y.; Ozin, G.A. Effect of precursor selection on the photocatalytic performance of indium oxide nanomaterials for gas-phase CO_2 reduction. *Chem. Mater.* **2016**, *28*, 4160–4168. [CrossRef]
30. Hu, B.; Guo, Q.; Wang, K.; Wang, X.T. Enhanced photocatalytic activity of porous In_2O_3 for reduction of CO_2 with H_2O. *J. Mater. Sci. Mater. Electron.* **2019**, *30*, 7950–7962. [CrossRef]

31. Ba-Abbad, M.M.; Takriff, M.S.; Benamor, A.; Mohammad, A.W. Size and shape controlled of α-Fe$_2$O$_3$ nanoparticles prepared via sol–gel technique and their photocatalytic activity. *J. Sol-Gel Sci. Technol.* **2016**, *81*, 880–893. [CrossRef]
32. Boumaza, S.; Kabir, H.; Gharbi, I.; Belhadi, A.; Trari, M. Preparation and photocatalytic H$_2$ -production on α-Fe$_2$ O$_3$ prepared by sol-gel. *Int. J. Hydrog. Energy* **2018**, *43*, 3424–3430. [CrossRef]
33. Liu, J.; Ke, J.; Li, D.; Sun, H.; Liang, P.; Duan, X.; Tian, W.; Tade, M.; Liu, S.; Wang, S. Oxygen vacancies in shape controlled Cu$_2$O/reduced graphene oxide/In$_2$O$_3$ hybrid for promoted photocatalytic water oxidation and degradation of environmental pollutants. *ACS Appl. Mater. Interfaces* **2017**, *9*, 11678–11688. [CrossRef]
34. Liu, J.; Zhao, Y.; Zhang, J.-N.; Ye, J.-H.; Ma, X.-N.; Ke, J. Construction of Cu$_2$O/In$_2$O$_3$ hybrids with p-n heterojunctions for enhanced photocatalytic performance. *J. Nanosci. Nanotechnol.* **2019**, *19*, 7689–7695. [CrossRef] [PubMed]
35. Li, F.; Dong, B. Construction of novel Z-scheme Cu$_2$O/graphene/α-Fe$_2$O$_3$ nanotube arrays composite for enhanced photocatalytic activity. *Ceram. Int.* **2017**, *43*, 16007–16012. [CrossRef]
36. Lakhera, S.K.; Watts, A.; Hafeez, H.Y.; Neppolian, B. Interparticle double charge transfer mechanism of heterojunction α-Fe$_2$O$_3$/Cu$_2$O mixed oxide catalysts and its visible light photocatalytic activity. *Catal. Today* **2018**, *300*, 58–70. [CrossRef]
37. Shen, H.; Liu, G.; Yan, X.; Jiang, J.; Hong, Y.; Yan, M.; Mao, B.; Li, D.; Fan, W.; Shi, W. All-solid-state Z-scheme system of RGO-Cu$_2$O/Fe$_2$O$_3$ for simultaneous hydrogen production and tetracycline degradation. *Mater. Today Energy* **2017**, *5*, 312–319. [CrossRef]
38. Li, P.; Jing, H.; Xu, J.; Wu, C.; Peng, H.; Lu, J.; Lu, F. High-efficiency synergistic conversion of CO$_2$ to methanol using Fe$_2$O$_3$ nanotubes modified with double-layer Cu$_2$O spheres. *Nanoscale* **2014**, *6*, 11380–11386. [CrossRef]
39. Wang, J.-C.; Zhang, L.; Fang, W.-X.; Ren, J.; Li, Y.-Y.; Yao, H.-C.; Wang, J.-S.; Li, Z.-J. Enhanced photoreduction CO$_2$ activity over direct Z-scheme α-Fe$_2$O$_3$/Cu$_2$O heterostructures under visible light irradiation. *ACS Appl. Mater. Interfaces* **2015**, *7*, 8631–8639. [CrossRef]
40. Heinemann, M.; Eifert, B.; Heiliger, C. Band structure and phase stability of the copper oxides Cu$_2$O, CuO, and Cu$_4$O$_3$. *Phys. Rev. B* **2013**, *87*, 87. [CrossRef]
41. Biesinger, M.C.; Lau, L.W.; Gerson, A.R.; Smart, R.S. Resolving surface chemical states in XPS analysis of first row transition metals, oxides and hydroxides: Sc, Ti, V, Cu and Zn. *Appl. Surf. Sci.* **2010**, *257*, 887–898. [CrossRef]
42. Biesinger, M.C. Advanced analysis of copper X-ray photoelectron spectra. *Surf. Interface Anal.* **2017**, *49*, 1325–1334. [CrossRef]
43. X-ray Photoelectron Spectroscopy (XPS) Reference Pages: Indium. Available online: http://www.xpsfitting.com/search/label/Indium (accessed on 9 May 2020).
44. Barr, T.L.; Ying, L.L. An x-ray photoelectron spectroscopy study of the valence band structure of indium oxides. *J. Phys. Chem. Solids* **1989**, *50*, 657–664. [CrossRef]
45. Grosvenor, A.; Kobe, B.A.; Biesinger, M.C.; McIntyre, N.S. Investigation of multiplet splitting of Fe 2p XPS spectra and bonding in iron compounds. *Surf. Interface Anal.* **2004**, *36*, 1564–1574. [CrossRef]
46. Radu, T.; Iacovita, C.; Benea, D.; Turcu, R. X-ray photoelectron spectroscopic characterization of iron oxide nanoparticles. *Appl. Surf. Sci.* **2017**, *405*, 337–343. [CrossRef]
47. Kraut, E.A.; Grant, R.W.; Waldrop, J.R.; Kowalczyk, S.P. Precise determination of the valence-band edge in x ray photoemission spectra. *Phys. Rev. Lett.* **1980**, *44*, 1620–1623. [CrossRef]
48. Kraut, E.A.; Grant, R.W.; Waldrop, J.R.; Kowalczyk, S.P. Semiconductor core-level to valence-band maximum binding-energy differences: Precise determination by x-ray photoelectron spectroscopy. *Phys. Rev. B* **1983**, *28*, 1965–1977. [CrossRef]
49. Chambers, S.A.; Droubay, T.C.; Kaspar, T.; Gutowski, M. Experimental determination of valence band maxima for SrTiO$_3$, TiO$_2$ and SrO and the associated valence band offsets with Si(001). *J. Vac. Sci. Technol. B Microelectron. Nanometer Struct.* **2004**, *22*, 2205. [CrossRef]
50. Yang, Y.; Xu, D.; Wu, Q.; Diao, P. Cu$_2$O/CuO bilayered composite as a high-efficiency photocathode for photoelectrochemical hydrogen evolution reaction. *Sci. Rep.* **2016**, *6*, 35158. [CrossRef]
51. Wang, Y.; Lany, S.; Ghanbaja, J.; Fagot-Revurat, Y.; Chen, Y.P.; Soldera, F.; Horwat, D.; Mücklich, F.; Pierson, J.-F. Electronic structures of Cu$_2$O, Cu$_4$O$_3$, and CuO: A joint experimental and theoretical study. *Phys. Rev. B* **2016**, *94*, 245418. [CrossRef]

52. Xu, Y.; Schoonen, M.A. The absolute energy positions of conduction and valence bands of selected semiconducting minerals. *Am. Miner.* **2000**, *85*, 543–556. [CrossRef]
53. Erhart, P.; Klein, A.; Egdell, R.G.; Albe, K. Band structure of indium oxide: Indirect versus direct band gap. *Phys. Rev. B* **2007**, *75*, 75. [CrossRef]
54. Dixon, S.C.; Scanlon, D.O.; Carmalt, C.J.; Parkin, I.P. n-Type doped transparent conducting binary oxides: An overview. *J. Mater. Chem. C* **2016**, *4*, 6946–6961. [CrossRef]
55. Piccinin, S. The band structure and optical absorption of hematite (α-Fe_2O_3): A first-principles GW-BSE study. *Phys. Chem. Chem. Phys.* **2019**, *21*, 2957–2967. [CrossRef] [PubMed]
56. Temesghen, W.; Sherwood, P. Analytical utility of valence band X-ray photoelectron spectroscopy of iron and its oxides, with spectral interpretation by cluster and band structure calculations. *Anal. Bioanal. Chem.* **2002**, *373*, 601–608. [CrossRef]
57. Luan, P.; Xie, M.; Liu, D.; Fu, X.; Jing, L. Effective charge separation in the rutile TiO_2 nanorod-coupled α-Fe2O3 with exceptionally high visible activities. *Sci. Rep.* **2014**, *4*, 6180. [CrossRef]
58. Dolgonos, A.; Mason, T.O.; Poeppelmeier, K.R. Direct optical band gap measurement in polycrystalline semiconductors: A critical look at the Tauc method. *J. Solid State Chem.* **2016**, *240*, 43–48. [CrossRef]
59. Zanatta, A.R. Revisiting the optical bandgap of semiconductors and the proposal of a unified methodology to its determination. *Sci. Rep.* **2019**, *9*, 1–12. [CrossRef]

© 2020 by the authors. Licensee MDPI, Basel, Switzerland. This article is an open access article distributed under the terms and conditions of the Creative Commons Attribution (CC BY) license (http://creativecommons.org/licenses/by/4.0/).

Article

Evaluation of CO₂ Hydrogenation in a Modular Fixed-Bed Reactor Prototype

Heather D. Willauer [1,*], Matthew J. Bradley [2], Jeffrey W. Baldwin [3], Joseph J. Hartvigsen [4], Lyman Frost [4], James R. Morse [1], Felice DiMascio [5], Dennis R. Hardy [6] and David J. Hasler [5]

1. Naval Research Laboratory, Materials Science & Technology Division, Washington, DC 20375, USA; James.Morse@nrl.navy.mil
2. ASEE Postdoctoral Research Associate, Naval Research Laboratory, Materials Science & Technology Division, Washington, DC 20375, USA; mbradley757@gmail.com
3. Naval Research Laboratory, Acoustics Division, Washington, DC 20375, USA; jeffrey.baldwin@nrl.navy.mil
4. OxEon Energy, 257 S Riverbend Way Suite 300, North Salt Lake, UT 84054, USA; jjh@oxeonenergy.com (J.J.H.); Lyman.Frost@oxeonenergy.com (L.F.)
5. Office of Naval Research, Arlington, VA 22203, USA; felixdim@aol.com (F.D.); david.hasler@navy.mil (D.J.H.)
6. NOVA Research Inc., 1900 Elkin Street, Alexandria VA 22308, USA; kashardy@gmail.com
* Correspondence: heather.willauer@nrl.navy.mil

Received: 26 June 2020; Accepted: 18 August 2020; Published: 26 August 2020

Abstract: Low-cost iron-based CO_2 hydrogenation catalysts have shown promise as a viable route to the production of value-added hydrocarbon building blocks. It is envisioned that these hydrocarbons will be used to augment industrial chemical processes and produce drop-in replacement operational fuel. To this end, the U.S. Naval Research Laboratory (NRL) has been designing, testing, modeling, and evaluating CO_2 hydrogenation catalysts in a laboratory-scale fixed-bed environment. To transition from the laboratory to a commercial process, the catalyst viability and performance must be evaluated at scale. The performance of a Macrolite®-supported iron-based catalyst in a commercial-scale fixed-bed modular reactor prototype was evaluated under different reactor feed rates and product recycling conditions. CO_2 conversion increased from 26% to as high as 69% by recycling a portion of the product stream and CO selectivity was greatly reduced from 45% to 9% in favor of hydrocarbon production. In addition, the catalyst was successfully regenerated for optimum performance. Catalyst characterization by X-ray diffraction (XRD) and X-ray photoelectron spectroscopy (XPS), along with modeling and kinetic analysis, highlighted the potential challenges and benefits associated with scaling-up catalyst materials and processes for industrial implementation.

Keywords: carbon dioxide; hydrogenation; catalyst; gas hourly space velocity (GHSV); fixed-bed reactor

1. Introduction

Operating in a littoral and marine environment provides the U.S. Navy with unique access to a vast environmental resource of carbon. The world's oceans are the largest carbon reservoirs containing approximately 38,000 gigatons [1]. Carbon and hydrogen are the principal building blocks needed to synthesize hydrocarbons. It is envisioned that these hydrocarbons may one day be used to produce operational fuel. Synthesizing "drop-in" replacement fuel at or near the point of use, translates into "Freedom of Action for the Warfighter" and potential long-term cost savings and strategic advantages for the Department of Defense (DOD) [2–4]. If the energy required for the process is nuclear or renewable, the entire low carbon fuel process could be considered CO_2 neutral [5–7].

The future capability of producing fuel from inorganic carbon (CO_2) and H_2 in seawater is dependent on the development of processes and technologies specifically designed for such applications. The U.S. Navy has recently patented a process and an apparatus for the simultaneous extraction of CO_2 and production of H_2 from seawater [8–10]. However, the primary limitations in using the CO_2 and H_2 as building blocks for the synthesis of hydrocarbons are the high energy barrier for the redox and polymerization reactions necessary to synthesize longer chain molecules to be used as fuel [5–7,11]. While electrochemical and photochemical CO_2 conversion processes in water continue to improve in efficiency, challenges remain for these fuel synthesis approaches. These challenges include low hydrocarbon yields, catalyst stability, and difficulty in scaling-up the processes [5,7,11]. Two-step thermochemical approaches are one of the few proven scalable methods for the production of liquid hydrocarbons ranging from C_6–C_{17} from CO_2 and H_2 [6,12–14]. Step 1 involves the conversion of CO_2 and H_2 to intermediates (methanol, olefins, CO) [6,12–17]. Step 2 processes these intermediates to C_6–C_{17} hydrocarbons. Commercially, methanol and CO intermediates have both been successfully utilized in this two-step thermochemical approach [13,14], whereas the synthesis of olefin intermediates has only been extensively studied and demonstrated at the laboratory scale [6,17]. In order to evaluate the feasibility of directly synthesizing olefin intermediates as the first step towards operational fuel production for military and commercial applications, the process has to be scaled-up and demonstrated in thermochemical reactor platforms that will be relevant to off-shore and remote synthetic fuel production applications [18,19]. An additional advantage to scaling the chemical conversion of CO_2 to light olefin intermediates is that these intermediates serve as key building blocks in the chemical industry [6,12,17].

Commercial-scale, low-cost, modular fixed-bed reactors are being designed and evaluated for remote Fischer-Tropsch synthesis (FTS) processes that use natural gas as the starting material [18]. These advantages could also make commercial-scale fixed-bed reactors ideal for the scale-up of CO_2 hydrogenation technologies. Since catalyst physical properties and the reactor type are known to influence the product selectivity, mass transfer, and conversion of hydrogenation reactions [17–21], they are important parameters to consider upon transitioning from the laboratory to commercial scale.

In previous work, highly active Fe-Mn-K/supported CO_2 hydrogenation catalysts were characterized and evaluated at the laboratory scale in both a continuously stirred tank/thermal reactor (CSTR) [22] and a fixed-bed reactor [17]. The catalyst materials were demonstrated to be capable of functioning as an effective catalyst to convert CO_2 to short-chain olefins. In the present work, the synthesis of Fe-Mn-K-based catalyst is scaled-up 300 times and operated in a commercial-scale prototype modular fixed-bed reactor that is 176 times larger by volume than previous laboratory scale studies. The findings of this paper show how catalyst and reactor scale-up along with recycling a portion of the product stream significantly enhance CO_2 conversion efficiency and dramatically change product selectivity. These results are used to expand modeling efforts to bridge the gap between bench-scale research and the development and implementation of a commercial process.

2. Results and Discussion

Commercial tubular fixed-bed reactors for FTS offer scalable solutions for off-shore and remote synthetic fuel production applications [18,19]. The potential challenges associated with using the reactors for FTS are the likely high-pressure drops, low catalyst utilization, and insufficient removal of the heat generated during the exothermic reaction [18]. OxEon Energy developed the reactor used in this study (Figure 1) along with the cooling fin (Figure 2) after having several years of experience working with the hydrogenation of syngas (CO and H_2) to hydrocarbons using cobalt- and iron-based catalysts [18]. Industry standard FT tube size is typically under 1". OxEon Energy is able to use a larger tube size because of the thermal management structure shown in Figures 1 and 2 that distributes the heat throughout the catalyst bed produced by the exothermic FT reaction. Optimizing the hydrogenation reactions in the larger reactor tube size will drastically reduce the number of tubes required for a process. This was intended to reduce long-term fabrication and catalyst servicing costs [18]. The reactor and the skid that supports the hydrogenation reactions in this test series are pictured in Figure 3.

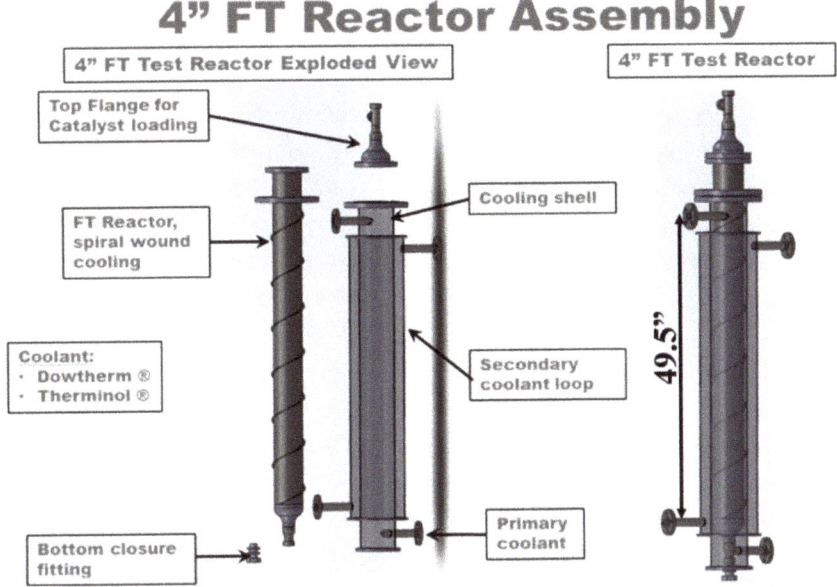

Figure 1. Schematic of OxEon Energy commercial-scale fixed-bed reactor prototype and dimensions.

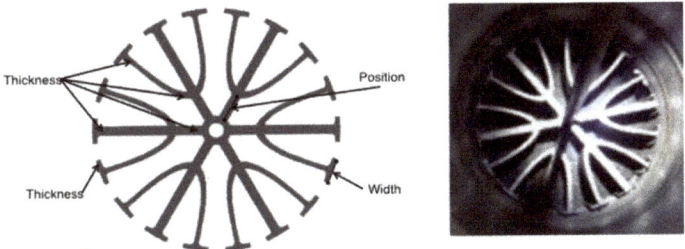

Figure 2. Schematic and picture of internal cooling fin.

The synthesis of light olefins from CO_2 in the fixed-bed environment presents even more challenges than traditional FTS. Equation (1) describes the reverse water–gas shift reaction (RWGS), which is endothermic and produces carbon monoxide (CO). Modeling and kinetic analysis of this reaction on the laboratory scale indicates the RWGS reaction rate is highest at the top of the reactor bed and decreases over the length of the catalyst bed [17]. The CO produced is carried forward in an exothermic FT step (Equation (2)) to produce predominantly monounsaturated hydrocarbons (Equation (3)).

$$nCO_2 + nH_2 \rightleftarrows nCO + nH_2O \qquad \Delta_R H_{300\,°C} = +38 \text{ kJ/mol} \qquad (1)$$

$$nCO + 2nH_2 \rightarrow (CH_2)_n + nH_2O \qquad \Delta_R H_{300\,°C} = -166 \text{ kJ/mol} \qquad (2)$$

$$nCO_2 + 3nH_2 \rightarrow (CH_2)_n + 2nH_2O \qquad (3)$$

A thermodynamically favorable side reaction associated with CO_2 hydrogenation is the highly competitive methanation reaction (Equation (4)) [23].

$$CO_2 + 4H_2 \rightarrow CH_4 + 2H_2O \qquad (4)$$

Another competing side reaction is the Boudouard reaction (Equation (5)) [23].

$$2CO \rightleftarrows C + CO_2 \qquad (5)$$

The water formed in the primary reactions shown in Equations (1)–(3) is twice the amount of water produced in traditional FTS. As this water accumulates along the fixed catalyst bed during the hydrogenation reaction, the catalyst will likely be more susceptible to re-oxidizing [17]. Even though the overall reaction is exothermic in nature, the role the RWGS reaction has on the overall process and temperatures along the catalyst bed can be better elucidated at the commercial-scale. The larger scale allows for more precise monitoring of the reaction conditions across the reactor bed.

Figure 3. The OxEon Energy commercial-scale fixed-bed reactor prototype and skid that supports the reactor assembly.

Five K profile probe thermocouples purchased from Omega (31 Stainless steel, 72) were positioned approximately 12 inches from one another along the catalyst bed starting from the top of the bed. The thermocouples were measured by Labview to monitor the exothermic behavior of the reaction as it proceeds down the reactor bed. During a typical commercial-scale reaction, the top of the catalyst bed measured an average temperature of 555 K, while the remainder of the catalyst bed was measured at 574 K and never rose above 575 K in a single pass of the feedstock. Since the feedstock gases are preheated before they enter the top of the reactor bed, the 19 K difference in temperature is attributed to the endothermic nature of the RWGS reaction (Equation (1)) and slow heat transfer during preheating the feedstock entering the bed. This temperature difference is not common in the FTS process that is highly exothermic (Equation (2)) and requires the thermal management structure shown in Figure 2 to distribute the heat throughout the catalyst bed, preventing any risk of thermal runaway. The thermocouple data supports findings from a three-dimensional fixed-bed computational model that shows how the RWGS reaction rate is high near the entrance of the catalyst bed and is always more than the FT reaction rate throughout the length of the catalyst bed [17]. This explains why the temperature should be cooler at the beginning of the reaction and suggests there is a lower risk of thermal runaway or the need for additional thermal management structures for these types of CO_2 hydrogenation reactions.

Prior to choosing a catalyst to measure performance in the commercial-scale fixed-bed reactor environment, the composition of Fe:Mn:K-based catalysts on gamma alumina and Macrolite supports were well characterized in a laboratory scale fixed-bed environment [17]. Since gamma alumina supports are susceptible to hydroxylation in fixed-bed environments [17], there was motivation to replace the gamma alumina support of the iron-based catalysts with the engineered chemically inert ceramic aluminosilicate material, Macrolite®, (M2). After the successful demonstration of commercially prepared 100:3.93:2.36:3.23 M2: Fe:Mn:K catalyst (Fe:M2-1, Table 1) at the laboratory scale, Water Star Inc. (Newbury, OH, USA) prepared a small batch (~500 to 1000 g) of higher concentrated iron-based catalyst on the Macrolite® support (support 100:17:12:16.5 ratio of M2:Fe:Mn:K) (Fe:M2-2, Table 1). The Fe:M2-2 catalyst composition closely resembled previously published iron-based catalyst compositions loaded on gamma alumina support [22]. Fe:M2-2 was characterized and tested at the laboratory scale to ensure that the CO_2 hydrogenation performance properties were similar or better than those measured for the gamma alumina supported catalyst [17] under similar reaction conditions. Table 1 (Row 2) shows a CO_2 conversion of 41% and an O/P ratio of almost 5. Based on the laboratory-scale results, Water Star Inc. scaled up the catalyst synthesis of the 100:17:12:16.5 support M2: iron: manganese: potassium catalyst (Fe:M2-3) to kilogram quantities for testing in the commercial-scale fixed-bed reactor environment that are presented in this study.

The single pass of the feedstock over the commercial-scale catalyst bed is the closest reaction conditions to those run at the laboratory scale, as there is currently no mechanism in place to recycle the product stream at the laboratory scale. Data were collected on the conversion of CO_2 and H_2 over the Fe:M2-3 catalyst for a 96 h period. CO_2 conversion, product selectivity, and olefin/paraffin ratio as a function of GHSV, product recycling, and catalyst regeneration are reported in Table 1, along with data for other catalyst blends and reactor scales. When the Fe:M2-3 catalyst was reacted in the commercial-scale fixed-bed reactor at the lowest total GHSV of 4.6×10^{-4} L/s-g for 96 h under single pass feedstock conditions (Table 1, Rows 4–7), the average CO_2 conversion was 26% with a CO selectivity of 46% and methane selectivity of 6%. The olefin/paraffin (O/P) ratio increased from 3.1 to 4.4 in the first 48 h of operation. The hydrocarbon selectivity (percent conversion of CO_2 to C2–C5+ hydrocarbons) was 48% on average. The percent yield values in Table 1 were calculated by multiplying the C2–C5+ selectivity values by the CO_2 conversion given in the Table. The average C2–C5+ yield measured is approximately 12% by weight on a carbon basis at a GHSV of 4.6×10^{-4} L/s-g.

Loosely comparing the average commercial-scale hydrogenation results of Fe:M2-3 (Table 1, Rows 4–7) to the laboratory-scale operating conditions performed with the small-scale commercially prepared catalyst, Fe:M2-2, at a higher GHSV of 9.6×10^{-4} L/s-g (Table 1, Row 2), a 37% decrease in CO_2 conversion and an 86% increase in CO selectivity is observed on the commercial scale for the Fe:M2-3 catalyst. This was an unexpected result, as the lower GHSV tested at the commercial scale was anticipated to yield higher CO_2 conversions. This significant difference in conversion and selectivity at the two reactor scales prompted the implementation of a regeneration process before changing the commercial-scale reactor operating conditions and studying the catalyst performance. Mechanisms for catalyst regeneration and the long-term catalyst stability are important for the commercial feasibility of any catalyst process [24]. Fe:M2-3 was regenerated by flowing hydrogen over the catalyst bed at until methane was not detected by the inline GC equipped with a TCD detector. Then nitrogen with air to make 1% O_2 was flowed over the catalyst bed. The O_2 content was gradually increased until no further CO_2 was measured by the GC. This method was used to remove all hydrocarbon and carbon remaining on the catalyst and also dried the catalyst bed.

Table 1. Summary of product selectivity, olefin/paraffin ratio, CO_2 conversion, and ASF values over Fe-based CO_2 hydrogenation catalysts in both laboratory and commercial single-channel scalable modular thermochemical reactor units.

Row	Reactor Scale/Catalyst Scale (g/kg)	Laboratory Reactor Scale (LS)/Commercial Reactor Scale (CS)	Fresh GHSV (L/s-g)	Recycled GHSV (L/s-g)	Total GHSV (L/s-g)	Mass (g)	Conversion CO_2 (%)	Product Selectivity (%) C1	C2–C5+	CO	O/P	α C1–C6	Hydrocarbon Yield (%) C2–C5+
\multicolumn{14}{l}{Laboratory reactor scale studies}													
1	LS/g 100:3.93:2.36:3.23 M2:Fe:Mn:K	Fe:M2-1 [17]	9.6×10^{-4}	-	9.6×10^{-4}	2	26	22	68	9.8	3.7	0.6	18
2	LS/g 100:17:12:16.5 M2:Fe:Mn:K	Fe:M2-2	9.6×10^{-4}	-	9.6×10^{-4}	2	41	21	73	6.3	4.8	0.7	30
3	LS/kg 100:17:12:16.5 M2:Fe:Mn:K	Fe:M2-3	4.6×10^{-4}	-	4.6×10^{-4}	2	11	5	75	19.7	4.4	0.9	8
\multicolumn{14}{l}{Commercial reactor scale studies using 100:17:12:16.5 M2:Fe:Mn:K Fe:M2-3}													
4	CS/kg	Fe:M2-3	4.6×10^{-4}	-	4.6×10^{-4}	1888	26	7	48	45	3.1	0.7	12
5	CS/kg	Fe:M2-3	4.6×10^{-4}	-	4.6×10^{-4}	1888	26	6	48	45	4.4	0.6	12
6	CS/kg	Fe:M2-3	4.6×10^{-4}	-	4.6×10^{-4}	1888	26	6	49	45	-	-	13
7	CS/kg	Fe:M2-3	4.6×10^{-4}	-	4.6×10^{-4}	1888	25	6	46	48	-	-	11
8	CS/kg	Fe:M2-3	3.5×10^{-4}	9.2×10^{-4}	1.3×10^{-3}	1888	55	9	76	15	4	0.6	42
9	CS/kg	Fe:M2-3	3.5×10^{-4}	9.2×10^{-4}	1.3×10^{-3}	1888	62	9	78	12	-	-	48
10	CS/kg	Fe:M2-3	3.5×10^{-4}	9.2×10^{-4}	1.3×10^{-3}	1888	66	9	78	13	4	0.5	51
11	CS/kg	Fe:M2-3	3.5×10^{-4}	9.2×10^{-4}	1.3×10^{-3}	1888	64	9	78	12	4	0.5	50
12	CS/kg	Fe:M2-3	4.6×10^{-4}	9.4×10^{-4}	1.4×10^{-3}	1888	54	26	63	11	-	-	34
13	CS/kg	Fe:M2-3	4.6×10^{-4}	9.4×10^{-4}	1.4×10^{-3}	1888	54	10	78	12	-	-	42
14	CS/kg	Fe:M2-3	4.6×10^{-4}	9.4×10^{-4}	1.4×10^{-3}	1888	52	17	65	17	-	-	34
15	CS/kg	Fe:M2-3	2.2×10^{-4}	5.8×10^{-4}	8.0×10^{-4}	1888	69	12	79	9	3	0.5	54
16	CS/kg	Fe:M2-3	2.2×10^{-4}	5.8×10^{-4}	8.0×10^{-4}	1888	68	12	79	9	3	0.5	54

After catalyst regeneration, the reaction was resumed at a total GHSV of 1.3×10^{-3} L/s-g with the recycling feature applied to the experimental runs. The water in the recycle feed was condensed and removed from the feed. The dry recycle feed at 9.2×10^{-4} L/s-g was blended with fresh feed at 3.5×10^{-4} L/s-g back into the reactor at a 2.6:1 recycle feed to fresh feed Table 1, Rows 8–11). It is worth noting that the top of the catalyst bed measured an average temperature of 540 K. This is, on average, 15 K cooler than the single pass temperatures measured at the top of the reactor. This additional drop in temperature is believed to be the result of the recycle feed being treated for removal of liquid and water by passing it through a cold trap (Figure 4). The temperatures in the remainder of the catalyst bed reached an average of 575 K, which is similar to those measured during the single pass conditions.

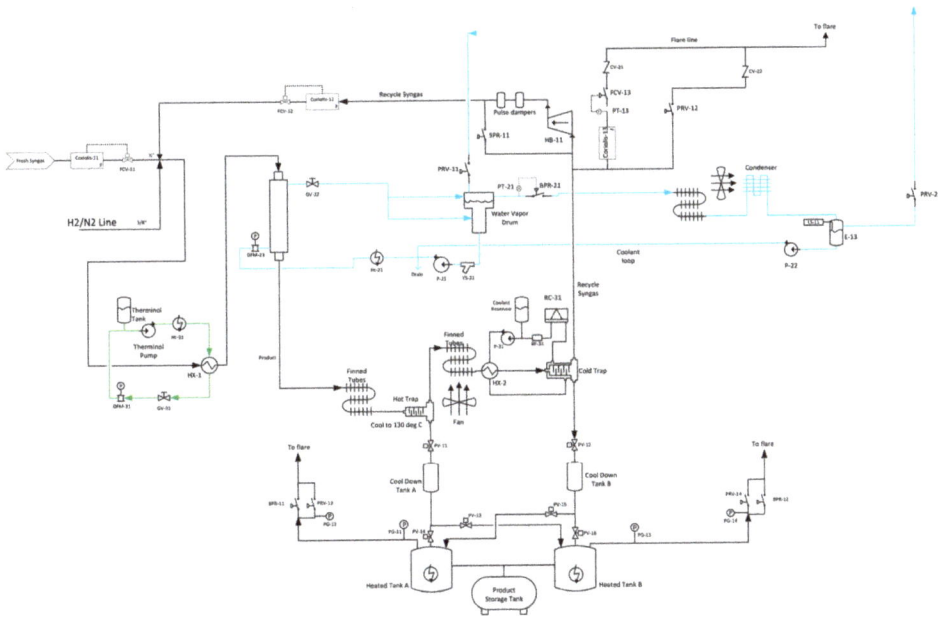

Figure 4. OxEon Energy commercial-scale fixed-bed reactor prototype process flow diagram.

After 96 h on stream, under these recycling conditions, the average CO_2 conversion was measured to be 62% (Table 1, Rows 8–11), far greater than the 26% CO_2 conversion originally measured under the single pass conditions (Table 1, Rows 4–7). Under the recycling conditions, methane selectivity increased from 6% to 9% and C2–C5+ selectivity increased from 48% to 76%. CO selectivity was reduced from 45% to 15%, while the O/P remained at 4. This reduction in CO selectivity also corresponds well with the 75% increase in average C2–C5+ yield, indicating the CO produced during the first pass over the catalyst is hydrogenated to higher hydrocarbons during the second pass over the catalyst bed.

Increasing the total GHSV from 1.3×10^{-3} to 1.4×10^{-3} L/s-g by the addition of 1.1×10^{-4} L/s-g of fresh feed and 2.0×10^{-5} L/s-g of recycled feed over a 72 h period significantly affects the selectivity of the catalyst (Table 1, Rows 12–14). Since the GHSV was only increased by 7.7%, the change in CO_2 conversion and product selectivity can be attributed more to the lower ratio of recycle feed 2:1 to fresh feed. Reducing the amount of unreacted CO by reducing the amount of recycle feed on each pass over the catalyst bed lowered the average CO_2 conversion from 62% (Table 1, Rows 8,9) to 53% (Table 1, Rows 12–14) and hydrocarbon selectivity from 78% to 69%. This is further substantiated at lower GHSV. While GHSV has been shown to have little effect on the chain growth probability of C1–C5+ at the laboratory scale, it has had a significant impact on CO_2 and CO conversion. In particular, their conversion can be significantly improved by lowering the GHSV and thus increasing the residence time of the reagents CO_2, CO, and H_2 in contact with the active catalyst bed. When the GHSV is lowered to 8.0×10^{-4} L/s-g and the ratio of recycle feed to fresh feed is returned to 2.6:1, the highest average CO_2 conversion and hydrocarbon selectivity are reported at 68.5% and 79 (Table 1, Rows 15,16). In addition, the lower GHSV reduces CO selectivity to 9% in favor of greater C2–C5+ yield at 54%.

A detailed kinetic analysis at the commercial prototype modular fixed-bed reactor scale is challenging from a cost, time, and materials perspective. In this study, the single pass feedstock conditions and the ability to recycle a portion of the product stream in the commercial-scale prototype reactor are modeled using reaction kinetics ascertained at the laboratory scale [17]. The single pass feedstock reactor conditions were modeled at a GHSV of 4.6×10^{-4} L/s-g, a gas mass fraction composition of 82.2% CO_2, 10.9% H_2, and 6.9%N_2, inlet temperature of 553 K, and 2 Mpa (20 Bar) pressure. The single pass feedstock reactor model yielded results with a conversion of 22% CO_2 and product selectivities of 52% CO, 43% C_3H_6, and 5% CH_4 with an outlet temperature of 575 K (Table 2). The model suggests that if the reactor outlet temperatures are raised to 608 K, the CO_2 conversion will increase to 26% and the CO and CH_4 selectivities will decrease (45% and 6%) in favor of higher hydrocarbon formation, 49% C_3H_6 (Table 2). Both of these sets of model results (575 and 608 K outlet temperature) are very similar to those reported by Riedel et al. [24]. The model results also fall in a similar range as the averaged prototype reactor data measured and are shown in Tables 1 and 2.

Table 2. 3D kinetic modeling predictions for product selectivity and CO_2 conversion under single pass feedstock conditions and recycling conditions compared to the average measured for the rector prototype. Single pass feedstock conditions at GHSV of 4×10^{-4} L/s-g and recycling conditions fresh feed GHSV of 4×10^{-4} L/s-g and 9.4×10^{-4} L/s-g of recycled feed.

	Model	Model	Test	Model	Test
	Ceramatec Pilot Reactor	Ceramatec Pilot Reactor	Ceramatec Pilot Reactor	Ceramatec Pilot Reactor	Ceramatec Pilot Reactor
	No Recycle	No Recycle	No Recycle	Recycle	Recycle
Outlet Temperature (K)	575	608	~573	574	~573
% CO_2 Conversion	22	26	25.9 ± 0.4	53	53 ± 1
% CO Selectivity	52	45	45.9 ± 1.6	21	14 ± 3
% C_3H_6 Selectivity	43	49	47.8 ± 1.5	65	69 ± 8
% CH_4 Selectivity	5	6	6.3 ± 0.4	14	18 ± 8

Recycling conditions were incorporated into the model using a feed gas consisting of a combined fresh (4.6×10^{-4} L/s-g GHSV and recycled 9.4×10^{-4} L/s-g GHSV) gas mass fraction composition of 81.0% CO_2, 10.6% H_2, 1.0% CO, 1.2% C_3H_6, 0.2% CH_4, 6.0% N_2, and 0.0% H_2O. The model provided results indicating 53% conversion of CO_2 and product selectivities of 21% CO, 65% C_3H_6, and 14% CH_4 with an outlet temperature of 574 K (Table 2). The model results align well with the measured data for the recycled conditions as provided in Tables 1 and 2.

Any differences in the modeling results may be attributed to the complexities of the commercial-scale reactor configuration and the axial location of the thermocouples. Temperature excursions within the prototype reactor could have risen above the 573 K without being measured by the thermocouples due to the finned nature of the reactor (Figure 2). The internally finned reactor creates separate channels of flow all separated by an aluminum wall, which causes multiple "micro" reactors within the large reactor resulting in potential differences in temperatures. The model supports this possible temperature variation by providing a better fit at reactor outlet temperatures that are 35 K higher than those measured during the reactions. Neither of the reactor models indicated a radial heat transfer limitation. However, a primary limitation in CO_2 conversion is the heat required by the endothermic RWGS reaction that initially drops the temperature in the first section of the bed followed by the subsequent exothermic FT reaction that begins to release heat that needs to be removed. This further substantiates previous modeling and kinetic analysis that indicate that the FT reaction is the rate-limiting step in the use of these supported iron-based catalysts [17].

After the commercial-scale testing was completed, significant differences in catalyst activity and selectivity were observed for the commercially prepared small-batch catalyst Fe:M2-2 tested at the laboratory scale (Table 1, Row 2) and the large batch Fe:M2-3 prepared and measured at the commercial scale. Fe:M2-3 activity and selectivity with respect to CO_2 conversion, O/P, and hydrocarbon yields align more closely with the Fe:M2-1 that was prepared commercially with a lower iron concentration (100:3.93:2.36:3.23 M2: Fe:Mn:K) [17]. These observations motivated further characterization of both catalysts. The activity of the Fe:M2-3 catalyst was measured at a GHSV of 4.6×10^{-4} L/s-g in the NRL laboratory scale fixed-bed reactor, while both the Fe:M2-2 and Fe:M2-3 were characterized and compared by XPS and XRD shown in Table 3. When the hydrogenation results of Fe:M2-3 are compared at the different reactor scales and GHSV of 4.6×10^{-4} L/s-g, the results in Table 1 (LS/kg) show a 57% loss in CO_2 conversion at the laboratory scale to 11%. CO selectivity was reduced by 56% in favor of C2–C5+ formation at 36%, and the methane selectivity was similar at 6%.

Table 3. Comparing XPS data for two commercially prepared catalyst formulations.

Element	Commercially Synthesized Fe:M2-2 Small Batch Catalyst (10 kg)			Commercially Synthesized Fe:M2-3 Large Batch Catalyst (500 to 1000 g)		
	XPS (mol%)	Atomic Weight (g/mol)	Wt%	XPS (mol%)	Atomic Weight (g/mol)	Wt%
Mn	6.4	54.9	25.5	16.2	54.9	47.7
K	7.5	39.1	21.3	13.9	39.1	29.0
Fe	6.2	55.8	25.1	3.8	55.8	11.4
N	8.5	14.0	8.7	6.4	14.0	4.7
Si	6.7	25.1	12.3	0	25.1	0
Ca	2.4	40.1	7.1	0	40.1	0
Cl	0	35.5	0	3.8	35.5	7.2
Sum	37.7	264.5	100.0	44.1	264.5	100.0

Elemental analysis using XPS (Table 3) reveals that Fe:M2-3 catalyst batch contains twice the concentration of manganese and less than half the iron concentration of Fe:M2-2 catalyst. The effects of iron loading on CO_2 conversion and product selectivity have been well established [22]. Table 1 provides data taken in the laboratory-scale reactor for the well-characterized iron/manganese/potassium/Macrolite® catalysts with a relative iron loading that is 3.93% by weight of the M2 support, Fe:M2-1. The CO_2 conversion observed is 26% compared to 41% for the higher iron-loaded catalyst, Fe:M2-2 under laboratory-scale reactor conditions. These results are in line with similar reports documenting the change in CO_2 conversion and product selectivity as a function of increase in the iron [22] and further supports the findings that the Fe:M2-3 catalyst was not made to the specified metal loadings. Contributions of aluminum, oxygen, and carbon were factored out of the XPS analysis to reduce any potential impact of surface oxygen and advantageous carbon species. There were also various trace elements that only existed in one catalyst or the other, including silicon, calcium, nitrogen, and chlorine.

X-ray diffraction analysis in Figure 5 revealed that both Fe:M2-2 and Fe:M2-3 contain the same iron oxide crystal phase (Fe_2O_3 hematite), but the support phases between the two catalyst are clearly different. Yellow tick marks have been added to the pattern to indicate discrepancies between the Fe:M2-2 and Fe:M2-3 catalysts. When compared to the hematite simulated pattern, it is evident that both catalysts contain the iron oxide phase, leading to the conclusion that the support is the primary difference between the two catalysts. Unfortunately, no one alumina or aluminosilicate phase we examined accurately matches the supports in question. In summary, it is clear that these catalysts are different compositionally and crystallographically from one another. After providing further evidence to the supplier on the differences in catalyst support, the supplier revealed that different supports were in fact used between the small-scale catalyst batch Fe:M2-2 and the large-scale catalyst batch Fe:M2-3. While it was believed that little difference in chemical activity existed between the two supports, this difference ultimately led to the decrease in activity of Fe:M2-3 relative to the Fe-M2-2 small-scale catalyst test batch measured at the laboratory scale.

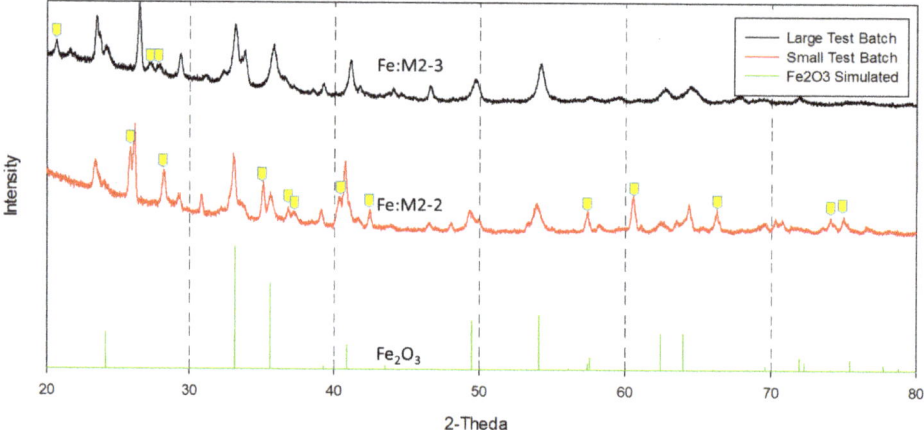

Figure 5. X-ray diffraction analysis of commercially prepared catalyst formulations, large test batch (Fe:M2-3), and small test batch (Fe:M2-2).

3. Experimental Methods and Methodology

3.1. Commercial Large Scale Catalyst Preparation

Approximately 10 kg of iron-based metal catalyst (consisting of the following ratio: 100:17:12:16.5 of support: iron: manganese: potassium) was synthesized by Water Star Inc (Newbury, OH, USA) using typical incipient wetness impregnation (IWI) methods. The support was a Macrolite® engineered ceramic media (M2) (Fairmount Water Solutions). This catalyst is denoted in Table 1 as Fe:M2-3. The catalyst particles ranged from 0.6 mm to 2 mm in diameter. Before the catalyst was loaded into the reactor, the catalyst was reduced in hydrogen, passivated, and soaked in mineral oil for 12 h. Approximately 9 mL of mineral oil was used for every 20 g of catalyst. This is similar to the amounts of mineral oil used in the laboratory scale studies [17].

3.2. Commercial-Scale Fixed-Bed Reactor Prototype Setup

Figure 1 provides a schematic of the commercial-scale fixed-bed reactor, which was designed and assembled by OxEon Energy (formally known as Ceramatec) located in Salt Lake City, Utah, US. Approximately 6090 g of catalyst was loaded into a 4" (100 mm) nominal tube size, single tube reactor. The reactor has a 4.5" outer diameter (OD), 0.12" wall, and an inner diameter (ID) of 4.26". The catalyst occupies a length of approximately 49.5" of the reactor tube assembly. The center of the tube contains an internal cooling fin to disseminate the heat throughout the catalyst bed that is generated during exothermic reactions (Figure 2). The reactor and the skid that supports the reactor used in this test series are pictured in Figures 3 and 4 and provide the process flow diagram of the commercial-scale OxEon Energy test facility. Following the process flow diagram shown in Figure 4, a 3:1 ratio of H_2/CO_2

was flowed over the Fe:M2-3 catalyst at GHSV ranging between 8.0×10^{-4} L/s-g and 1.4×10^{-3} L/s-g. Nitrogen was used as the internal standard and all experiments were carried out at 300 °C and 20 bar.

The effluent fixed gases (H_2, CO_2, CO, N_2) and CH_4 from the reactor were analyzed in real time using an inline GC (Inficon) equipped with a TCD detector. Figure 4 shows that the system was designed such that when the products exit the reactor, they go to a hot trap and heavy hydrocarbons such as wax (typically seen if FT reactions using cobalt-based catalyst) are drained; the remainder of the hydrocarbons and non-converted synthesis gas goes to the fin tube to cool to room temperature. The gases then proceed through a chiller to a cold trap (cold trap is maintained at −4°C) where the light oil and water are condensed, captured, and separated periodically. The non-condensate gases pass through a tee and are split such that part of the stream goes to flare and the other part goes to the recycle compressor. The composition of both gas streams are identical and the recycled feed is then blended back into the reactor at a ratio of 2:1 recycle feed to fresh feed (CO_2 and H_2). The flaring/removal of 1/3 of the gas stream is important to keep methane, an unwanted byproduct of this reaction, to reasonable levels. Labview is used for the inline GC analysis to ensure that the GHSV and CO_2:H_2 ratio remained constant during the recycling process. Analysis provided in Table 1 occurred approximately every 24 h for a given GHSV.

The long chain effluent from the reaction (i.e., hydrocarbons greater than C-2) was analyzed separately using a GC (Bruker 456, (Billerica, MA, USA) equipped with a flame ionization detector. It is important to note that all selectivities and yields are reported on a per carbon atom consumed basis and not per mole of product (i.e., propane selectivity is weighted by three due to the fact that it accounts for three carbon atoms per molecule). Five thermocouples (type and company) were positioned along the reactor bed and monitored by Labview to monitor the exothermic behavior of the reaction as it proceeds down the reactor bed.

3.3. Computational Modeling for Fixed-Bed Reactors

The OxEon Energy commercial-scale fixed-bed reactor prototype results were modeled according to previous reaction kinetics based on Willauer et al. but factored by 100 for the reverse water–gas shift (rWGS) reaction, 6.75 for the Fischer-Tropsch reaction, and 6 for the methanation reaction [17,24]. The same reaction kinetics were used to model both the single pass and recycling capability. The single pass of the feedstock over the reactor along with the recycle capability were modeled with commercial CFD software CFX© on the Department of Defense High Performance Computing Modernization Program servers. The number of nodes used in the model was 60,975 with 305,050 elements. The k-ε turbulence model was used with the scalable wall function, compressible gas flow, and finite rate with equilibrium chemical kinetics. The reactor was modeled as a packed bed of length 1.2573 m (49.5 inch) and inner diameter of 0.1082 m (4.26 inch), a bed porosity of 40%, and a permeability of 1.9×10^{-9} m²/s. A total catalyst weight of 6090 g with 1888 g of active metal was used as the basis for the reaction kinetics. The model outlet pressures and imposed wall temperatures were variables used to converge on the reported pilot test results for outlet temperature and mass flows. For both of the reactor configurations, heat was removed from the reactor, which is similar to the porotype reactor configuration.

3.4. Laboratory-Scale Fixed-Bed Reactor Setup

The laboratory-scale fixed-bed plug flow thermochemical reactor process requires 20 g of catalyst for evaluation in a 9 to 12" long stainless steel tube (3/8" ID, 1/2" OD). The catalyst bed occupies 8 to 10" length of the stainless steel tube. The catalyst quantities required for the evaluations were synthesized commercially by Water Star Inc. Water Star prepared two small-scale iron-based metal catalysts for evaluation on the laboratory scale that differed in wt.% metal loading by IWI methods. The first catalyst, 100:3.93:2.36:3.23 M2:Fe:Mn:K, has been studied extensively [17] and is referred to as Fe:M2-1 in Table. The second catalyst was prepared at a higher catalyst loading of 100:17:12:16.5 M2:Fe:Mn:K and is listed in Table 1 as Fe:M-2. Each small-scale batch consisted of 500 g to 1000 g of catalyst.

The catalysts were recovered after the hydrogenation reactions, rinsed with hexanes, vacuum-dried, and then characterized by XRD and XPS.

In a typical CO_2 hydrogenation experiment, 10 mL of mineral oil is added to the catalyst and the catalyst was reduced in situ by flowing 100 mL/min H_2 at 300 °C and 20 bar for 18 h. Three mass flow controllers (Brooks Instruments, Hatfield, PA, USA) were used to control the flows of CO_2, H_2, and N_2 into the reactor. Immediately following reduction, hydrogenation of CO_2 was conducted at 20 bar and 300 °C with a H_2/CO_2 ratio of 3:1 and a 10 mL/min N_2 internal standard at 20 bar and gas hourly space velocity (GHSV) of 9.6×10^{-4} L/s-g. The GHSV is defined as standard liters per second of total CO_2 and H_2 flow divided by the grams of total elemental Fe, Mn, and K metals in the reactor. The effluent gases were passed through a cold trap, at 10 °C to condense the water vapor and any heavy liquid hydrocarbons formed in the reactor. The effluent gases were analyzed in real time using an inline gas chromatograph (GC) (Agilent Technologies, Fast RGA analyzer, Santa Clara, CA, USA. Hydrocarbons were separated using an HP-Al/S column (Agilent Technologies) 19091P-512, 25 μm × 320 μm × 8 μm) and detected with an FID detector. Fixed gases (H_2, CO_2, CO, N_2) were separated on a Unibead IS column (4 ft, 60/80 mesh in UltiMetal, Agilent Technologies) and a 5Å molecular sieve column (8 ft, 60/80 mesh) and detected with a TCD detector. The GC was calibrated using a mixture of gases with known molar ratio (MESA Specialty Gas, Santa Ana, CA, USA). Time-on-stream (TOS) for the catalyst was 48 h.

3.5. Catalyst Characterization

Powder X-ray diffraction analysis was performed using a Rigaku Smartlab X-ray Diffractometer (Austin, TX, USA) using a Cu_Kα source and collected between 20 and 80 2Θ. The XPS measurements were performed using a Thermo Scientific K-alpha equipped with a monochromatic Al Kα source and 180° double focusing hemispherical analyzer with 128-channel detector was used to collect X-ray photoelectron spectroscopy data. The nominal XPS spot size and analyzer field of view were 100 μm^2. Charge compensation was necessary.

4. Conclusions

The commercial large-scale synthesis of a well-characterized NRL iron-based catalyst and its performance in a commercial-scale modular fixed-bed reactor prototype was evaluated in this test series. A significant reduction in CO_2 conversion and change in product selectivity at the large scale suggested the commercially scaled-up synthesized catalyst (Fe:M2-3) was significantly less active than previous catalysts synthesized and tested at the laboratory scale (Fe:M2-2). The reduced catalyst activity was further substantiated spectroscopically by the confirmation of lower iron loadings and differences in catalyst support characteristics used in the large-scale synthesis of the catalyst. The ability to recycle a portion of the product stream at this scale provided a solution to overcome challenges associated with the lower catalyst activity. CO_2 conversion increased from 26% to as high as 69% and the product selectivity shifted from 45% CO to 9% CO in favor of C2–C5+ hydrocarbon production upon recycling the effluent product stream.

The results presented serve to highlight the potential challenges associated with scaling-up materials and processes for commercial implementation. They also suggest that the iron-based catalyst with the higher metal loading will have even better performance characteristics in the commercial-scale prototype reactor. The modeling and kinetic analysis using kinetics established at the laboratory support trends associated with reaction temperatures and FT reaction rates. The higher loading catalyst and the endothermic cooling associated with the RWGS reaction will be the subject of further evaluations as NRL pursues determination of the feasibility of producing olefin intermediates from CO_2 and H_2 as the first step of a two-step thermochemical process to produce operational fuel for the military and commercial applications.

Author Contributions: All authors assisted in the collection of the data, the data analysis, and the preparation of the manuscript. All authors have read and agreed to the published version of the manuscript.

Funding: This work was supported by the Office of Naval Research both directly and through the Naval Research Laboratory and OPNAV N45.

Conflicts of Interest: The authors declare no conflict of interest.

References

1. United States. *DOE Genomics: GTL Roadmap: Systems Biology for Energy and Environment*; Department of Energy Office of Science: Washington, DC, USA, 2005.
2. Summary of the 2018 National Defense Strategy of the United States of America. Available online: https://dod.defense.gov/Portals/1/Documents/pubs/2018-National-Defense-Strategy-Summary.pdf (accessed on 10 January 2020).
3. Department of Defense. Operational Energy Strategy. 2016. Available online: http://www.acq.osd.mil/eie/Downloads/OE/2016%20DoD%20Operational%20Energy%20Strategy%20WEBc.pdf (accessed on 10 January 2020).
4. Naval S&T Strategy 2015, Office of Naval Research. Available online: https://www.navy.mil/strategic/2017-Naval-Strategy.pdf (accessed on 10 January 2020).
5. Burkart, M.D.; Hazari, N.; Tway, C.L.; Zeitler, E.L. Opportunities and challenges for catalysis in carbon dioxide utilization. *ACS Catal.* **2019**, *9*, 7937–7956. [CrossRef]
6. Choi, Y.H.; Jang, Y.J.; Park, H.; Kim, W.Y.; Lee, Y.H.; Choi, S.H.; Lee, J.S. Carbon dioxide fischer-tropsch synthesis: A new path to carbon-neutral fuels. *Appl. Catal. B* **2017**, *202*, 605–610. [CrossRef]
7. Song, J.T.; Song, H.; Kim, B.; Oh, J. Towards higher rate electrochemical CO_2 conversion: From liquid-phase to Gas-phase systems. *Catalysts* **2019**, *9*, 224. [CrossRef]
8. DiMascio, F.; Hardy, D.R.; Lewis, M.K.; Willauer, H.D.; Williams, F.W. Extraction of Carbon Dioxide and Hydrogen from Seawater and Hydrocarbon Production Therefrom. U.S. Patent 9,303,323, 5 April 2016.
9. DiMascio, F.; Willauer, H.D.; Hardy, D.R.; Williams, F.W.; Lewis, M.K. Electrochemical Module Configuration for the Continuous Acidification of Alkaline Water Sources such as Seawater and Recovery of CO_2 with Continuous Hydrogen Production. U.S. Patent 9,719,178, 1 August 2017.
10. DiMascio, F.; Willauer, H.D.; Hardy, D.R.; Lewis, M.K.; Williams, F.W. Electrochemical Module Configuration for the Continuous Acidification of Alkaline Water Sources such as Seawater and Recovery of CO_2 with Continuous Hydrogen Gas Production. U.S. Patent 10,450,661, 22 October 2019.
11. Chang, K.; Zhang, H.; Cheng, M.-J.; Lu, Q. Application of ceria in CO_2 conversion catalysis. *ACS Catal.* **2020**, *10*, 613–631. [CrossRef]
12. Porosoff, M.D.; Yan, B.; Chen, J.G. Catalytic reduction of CO_2 by H_2 for synthesis of CO, methanol and hydrocarbons: Challenges and opportunities. *Energy Environ. Sci.* **2016**, *9*, 62–73. [CrossRef]
13. Olah, G.A.; Goeppert, A.; Prakash, G.K.S. *Beyond Oil and Gas: The Methanol Economy*; Wiley-VCH Verlag GmbH & Co. KGaA: Weinheim, Germany, 2006.
14. MacDonald, F. Science Alert 27 Apr. 2015. Available online: http://www.sciencealert.com/audi-have-successfully-made-diesel-fuel-from-air-and-water (accessed on 3 June 2020).
15. Porosoff, M.D.; Yang, X.; Bosocoboinik, J.A.; Chen, J.G. Molydenum carbide as alternative catalysts to precious metals for highly selective reduction of CO_2 to CO. *Angew. Chem. Int. Ed.* **2014**, *53*, 6705–6709. [CrossRef] [PubMed]
16. Porosoff, M.D.; Baldwin, J.W.; Peng, X.; Mpourmpakis, G.; Willauer, H.D. Potassium-promoted molybdenum carbide as a highly active and selective catalyst for CO_2 conversion to CO. *ChemSusChem* **2017**, *10*, 1–9. [CrossRef] [PubMed]
17. Bradley, M.J.; Ananth, R.; Willauer, H.D.; Baldwin, J.W.; Hardy, D.R.; DiMascio, F.; Williams, F.W. The role of catalyst environment on CO_2 hydrogenation in a fixed-bed reactor. *J. CO2 Utiliz.* **2017**, *17*, 1–9. [CrossRef]
18. Frost, L.; Elangovan, E.; Hartvigsen, J. Production of synthetic fuels by high-temperature co-electrolysis of carbon dioxide and steam with Fischer-Tropsch synthesis. *Can. J. Chem. Eng.* **2016**, *94*, 636–641. [CrossRef]
19. LeViness, S.; Deshmukh, S.R.; Richard, L.A.; Robota, H.J. Velocys Fischer-Tropsch synthesis technology-new advances on state-of-the art. *Top. Catal.* **2014**, *57*, 518–525. [CrossRef]

20. Davis, B.H. Fischer-Tropsch synthesis: Overview of reactor development and future potentialities. *Top. Catal.* **2005**, *32*, 143–168. [CrossRef]
21. Espinoza, R.L.; Steynberg, A.P.; Jager, B.; Vosloo, A.C. Low temperature Fischer-Tropsch synthesis from a Sasol perspective. *Appl. Catal. A* **1999**, *186*, 13–16. [CrossRef]
22. Dorner, R.W.; Hardy, D.R.; Williams, F.W.; Willauer, H.D. K and Mn doped iron-based CO_2 hydrogenation catalysts: Detection of KAlH4 as part of the catalyst's active phase. *Appl. Catal. A* **2010**, *373*, 112–121. [CrossRef]
23. Riedel, T.; Schaub, G.; Jun, K.-W.; Lee, K.-W. Kinetics of CO_2 hydrogenation on a K-promoted Fe catalyst. *Ind. Eng. Chem. Res.* **2001**, *40*, 1355–1363. [CrossRef]
24. Moulijn, J.A.; van Diepen, A.E.; Kapeteijn, F. Catalyst deactivation: Is it predictable? What to do? *Appl. Catal. A* **2001**, *212*, 3–16. [CrossRef]

© 2020 by the authors. Licensee MDPI, Basel, Switzerland. This article is an open access article distributed under the terms and conditions of the Creative Commons Attribution (CC BY) license (http://creativecommons.org/licenses/by/4.0/).

Review

Utilization of CO_2-Available Organocatalysts for Reactions with Industrially Important Epoxides

Tomáš Weidlich * and Barbora Kamenická

Chemical Technology Group, Institute of Environmental and Chemical Engineering, Faculty of Chemical Technology, University of Pardubice, Studentská 573, 532 10 Pardubice, Czech Republic; barbora.kamenicka@student.upce.cz
* Correspondence: tomas.weidlich@upce.cz; Tel.: +420-46-603-8049

Abstract: Recent knowledge in chemistry has enabled the material utilization of greenhouse gas (CO_2) for the production of organic carbonates using mild reaction conditions. Organic carbonates, especially cyclic carbonates, are applicable as green solvents, electrolytes in batteries, feedstock for fine chemicals and monomers for polycarbonate production. This review summarizes new developments in the ring opening of epoxides with subsequent CO_2-based formation of cyclic carbonates. The review highlights recent and major developments for sustainable CO_2 conversion from 2000 to the end of 2021 abstracted by Web of Science. The syntheses of epoxides, especially from bio-based raw materials, will be summarized, such as the types of raw material (vegetable oils or their esters) and the reaction conditions. The aim of this review is also to summarize and to compare the types of homogeneous non-metallic catalysts. The three reaction mechanisms for cyclic carbonate formation are presented, namely activation of the epoxide ring, CO_2 activation and dual activation. Usually most effective catalysts described in the literature consist of powerful sources of nucleophile such as onium salt, of hydrogen bond donors and of tertiary amines used to combine epoxide activation for facile epoxide ring opening and CO_2 activation for the subsequent smooth addition reaction and ring closure. The most active catalytic systems are capable of activating even internal epoxides such as epoxidized unsaturated fatty acid derivatives for the cycloaddition of CO_2 under relatively mild conditions. In case of terminal epoxides such as epichlorohydrin, the effective utilization of diluted sources of CO_2 such as flue gas is possible using the most active organocatalysts even at ambient pressure.

Keywords: cycloaddition; ionic liquid; deep eutectic solvents; onium salt; homogeneous catalysts

Citation: Weidlich, T.; Kamenická, B. Utilization of CO_2-Available Organocatalysts for Reactions with Industrially Important Epoxides. *Catalysts* **2022**, *12*, 298. https://doi.org/10.3390/catal12030298

Academic Editors: Javier Ereña and Ainara Ateka

Received: 16 February 2022
Accepted: 4 March 2022
Published: 6 March 2022

Publisher's Note: MDPI stays neutral with regard to jurisdictional claims in published maps and institutional affiliations.

Copyright: © 2022 by the authors. Licensee MDPI, Basel, Switzerland. This article is an open access article distributed under the terms and conditions of the Creative Commons Attribution (CC BY) license (https://creativecommons.org/licenses/by/4.0/).

1. Introduction

Carbon dioxide capture and utilization (CCU technologies) has been recognized as a possible and cost-effective way to reduce worldwide greenhouse gas emissions [1–10]. The use of CO_2 as a raw material in chemical synthesis is a research area of great scientific, economic and ecological interest [1–14]. The abundance and benignity of carbon dioxide, which is cheap, nontoxic and nonflammable, makes it a very attractive low-cost C1-synthon in organic chemistry. Moreover, the mitigation of CO_2 emission from industrial processes in order to reduce CO_2 causing the greenhouse effect encourages chemists to carry out research in this area.

CO_2 is thermodynamically stable ($\Delta G^0 = -394.228$ kJ/mol) [10], however, and needs catalytic activation and a corresponding reactive agent for its possible fixation into the organic molecules. Thermodynamically non-stable three-membered heterocyclic rings such as epoxides serve as ideal reactants for CO_2 fixation. The reaction of epoxides and carbon dioxide to produce cyclic carbonates is attractive because CO_2 can be incorporated in the epoxide molecule without the formation of any side products (with 100% atom-economy) (Scheme 1) [1,2].

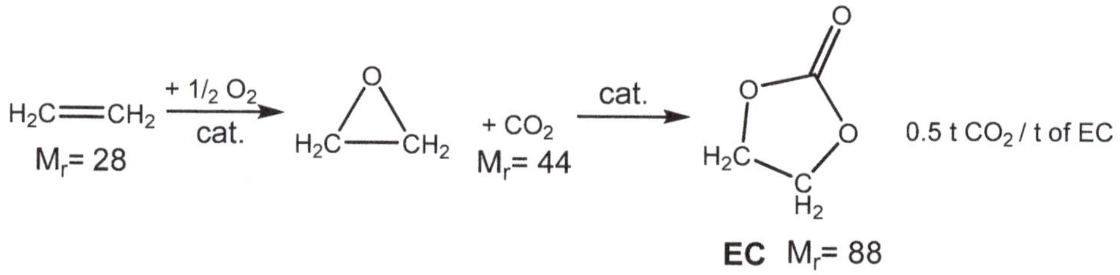

Scheme 1. CO_2 consumption for ethylene carbonate (EC) production applying the carbonation of oxirane produced from ethylene [2].

The epoxides available for carbonation can be classified according to their bound substituents as terminal epoxides (containing at least one unsubstituted CH_2 group in the oxirane ring) and internal epoxides (containing substituents on both carbon atoms of the oxirane ring, Figure 1).

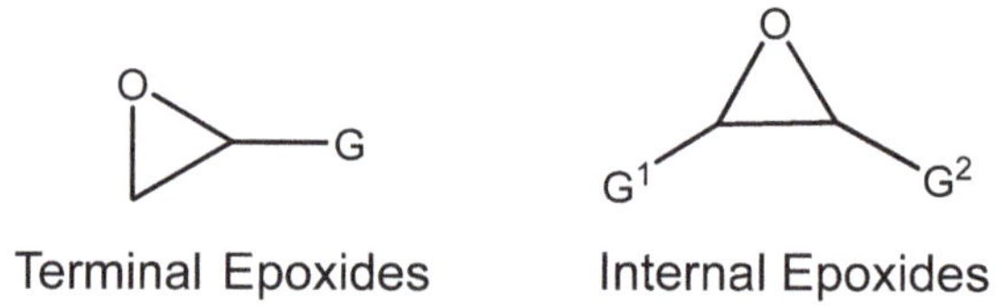

Figure 1. Epoxides available for cycloaddition reactions.

The reaction conditions for the efficient carbonation of epoxides differ significantly, utilizing terminal (for example, glycidyl derivatives such as epichlorohydrin or glycidol and corresponding glycidyl ethers) [1–3,5–14]) or sterically, much more hindered internal epoxides (for example, epoxidized fatty acids and their derivatives [4,8–14]). From the sustainability point of view, bio-based epoxides ideally serve as suitable reactants for the production of cyclic carbonates utilizing waste CO_2.

In particular, the connection of direct air oxidation and carbonation of ethylene oxide serves as a simple and effective technology that is useful for the subsequent efficacious utilization of anthropogenic CO_2 (Scheme 1). In case of ethylene carbonate (EC) produced from ethylene, 44 g of CO_2 is utilized per mol of produced EC (approximately 2/3 of EC molecular weight creates CO_2) according to the stoichiometry of this reaction. On the other hand, CO_2 is emitted during the production of ethylene using conventional technologies such as steam cracking (discussed in Section 2). According to the life cycle assessment (LCA) methodology, technology based on the carboxylation of petrochemical ethylene via catalytic air oxidation emits only 0.92 t CO_2/t EC [2].

The catalytic carboxylation of epoxides may afford either cyclic carbonates (Scheme 2, Path a) or eventually polycarbonates (Scheme 2, Path b) [4,7,9], depending on the used catalyst and the reaction conditions.

The nucleophile-based ring opening of oxirane activated by the catalyst with the subsequent addition of CO_2 and five-membered ring closure accompanied by the release of nucleophile is described in Scheme 1, Path a. The catalyst that activates oxirane (H-bond donor or Lewis acid) decreases the highest reaction barrier of the ring opening (usually the rate-determining step) or/and stabilizes the alkoxide produced by the ring-opening, thus promoting the cycloaddition reaction. Different Lewis bases act as nucleophiles, preferentially bromide or iodide.

Path b. describes the insertion of another oxirane molecule followed by alternate copolymerization with carbon oxide and epoxide. The polymerization occurs in the case of insertion of another epoxide molecule competes with the ring-closure process. Generally, the utilization of metal-based catalysts is necessary for the direct formation of polycarbonates starting from CO_2 and epoxides via ring-opening polymerization.

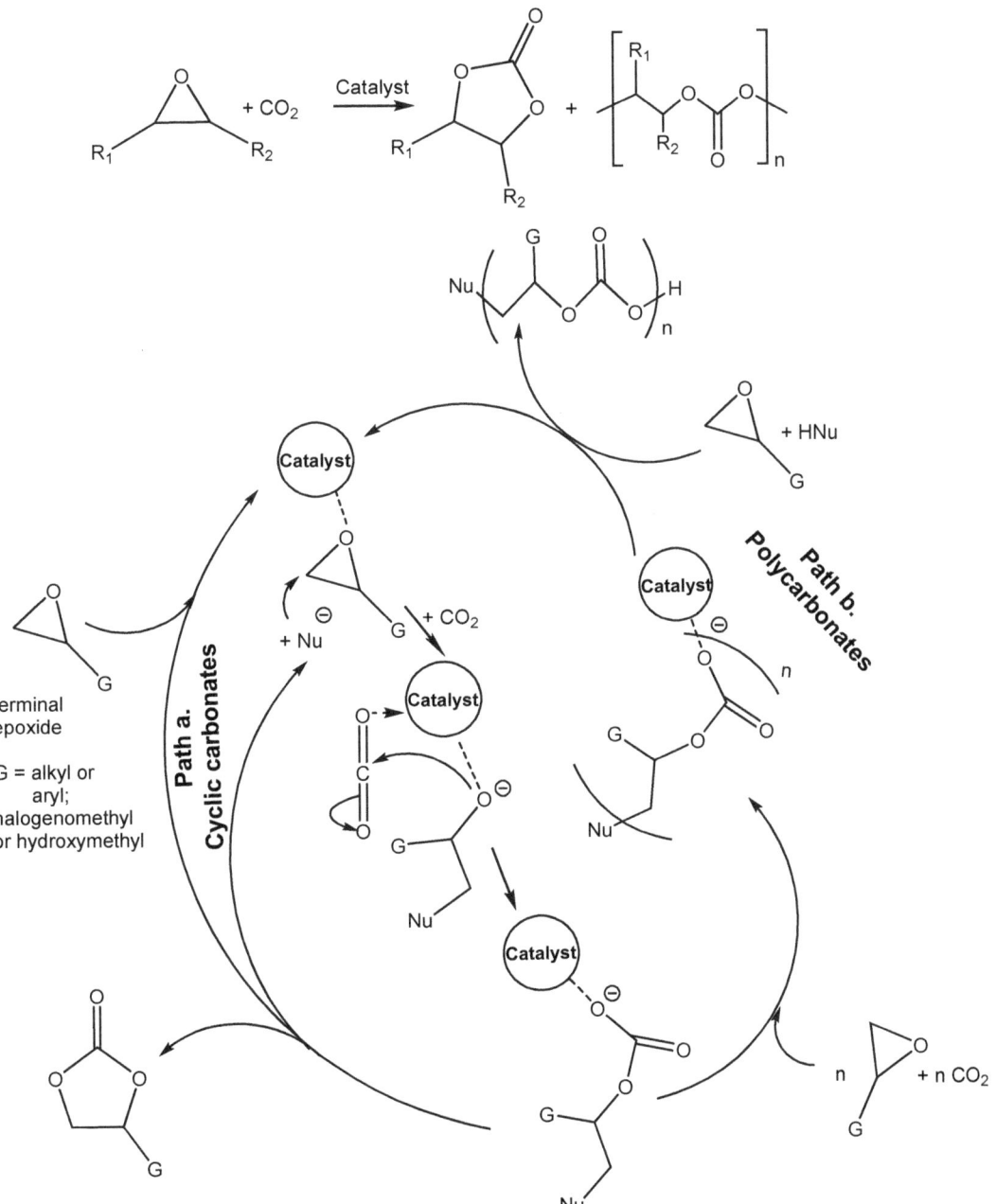

Scheme 2. Reaction of CO_2 with epoxides producing cyclic carbonates and/or polycarbonates [11].

Utilizing CO_2 via cycloaddition to epoxides is the exothermic process that can be carried out under mild reaction conditions using a broad spectrum of catalysts [11,12].

Cyclic carbonates are generally stable liquids, which enables the long-term sequestration of CO_2, especially when cyclic carbonates are subsequently used for polymerization. Although the production of cyclic carbonates from CO_2 and epoxides has been industrialized since 1958 and uses inexpensive catalysts such as KI, the current production processes still suffer from major disadvantages, such as high reaction temperatures (180–200 °C), high pressure (5–8 MPa) and stoichiometric amounts of activating reagents [13,14].

It is known that the increasing of the reaction rate with the increase in CO_2 pressure is not only counterproductive due to the high energy consumption, but is even limited. The high increase in CO_2 pressure (and the overrunning at a concentration of 0.47 g CO_2/mL) is accompanied by sudden decrease in reaction rate due to the dilution effect causing the reduction in epoxide and catalyst concentrations in the reaction mixture [11].

Cyclic carbonates are used for polymer production, as electrolytes in lithium-ion batteries, polar aprotic (ethylene or propylene carbonate) or protic solvents (glycerol carbonate, etc.) and as chemical intermediates in organic synthesis [1–14] (Scheme 3).

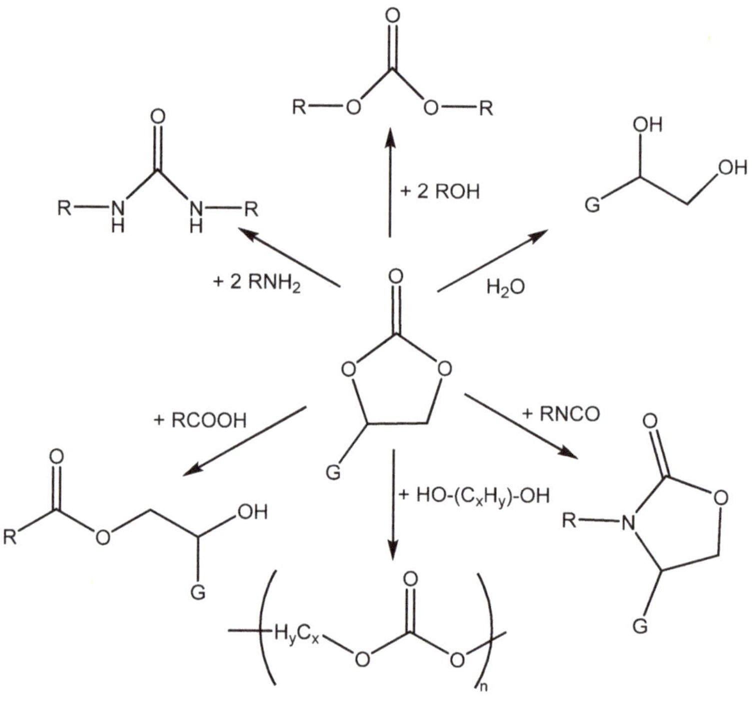

Scheme 3. Possibilities for the chemical transformation of cyclic carbonates to linear (poly)carbonates, esters, ureas, ethyleneglycols and urethanes (carbamates) [1,2,5,13].

Alluding to cyclic carbonates used as solvents, a grade deal of attention has been given to the application of glycerol carbonate as it possesses low toxicity and good biodegradability, and has a high boiling point and simple availability from lipids and CO_2, giving it many applications [10]. Generally, the direct fixation of CO_2 in cyclic carbonates and their products is regarded as a greener approach than the existing practices. As North et al. have mentioned, only two reactions of CO_2, the dry reforming of methane (for fuel production) and cyclic carbonate chemical production, could consume up to 25% of the anthropogenic CO_2 produced annually [12].

The cyclic carbonate formation based on the cycloaddition of CO_2 requires oxirane ring opening in the first reaction step, which is followed by insertion of CO_2 in the second reaction step and ring closure to form cyclic carbonate. The general mechanisms of this reaction are illustrated in Scheme 4 [3,12].

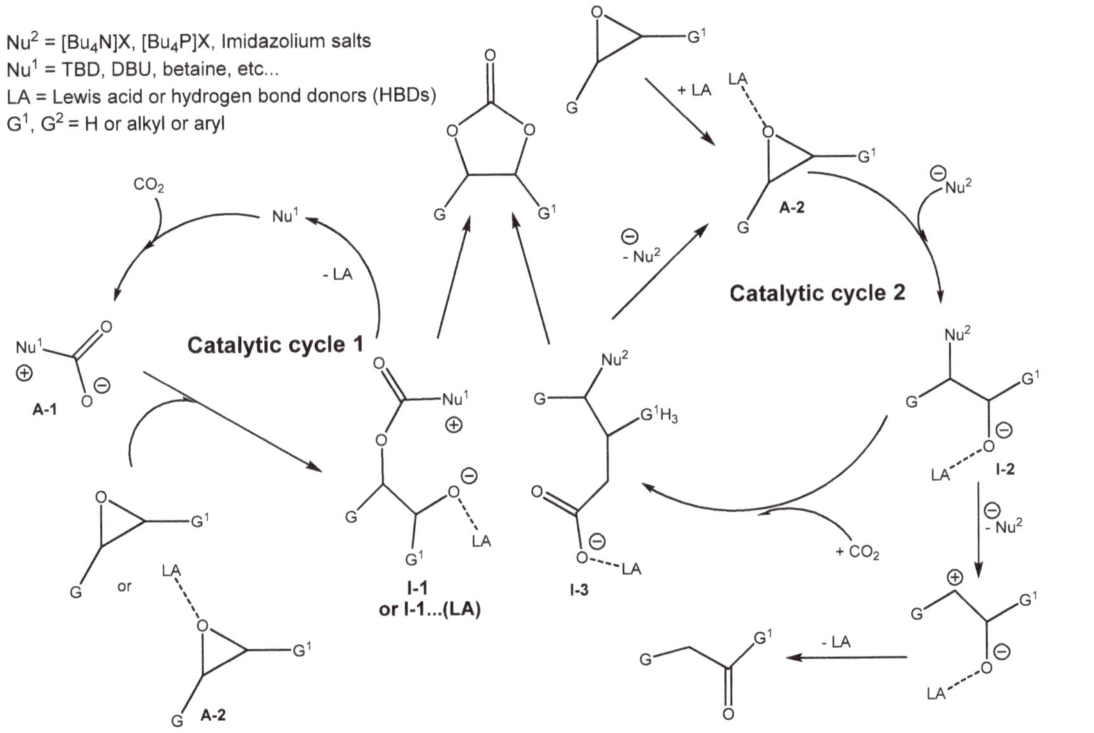

Nu² = [Bu₄N]X, [Bu₄P]X, Imidazolium salts
Nu¹ = TBD, DBU, betaine, etc...
LA = Lewis acid or hydrogen bond donors (HBDs)
G¹, G² = H or alkyl or aryl

Scheme 4. Generally accepted mechanisms of cyclic carbonate formation via the cycloaddition of epoxide on CO_2 [3,12].

According to the computations, the epoxide ring-opening has the highest energetic barrier. Due to this reason, the applicable catalysts decrease the reaction barrier, acting as Lewis bases (nucleophile, Nu) that are capable of nucleophilic attack and opening the oxirane ring and/or stabilizing intermediates. In addition, the oxirane can be activated by hydrogen bond formation interacting with a Lewis acid or a hydrogen bond donor (HBD), thereby decreasing the energetic barrier of the epoxide ring opening. In order to catalyze cyclic carbonate formation most effectively, the best catalysts often contain a combination of both Lewis base and Lewis acid components or Lewis base/HBD parts. If the catalysis influencing the ring opening is very effective, then the ring closure step can be the rate-determining step with a higher energetic barrier [15].

The action of some described effective catalysts is based even on the activation of CO_2, as is assumed. The activation of CO_2 comprises formation of carbamate or carbonate, which may also take part in the epoxide ring-opening step [3,5,10–12].

Catalytic cycle 1 in Scheme 4 (CO_2 activation mechanism) describes the nucleophilic activation of CO_2 using nucleophiles such as tertiary amines, amidines, guanidines or carbenes (Nu¹) to form an intermediate A-1 (carbamate or carboxylate), which subsequently promotes oxirane ring opening, leading to the second intermediate (I-1). A-1 attacks oxirane or oxirane activated by Lewis acid (A-2). Cyclization then occurs to produce a cyclic carbonate while the nucleophile (Nu¹) is recycled for a new catalytic cycle. The CO_2 activation mechanism requires a catalyst nucleophilic towards CO_2, but not for epoxide [4,12].

In addition, catalytic cycle 2 demonstrates a more common carbonation pathway through the activation of epoxide. The hydrogen bond donor(s) (HBDs) coordinate(s) to the epoxide ring (formation of intermediate A-2). Subsequently, the Lewis base, for instance, onium halide salts, acting as nucleophile (Nu^2) attacks the oxirane to enable the ring opening, producing an intermediate (I-2). Subsequently, upon ring opening, the CO_2 insertion occurs in the O-H bond of the intermediate (I-2), producing a new intermediate (I-3). The cyclization of the intermediate (I-3) accompanied by cleavage of the leaving group (Nu^2), produces cyclic carbonate as the final product, and nucleophile is recycled for a further reaction. Alternatively, cleavage of nucleophile Nu^2 from I-2 before CO_2 addition causes the formation of alternative products such as corresponding ketones via a Meinwald rearrangement.

2. Sources of Epoxides

Epoxides are generally accessible either via the oxidation of the C=C double bond in alkenes (using peroxides or O_2/Ag-based reaction) or via the dehydrohalogenation of geminal halogenoethanols [5].

Terminal epoxides such as epichlorohydrin (EPIC), glycidol (GL), methyloxirane (propylene oxide, PO) and oxirane (ethylene oxide, EO), and their functional derivatives, are the most studied epoxides for the CO_2/epoxide coupling reaction.

EPIC and GL are chemicals that were recently produced from bio-based glycerol obtained as a by-product from the chemical utilization of lipids (Scheme 5). Transesterification or hydrolysis of lipids is the main pathway for the production of bio-based glycerol and fatty acid esters (for biodiesel) and soap (sodium or potassium salts of fatty acids are used as surfactants) [5] (Scheme 6).

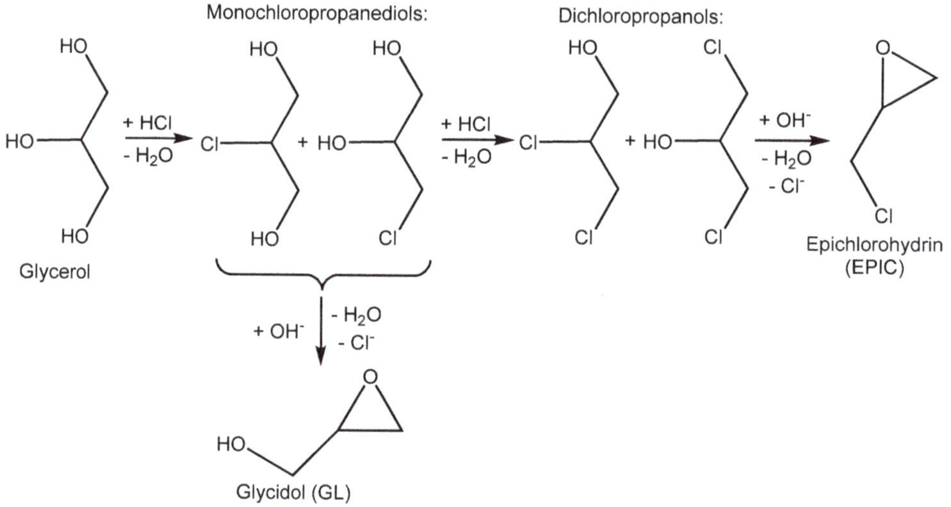

Scheme 5. Production of glycidol and epichlorohydrin starting from bioglycerol [16–20].

The nucleophilic substitution of OH groups in the glycerol structure with Cl via the action of gaseous HCl enables the formation of monochloropropanediols or dichloropropanols, in the presence of a catalyst, usually a carboxylic acid [2,5,16–18], as the crucial feedstock for production.

The subsequent ring closure induced by an inorganic base such as calcium or sodium hydroxide enables epoxide formation via a dehydrochlorination reaction with inorganic chloride (NaCl or $CaCl_2$) being obtained as a co-product [19,20] (Scheme 5).

Similarly, starting from chloropropanediols, glycidol is produced by basic hydrodechlorination [20].

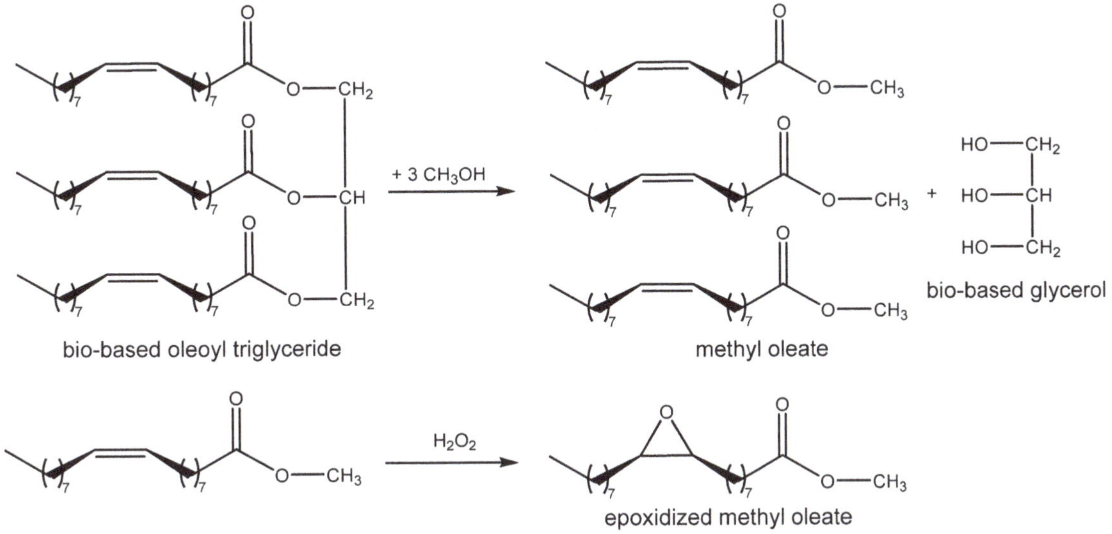

Scheme 6. Synthesis of glycerol and epoxidized methyl oleate from triglyceride of oleic acid (vegetable oil).

EO and PO are still mainly derived from ethylene and propylene produced by the cracking of petrochemical feedstock with subsequent silver-catalyzed direct air oxidation [2].

The development of appropriate sustainable alkenes production is based on the utilization of bio-based syngas (CO + H_2) obtained by the gasification of waste biomass [21]. The most promising pathway for alkene production exploits methanol as a crucial intermediate simply available from syngas with its subsequent dehydrogenation/coupling catalyzed by zeolites (methanol-to-olefins technology, MTO) [21]. Apart from the above-mentioned, the direct catalytic cracking of vegetable oils (lipids) may produce propylene [21].

The other sustainable pathway for ethylene or butylene production is based on the dehydrogenation of bio-based alcohols (bioethanol and biobutanol) [22] accessible by fermentation of oligosaccharides. Brasco Co. produces bio-ethylene, for example, through the dehydration of bioethanol for polyethylene production in Brazil [23].

In addition, other types of epoxide can be produced from waste biomass, such as limonene oxide and limonene dioxide, which can be synthesized via the promising epoxidation of bio-based limonene obtained from citrus peels, oak and pine tree, under solvent-free conditions with hydrogen peroxide and a tungsten-based catalyst [24].

In addition, vegetable oil-derived triglycerides and fatty acids contain double bond(s) (Scheme 6), which can undergo epoxidation with peroxides [2]. The epoxidized fatty acid derivatives can be subsequently exploited as the starting materials for cycloaddition or even a subsequent one-pot polymerization reaction with CO_2 [25,26]. Scheme 6 shows the transesterification of natural triglycidyl oleate producing methyl oleate and bioglycerol, and the subsequent epoxidation of methyl oleate using hydrogen peroxide.

It is worth noting, however, that epoxidation is a highly exothermic reaction (as $\Delta H = -55$ kcal/mol for each double bond); thus, H_2O_2 is slowly added or added in a stepwise manner in semi-batch operations, and requires a long reaction time. In order to avoid problems with heat and mass transfer, process intensification using a microreactor was proposed [27]. The microreactor could significantly decrease the reaction time with higher epoxide selectivity.

3. Homogeneous Metal-Free Catalysts

This review focuses on the recent development of homogeneous organocatalysts for the cycloaddition reaction of CO_2 and epoxides (insertion of CO_2 into the oxirane ring) with the aim of producing cyclic carbonates. Homogeneous catalytic systems (catalyst and reagent dissolved in the same phase) usually display the highest conversion and selectivity and are usually significantly cheaper compared with heterogeneous ones. Metal-free organocatalysts are usually readily available even from renewable sources, non-recalcitrant, biodegradable, affordable, and inert towards air and moisture. They may, however, be more difficult to separate and recycle from the produced cyclic carbonates. On the other hand, after the discovery of a new effective homogeneous catalysts, subsequent attempts to immobilize it on the insoluble surface have followed in order to solve problems with catalyst separation and recycling [12].

3.1. Catalytically Active Amines and Their Salts

Generally, the effective absorption of CO_2 into the liquid phase and its subsequent activation of chemisorbed CO_2 are important steps for the subsequent cycloaddition reaction of CO_2 with epoxides.

The alkaline absorbents described in the literature as being capable of effectively absorbing CO_2 are different amines [28], including amidines such as 1,8-diazabicyclo [5.4.0]undec-7-ene (DBU) [28–30], guanidines such as 1,5,7-triazabicyclo[4.4.0]dec-5-enium (TBD) [28,29], and azaheterocycles such as pyridines and imidazoles [31,32], which are efficient for the chemisorption and activation of CO_2.

The correlation between the structures of different organic amines and their catalytic activity in the cycloaddition of CO_2 was studied by Yu et al. [28]. Their article compares the catalytic activity of a variety of nucleophilic aliphatic amines (ethanolamine, bis-(3-aminopropyl)-amine, oleylamine), basic aminoacid arginine, nucleophilic aromatic amines (1,2-phenylenediamine, 2-aminobenzylamine) and non-nucleophilic cyclic amines such as TBD and DBU for the cycloaddition of CO_2 to methyloxirane (propylene oxide, PO) [28].

It is well known that the chemisorption ability of CO_2 (formation of carbamate) increases with the increasing basicity of amine [33,34]. Yu et al. observed no apparent relationship of the pK_a value of conjugated acids of the tested amines with respect to their catalytic activity in the case of propylene carbonate (PC) formation (Figure 2, Table 1). The authors observed the lowest propylene carbonate conversion when applying aliphatic amines.

Low conversion was even observed in the case of amine forming intramolecular hydrogen bonds (1,2-phenylenediamine and 2-aminobenzylamine).

In contrast, aromatic amines (such as 1,4-phenylenediamine) and amidines or guanidines involving conjugated "N=C–N" structures (arginine, TBD) were found to be the most active ones [28].

According to the published information [35,36], tertiary amines are often higher in activity than primary or secondary ones (Table 2, Entries 5, 10–12). It has been proven that amidines containing the "N=C–N" bond in their structures are particularly favorable for the cycloaddition of CO_2 with epoxides (Table 1, Entries 6–8, 17 and Table 2, Entries 3 and 12) [2–4,6,32]. The above-mentioned observations are in agreement with the statement of North et al. that for the efficient activation of CO_2, compounds that nucleophilically attack CO_2 but not the epoxide ring are sought out [12].

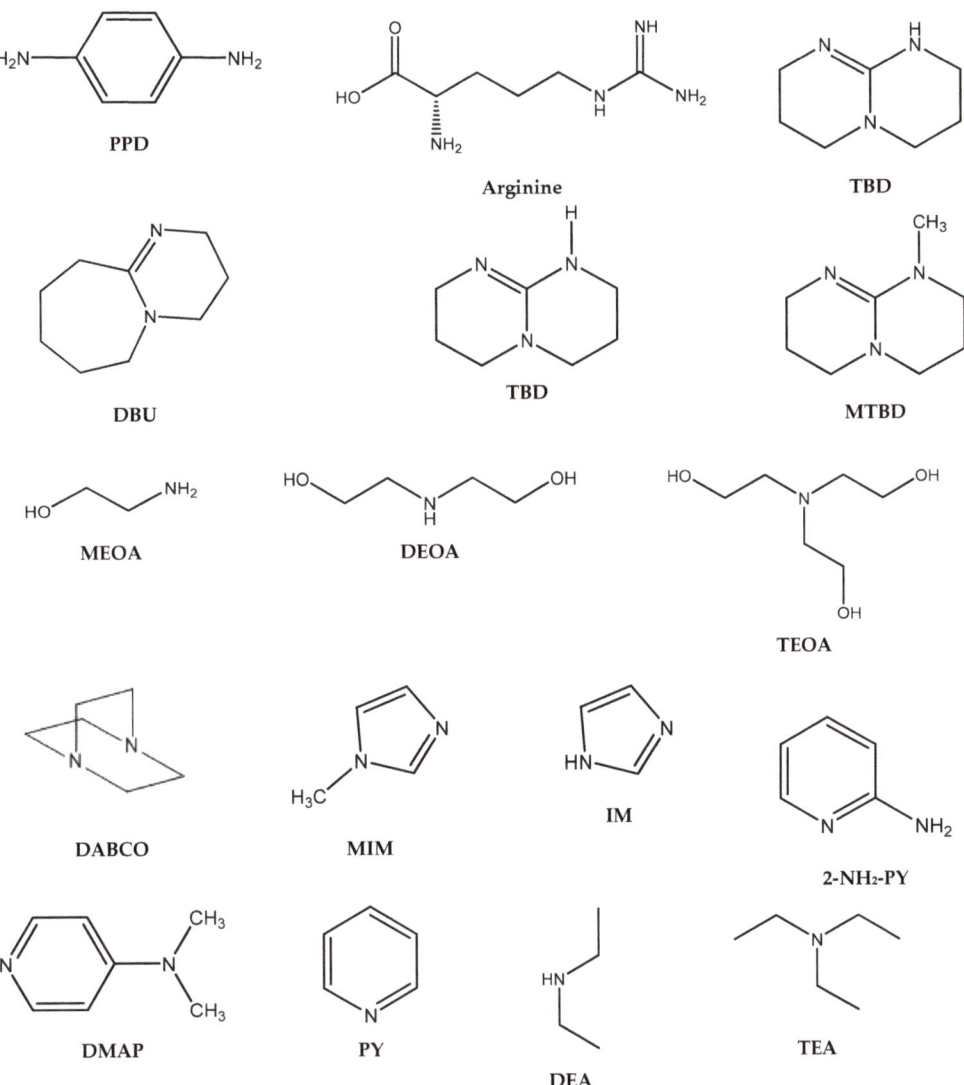

Figure 2. Structures and abbreviations of the tested amines.

Azzouz et al. demonstrated the effective utilization of 2-aminopyridine (2-NH$_2$-PY) as a catalyst (using 10 molar % of 2-NH$_2$-PY) for the carbonation of different terminal epoxides at 60–85 °C and 1 MPa of CO$_2$ [35]. This carbonation was performed even at pilot scale.

Interestingly, corresponding salts with protic (Bronsted) acids (base.HA) of the above-mentioned non-nucleophilic amines (DBU.HA, N-methylimidazole (MIM.HA), N,N-dimethylaminopyridine (DMAP.HA)) and, alongside these, even triethanolamine, pyridine and caffeine (Figure 3), are quite catalytically active in the cycloaddition of CO$_2$ with epoxides (Table 3, Entries 3–12, 15–24). Hydrogen halides in particular are the most active in comparison with corresponding free bases [29,30,32,37–39] (Table 3, Entries 3–6, 8–12). The most active seems to be hydroiodides of the corresponding amines (Table 3, Entries 10, 18–24, 26–27). For the above-mentioned catalytic action of amine salts, the reaction mechanism based on the activation of epoxide via the protonation with amine salt in the role of Bronsted acid (HA) is proposed with subsequent anion-based epoxide ring opening.

Table 1. Effect of amine structure on the cycloaddition of CO_2 to propylene oxide (PO) producing propylene carbonate (PC).

Entry	Catalyst	Catalyst Amount (mol%)	CO_2 Pressure (MPa)	Temperature (°C)	Reaction Time (h)	PO Conversion/Yield (%)	Selectivity (%)	Ref.
1	Monoethanolamine	3.2	5	150	24	48.8	not specified	[31]
2	bis-(3-amonipropyl)-amine	3.2	5	150	24	55.8	not specified	[28]
3	o-phenylene-diamine	3.2	5	150	24	56.6	not specified	[28]
4	oleylamine	3.2	5	150	24	67.1	not specified	[28]
5	2-amino-benzylamine	3.2	5	150	24	70.2	not specified	[28]
6	Arginine	3.2	5	150	24	79.9	not specified	[28]
7	PPD	3.2	5	150	24	93.8	not specified	[28]
8	TBD	3.2	5	150	24	100	not specified	[28]
9	Pyridine	1.2	1.2	120	3	34	99	[32]
10	Pyridine	1.2	2.5	120	3	77	99	[32]
11	Imidazole	1.2	1.2	120	3	28	99	[32]
12	Imidazole	1.2	2.5	120	3	63	99	[32]
13	DMAP	12.5	2	120	2	76	96	[31]
14	DMAP	1.2	2.5	120	3	71	99	[32]
15	DMAP	1.2	1.2	120	3	30.7	not specified	[39]
16	DMAP	1.2	1.2	120	3	31	99	[32]
17	DBU	9	2.0	120	2	79.2	not specified	[39]
18	DBU	1	1	140	2	2	3	[29]
19	DBU	10.5	2	120	2	80	90	[31]
20	No catalyst	0	1	140	2	0	0	[29]

Abbreviations: PPD—p-Phenylenediamine.

Table 2. The effect of organic amines on the conversion of PO [36].

Entry	Catalyst	pKa	Catalyst Amount (mol%)	CO_2 Pressure (MPa)	Temperature (°C)	Reaction Time (h)	Conversion (%)	Selectivity (%)
1	TBD	26	11.3	2	120	2	8	96
2	MTBD	25.5	10.2	2	120	2	15	99
3	DBU	24.3	10.5	2	120	2	93	99
4	TEA	18.8	15.5	2	120	2	41	95
5	DMAP	18	12.5	2	120	2	87	99
6	PY	12.5	19.9	2	120	2	30	92
7	DEA	11	21.5	2	120	2	34	92
8	MEOA	9.5	25.7	2	120	2	64	88
9	DEOA	8.9	15	2	120	2	74	94
10	DABCO	8.7	14.1	2	120	2	80	98
11	TEOA	7.8	10.6	2	120	2	81	93
12	MIM	7.1	19.2	2	120	2	74	99
13	IM	7	23.1	2	120	2	10	93

Table 3. Comparison of the catalytic effects of amines and their salts on the carbonation of PO.

Entry	Catalyst	Catalyst Amount (mol%)	CO_2 Pressure (MPa)	Temperature (°C)	Reaction Time (h)	Conversion/Yield (%)	Selectivity (%)	Ref.
1	No catalyst	0	1	140	2	0	0	[29]
2	DBU	1	1	140	2	2	3	[29]
3	[HDBU]OAc	1	1	140	2	86	90	[29]
4	[HDBU]Cl	1	1	140	2	97	>99	[29]
5	[HTBD]Cl	1	1	140	2	86	93	[29]
6	[HMIM]Cl	1	1	140	2	83	90	[29]
7	[HPY]Cl	1	1	140	2	59	87	[29]
8	CAFH.Br	7	0.7	70	16	93	not specified	[37]
9	CAFH.Br	7	0.7	70	6	>99	not specified	[37]
10	[HTEA]I	1.88	2	110	6	91	not specified	[37]
11	IM + CH_3COOH	1.2	1.2	120	3	95	98	[39]
12	PY + CH_3COOH	1.2	1.2	120	3	94	98	[39]
13	No catalyst	0	2	110	6	trace	not specified	[38]
14	TEOA	1.88	2	110	6	trace	not specified	[38]
15	[HTEOA]F	1.88	2	110	6	4	99	[38]
16	[HTEOA]Cl	1.88	2	110	6	5	99	[38]
17	[HTEOA]Br	1.88	2	110	6	6	99	[38]
18	[HTEOA]I	1.88	2	110	6	91	99	[38]
19	[HMEOA]I	1.88	2	110	6	65	99	[38]
20	[HDEOA]I	1.88	2	110	6	90	99	[38]
21	PANI-HI	1.88	2	110	6	99	not specified	[38]
22	[HHMTA]Cl	1.88	2	110	6	33	48	[38]
23	[Hbet]I	1.88	2	110	6	98	not specified	[38]
24	[HMIM]I	1.88	2	110	6	91	99	[38]

Abbreviations: [HMIM]Cl—1-methylimidazolium chloride; CAFH.X—caffeinium halide; [Hbet]I—$Me_3NCH_2COO.HI$; PANI-HI—Polyaniline hydrogen iodide; [HHMTA]Cl—urotropin hydrogen chloride.

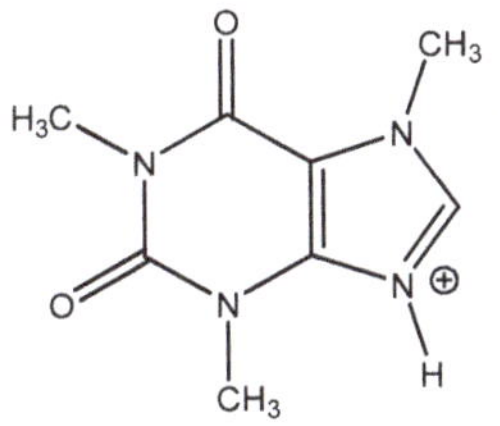

Figure 3. Structure of catalytically active caffeine hydrohalides [37].

Sun et al. reported very effective carbonation using triethanolamine hydroiodide, including the simple recyclability of this catalyst without loss of activity even after four recycling steps [38].

The published results indicate that the synergetic effect of hydroxyl groups from protonated aminoalcohol in the role of HBD, together with naked bromide or iodide, significantly influences the cycloaddition of CO_2 to the studied epoxides and makes possible the application of this reaction even at ambient conditions (Table 3, Entry 17–18).

Catalytically active ammonium halide activates not only the epoxide ring for opening through the H-bond with hydroxyl groups of triethanolammonium cation and the subsequent addition of intermediate 2-halogenoalkoxide to CO_2, but even the next ring closure caused by the facile withdrawal of halide from the produced 2-halogenocarbonate [38] (Schemes 3 and 4).

Apart from the above-mentioned, in the case of caffeine hydrobromide, potassium halides added to the reaction mixture as an additional source of nucleophiles were successfully tested. With the same reaction conditions, the enhanced efficiency of cyclic carbonate formation was observed utilizing equimolar quantities of KX and caffeine.HBr or even

2: 1 KX: caffeine.HBr using DMSO as the reaction solvent at 70 °C and 0.7 MPa CO_2 pressure [29,37]. The yield of cyclic carbonate increased following the trend KF < KCl < KBr < KI, which was in good agreement with nucleophilicity and nucleofugacity of the corresponding halide anions (Table 4, Entries 17–25).

Table 4. Catalyst screening for the cycloaddition of EPIC.

Entry	Catalyst	Catalyst Amount (mol%)	CO_2 Pressure (MPa)	Temperature (°C)	Reaction Time (h)	Conversion (%)	Reference
1	PY	5	0.1	25	24	3	[40]
2	PY	5	0.1	60	24	39	[40]
3	PY-CH$_2$OH	5	0.1	25	24	18	[40]
4	PY-CH$_2$OH	5	0.1	60	24	73	[40]
5	Ph CH$_2$OH	5	0.1	60	24	0	[40]
6	CAFH.Br	7	0.7	70	16	>99	[37]
7	CAFH.I	7	0.7	70	16	95	[37]
8	CAFH.Cl	7	0.7	70	16	81	[37]
9	CAF	7	0.7	70	16	9	[37]
10	HBr	7	0.7	70	16	78	[37]
11	HI	7	0.7	70	16	42	[37]
12	HCl	7	0.7	70	16	26	[37]
13	1-MIM.HBr	7	0.7	70	16	99	[37]
14	1,2-MMIM.Br	7	0.7	70	16	93	[37]
15	1-MIM.HBr	7	0.7	70	16	77	[37]
16	1,2-MMIM.HBr	7	0.7	70	16	67	[37]
17	CaFH.Br	7	0.7	70	6	14	[37]
18	CaFH.Br/KI (1:1)	7	0.7	70	6	94	[37]
19	CaFH.Br/KI (1:2)	7	0.7	70	6	>99	[37]
20	CaFH.Br/KBr (1:1)	7	0.7	70	6	87	[37]
21	CaFH.Br/KBr (1:2)	7	0.7	70	6	95	[37]
22	CaFH.Br/KCl (1:1)	7	0.7	70	6	70	[37]
23	CaFH.Br/KCl (1:2)	7	0.7	70	6	81	[37]
24	CaFH.Br/KF (1:1)	7	0.7	70	6	38	[37]
25	CaFH.Br/KF (1:2)	7	0.7	70	6	51	[37]
26	TEA.HI	10	0.1	40	24	87	[41]

In Table 5, the increase in carbonate yields using CAFH.Br/KI in the case of carbonation of terminal epoxides is documented. In the case of internal epoxide (limonene oxide), however, no carbonation was observed using caffeine hydrobromide or even its mixture with KI (Table 5, Entry 5).

Roshan et al. came to the conclusion that even the addition of a low quantity (a catalytic amount) of H_2O significantly enhances CO_2-based formation of PC over tertiary heterocyclic amines such as IM, PY and DMAP, giving over 98% selectivity of PC formation (at 120 °C, 1.2 MPa, 3 h). The observed results were evaluated by a DFT study comparing energy profiles for free amine in comparison with corresponding amine hydrogencarbonate-mediated cycloadditions of CO_2 to propylene oxide. Ammonium hydrogencarbonates produced in situ from heterocyclic amine, H_2O and CO_2 works similarly to the above-mentioned amine salts as activators of the epoxide ring. As was argued, the HCO_3^- anion generated in the water-CO_2-base reaction was the key active species that gave the higher activity of the base-water systems rather than the carbamate salt (produced by a reaction of R$_3$N with CO$_2$) [32].

Table 5. Reaction of various epoxides with CO$_2$ [37].

Entry	Abbreviation of Used Epoxide	Cyclic Carbonate	Structure	Conversion (%) CAFH.Br	Conversion (%) CAFH.Br/KI
1	EPIC	CMEC		>99	>99
2	EPIB	BMEC		98	98
3	PO	MEC		93	99
4	SO	PEC		28	53
5	LO	LC		no conversion	no conversion

Reaction conditions: 7 mol% catalyst at 70 °C and 0.7 MPa; reaction time, CAFH.Br = 16 h and CAFHBr/KI = 6 h.

It should be mentioned that low-melting salts (melting point below 100 °C) obtained by the neutralization of organic bases with organic or inorganic acids embodies are called protic ionic liquids (PILs). The melting of PILs could enhance the miscibility of catalytically active PILs with reacting epoxide and CO$_2$ compared with solid catalysts, as was published by Zhang et al. in the case of DMAP hydroiodide [42] or by Kumatabara et al. using triethylamine hydroiodide [41] at ambient pressure (Table 3, Entry 10 and Table 4 Entry 26).

Zhang et al. published results obtained even by means of the capture and utilization of CO$_2$ for the cycloaddition into SO using PIL (DMAP hydrobromide) at ambient pressure and 120 °C [42]. This PIL has superior catalytic effect compared with other hydrogenhalides of tertiary bases such as DBU, MIM, DABCO or tetramethylguanidine (Table 6). DMAP.HBr is well reusable with no drop in activity after five recycling steps. DMAP.HBr is able to carbonate even internal epoxides such as ChO, although this cycloaddition is quite sluggish.

Table 6. Synthesis of SC catalyzed by different PILs [42].

Entry	Catalyst	Catalyst Amount (mol%)	CO_2 Pressure (MPa)	Temperature (°C)	Reaction Time (h)	SO Conversion (%)	SC Selectivity (%)
1	none	1	0.1	120	4	n.d.	n.d.
2	DMAP/NaBr	1	0.1	120	4	48	99
3	[DMAPH]Br	1	0.1	120	4	96	99
4	[DMAPH]Br	0.5	0.1	120	4	76	99
5	[DMAPH]Br	0.7	0.1	120	4	85	99
6	[DMAPH]Br	1	0.1	120	4	96	99
7	[4-MeNH-PyH]Br	1	0.1	120	4	78	99
8	[4-NH_2-PyH]Br	1	0.1	120	4	60	99
9	[4-OH-PyH]Br	1	0.1	120	4	47	99
10	[DBUH]Cl	1	0.1	120	4	80	99
11	[DBUH]Br	1	0.1	120	4	85	99
12	[HMIM]Br	1	0.1	120	4	84	99
13	[HTMG]Br	1	0.1	120	4	71	99
14	[DABCOH]Br	1	0.1	120	4	61	99
15	[DMAPH]Br	1	0.1	80	4	36	99
16	[DMAPH]Br	1	0.1	80	12	76	99
17	[DMAPH]Br	1	0.1	120	14	92	99

Abbreviations: [HMIm]Br—3-methylimidazolium bromide; [4-MeNH-PyH]—4-methylaminopyridine; [HTMG]Br—N,N,N',N'-tetramethylguanidinium hydrogen bromide.

3.2. Two Components Catalysts Based on a Combination of Organic Base and Hydrogen Bond Donor

As was mentioned in Section 3.1, the combined action of protonated amine with nucleophilic anion positively influences the efficiency of cycloaddition. This synergic action between the Lewis base, such as the amine, and hydrogen bond donors (HBDs) was reported in the literature [2,4,6,32,36,43].

The possible synergic effects of alcohols (glycerol, glycidol, 1,2-propylene glycol (PG), poly(ethylene glycol)-600 (PEG600), poly(ethylene glycol)-400 (PEG400), cellulose, chitosan and β-cyclodextrin (β-CD)) known as HBDs was explored in CO_2-based cyclic carbonates synthesis catalyzed by amines, as mentioned in Section 3.1 [36]. For this purpose, the most catalytically active DBU and DMAP were tested in relation to the co-action of the chosen HBDs (Table 7).

Out of the set of experiments, cellulose was recognized as the most effective HBD in the addition of CO_2 to propylene oxide [36].

The effective quantity of cellulose used as HBD in the case of the DBU catalysis of the chemical fixation of CO_2 into propylene carbonate is very low with respect to the optimal quantity of DBU (15 mg of cellulose + 300 mg of DBU per mL of PO). The effect of the DBU excess on the yield of PC in the DBU-cellulose reaction system was studied. Generally, the yield rises with the increasing of the DBU: cellulose ratio with the maximum conversion and selectivity reached at a mass ratio of 25–30:1 [36]. The high catalytical activity of cellulose was, in all probability, explained by Khiari et al. and Gunnarson et al. [44,45]. Cellulose reacts in co-action with a significant excess of non-nucleophilic DBU, with CO_2 producing carbonate by means of a reaction similar to that of cellulose xanthate formation during the production of viscose utilizing a sulfur analogue of CO_2, carbon disulfide. The produced carbonate should be a nucleophilic agent that attacks and opens the epoxide ring rather than the known non-nucleophilic DBU.

Table 7. Catalyst screening for the base catalyzed cycloaddition reaction of PO; effects of different HBDs [36].

Entry	Catalyst	HBD	Catalyst Amount (mol%)	CO_2 Pressure (MPa)	Temperature (°C)	Reaction Time (h)	Conversion (%)	Selectivity (%)
1	DBU	none	9.5	2	120	2	80	99
2	DBU	H_2O	9.5	2	120	2	51	89
3	DBU	PG	9.5	2	120	2	89	98
4	DBU	Chitosan	9.5	2	120	2	88	99
5	DBU	β-CD	9.5	2	120	2	83	99
6	DMAP	none	11.1	2	120	2	76	96
7	DMAP	PG	11.1	2	120	2	86	97
8	DMAP	Chitosan	11.1	2	120	2	83	98
9	DMAP	β-CD	11.1	2	120	2	80	98
10	DBU	Cellulose	9.5	2	120	2	93	99
11	DBU	PEG600	9.5	2	120	2	83	99
12	DBU	PEG400	9.5	2	120	2	86	99
13	DBU	Glycerine	9.5	2	120	2	87	98
14	DBU [a]	Cellulose	9.5	2	120	2	92	99
15	DBU [b]	Cellulose	9.5	2	120	2	89	99
16	DBU [c]	Cellulose	9.5	2	120	2	85	99

[a] 2nd reuse; [b] 3rd reuse; [c] 4th reuse.

Aoyragi et al. described the markedly increased formation of cyclic carbonates in isopropylalcohol using triphenylphosphine hydroiodide as a catalyst. 1H NMR spectra documented the formation of H-bonds between the used isopropylalcohol and the starting epoxide [46]. The high activity of the above-mentioned hydroiodide (compared with other HXs) was explained by both the high nucleophilicity and even the high leaving ability (nucleofugality) of iodide ion.

Section 3.1 mentioned the significant catalytic activity of triethanolamine [36], which could be explained by the synergy of HBDs (bound alcoholic OH groups) and the Lewis basicity of the tertiary amine.

More advanced catalysts such as 2-hydroxymethylpyridine (2-PY-CH_2OH) and 2,6-bis(hydroxymethyl)pyridine (2,6-PY-CH_2OH)$_2$ were developed for the high-efficiency cycloaddition of CO_2 with epichlorohydrin (EPIC) under a slightly elevated temperature and ambient pressure (T = 25–60 °C, 0.1 MPa of CO_2; see Table 4, Entry 4) [40]. The high catalytic effect was demonstrated by ^{1}H NMR spectroscope observing the formation of a stable H-bond between the PY-CH_2OH and oxygen of epichlorohydrin. The authors demonstrated that the tested compounds with either heterocyclic nitrogen (benzylalcohol PhCH_2OH) or hydroxymethyl (CH_2OH) groups (PY) catalyzed EC formation only sparingly (PY) or not at all (PhCH_2OH) (Table 4, Entries 1–5).

The catalytic activity of nitrogen-doped charcoal for CO_2 cycloaddition reactions could be explained by the co-operation of OH groups working as HBDs and tertiary amines bound in the graphitic structure of specially prepared N-doped carbons together with the ability of active carbon to adsorb CO_2 [47].

3.3. Aminoacids (AAs) as Catalysts

AAs contain amino and carboxylic groups in their structures. Amino groups can react with CO_2 to form N-COO$^-$ (carbamate) products with low binding energy, which can catalyze the transfer of CO_2 to the 2-halogenoalcoholate produced by the halide-based opening of the epoxide ring. The carboxylate group (–COOH) can catalyze the oxirane ring opening as effective HBD analogously to the amino group (–NHR) and hydroxyl (–OH). Some AA salts have been successfully tested in the capture of CO_2 from flue gas. In addition, the amino groups can also be utilized in quaternization with the aim of introducing halide as a nucleophile into the engineered catalytically active molecules (Table 8).

Table 8. PO catalyzed by AAs.

Entry	Catalyst	Cocatalyst	Cat./Cocat. (mol%)	CO_2 Pressure (MPa)	Temperature (°C)	Reaction Time (h)	Yield (%)	Ref.
1	L-His	-	0.8/-	8	130	48	100	[48]
2	L-His	H_2O	0.44/12.82	1.2	120	3	82	[49]
3	QGly	-	2.15/-	1.2	120	2	84	[50]
4	L-His	KI	0.2/0.2	1	120	3	98	[51]
5	L-His	DBU	2/10	2	120	2	96	[52]
6	-	DBU	-/10	2	120	2	95	[52]

Abbreviations: L-His—L-Histidine; QGly—glycine quaternized using MeI under 10 min of microwave irradiation.

The yield of PO is dependent on the AA structure; basic AAs such as L-histidine (L-His) and proline (Pro) containing basic additional groups (imidazolium and amino, respectively) provided higher yields than acidic aspartic or glutamic acids (Figure 4).

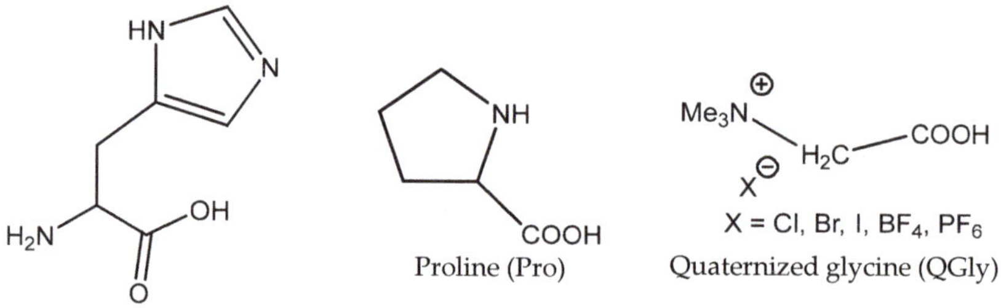

Figure 4. Structures of highly catalytically active AAs.

In addition, the combination of an amino acid with an HBD results in a higher catalytic activity under milder reaction conditions than in the absence of an HBD. The binary catalytic systems formed with the amino acid and H_2O produced, for example, more active systems for the synthesis of PO than in the absence of H_2O [53]. In the case of L-His, the time taken for the nearly total conversion was reduced from 48 h to 3 h by adding H_2O as an HBD (L-His:H_2O ratio = 1:29) under similar conditions (Table 8, Entries 1 and 2). The low H_2O concentration was used to avoid the hydrolysis of the produced PC [49,53].

Roshan et al. showed that the combination of halide ions as nucleophiles (added in the form of KI, for example) with L-histidine produced highly active catalytic systems for the cycloaddition of CO_2 to epoxides [51] (Table 8, Entry 4).

The most effective binary catalytic systems contained KI with basic AAs such as His/KI (Table 8, Entry 4). In a related work, Yang et al. proved the high stability of AA/KI catalytic systems, namely L-Trp/KI, for the cycloaddition of CO_2 to PO to form propylene carbonate. After carbonate separation conducted by means of distillation, the catalytic system was reused five times without loss of activity [54].

The catalytic activity of the different KX salts followed the order Cl < Br < I corresponding to the increasing nucleophilicity and leaving group ability [54]. This confirmed the role of these anions in the opening of the epoxide ring [54].

3.4. Onium Salts as Catalysts

Quaternary ammonium (most often tetrabutylammonium bromide, TBAB, and iodide (TBAI), phosphonium and sometimes even sulfonium salts [55]) are common catalysts for the formation of cyclic carbonates from epoxides and CO_2 [5,6,12] (Figure 5).

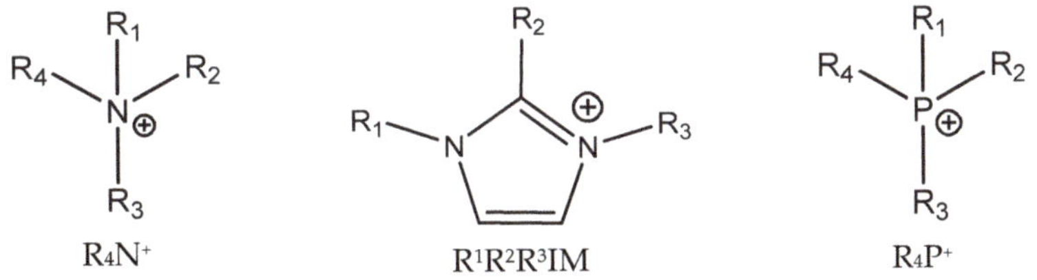

Figure 5. Structures of cations of commonly used ILs.

The low-melting (below 100 °C) quaternary ammonium salts are called ionic liquids (ILs). In addition, ILs show good solvating ability including CO_2, variable polarity and negligible vapor pressure [56]. ILs are considered to be sustainable ("green") solvents because of their properties such as relatively high thermal stability and negligible vapor pressure, high chemical stability and simple separability, and the modularity of their properties changing the structure of anions and cations.

The first published article that mentioned the formation of cyclic carbonates via the cycloaddition of CO_2 in epoxides catalyzed solely by ILs was published by Deng and Peng [57]. The authors studied different 1-butyl-3-methylimidazolium (BMIM) and N-butylpyridinium (BPY) salts varying in anions (chloride and tetrafluoroborate (BF_4^-, hexafluorophosphate (PF_6^-)) and observed the highest activity in the case of $BMIMBF_4$ salt. The catalytic activity increase was in the order: $BMIMPF_6 < BPYBF_4 < BMIMCl < BMIMBF_4$. This observation is in agreement with the solubility of CO_2 in these Ils; the highest solubility of CO_2 was determined in $BMIMBF_4$ [58].

Dyson et al. and Wang et al. studied the abilities of different ILs that differed in terms of the cations (alkylated imidazolium and tetraalkylammonium) and anions (halides) used [59,60]. Interestingly, cheap Bu_4NCl was found to be a very active catalyst compared with the much more expensive alkylated imidazolium halides. Based on the experimental results obtained at ambient pressure and 50 °C, they hypothesized that the balance between the nucleophilicity of anions and the acidity of hydrogens bound in the imidazolium ring of used IL is most important.

Yang et al. published an even higher catalytic activity of corresponding bromides when comparing them with tetrafluoroborates (Table 9, Entries 10 and 11) [29].

An increase in catalytical activity was observed with the increasing of the lipophilic alkyl chain length in RMIMXs when comparing 1-butyl-3-methylimidazolium and 1-octyl-3-methylimidazolium bromide [29].

Based on this idea, Akhdar et al. successfully tested the carbonation of internal epoxide produced via the epoxidation of methyl oleate [61–63]. The catalysts were prepared by means of the alkylation of N-alkylimidazoles with oligoethylene iodide and modified by ion-exchange to the corresponding bromide. N'-Oligoethylene-N-butylimidazolium bromide was recognized as the most active catalyst enabling the carbonation of epoxidized methyl oleate in 96% yield even at 100 °C and 2 MPa of CO_2.

Table 9. Effect of the ionic liquid structure on the cycloaddition reaction of PO.

Entry	Catalyst	Catalyst Amount (mol%)	CO_2 Pressure (MPa)	Temperature (°C)	Reaction Time (h)	Conversion (%)	Selectivity (%)	Ref.
1	[BMIm]PF_6	1.5	2	110	6	11.3	100	[57]
2	[BPy]BF_4	1.5	2	110	6	25.3	100	[57]
3	[BMIm]Cl	1	2	110	6	63.8	100	[57]
4	[BMIm]BF_4	0.75	2	110	6	67.4	100	[57]
5	[BMIm]BF_4	1.5	2	110	6	80.2	100	[57]
6	[BMIm]BF_4	2	2	110	6	90.3	100	[57]

Table 9. Cont.

Entry	Catalyst	Catalyst Amount (mol%)	CO$_2$ Pressure (MPa)	Temperature (°C)	Reaction Time (h)	Conversion (%)	Selectivity (%)	Ref.
7	[BMIm]BF$_4$	2.5	2	110	6	100	100	[57]
8	[BMIm]BF$_4$ + [BMIm]Cl	1.5 + 0.2	2	110	6	90.5	100	[57]
9	[BMIm]PF$_6$ + [BMIm]Cl	1.5 + 0.2	2	110	6	45.3	100	[57]
10	[BMIm]Br	1	1	140	2	54	70	[29]
11	[OMIm]Br	1	1	140	2	85	91	[29]
12	[BMIm]BF$_4$	1	1	140	2	7	29	[29]
13	[OMIm]BF$_4$	1	1	140	2	1	5	[29]
14	[p-ClBzMIM]Cl	0.25	2	130	4	86.34	not specified	[60]
15	[o-MeBzMIM]PF$_6$	0.25	2	130	4	31.82	not specified	[60]
16	[o-MeBzMIM]BF$_4$	0.25	2	130	4	33.03	not specified	[60]
17	[BMIm]Br	9	4	70	22	85	>99	[64]
18	[OMIm]Br	9	4	70	22	88	>99	[64]
19	[BBzIm]Br	9	4	70	22	77	>99	[64]
20	[OBzIm]Br	9	4	70	22	73	>99	[64]

The catalytic activity of ILs cannot be known based on a comparison of the anion effects using one type of cation. In the case of N-benzyl-N′-methylimidazolium salts, the most active one is N-(o-methyl)benzyl-N′-methylimidazolium chloride o-MeBzMIMCl; the catalytic activity of the corresponding o-MeBzMIMBF$_4^-$ and o-MeBzMIMPF$_6^-$ salts is lower (Table 9, Entries 14–16) [60].

Anthofer et al. studied the relationship between the structure, affinity to CO$_2$ and catalytic activity of different low-melting N,N′-dialkylimidazoles in detail [64] (Figure 6).

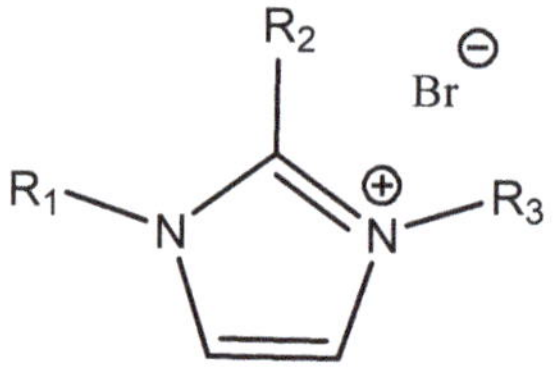

Figure 6. Structures of tested N,N′-dialkyl IM-based ILs in the carbonation of PO [64]. R$_1$ = Me; R$_3$ = Bu; R$_2$ = H = 1-methyl-3-butyl imidazolium bromide [BMIm]Br; R$_1$ = Me; R$_3$ = Bu; R$_2$ = Me = 1,2-dimethyl-3-butylimidazolium bromide [BM$_2$Im]Br; R$_1$ = Me; R$_3$ = Bu; R$_2$ = Et = 1-methyl-2-ethyl-3-butylimidazolium bromide [BEMIm]Br; R$_1$ = Me; R$_3$ = Oct; R$_2$ = H = 1-methyl-3-octylimidazolium bromide [OMIm]Br; R$_1$ = Me; R$_3$ = Oct; R$_2$ = Me = 1,2-dimethyl-3-octylimidazolium bromide [OM$_2$Im]Br; R$_1$ = Me; R$_3$ = Oct; R$_2$ = Et = 1-methyl-2-ethyl-3-octylimidazolium bromide [OEMIm]Br; R$_1$ = Bz; R$_2$ = H; R$_3$ = Bu = 1-benzyl-3-butylimidazolium bromide [BBzIm]Br; R$_1$ = Bz; R$_2$ = H; R$_3$ = Oct = 1-benzyl-3-octylimidazolium bromide [OBzIm]Br; R$_1$ = CH$_2$C$_6$F$_5$; R$_2$ = H; R$_3$ = Bu = 1-(2,3,4,5,6-pentafluoro)benzyl-3-butylimidazolium bromide [BBzF5Im]Br; R$_1$ = CH$_2$C$_6$F$_5$; R$_2$ = H; R$_3$ = Oct = 1-(2,3,4,5,6-pentafluoro)benzyl-3-octylimidazolium bromide [OBzF5Im]Br.

As could be seen, the most active ILs were those that contained bulk lipophilic substituents (N-octyl-N′-pentafluorophenyl-, N-butyl-N′-pentafluorophenyl- or N-methyl-N′-octyl-imidazole). The observed catalytic activity correlates well with the measured sorption of CO$_2$ in the studied ILs under the reaction conditions of this study. The substitution of the 2-position also significantly reduced the activity of the tested ILs [64].

The authors detected an interaction (hydrogen bond formation) between the acidic hydrogen of the used imidazole bromides (bound in position 2) and the oxygen atom of PO using FT-IR spectroscopy [64]. The mentioned catalyst was very effective even for the carbonation of internal epoxides such as cyclohexene oxide (ChO).

As could be seen, the most active ILs were those that contained bulk lipophilic substituents (N-octyl-N'-pentafluorophenyl-, N-butyl-N'-pentafluorophenyl- or N-methyl-N'-octyl-imidazole). The observed catalytic activity correlates well with the measured sorption of CO_2 in the studied ILs under the reaction conditions of this study. The substitution of the 2-position also significantly reduced the activity of the tested ILs [64] (Table 10).

Table 10. Synthesis of PC from CO_2 and PO using IL catalysts [64].

Entry	Catalyst	Catalyst Amount (mol%)	Mole Fraction CO_2 to IL	CO_2 Pressure (MPa)	Temperature (°C)	Reaction Time (h)	Conversion (%)	Selectivity (%)
0	-	0	-	0.4	70	22	0	-
1	[BMIm]Br	9.1	-	0.4	70	22	85	>99
2	[BM$_2$Im]Br	9.1	-	0.4	70	22	64	>99
3	[BEMIm]Br	9.1	-	0.4	70	22	69	>99
4	[OMIm]Br	9.1	0.058 ± 0.002	0.4	70	22	88	>99
5	[OM$_2$Im]Br	9.1	0.077 ± 0.003	0.4	70	22	80	>99
6	[OEMIm]Br	9.1	0.109 ± 0.001	0.4	70	22	71	>99
7	[BBzIm]Br	9.1	-	0.4	70	22	77	>99
8	[OBzIm]Br	9.1	0.096 ± 0.004	0.4	70	22	73	>99
9	[BBz^{F5}Im]Br	9.1	-	0.4	70	22	86	>99
10	[OBz^{F5}Im]Br	9.1	0.102 ± 0.011	0.4	70	22	71	>99

The authors detected an interaction (hydrogen bond formation) between the acidic hydrogen of the used imidazole bromides (bound in position 2) and the oxygen atom of PO using FT-IR spectroscopy [64]. The mentioned catalyst was very effective even for the carbonation of internal epoxides such as ChO (Table 11).

Table 11. Cycloaddition of different epoxides catalyzed by [OBzF5Im]Br; effect of epoxide structure [64].

Entry	Substrate	Substrate Abbreviation	Conversion (%)
1	Me-epoxide	PO	91
2	Cl-epoxide	EPIC	99
3	HO-epoxide	GL	98
4	hexyl-epoxide	OO	72
5	Ph-CH$_2$-epoxide	SO	96
6	cyclohexene oxide	ChO	78

Reaction conditions: catalyst: 1-(2,3,4,5,6-pentafluoro)benzyl-3-octylimidazolium bromide (10 mol%), 0.4 MPa, 70 °C, 22 h.

As was published by Yang et al., in the case of butylated DABCO (BuDABCO), the corresponding bromides, chlorides and hydroxides were recognized as the most effective cycloaddition catalysts [29] (Table 12, Entries 1–4). In contrast, non-nucleophilic anions such as NTf_2^-, PF_6^- or BF_4^- caused the loss of the catalytical activity of the studied BuDABCO salts.

Table 12. PC synthesis catalyzed by DABCO-based Lewis basic ionic liquids [29].

$X = Br, Cl, OH, BF_4, PF_6$
Alk = butyl or octyl

Entry	Catalyst	Catalyst Amount (mol%)	CO_2 Pressure (MPa)	Temperature (°C)	Reaction Time (h)	Conversion (%)	Selectivity (%)
1	[BuDABCO]OH	1	1	140	2	88	97
2	[BuDABCO]Cl	1	1	140	2	81	96
3	[BuDABCO]Br	1	1	140	2	79	>99
4	[BuDABCO]Br	1	1	140	2	78	98
5	[BuDABCO]NTf$_2$	1	1	140	2	9	50
6	[BuDABCO]PF$_6$	1	1	140	2	4	9
7	[BuDABCO]BF$_4$	1	1	140	2	2	6

Surprisingly, when ILs containing activated CO_2 in their structures in N,N'-di(alkyl)imidazolium-2-carboxylates were tested as CO_2 cycloaddition catalysts by Kayaki et al. [65], the observed activity was quite low and a high CO_2 pressure was required to obtain satisfactory conversion (Table 13). The hydrogen in position 2 on the imidazolium ring is, in all probability, important as an HBD for epoxide ring activation and substitution with $-COO^-$ causes a decrease in catalytic activity (Table 13).

Table 13. Cycloaddition of CO_2 to epoxides catalyzed by 1,3-di-tert-butylimidazolium-2-carboxylate [65].

Entry	R	Catalyst	Catalyst Amount (mol%)	CO_2 Pressure (MPa)	Temperature (°C)	Reaction Time (h)	Conversion (%)
1	C_6H_5		5	4.5	100	24	89
2	$C_6H_5OCH_2$		5	2.5	100	15	81
3	$ClCH_2$		5	2.5	100	18	87
4	n-C_4H_9		5	2.5	100	40	71

Interestingly, some of the attempts to boost the catalytic activity of onium salts constructing di- or tricationic ILs (Scheme 7) often fall flat (Figure 7) [66–68].

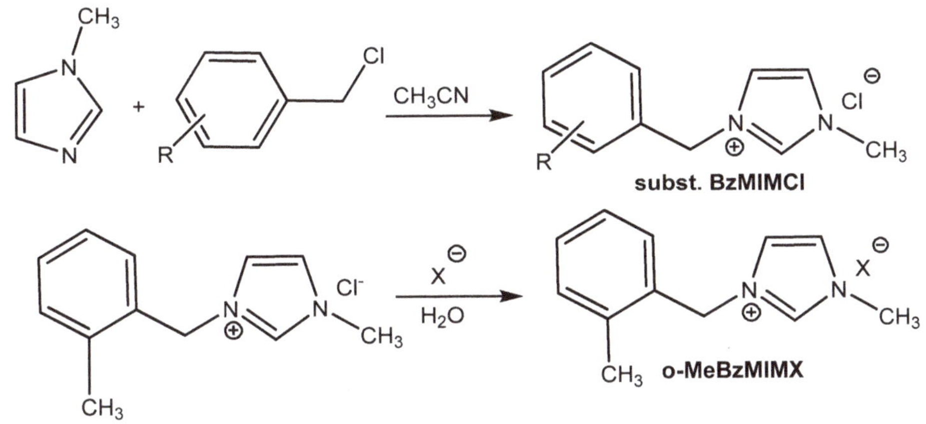

Scheme 7. Preparation of tested ILs (substituted BzMIMs) described by Yang et al. [60]. R = H = [BzMIM]Cl; R = 4-CH$_3$ = [*p*-MBzMIM]Cl; R = 2-CH$_3$ = [*o*-MeBzMIM]Cl; R = 4-NO$_2$ = [*p*-NBzMIM]Cl; R = 2-Cl = [*o*-ClBzMIM]Cl; R = 4-Cl = [*p*-ClBzMIM]Cl; X = PF$_6$ = *o*-MeBzMIMBF$_4^-$; X = BF$_4$ = *o*-MeBzMIMPF$_6$.

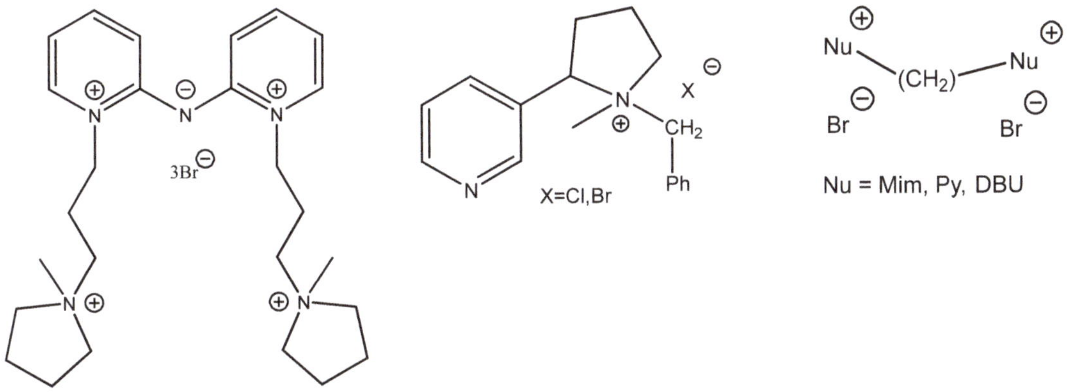

Figure 7. Structures of studied task-specific dicationic ILs (amino-pyridinium-pyrrolidinium bromide [66], Quaternized nicotine based ammonium ILs [67] and CH$_2$-bridged tertiary amines [68].

Isothiouronium salts (Scheme 8) were also chosen for the testing of catalytic activity for CO$_2$ addition, providing encouraging results (over 90% yield with selectivity over 99%) at 2 MPa pressure of CO$_2$ and 140 °C after 2 h of action using 1 molar % of catalyst. The corresponding thiourea was practically nonactive [55].

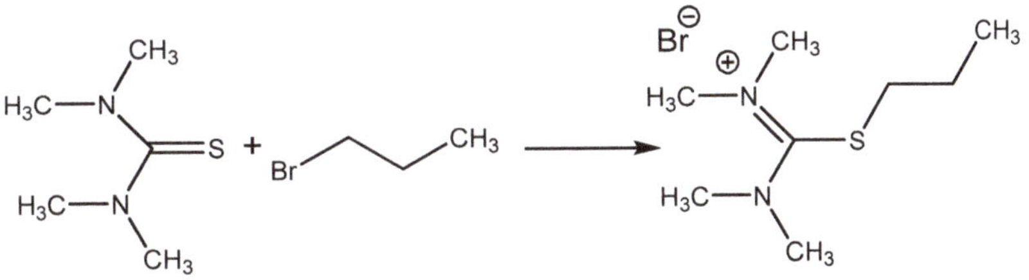

Scheme 8. Preparation of catalytically active S-alkylisothiouronium salt [55].

Apart from ammonium and sulfonium salts, the methyl-trioctylphosphonium-based ILs with organic anions were studied as cycloaddition catalysts. Their catalytic activity was remarkable even for cycloaddition reaction of less reactive styrene oxide (SO) with CO_2 at ambient pressure [46].

Wilhelm et al. compared the action of different aromatic or heterocyclic alcoholates (phenolates or anions of hydroxypyridine regioisomers, Figure 8) used as anions in combination with tetrabutylphosphonium, tetrabutylammonium and N-ethyl-DBU-based cations [69] (Table 14). The authors discovered the cooperative effect of the alcoholate anion of 2-hydroxypyridine with the tetrabutylphosphonium cation in the case of a reaction of CO_2 with epichlorohydrin. The catalytic activity, however, of other ILs containing Bu_4N^+ or Et-DBU$^+$ cations was slender (Table 14).

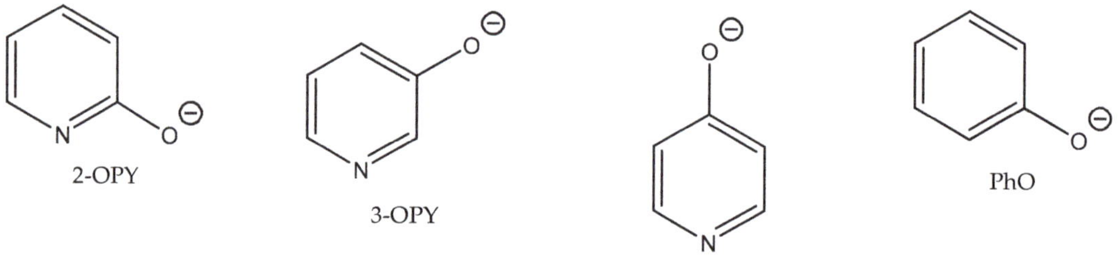

Figure 8. Structures of hydroxypyridine anion OPYs [69].

Table 14. Cycloaddition of CO_2 with EPIC catalyzed ILs [69].

Entry	Catalyst	Catalyst Amount (mol%)	CO_2 Pressure (MPa)	Temperature (°C)	Reaction Time (h)	Conversion (%)
1	-	-	0.1	30	20	0
2	[Bu$_4$P] 2-OPY	10	0.1	30	20	90
3	[Bu$_4$P] 3-OPY	10	0.1	30	20	77
4	[Bu$_4$P] 4-OPY	10	0.1	30	20	70
5	[Bu$_4$P] PhO	10	0.1	30	20	55
6	[Bu$_4$P] NO$_3$	10	0.1	30	20	10
7	[Bu$_4$N] 2-OPY	10	0.1	30	20	61
8	[Et-DBU] 2-OPY	10	0.1	30	20	42
9	[Bu$_4$P] 2-OPY	10	0.1	30	20	25
10	[Bu$_4$P] 2- OPY	10	0.1	80	4	24
11	[Bu$_4$P] 2-OPY	10	2	80	4	98

The authors suggested the mechanism of this reaction based on activation of the epoxide ring with the tetrabutylphosphonium cation as the Lewis acid with the simultaneous activation of CO_2 and phenolate [69]. The most active [Bu$_4$P] 2-OPY was tested at ambient pressure for the carbonation of different terminal epoxides with satisfactory results, using 50 molar % quantity of [Bu$_4$P] 2-OPY to an appropriate epoxide (Table 15).

Table 15. Carbonation of different epoxides catalyzed by Bu$_4$P.2-OPY at 0.1 MPa of CO$_2$ and at 30 °C [69].

Entry	Substrate		Product		Conversion (%)
1 [a]	EPIC		CMEC		90
2 [a]	EPIB		BMEC		86
3 [b]	AGL		AMEC		84
4 [b]	BO		EEC		80
6 [b]	SO		PEC		59

Reaction conditions: [a] catalyst (10 mol%), 0.1 MPa, 30 °C, 20 h; [b] catalyst (50 mol%).

Wang et al. prepared ammonium salts in situ by alkylating tertiary amides (N,N-dimethylformamide (DMF), N,N-dimethylacetamide (DMAc), N-formylmorpholine, N-methylpyrrolidone, tetramethylurea and N-formylpiperidine) with benzyl halogenides. The prepared ammonium salts enabled the formation of cyclic carbonates even at an ambient pressure, especially those prepared from DMF using benzylbromide [70] (Table 16).

Table 16. The effect of organic bases on cycloaddition (compared with DBU in co-action with alkyl halides) [71].

Entry	Catalyst	Catalyst Amount (mol%)	CO$_2$ Pressure (MPa)	Temperature (°C)	Reaction Time (h)	Conversion (%)
1	PhCH$_2$Br/DBU	5	0.1	65	22	95
2	PhCH$_2$Br/DBN	5	0.1	65	22	91
3	PhCH$_2$Br/DMAP	5	0.1	65	22	83
4	PhCH$_2$Br/DABCO	5	0.1	65	22	54
5	PhCH$_2$Br/Py	5	0.1	65	22	85
6	PhCH$_2$Br/TEA	5	0.1	65	22	84
7	PhCH$_2$Br/MIm	5	0.1	65	22	81
8	PhCH$_2$Br/Im	5	0.1	65	22	75
9	PhCH$_2$Cl/DBU	5	0.1	65	22	71
10	p-BuPhCH$_2$Br/DBU	5	0.1	65	22	92
11	4-Nitrobenzyl bromide/DBU	5	0.1	65	22	83

Table 16. Cont.

Entry	Catalyst	Catalyst Amount (mol%)	CO_2 Pressure (MPa)	Temperature (°C)	Reaction Time (h)	Conversion (%)
12	α-Bromodiphenyl-methane/DBU	5	0.1	65	22	86
13	$CH_3CH_2CH_2CH_2Br$/DBU	5	0.1	65	22	68
14	$CH_3CH_2CH_2CH_2I$/DBU	5	0.1	65	22	27
15	[nBu_4N]Br	5	0.1	65	22	72

Wang et al. attributed the high activity of the DMF + BnBr mixture in particular to the activation of the oxirane ring by benzyl cations and the contemporary nucleophilic activation of CO_2 by DMF [70] (Scheme 9).

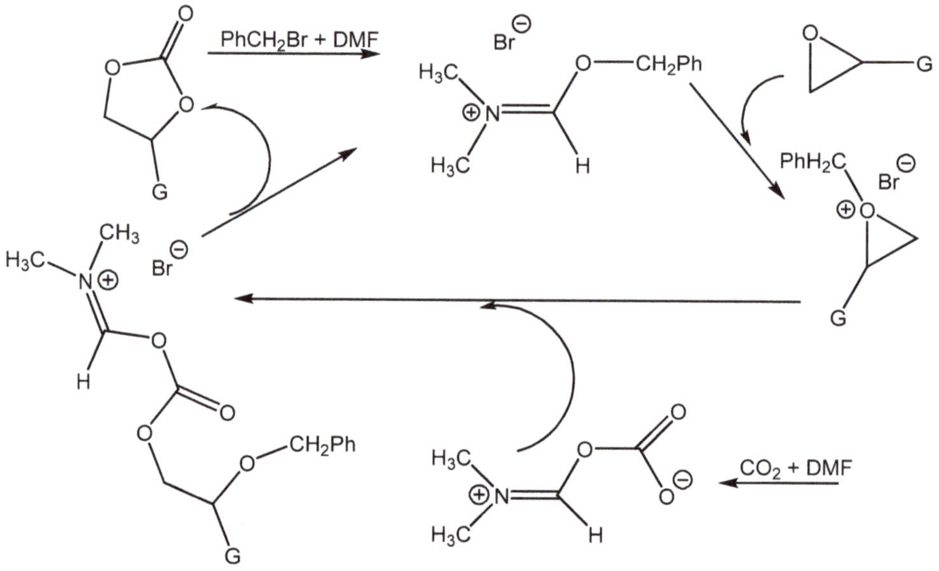

Scheme 9. Proposed reaction pathway for the formation of cyclic carbonates catalyzed by benzylbromide/DMF [71].

Similarly, the effectiveness tertiary amines described earlier as active catalysts (see Section 3.1) was satisfactorily proven for the reaction of epoxides with CO_2 at an ambient pressure after in situ quaternization via benzylation [71] (Table 16).

For a comparison of the action with the ammonium salts formed using arylmethylbromide derivatives, Bu_4NBr was employed as the bromide source using the same reaction conditions as the model reaction (Table 16, Entry 15).

As could be seen, the yield was lower than that using benzyl bromide as the bromide anion source, which was presumably due to the electrostatic interaction between the bromide anion and the ammonium center decreasing with the bulkiness of the cation. The authors stated, based on above-mentioned results, that the nucleophilicity of the bromide anion is weaker for Bu_4NBr than for the salts (Bn-$DBU^+.Br^-$).

The successful utilization of tetrabutylammonium halides, especially bromide and iodide, in CO_2 cycloaddition reactions was reported by Calo [72] (Table 17). The higher reactivity of Bu_4NBr/Bu_4NI in comparison with RMIMBr or RPYBr salts was explained by less coordination of halide with the bulkier Bu_4N^+ cation [72]. In addition, the catalytic activity of cheap and commercially available Bu$_4$NXs is high and quite comparable with much more expensive PPNXs salts (Figure 9).

Table 17. Cycloaddition of CO$_2$ to glycidol producing hydroxymethyl ethylene carbonate (HMEC) [74].

Entry	Catalyst	Catalyst Amount (mol%)	CO$_2$ Pressure (MPa)	Temperature (°C)	Reaction Time (h)	Conversion (%)
1	[PPN]Cl	1	0.1	80	1	70
2	[Bu$_4$N]Cl	1	0.1	80	1	80
3	[Bu$_4$N]I	1	0.1	80	1	82
4	[Bu$_4$N]Br	1	0.1	80	1	85
5	[Bu$_4$N]Br	1	0.1	60	1	51
6	[Bu$_4$N]Br	1	0.1	40	1	12
7	[Bu$_4$N]Br	1	0.1	20	1	2
8	[Bu$_4$N]Br	3	0.1	40	1	30
9	[Bu$_4$N]Br	5	0.1	40	1	40
10	[Bu$_4$N]Br	5	0.1	40	3	87
11	[Bu$_4$N]Br	5	0.1	60	3	>99
12	[Bu$_4$N]Br	5	0.1	40	24	52
13 [a]	[Bu$_4$N]Br	5	0.1	60	1	12
14 [b]	[Bu$_4$N]Br	5	0.1	60	1	4

[a] Methyl glycidyl ether as substrate; [b] PO as substrate; [PPN]Cl—bis(-triphenylphosphine)iminium chloride.

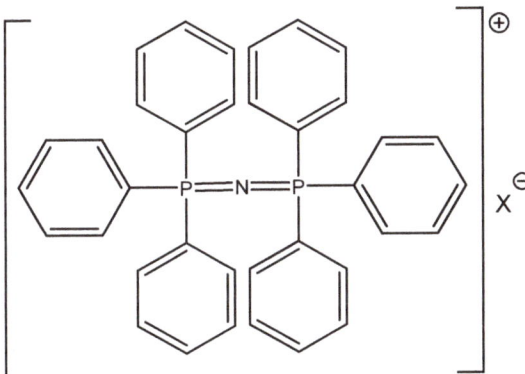

Figure 9. Structure of bis(triphenylphosphoranylidene)ammonium halide (PPNX) [73].

3.5. Two Component Catalysts Containing HDBs and Onium Salts

The cheap and easily available quaternary ammonium halides TBAB and TBAI are often combined with different HDBs with the aim of boosting catalytic activity for the insertion of CO$_2$ in the oxirane ring.

It was observed that even the addition of glycidol to the Bu$_4$NX significantly increased the yield of PO compared with Bu$_4$NX used alone [74] (Table 18, Entries 4, 7–9).

Some mixtures of onium salts with HBDs produce low-melting eutectic solvents (DESs) that readily dissolve both CO$_2$ and epoxide, enabling cycloaddition even at ambient pressure and low temperature [75]. DES is defined as a mixture of two or more compounds that are typically solid at room temperature, but when combined at a particular molar ratio, changes into liquid at room temperature [76].

Table 18. Comparison of the catalytic activity of various PILs and DESs for the carbonation of SO.

Entry	Catalyst	Catalyst Amount (mol%)	CO_2 Pressure (MPa)	Temp. (°C)	Reaction Time (h)	Yield (%)	Sel. (%)	Ref.
1	Glycidol/[Bu$_4$N]Br	1/1	1	50	24	15	n.d.	[74]
2	Glycidol/[Bu$_4$N]Br	1/1	1	60	24	40	n.d.	[74]
3	Glycidol/[Bu$_4$N]Br	1/1	1	70	24	66	n.d.	[74]
4	Glycidol/[Bu$_4$N]Br	1/1	1	80	24	83	n.d.	[74]
5	Glycidol/[Bu$_4$N]Br	-/1	1	80	24	30	n.d.	[74]
6	Glycidol/[Bu$_4$N]Br	3/3	1	60	24	63	n.d.	[74]
7	Glycidol/[Bu$_4$N]Br	5/5	1	60	24	84	n.d.	[74]
8	Glycidol/[Bu$_4$N]Br	-/5	1	60	24	16	n.d.	[74]
9	Glycidol/[Bu$_4$N]Br	5/5	1	60	24	85	n.d.	[74]
10	[DBUH]Br/EDA (2:1)	20	0.1	25	48	94	99	[75]
11	[DBUH]Br/DEA (2:1)	20	0.1	25	48	97	>99	[75]
12	DEA	20	0.1	25	48	0	0	[75]
13	[DBUH]Br	20	0.1	25	48	79	>99	[75]
14	[TMGH]Br/DEA (2:1)	20	0.1	25	48	97	>99	[75]
15	[DMAPH]Br/DEA (2:1)	20	0.1	25	48	92	>99	[75]
16	[Et$_3$NH]Br/DEA (2:1)	20	0.1	25	48	92	>99	[75]
17	[DBUH]Br/DEA (2:1)	20	0.1	25	48	81	>99	[75]
18	[DBUH]Br/DEA (1:1)	20	0.1	25	48	92	>99	[75]
19	[DBUH]Br/DEA (2:1)	20	0.1	60	5	>99	>99	[75]
20 [a]	[DBUH]Br/DEA (2:1)	20	0.1	25	48	10	>99	[75]
21	[DBUH]Br/DEA (2:1)	20	0.1	60	48	43	n.d.	[75]
22	[HMIM]Br	1	1.5	120	2	77.1	n.d.	[75,77]
23	[Et$_3$NH]Br	10	0.1	35	24	20	n.d.	[41,75]
24	[DBUH]I	10	0.1	70	4	85	n.d.	[30,75]
25	[DMAPH]Br	1	0.1	120	4	95	n.d.	[42,75]
26	ChCl/PEG 400 (1:2)	2	1.2	130	5	89.4	n.d.	[75,78]
27	ChCl/PEG 200 (1:2)	2	0.8	150	5	99.1	n.d.	[75,78]
28	[Bu$_4$P]Br/2-Aminophenol (1:2)	4.5	0.1	80	1	96	n.d.	[75,79]
29	ChI/NHS (1:2)	6	1	30	10	96	n.d.	[75,80]
30	[DBUH]Br/DEA (2:1)	20	0.1	25	48	97	n.d.	[75]
31	[DBUH]Br/DEA (2:1)	20	0.1	60	5	99	n.d.	[75]

[a] 15% of CO_2 and 85% N_2. Abbreviations: TMGH—N,N,N′,N′-tetramethylguanidinium; DEA—diethanolamine; Ch—choline.

They not only possessed comparable physicochemical properties to traditional ILs (designability, non-volatility and high thermal stability), but also had advantages such as low cost and a simple preparation process (mixing and melting) without the need for purification.

DESs prepared via the mixing of tertiary amines hydrogen halides (R$_3$N.HX) and ethylene diamine or different aminoethanols were compared in the carbonation of SO at an ambient pressure, obtaining intriguing yields and selectivities of SC even in the case of DES prepared from hydrobromide of cheap triethyl amine and diethanolamine [75] (Table 18, Entries 10–13). DBU hydrobromide mixed with diethanolamine at a molar ratio of 2:1 was recognized as the most catalytically active. Testing this most effective DES, high yields of different carbonates were determined by GC-MS even at an ambient pressure and room temperature of CO_2 after 48 h using 20 molar % of this DES. Testing the carbonation of internal ChO, the yield of CC was 43%. Applying a mixture of 15% CO_2 with nitrogen (simulated flue gas) drops, however, the yield decreased from 92% (100% CO_2 at an ambient pressure after 48 h at room temperature) to 10% (using 15%CO_2 in nitrogen under the same reaction conditions, Table 18, Entries 14–21).

Comparing the catalytic activity of different DESs with various protic ILs, the DES (2 DBU.HBr + 1 DEA) is much more active than PIL DBU hydroiodide (DBU.HI) alone. The observed high catalytic activity was explained by the synergistic action of DEA (as HBD) and DBU.HBr as a source of highly nucleophilic naked bromide [75] (Table 18, Entries 16–21 and 24).

Similarly, high activity was observed using DES prepared from Bu_4PBr with 2-aminophenol for the carbonating of terminal epoxides. The carbonation of internal epoxide CO to CC was, however, very slow [79] (Table 18, Entry 28).

Pentaerythritol as an aliphatic polyol-based HDB was recognized as effective for the carbonation of PO at elevated pressure [80]. Although completely inactive used alone or with KI, in a mixture with Bu_4N^+ bromide or iodide, it is very active, obtaining a 97% yield of PC at 70 °C after 22 h of CO_2 (0.4 MPa) action (Table 19).

Table 19. Comparison of catalytic activities of DESs based on pentaerythritol (PETT) and Bu_4NXs for the preparation of PC [80].

Entry	Catalyst	Catalyst Amount (mol%)	CO_2 Pressure (MPa)	Temp. (°C)	Reaction Time (h)	Sel. (%)	Yield of PC (%)
1	PETT	5	0.4	70	22	>99	0
2	[Bu$_4$N]I	5	0.4	70	22	>99	10
3	PETT/[Bu$_4$N]I	5	0.4	70	22	>99	96
4	PETT/[Bu$_4$N]Br	5	0.4	70	22	>99	97
5	PETT/[N-methyl-N'-octyl imidazolium Br]	5	0.4	70	22	>99	79
6	PETT/[N-methyl-N'-octyl imidazolium Br]	5	0.4	70	22	>99	86
7	PETT/[N-methyl-N'-octyl imidazolium Br]	5	0.4	70	22	>99	87
8	PETT/KI	5	0.4	70	22	>99	6

PETT—pentaerythritol: C(CH$_2$OH)$_4$

Choline iodide together with N-hydroxysuccinimide forms DES, enabling the high-yield carbonation of PO to SC at 30–80 °C and 1 MPa pressure of CO_2. Instead of choline iodide, Bu_4NX in a mixture with N-hydroxysuccinimide is applicable [81]. Using a 2 MPa pressure of CO_2, a high yield of CC from internal ChO was obtained at 70 °C after 10 h of reaction (Table 18, Entry 29).

A broad set of aliphatic and aromatic alcohols in terms of their role as potential HDBs was studied by Alves et al. [82] in the co-action of Bu_4NBr using pressurized CO_2 (2 MPa) for PO carbonation. The authors observed that the most active HDBs were low-polar polyfluorinated secondary alcohols such as tertiary alcohols HFTI or 1,3-bis-HFAB (Figure 10).

Aromatic polyols such as pyrocatechol, pyrogallol and gallic acid were less catalytically active. Aliphatic alcohols exhibited low cooperative activity in the case of the tested Bu_4NBr, which was practically comparable with the catalytic effect of sole Bu_4NBr. Interestingly, some of tested alcohols even exhibited inhibition effects.

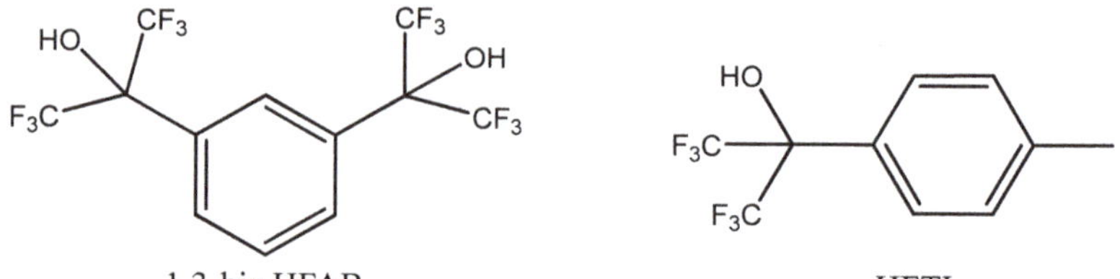

Figure 10. Structures of the most effective HBDs in co-action with Bu$_4$NBr for the carbonation of PO [81].

The high catalytical activity of RMIMs/phenols-based DESs was published by Liu et al. even at an ambient pressure of CO$_2$ and at room temperature for SO [83]. Especially *N*-ethyl-*N*′-methylimidazolium iodide (EMIMI) was recognized as a very suitable part of DES in co-action with phenols substituted with electron-donating groups such as –NH$_2$, –C(CH$_3$)$_3$ and –Cl, –OH. The most effective DES contained EMIMI (2 mol) and resorcinol (1 mol). The authors explained its high catalytic activity as multifunctional HBD-based activation by acidic hydrogen bound in position 2 of EMIM salt together with hydrogen from the –OH group of resorcinol (Figure 11) and the subsequent action of iodide as a nucleophile. Interestingly, comparing the activity of (2 EMIMI + 1 resorcinol) DES with the much cheaper (2 Bu$_4$NI + 1 resorcinol) binary system for SO carbonation, the obtained yields of PEC were very similar [83]. SO was the single epoxide studied, however, in this article. Another catalytically very effective DESs containing mixture of choline chloride and malic acid or choline iodide and glycerol published Vagnoni et al. [84].

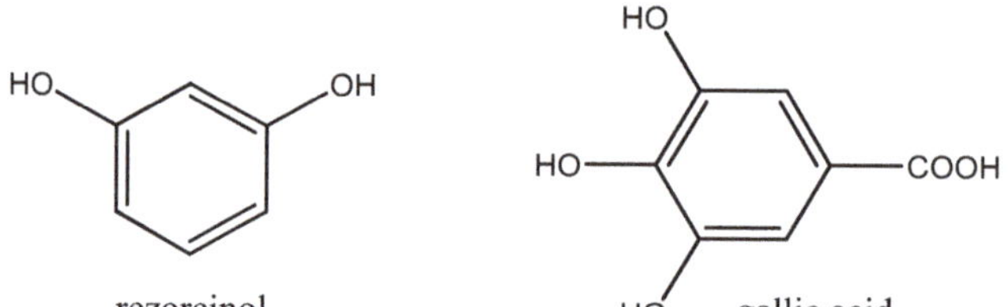

Figure 11. Structures of resorcinol and gallic acid.

Additional very effective DESs were obtained as catalysts by mixing 2-hydroxymethylpyridine or 2,6-hydroxymethylpyridine with Bu$_4$NI [40]. These DESs were able to catalyze the carbonation of EPIC to chloromethyl-ethylenecarbonate (CMEC) even at room temperature and ambient CO$_2$ pressure. The carbonation of internal epoxide ChO was very slow, however, under ambient conditions even after 20 h using 8 molar % of catalyst [40].

Gallic acid (Figure 11), as a green, biobased and biodegradable HDB, was discovered by Sopena et al. as a more effective alternative of resorcinol in a binary Bu$_4$NI + gallic acid catalytic system dissolved in 2-butanone [85]. Even internal epoxide was carbonated with a high yield at 80 °C and 1 MPa pressure of CO$_2$ after 18 h [85].

Polycarboxylic acids such as citric acid were effectively applied as the HDB part of DES together with choline iodide [86]. The other tested carboxylic acids were less active HDBs compared with citric acid. Additionally, it was observed that the molar ratio of the used HBA and HDB is crucial. For DES obtained by the melting of choline iodide, citric acid at a molar ratio of 2:1 (excess of iodide source) is highly active. Changing the molar ratios significantly decreased the reaction yield (but not selectivity). ChO tested as an internal epoxide at 70 °C and 0.5 MPa of CO$_2$ produced only 36% CC after 6 h [86].

The attempts to substitute ILs-based iodides or bromides as key parts of DESs were described by Wang et al. [87]. Applying boric and glutaric acids, together with BMIMCl, the authors described significant catalytic activity even without the presence of bromide or iodide ions for the carbonation of terminal epoxides at 0.8 MPa of CO_2 and 70 °C. [87]. The carbonation of internal ChO was below 40% after 7 h of CO_2 action.

The most active HDB described until this time for the catalysis of epoxides' carbonation is ascorbic acid in co-action with Bu_4NI [15] (Table 20). This mixture was effective even for the carbonation of internal epoxides, even at an elevated temperature (100 °C) and 2 MPa CO_2 pressure [15] (Table 21).

Table 20. Comparison of various HBD/Bu_4NI catalytic systems for the carbonation of EPIC.

Entry	Catalyst	Catalyst Amount (mol%)	CO_2 Pressure (MPa)	Temp. (°C)	Reaction Time (h)	Conv. (%)	Ref.
1	L-Ascorbic acid	2	0.1	25	23	70	[15]
2	L-Ascorbic acid	2	0.1	40	23	94	[15]
3	APAA [b]	2	0.1	25	23	58	[15]
4	Lactic acid	2	0.1	25	23	59	[15]
5	D-Glucose	2	0.1	25	23	45	[15]
6	Erythritol	2	0.1	25	23	54	[15]
7	Pentaerythritol	2	0.1	25	23	41	[15]
8	2-Pyridinemethanol	2	0.1	25	23	78	[15,84]
9	Dinaphtyl Si-diol	2	0.1	25	23	93	[15,88]
10 [a]	DBU/$PhCH_2Br$	2	0.1	25	23	93	[15,70]
11	Tetraethylene glycol/KI	2	0.1	40	23	92	[15,89]
12	P-ylide-CO_2-adduct	2	0.1	25	23	90	[15,90]
13	[Bu_4N]I	4	0.1	25	23	31	[15]

[a] using SO as a substrate. [b] APAA—acetal protected ascorbic acid.

Table 21. Carbonation of methyl oleate using ascorbic acid (HBD) and different sources of nucleophile (Bu_4NX) [91].

Entry	Catalyst	Catalyst Amount (mol%)	CO_2 Pressure (MPa)	Temp. (°C)	Reaction Time (h)	Conversion (%)	Selectivity (%)
1	[Bu_4N]I	5	0.5	100	24	70	59
2	[Bu_4N]Br	5	0.5	100	24	83	87
3	[Bu_4N]Cl	5	0.5	100	24	44	>99
4	L-ascorbic acid/[Bu_4N]I	0.5/5	0.5	100	24	>99	20
5	L-ascorbic acid/[Bu_4N]Br	0.5/5	0.5	100	24	91	83
6	L-ascorbic acid/[Bu_4N]Cl	0.5/5	0.5	100	24	61	>99
7	L-ascorbic acid/[Bu_4N]Cl	1/5	0.5	100	24	62	>99
8	L-ascorbic acid/[Bu_4N]Cl	1.5/5	0.5	100	24	69	>99
9	L-ascorbic acid/[Bu_4N]I	1.5/5	0.5	100	24	>99	74
10	L-ascorbic acid/[Bu_4N]Br	1.5/5	0.5	100	24	98	>99
11	L-ascorbic acid/[Bu_4N]Cl	1.5/5	0.5	100	24	69	>99
12	L-ascorbic acid/[Bu_4N]Cl	1.5/5	0.5	100	24	59	>99
13	L-ascorbic acid/[Bu_4N]Cl	1.5/5	0.5	100	24	57	>99
14	L-ascorbic acid/[Bu_4N]Cl	2/5	0.5	100	24	49	>99
15	L-ascorbic acid/[Bu_4N]Cl	5/5	0.5	100	24	15	>99
16	L-ascorbic acid/[Bu_4N]Cl	1.5/5	0.5	100	48	>99	>99
17 [a]	L-ascorbic acid/[Bu_4N]Cl	1.5/5	0.5	100	24	38	>99
18	L-ascorbic acid/[Bu_4N]Cl	1.5/5	1	100	24	92	>99

[a] Using recovered catalysts.

Encouraged by the robustness of this catalytic system, Elia et al. tested the Bu_4NI/ascorbic acid system for the cycloaddition of CO_2 in epoxidized fatty acid esters [91]. Cyclic carbonates based on fatty acid esters seemed to be potential plasticizers for polyvinyl chloride instead of harmful phtalates, for example [92].

As can be seen in Table 21, the most effective catalytic mixture found contains Bu_4NCl/ascorbic acid. Bu_4NCl is superior because overly nucleophilic Bu_4NI causes undesirable Meinwald rearrangement producing ketones instead of cyclic carbonates, probably due to the sterical hindrance in the case of epoxidized oleic acid methyl ester [91]. In the case of epoxidized polyunsaturated fatty acid esters, allylic alcohols are produced as by-products using Bu$_4$NI [91].

3.6. Application of Bifunctional (or Multifunctional) Onium Salts

The functionalization of ILs involves an increase in catalytic activity owing the synergistic effect between the bound functional groups (nucleophilic iodide or bromide anions together with –NH$_2$, –COOH or –OH groups in the role of HBDs). The reached synergism enables a decrease in the quantity of the multifunctional catalyst and simpler separation in an optimal case [93] (Table 22, Figure 12). On the other hand, multifunctional ILs are more difficult to prepare and more expensive due to this reason.

Table 22. Carbonation of butylene oxide (BO) using bifunctional phosphonium salts and onium salts [93].

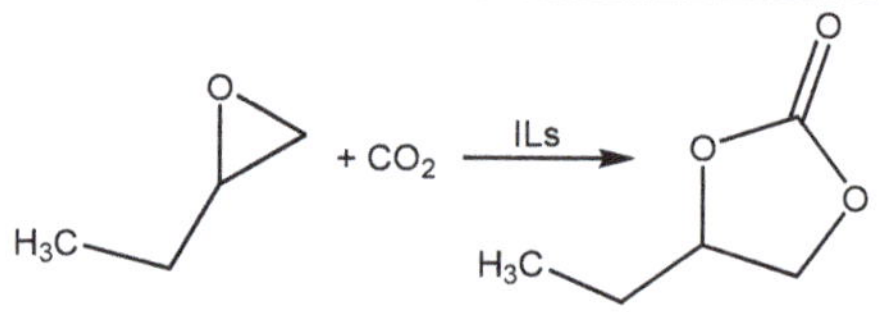

Entry	Catalyst	Catalyst Amount (mol%)	CO$_2$ Pressure (MPa)	Temp. (°C)	Reaction Time (h)	Yield (%)
1	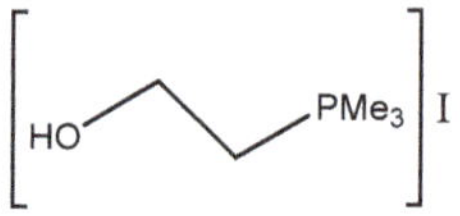 [HOCH$_2$CH$_2$PMe$_3$]I	2	1	90	2	21
2	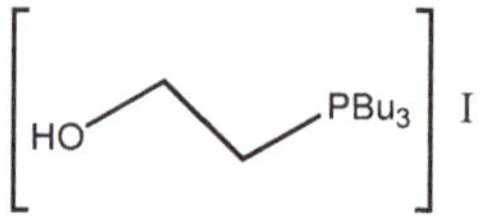 [HOCH$_2$CH$_2$PBu$_3$]I	2	1	90	2	95
3	 [HO(CH$_2$)$_4$PMe$_3$]I	2	1	90	2	78
4	[Bu$_4$P]Cl	2	1	90	2	36
5	[Bu$_4$P]Br	2	1	90	2	25
6	[Bu$_4$P]I	2	1	90	2	19
7	[Bu$_4$N]Cl	2	1	90	2	51
8	[Bu$_4$N]Br	2	1	90	2	24
9	[Bu$_4$N]I	2	1	90	2	19

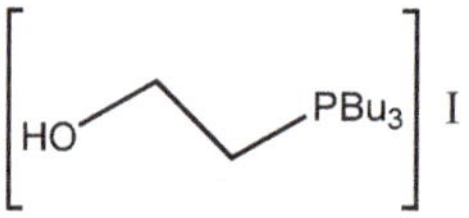

Figure 12. Structure of the most active bifunctional phosphonium salt tri-n-butyl-(2-hydroxyethyl) phosphonium iodide [93].

Bifunctional catalysis based on aromatic OH-functionalized onium salt was described by Tsutsumi et al. [94]. This research work demonstrates that it still possible to discover a very active and quite cost-effective bifunctional catalyst such as m-trimethylammonium phenolate, which is much more effective than more structurally complicated ones [94].

This catalyst was only tested, however, for the carbonation of terminal epoxides and still requires higher pressure of CO_2 at an elevated temperature [94] (Scheme 10, Table 23).

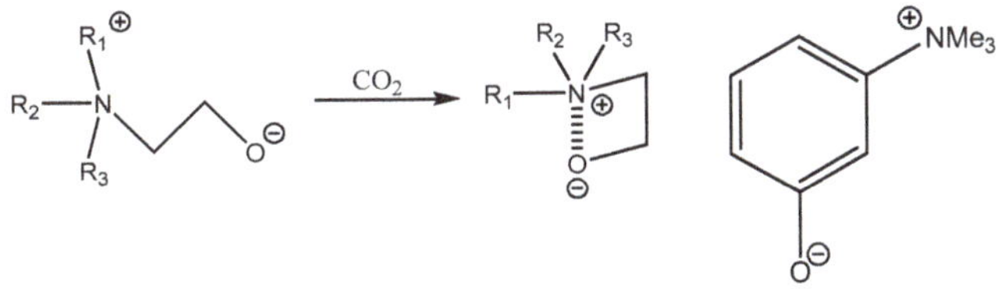

Scheme 10. CO_2 activation by ammonium betaine organocatalyst 3-trimethylammonium phenolate [94].

Table 23. Comparison of catalytic action of different substituted phenols on carbonation of hexylene oxide [94].

Entry	Catalyst	Catalyst Amount (mol%)	CO_2 Pressure (MPa)	Temp. (°C)	Reaction Time (h)	Yield (%)
1	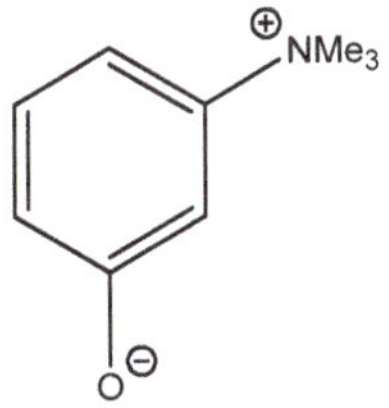 3-trimethylammonium phenolate	3	1	120	24	>99
2	3-trimethylammonium phenolate	2	1	120	24	>99
3	3-trimethylammonium phenolate	1	1	120	24	88
4	o-dimethylaminophenol	3	1	120	24	4
5	p-MeOC$_6$H$_4$OH/DMAP	3	1	120	24	61
6	PhN$^+$Me$_3$.PhO$^-$	3	1	120	24	15

Meng et al. published very promising results obtained by testing a series of OH-functionalized DBU-based ILs (Figure 13) [95]. They discovered not only a recyclable organocatalyst with excellent activity for carbonation of EPIC at 30 °C and an ambient pressure of CO_2, but also a compound suitable for the utilization of CO_2 from a simulated flue gas (Scheme 11). Its high activity is explained by the cooperative activation of the epoxide ring by protonated DBU in the role of HDB and the activation of CO_2 via carbonate formation utilizing alcoholate on a bridge-functionalized bis-DBU anion.

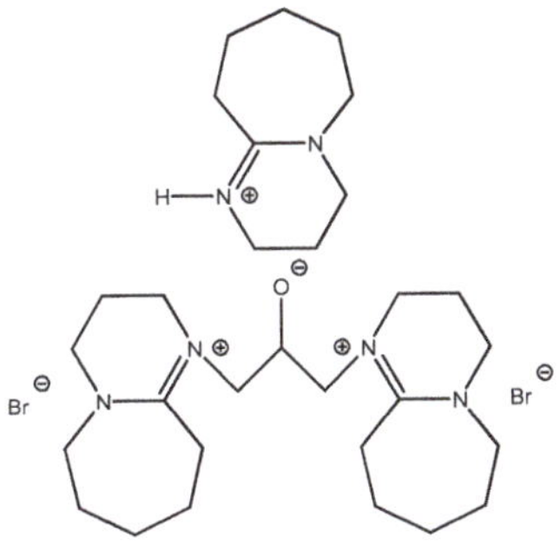

Figure 13. Structure of most effective OH-functionalized DBU-based IL [95].

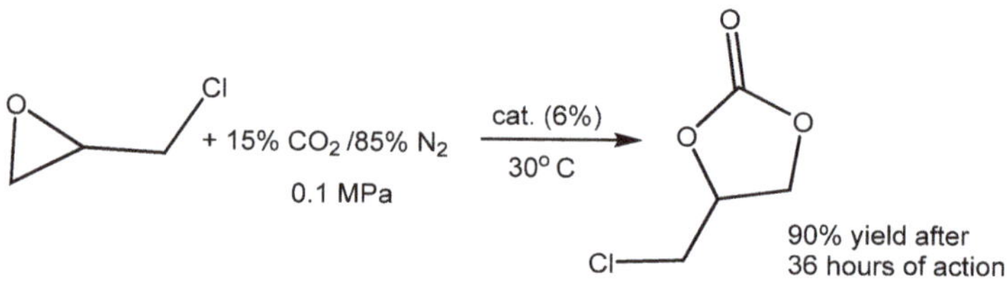

Scheme 11. Carbonation of EPIC using simulated flue gas (15% CO_2 in nitrogen) catalyzed by OH-functionalized DBU-based IL at ambient pressure and at 30 °C [95].

Another described group of catalytically active bifunctional catalysts consists of 1-alkyl-2-hydroxyalkyl pyrazolium salts [96]. The most active one from the broad group of tested derivatives was 1-ethyl-2-(2-hydroxy)ethylpyrazolium bromide (Figure 14). It was demonstrated that using 1 MPa pressure of CO_2 at 110 °C, even internal epoxide ChO was carbonated to CC with a considerable yield of over 60% after 4 h of action [96].

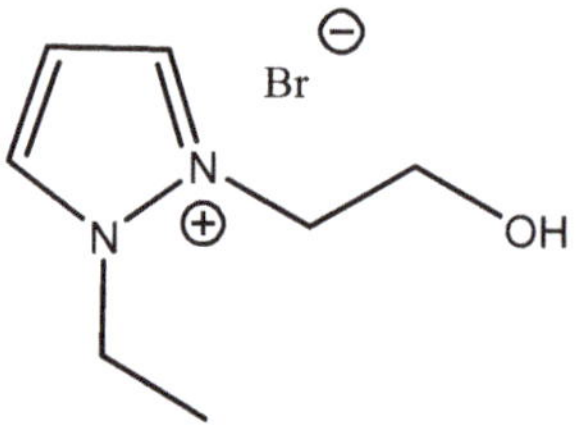

Figure 14. Structure of 1-ethyl-2-(2-hydroxy)ethylpyrazolium bromide, the most effective carbonation catalyst based on 1-alkyl-2-hydroxyalkyl pyrazolium salts [96].

Zhou et al. compared the catalytic activity of quaternized aminoacid glycine (betaine) and quaternized aminoethanol salts (choline salts, Scheme 12) [97].

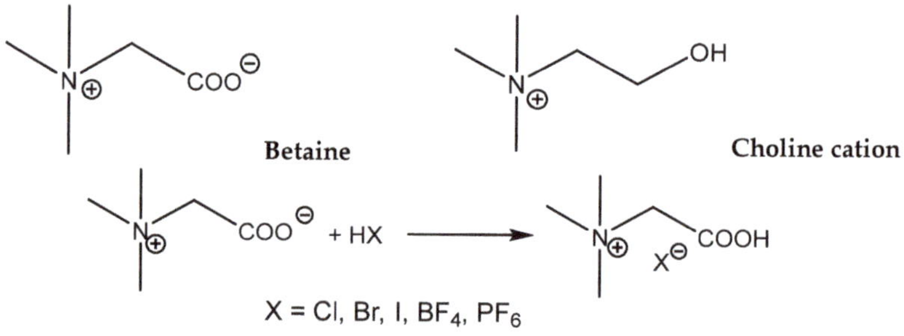

Scheme 12. Structures of tested bifunctional onium salts and scheme of preparation of betaine hydrohalides [97].

The main difference in the structures of these onium salts is the presence of a more acidic carboxylic group (a stronger HBD) in the betaine structure compared with the choline hydroxyl group. In addition, the effects of different anions in the case of protonated betaine were compared, and it was observed that the most active was the corresponding iodide salt. The catalytic activity of different betaine and choline salts decreased in the range: betaine.HI > betaine.HCl > choline.HCl > betaine.HBr > betaine.BF_4^-. The authors interpret the low betaine.HBr activity as the effect of its low solubility in the reaction mixture. The tested choline.HCl possesses an activity comparable with Bu_4NBr, which supports the positive effect of the hydroxyl group in the activation of the oxirane ring of PO. This positive effect could be both based on HBD action and/or the activation of CO_2 on account of carbonate formation. The considerable HDB effect of the carboxylic group of protonated betaines exceeds, however, the effect of the hydroxyl group in choline. The reaction conditions for effective carbonation even of terminal epoxides are, however, harsh (8 MPa CO_2, 140 °C) [97]. Due to the above-mentioned reasons, the research that focused on ILs functionalized with the carboxylic group(s) provided fruitful results.

Xiao et al. suggested that the influence of suitable acidity, even with the flexibility of the bound -$(CH_2)_n$-COOH chain in the IL structure, is crucial for the carbonation of epoxides due to the cooperation function of the ring-opening of oxiranes [98]. When 1-(2-carboxyethyl)-3-methylimidazolium bromide was used as bifunctional IL, the PC from PO was obtained with ca. 96% yield using pressurized CO_2 (1.5 MPa) at 110 °C after 2 h. The IL showed high thermal stability and could be recycled with a slight loss in activity, while the selectivity of the cyclic carbonates remained at over 98%. The catalytic activity of the described functionalized IL-based carboxylic acids is still not unique enough, however, and these types of catalysts still require elevated pressure of CO_2 to maintain a high conversion of epoxides to cyclic carbonates. In addition, the carbonation of internal epoxide is still quite slow even at the above-mentioned high pressure and elevated temperature [98].

The construction of bridge-functionalized bisimidazolium-based ILs improves the catalytic activity of acidic ILs, as was discovered by Kuhn et al. [99]. The most active catalyst is the most branched one, bis(imidazoyl)isobutyric acid derivative, N-arylated with hydrophobic mesitylene (Figure 15). It is well recyclable and active even using 0.4 MPa CO_2 at 70 °C. It is ineffective, however, in the case of internal epoxides' carbonation [99].

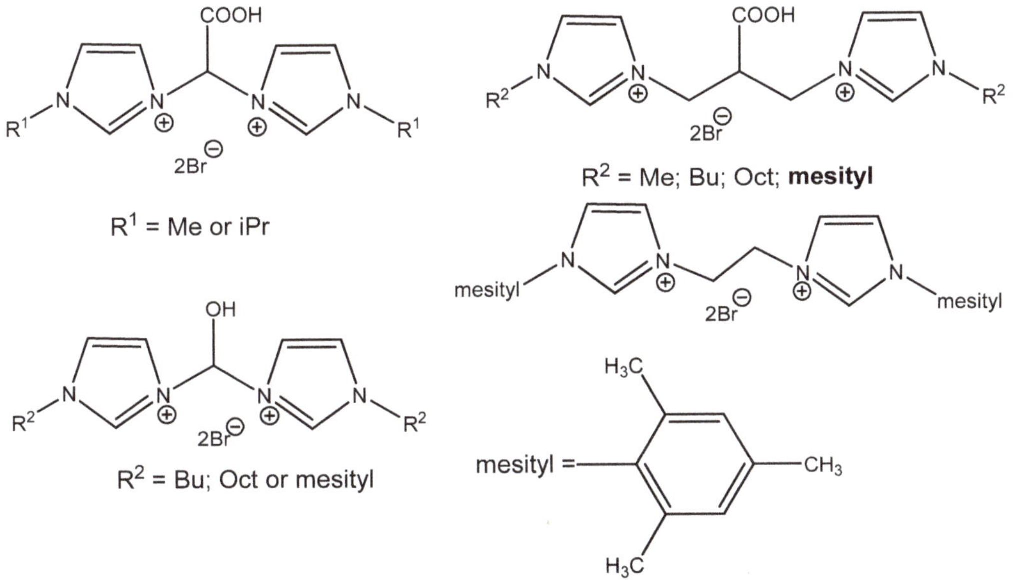

Figure 15. Multifunctional bisimidazolium bromides tested for carbonation of epoxides [99].

An additional direction of research related to the significant increasing of catalytic activity was discovered by Han et al. [100] and Wang et al. [101]. The Han and Wang research groups recognized the crucial role of ion pairs produced by a combination of acidic ILs with guanidinium cations. Using the same acidic ILs, neutralized by alkylated guanidines, enables an increase in activity, probably due to the distinctive activation of reacting CO_2. This catalytic system is active at 0.1 MPa CO_2 and 30 °C for the carbonation of EPIC, but fails even in the case of PO (Table 24). Additionally, the used ion pairs are simply separable from the produced cyclic carbonates by means of extraction with ethyl acetate, enabling simple recycling without a significant drop in conversion.

Table 24. Catalytic activity of multifunctional IM-based ILs on the synthesis of CPC by carbonation of EPIC [100,101].

Entry	Catalyst	Catalyst Amount (mol%)	CO_2 Pressure (MPa)	Temp. (°C)	Reaction Time (h)	Yield (%)
1	C[CMIm]$_2$	5	0.1	50	6	48.4
2	[$^-O_2$MMIm$^+$]$_2$[Br$^-$]$_2$[TMGH$^+$]	5	0.1	50	6	92.1
3	[$^-O_2$MMIm$^+$]$_2$[Br$^-$]$_2$[TMGH$^+$]	5	0.1	50	6	91.4
4	[TMGH$^+$][$^-O_2$MMIm$^+$][Br$^-$]	25	0.1	30	12	84

Bridged methylene(bis)imidazolium salts substituted on both N'-nitrogens by carboxymethyl groups are more active after neutralization with tetramethylguanidine [101] (Figure 16, Table 24). Even this catalytic system is active at 0.1 MPa CO_2 and 50 °C but the carbonation of internal ChO seems to be sluggish using the above-mentioned reaction conditions. Additionally, the used catalyst is simply separable from the produced cyclic carbonates and enables simple recycling without a significant drop in conversion.

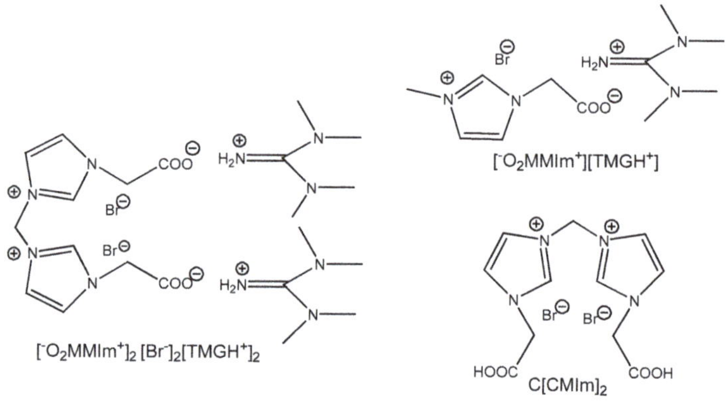

Figure 16. Structures of multifunctional IM-based ILs tested for the carbonation of EPIC [100,101].

The reverse activation of dual amino-functionalized ILs neutralized with acidic aminoacids, such as glutamic or aspartic acids, is possible, as was documented by Yue et al. [102]. This ion-pair-based catalytic system produces, however, high yields at 0.5 MPa and a temperature of 105 °C after 13 h of CO_2 action. It is recyclable without loss of activity and works well in the case of terminal epoxides [102] (Figure 17).

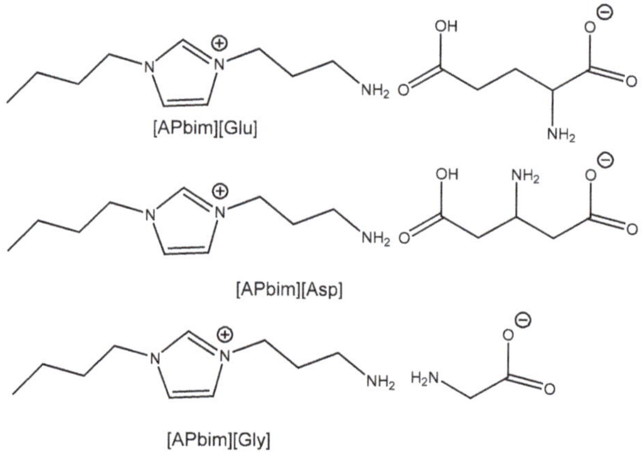

Figure 17. Structures of dual amino-functionalized IM-based ILs [102].

A very attractive alternative approach was published by Kumar et al. [103]. Their research was focused on the utilization of CO_2 from model flue gas (5–15% CO_2 in N_2 stream at atmospheric pressure) at 80 °C using task specific AA-based ILs (Scheme 13). The authors verified that tetrabutylammonium salt with histidine dissolved in dialkyl carbonate enables the capture and usage of CO_2 for the carbonation of terminal epoxides at the above-mentioned reaction conditions, with a high yield. This research work is one of the very infrequent examples of the direct capture and subsequent utilization of CO_2 from (model) flue gas. The authors documented that this catalyst is recyclable with no drop in efficiency after six recycling steps. This catalytic system was proved, however, only on terminal epoxides [103].

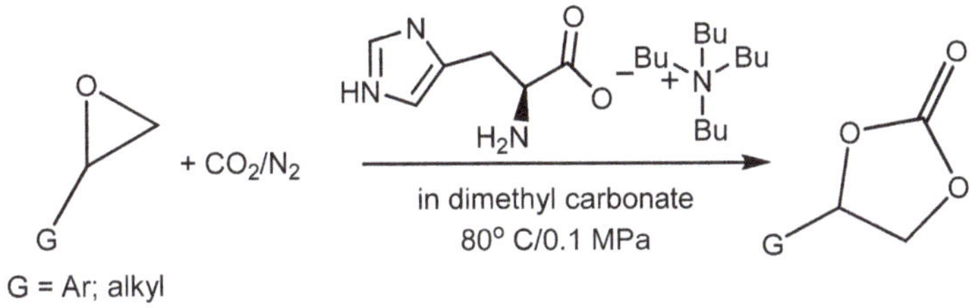

Scheme 13. Carbonation of terminal epoxides using model flue gas (5–15% CO_2 in nitrogen) [103].

A different approach was used for the preparation of highly catalytically active bifunctional ILs using allylation by Hui et al. [104] (Figure 18). They discovered tetramethylguanidine-based allylated IL with superior activity for the capture and utilization of CO_2 from simulated flue gas at 120 °C, ambient pressure and solventless conditions (Table 25). This catalyst is effective even for carbonation of ChO and reusable with low loss of activity after five recycling steps [104]. This type of catalyst seems to be very attractive even for the carbonation of other internal epoxides including eventually epoxidized fatty acid esters.

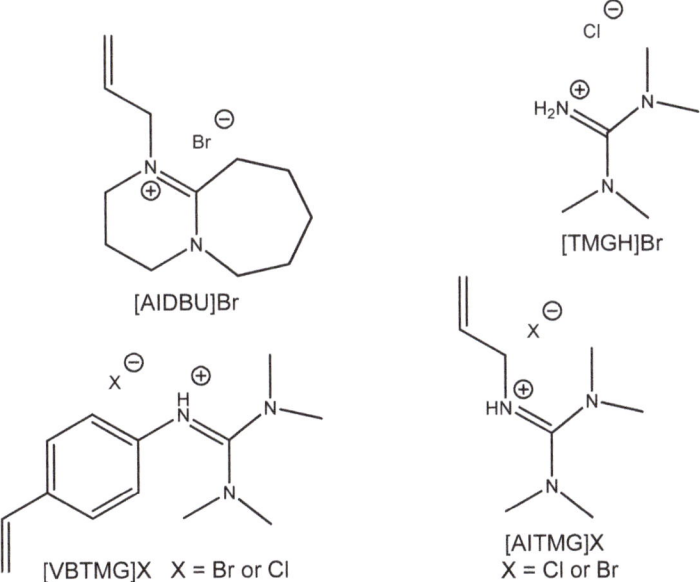

Figure 18. Structures of functionalized ILs tested as highly active CO_2 cycloaddition catalysts [104].

Table 25. Effect of different PILs and ILs on the carbonation of SO [71].

Entry	Catalyst	Catalyst Amount (mol%)	CO_2 Pressure (MPa)	Temp. (°C)	Reaction Time (h)	Yield (%)	Sel. (%)
1	[DBUH]Br	1.5	0.1	100	8	76	99
2	[AlDBU]Br	1.5	0.1	100	8	59	99
3	[DBUH]Cl	1.5	0.1	100	8	71	99
4	[MimH]Br	1.5	0.1	100	8	66	99
5	[DABCOH]Br	1.5	0.1	100	8	59	99
6	[TMGH]Br	1.5	0.1	100	8	80	99
7	[AlTMG]Br	1.5	0.1	100	8	99	99
8	[VBTMG]Cl	1.5	0.1	100	8	57	99
9	DBU	1.5	0.1	100	8	n.d.	n.d.
10	Mim	1.5	0.1	100	8	n.d.	n.d.
11	TMG	1.5	0.1	100	8	n.d.	n.d.
12	TMG + [Bu$_4$N]Br	1.5	0.1	100	8	32	99
13	[Bu$_4$N]Br	1.5	0.1	100	8	17	99
14	[AlTMG]Cl	1.5	0.1	100	8	79	99
15	[VBTMG]Br	1.5	0.1	100	8	62	99
16	none	1.5	0.1	100	8	n.d.	n.d.

4. Conclusions

The carboxylation of epoxides is a sustainable pathway for the fixation of CO_2 into valuable chemicals, considering the industrial utilization of cyclic and polymeric carbonates. The effect of homogeneous organocatalysts published in recent literature is presented. We hope that this review affords insights into the recent research and development of efficient metal-free homogeneous catalysts.

Hopefully, the next development of homogeneous catalysts, including organocatalysts (e.g., organic salt, ILs and DESs), will facilitate the expanding of the spectrum of available metal-free organocatalysts applicable for the reaction of CO_2 with terminal epoxides even at CO_2 pressures of 1 bar and reaction temperatures of less than around 50 °C. Except high catalytic activity, the simple catalyst separability should be profitable because of the necessity of high catalytic loading for the effective course of cycloaddition reaction. The bulky tetrabutylammonium or tetrabutylphosphonium cations in Bu$_4$NX or Bu$_4$PX enable the high nucleophilic activity of the appropriate naked anions of X^-, such as bromide or iodide, in most cases. Onium salts are widely applied as part of multicomponent catalytic systems in the research and development of epoxides' carbonation processes. Most of all, deep eutectic solvents constitute an important group of multicomponent low-cost homogeneous organocatalysts. In particular, DESs containing choline chloride and urea exhibit high catalytic activity [84,85]. Besides the above-mentioned onium halides, some ion pairs produced by the mixing of DBU with amidine-based alcohols are highly active [95]. The most effective halide free IL-based homogeneous catalyst was recognized to be the Bu$_4$N salt of histidine [103]. Several ion pairs based on N'-carboxymethylated MIM bromides neutralized with tetraalkylguanidines enable CO_2 cycloaddition at ambient pressure [100–102]. Searching for simple and cheap catalytic systems that are active at mild reaction conditions is attractive not only due to the environmental point of view (lower energy consumption) but even due to the thermodynamic reasons. As the formation the cyclic carbonate is exothermic, the lower reaction temperature affects the shifting of the reaction equilibrium in the products.

As we illustrate in this review, many simple molecules are known to act as effective mediators and/or catalysts, including Lewis and Bronsted acids such as water, ascorbic acid, cellulose, etc. Based on generally accepted mechanisms of carboxylation of epoxides, research focused on utilization of other HBDs such as bidentate nucleophiles could be profitable. The promising groups of simply available catalytically active compounds such as polyalkyl guanidines [100,101,104], enaminones [105], N-hydroxylamines [106] and amidoximes [107] should, in our opinion, be investigated in more detail.

The utilization of tandem reactions such as the one-pot production of cyclic carbonates starting directly from biobased unsaturated fatty acids esters [8], the one-pot production of ethylene carbonate from ethylene produced by low-energy-demanding methods [2,108] or the production of HMEC from chlorinated bio-based glycerol [109] seems to be very promising for effective CO_2 fixation.

It is evident from the recently published results that both the possible utilization of CO_2 from flue gas and the carbonation of internal bio-based epoxides such as epoxidized fatty acid esters are the main developing areas of research focused on CO_2 capture and utilization. However, the mild reaction conditions and lower catalytic loading are still challenging for both the carboxylation of internal epoxide substrates such as epoxidized fatty acid esters as well as for the direct utilization of waste CO_2 from power plant flue gas.

Author Contributions: T.W. conceived, designed and wrote the paper. B.K. provided technical support. All authors have read and agreed to the published version of the manuscript.

Funding: This research was funded by Faculty of Chemical Technology, University of Pardubice, with the support of excellent research.

Conflicts of Interest: The authors declare no conflict of interest.

References

1. Artz, J.; Müller, T.E.; Thenert, K. Sustainable Conversion of Carbon Dioxide: An Integrated Review of Catalysis and Life Cycle Assessment. *Chem. Rev.* **2018**, *118*, 434–504. [CrossRef]
2. Aresta, M.; Dibenedetto, A.; Angelini, A. Catalysis for the Valorization of Exhaust Carbon: From CO_2 to Chemicals, Materials, and Fuels. Technological Use of CO_2. *Chem. Rev.* **2014**, *114*, 1709–1742. [CrossRef]
3. Shaikh, R.R.; Pornpraprom, S.; D'Elia, V. Catalytic Strategies for the Cycloaddition of Pure, Diluted, and Waste CO_2 to Epoxides under Ambient Conditions. *ACS Catal.* **2018**, *8*, 419–450. [CrossRef]
4. Cui, S.; Borgemenke, J.; Liu, Z.; Li, Y. Recent advances of "soft" bio-polycarbonate plastics from carbon dioxide and renewable bio-feedstocks via straightforward and innovative routes. *J. CO_2 Util.* **2019**, *34*, 40–52. [CrossRef]
5. Kiatkittipong, K. Green Pathway in Utilizing CO_2 via Cycloadditon Reaction with Epoxide-A Mini Review. *Processes* **2020**, *8*, 548. [CrossRef]
6. Cahugule, A.A.; Tamboli, A.H.; Kim, H. Ionic liquid as a catalyst for utilization of carbon dioxide to production of linear and cyclic carbonate. *Fuel* **2017**, *200*, 316–332. [CrossRef]
7. Darensbourg, D.J. Making Plastics from Carbon Dioxide: Salen Metal Complexes as Catalysts for the Production of Polycarbonates from Epoxides and CO_2. *Chem. Rev.* **2007**, *107*, 2388–2410. [CrossRef]
8. Calmanti, R.; Sargentoni, N.; Selva, M.; Perosa, A. A One-Pot Tandem Catalytic Epoxidation—CO_2 Insertion of Monounsaturated Methyl Oleate to the Corresponding Cyclic Organic Carbonate Catalysts. *J. CO_2 Util.* **2021**, *11*, 1477. [CrossRef]
9. Coates, G.W.; Moore, D.R. Discrete Metal-Based Catalysts for the Copolymerization of CO_2 and Epoxides: Discovery, Reactivity, Optimization, and Mechanism. *Angew. Chem. Int. Ed.* **2004**, *43*, 6618. [CrossRef]
10. Gorbunov, D.N.; Nenasheva, M.V.; Terenina, M.V.; Kardasheva, Y.S.; Kardashev, S.V.; Naranov, E.R.; Bugaev, A.L.; Soldatov, A.V.; Maximov, A.L.; Karakhanova, E.A. Transformations of Carbon Dioxide under Homogeneous Catalysis Conditions (A Review). *Petroleum Chem.* **2022**, *62*, 1–39. [CrossRef]
11. Rehman, A.; Saleem, F.; Javed, F.; Ikhlaq, A.; Ahmad, S.W.; Harvey, A. Recent advances in the synthesis of cyclic carbonates via CO_2 cycloaddition to epoxides. *J. Environ. Chem. Eng.* **2021**, *9*, 105113. [CrossRef]
12. Guo, L.; Lamb, K.J.; North, M. Recent developments in organocatalyzed transformations of epoxides and carbon dioxide into cyclic carbonates. *Green Chem.* **2021**, *23*, 77–118. [CrossRef]
13. Sakakura, T.; Choi, J.C.; Yasuda, H. Transformation of Carbon Dioxide. *Chem. Rev.* **2007**, *107*, 2365. [CrossRef]
14. You, H.; Wang, E.; Cao, H.; Zhuo, C.; Liu, S.; Wang, X.; Wang, F. From Impossible to Possible: Atom-Economic Polymerization of Low Strain Five-Membered Carbonates. *Angew. Chem. Int. Ed.* **2022**, *61*, e202113152. [CrossRef]
15. Arayachukiat, S.; Kongtes, C.; Barthel, A.; Vummaleti, S.V.C.; Poater, A.; Wannakao, S.; Cavallo, L.; D'Elia, V. Ascorbic acid as a bifunctional hydrogen bond donor for the synthesis of cyclic carbonates from CO_2 under ambient conditions. *ACS Sustain. Chem. Eng.* **2017**, *5*, 6392–6397. [CrossRef]
16. Lari, G.M.; Pastore, G.; Mondelli, C.; Pérez-Ramírez, J. Towards sustainable manufacture of epichlorohydrin from glycerol using hydrotalcite-derived basic oxides. *Green Chem.* **2018**, *20*, 148–159. [CrossRef]
17. Vitiello, R.; Tesser, R.; Santacesaria, E.; Di Serio, M. New Production Processes of Dichlorohydrins from Glycerol Using Acyl Chlorides as Catalysts or Reactants. *Ind. Eng. Chem. Res.* **2016**, *55*, 1484–1490. [CrossRef]

18. Morodo, R.; Gerardy, R.; Petit, G.; Monbaliu, J.M. Continuous flow upgrading of glycerol toward oxiranes and active pharmaceutical ingredients thereof. *Green Chem.* **2019**, *21*, 4422–4433. [CrossRef]
19. Kubicek, P.; Sladek, P. Method of Preparing Dichloropropanols from Glycerine. U.S. Patent EP 1663924 B1, 23 August 2004.
20. Cespi, D.; Cucciniello, R.; Ricciardi, M.; Capacchione, C.; Vassura, I.; Passarini, F.; Proto, A. A simplified early stage assessment of process intensification: Glycidol as a value-added product from epichlorohydrin industry wastes. *Green Chem.* **2016**, *18*, 4559–4570. [CrossRef]
21. Phung, T.K.; Pham, T.L.M.; Vu, K.B.; Busca, G. (Bio)Propylene production processes: A critical review. *J. Environ. Chem. Eng.* **2021**, *9*, 105673. [CrossRef]
22. Zhang, M.; Yu, Y. Dehydration of ethanol to ethylene. *Ind. Eng. Chem. Res.* **2013**, *52*, 9505–9514. [CrossRef]
23. Broeren, M. *Production of Bio-Ethylene*; IEA-ETSAP: Paris, France; IRENA: Abu Dhabi, United Arab Emirates, 2013; pp. 1–20.
24. Resul, M.F.M.G.; López Fernández, A.M.; Rehman, A.; Harvey, A.P. Development of a selective, solvent-free epoxidation of limonene using hydrogen peroxide and a tungsten-based catalyst. *React. Chem. Eng.* **2018**, *3*, 747–756. [CrossRef]
25. Shaarani, F.W.; Bou, J.J. Synthesis of vegetable-oil based polymer by terpolymerization of epoxidized soybean oil, propylene oxide and carbon dioxide. *Sci. Total Environ.* **2017**, *598*, 931–936. [CrossRef]
26. Cui, S.; Qin, Y.; Li, Y. Sustainable Approach for the Synthesis of Biopolycarbonates from Carbon Dioxide and Soybean Oil. *ACS Sustain. Chem. Eng.* **2017**, *5*, 9014–9022. [CrossRef]
27. Phimsen, S.; Yamada, H.; Tagawa, T.; Kiatkittipong, W.; Kiatkittipong, K.; Laosiripojana, N.; Assabumrungrat, S. Epoxidation of methyl oleate in a TiO_2 coated-wall capillary microreactor. *Chem. Eng. J.* **2017**, *314*, 594–599. [CrossRef]
28. Yu, K.M.K.; Curcic, I.; Gabriel, J.; Morganstewart, H.; Tsang, S.C. Catalytic Coupling of CO_2 with Epoxide Over Supported and Unsupported Amines. *J. Phys. Chem. A* **2010**, *114*, 3863–3872. [CrossRef]
29. Yang, Z.Z.; He, L.N.; Miao, C.X.; Chanfreau, S. Lewis basic ionic liquids-catalyzed conversion of carbon dioxide to cyclic carbonates. *Adv. Synth. Catal.* **2010**, *352*, 2233–2240. [CrossRef]
30. Fanjul-Mosteirin, N.; Jehanno, C.; Ruiperez, F.; Sardon, H.; Dove, A. Rational study of DBU salts for the CO_2 insertion into epoxides for the synthesis of cyclic carbonates. *ACS Sustain. Chem. Eng.* **2019**, *7*, 10633–10640. [CrossRef]
31. Aoyagi, N.; Furusho, Y.; Endo, T. Cyclic amidine hydroiodide for the synthesis of cyclic carbonates and cyclic dithiocarbonates from carbon dioxide or carbon disulfide under mild conditions. *Tetrahedron* **2019**, *75*, 130781. [CrossRef]
32. Roshan, K.R.; Palissery, R.A.; Kathalikkattil, A.C.; Babu, R.; Mathai, G.; Lee, H.S.; Park, D.W. A computational study of the mechanistic insights into base catalysed synthesis of cyclic carbonates from CO_2: Bicarbonate anion as an active species. *Catal. Sci. Technol.* **2016**, *6*, 3997–4004. [CrossRef]
33. Lee, C.S.; Ong, Y.L.; Aroua, M.K.; Daud, W.M.A.W. Impregnation of palm shell based activated carbon with sterically hindered amines for CO_2 adsorption. *Chem. Eng. J.* **2013**, *219*, 558–564. [CrossRef]
34. Zelenak, V.; Halamova, D.; Gaberova, L.; Bloch, E.; Llewellyn, P. Amine-modified SBA-12 mesoporous silica for carbon dioxide capture: Effect of amine basicity on sorption properties. *Microporous Mesoporous Mater.* **2008**, *116*, 358–364. [CrossRef]
35. Azzouz, R.; Moreno, V.C.; Herasme-Grullon, C.; Levacher, V.; Estel, L.; Ledoux, A.; Bischoff, L. Efficient Conversion of Epoxides into Carbonates with CO_2 and a Single Organocatalyst: Laboratory and Kilogram-Scale Experiments. *Synlett* **2020**, *31*, 183–188. [CrossRef]
36. Sun, J.; Cheng, W.; Yang, Z.; Wang, J.; Xu, T.; Xin, J.; Zhang, S. Superbase/cellulose: An environmentally benign catalyst for chemical fixation of carbon dioxide into cyclic carbonates. *Green Chem.* **2014**, *16*, 3071–3078. [CrossRef]
37. Qaroush, A.K.; Hasan, A.K.; Hammad, S.B.; Feda'a, M.; Assaf, K.I.; Alsoubani, F.; Ala'a, F.E. Mechanistic insights on CO_2 utilization using sustainable catalysis. *New J. Chem.* **2021**, *45*, 22280–22288. [CrossRef]
38. Liu, M.; Li, X.; Liang, L.; Sun, J. Protonated triethanolamine as multi-hydrogen bond donors catalyst for efficient cycloaddition of CO_2 to epoxides under mild and cocatalyst-free conditions. *J. CO_2 Util.* **2016**, *16*, 384–390. [CrossRef]
39. Lan, D.H.; Fan, N.; Wang, Y.; Gao, X.; Zhang, P.; Chen, L.; Yin, S.F. Recent advances in metal-free catalysts for the synthesis of cyclic carbonates from CO_2 and epoxides. *Chin. J. Catal.* **2016**, *37*, 826–845. [CrossRef]
40. Wang, L.; Zhang, G.Y.; Kodama, K.; Hirose, T. An efficient metal- and solvent-free organocatalytic system for chemical fixation of CO_2 into cyclic carbonates under mild conditions. *Green Chem.* **2016**, *18*, 1229–1233. [CrossRef]
41. Kumatabara, Y.; Okada, M.; Shirakawa, S. Triethylamine hydroiodide as a simple yet effective bifunctional catalyst for CO_2 fixation reactions with epoxides under mild conditions. *ACS Sustain. Chem. Eng.* **2017**, *5*, 7295–7301. [CrossRef]
42. Zhang, Z.G.; Fan, F.J.; Xing, H.B.; Yang, Q.W.; Bao, Z.B.; Ren, Q.L. Efficient synthesis of cyclic carbonates from atmospheric CO_2 using a positive charge delocalized ionic liquid catalyst. *ACS Sustain. Chem. Eng.* **2017**, *5*, 2841–2846. [CrossRef]
43. Shen, Y.M.; Duan, W.L.; Shi, M. Phenol and organic bases co-catalyzed chemical fixation of carbon dioxide with terminal epoxides to form cyclic carbonates. *Adv. Synth. Catal.* **2003**, *345*, 337–340. [CrossRef]
44. Khiari, R.; Salon, M.C.B.; Mhenni, M.F.; Mauret, E.; Belgacem, M.N. Synthesis and characterization of cellulose carbonate using greenchemistry: Surface modification of Avicel. *Carbohydr. Polym.* **2017**, *163*, 254–260. [CrossRef] [PubMed]
45. Gunnarsson, M.; Bernin, D.; Hasani, M.; Lund, M.; Bialik, E. Direct evidence for reaction between cellulose and CO_2 from nuclear magnetic resonance. *ACS Sustain. Chem. Eng.* **2021**, *9*, 14006–14011. [CrossRef]
46. Aoyagi, N.; Furusho, Y.; Endo, T. Effective synthesis of cyclic carbonates from carbon dioxide and epoxides by phosphonium iodides as catalysts in alcoholic solvents. *Tetrahedron Lett.* **2013**, *54*, 7031–7034. [CrossRef]

47. Samikannua, A.; Konwara, L.J.; Mäki-Arvelab, P.; Mikkola, J.P. Renewable *N*-doped active carbons as efficient catalysts for direct synthesis of cyclic carbonates from epoxides and CO_2. *Appl. Catal. B Environ.* **2019**, *241*, 41–51. [CrossRef]
48. Qi, C.; Jiang, H. Histidine-catalyzed synthesis of cyclic carbonates in supercritical carbon dioxide. *Sci. China Chem.* **2010**, *53*, 1566–1570. [CrossRef]
49. Tharun, J.; Roshan, K.R.; Kathalikkattil, A.C.; Kang, D.H.; Ryu, H.M.; Park, D.W. Natural amino acids/H_2O as a metal- and halide-free catalyst system for the synthesis of propylene carbonate from propylene oxide and CO_2 under moderate conditions. *RSC Adv.* **2014**, *4*, 41266–41270. [CrossRef]
50. Tharun, J.; Mathai, G.; Roshan, R.; Kathalikkattil, A.C.; Bomi, K.; Park, D.W. Simple and efficient synthesis of cyclic carbonates using quaternized glycine as a green catalyst. *Phys. Chem. Chem. Phys.* **2013**, *15*, 9029–9033. [CrossRef]
51. Roshan, K.R.; Kathalikkattil, A.C.; Tharun, J.; Kim, D.W.; Won, Y.S.; Park, D.W. Amino acid/KI as multi-functional synergistic catalysts for cyclic carbonate synthesis from CO_2 under mild reaction conditions: A DFT corroborated study. *Dalton Trans.* **2014**, *43*, 2023–2031. [CrossRef]
52. Qi, Y.; Cheng, W.; Xu, F.; Chen, S.; Zhang, S. Amino acids/superbases as eco-friendly catalyst system for the synthesis of cyclic carbonates under metal-free and halide-free conditions. *Synth. Commun.* **2018**, *48*, 876–886. [CrossRef]
53. Claver, C.; Yeamin, M.B.; Reguero, M.; Masdeu-Bultó, A.M. Recent advances in the use of catalysts based on natural products for the conversion of CO_2 into cyclic carbonates. *Green Chem.* **2020**, *22*, 7665–7706. [CrossRef]
54. Yang, Z.; Sun, J.; Cheng, W.; Wang, J.; Li, Q.; Zhang, S. Biocompatible and recyclable amino-acid binary catalyst for efficient chemical fixation of CO_2. *Catal. Commun.* **2014**, *44*, 6–9. [CrossRef]
55. Dai, W.; Yang, W.; Zhang, Y.; Wang, D.; Luo, X.; Tu, X. Novel isothiouronium ionic liquid as efficient catalysts for the synthesis of cyclic carbonates from CO_2 and epoxides. *J. CO_2 Util.* **2017**, *17*, 256. [CrossRef]
56. Orhan, O.Y. Effects of various anions and cations in ionic liquids on CO_2 capture. *J. Mol. Liq.* **2021**, *333*, 115981. [CrossRef]
57. Peng, J.; Deng, Y. Cycloaddition of carbon dioxide to propylene oxide catalyzed by ionic liquids. *New J. Chem.* **2001**, *25*, 639–641. [CrossRef]
58. Neumann, J.G.; Stassen, H. Anion Effect on Gas Absorption in Imidazolium-Based Ionic Liquids. *J. Chem. Inf. Model.* **2020**, *60*, 661–666. [CrossRef]
59. Bobbink, F.D.; Vasilyev, D.; Hulla, M.; Chamam, S.; Menoud, F.; Laurenczy, G.; Katsyuba, S.; Dyson, P.J. Intricacies of Cation−Anion Combinations in Imidazolium SaltCatalyzed Cycloaddition of CO_2 Into Epoxides. *ACS Catal.* **2018**, *8*, 2589–2594. [CrossRef]
60. Wang, T.; Zheng, D.; Ma, Y.; Guo, J.; He, Z.; Ma, B.; Zhang, J. Benzyl substituted imidazolium ionic liquids as efficient solvent-free catalysts for the cycloaddition of CO_2 with epoxides: Experimental and Theoretic study. *J. CO_2 Util.* **2017**, *22*, 44–52. [CrossRef]
61. Akhdar, A.; Onida, K.; Vu, N.D.; Grollier, K.; Norsic, S.; Boisson, C.; Duguet, N. Thermomorphic Polyethylene-Supported Organocatalysts for the Valorization of Vegetable Oils and CO_2. *Adv. Sustain. Syst.* **2021**, *5*, 2000218. [CrossRef]
62. Buettner, H.; Grimmer, C.; Steinbauer, J.; Werner, T. Iron-based binary catalytic system for the valorization of CO_2 into biobased cyclic carbonates. *ACS Sustain. Chem. Eng.* **2016**, *4*, 4805–4814. [CrossRef]
63. Carrodeguas, L.P.; Cristòfol, À.; Fraile, J.M.; Mayoral, J.A.; Dorado, V.; Herrerías, C.I.; Kleij, A.W. Fatty acid based biocarbonates: Al-mediated stereoselective preparation of mono-, di-and tricarbonates under mild and solvent-less conditions. *Green Chem.* **2017**, *19*, 3535–3541. [CrossRef]
64. Anthofer, M.H.; Wilhelm, M.E.; Cokoja, M.; Markovits, I.I.; Pöthig, A.; Mink, J.; Kühn, F.E. Cycloaddition of CO_2 and epoxides catalyzed by imidazolium bromides under mild conditions: Influence of the cation on catalyst activity. *Catal. Sci. Technol.* **2014**, *4*, 1749–1758. [CrossRef]
65. Yoshihito, K.; Masafumi, Y.; Takao, I. *N*-Heterocyclic Carbenes as Efficient Organocatalysts for CO_2 Fixation Reactions. *Angew. Chem. Int. Ed.* **2009**, *48*, 4194–4197. [CrossRef]
66. Wong, W.L.; Chan, P.H.; Zhou, Z.Y.; Lee, K.H.; Cheung, K.C.; Wong, K.Y. A robust ionic liquid as reaction medium and efficient organocatalyst for carbon dioxide fixation. *ChemSusChem* **2008**, *1*, 67–70. [CrossRef]
67. Hajipour, A.R.; Heidari, Y.; Kozehgary, G. Nicotine-derived ammonium salts as highly efficient catalysts for chemical fixation of carbon dioxide into cyclic carbonates under solvent-free conditions. *RSC Adv.* **2015**, *5*, 61179–61183. [CrossRef]
68. Liu, M.; Liang, L.; Liang, T.; Lin, X.; Shi, L.; Wang, F. Cycloaddition of CO_2 and epoxides catalyzed by dicationic ionic liquids mediated metal halide: Influence of the dication on catalytic activity. *J. Mol. Catal. A Chem.* **2015**, *408*, 242. [CrossRef]
69. Yuan, G.; Zhao, Y.; Wu, Y.; Li, R.; Chen, Y.; Xu, D.; Liu, Z. Cooperative effect from cation and anion of pyridine-containing anion-based ionic liquids for catalysing CO_2 transformation at ambient conditions. *Sci. China Chem.* **2017**, *60*, 958–963. [CrossRef]
70. Wang, L.; Lin, L.; Zhang, G.; Kodama, K.; Yasutake, M.; Hirose, T. Synthesis of cyclic carbonates from CO_2 and epoxides catalyzed by low loadings of benzyl bromide/DMF at ambient pressure. *Chem. Commun.* **2014**, *50*, 14813–14816. [CrossRef]
71. Lin, W.; Koichi, K.; Takuji, H. DBU/benzyl bromide: An efficient catalytic system for the chemical fixation of CO_2 into cyclic carbonates under metal- and solvent- free conditions. *Catal. Sci. Technol.* **2016**, *6*, 3872–3877. [CrossRef]
72. Calo, V.; Nacci, A.; Monopoli, A.; Fanizzi, A. Cyclic carbonate formation from carbon dioxide and oxiranes in tetrabutylammonium halides as solvents and catalysts. *Org. Lett.* **2002**, *4*, 2561–2563. [CrossRef]
73. Sit, W.N.; Ng, S.M.; Kwong, K.Y.; Lau, C.P. Coupling reactions of CO_2 with neat epoxides catalyzed by PPN salts to yield cyclic carbonates. *J. Org. Chem.* **2005**, *70*, 8583–8586. [CrossRef] [PubMed]

74. Della Monica, F.; Buonerba, A.; Grassi, A.; Capacchione, C.; Milione, S. Glycidol: An hydroxyl-containing epoxide playing the double role of substrate and catalyst for CO_2 cycloaddition reactions. *ChemSusChem* **2016**, *9*, 3457–3464. [CrossRef] [PubMed]
75. Yang, X.; Zou, Q.; Zhao, T.; Chen, P.; Liu, Z.; Liu, F.; Lin, Q. Deep Eutectic Solvents as Efficient Catalysts for Fixation of CO_2 to Cyclic Carbonates at Ambient Temperature and Pressure through Synergetic Catalysis. *ACS Sustain. Chem. Eng.* **2021**, *9*, 10437–10443. [CrossRef]
76. Dai, Y.; van Spronsen, J.; Witkamp, G.-J.; Verpoorte, R.; Choi, Y.H. Ionic Liquids and Deep Eutectic Solvents in Natural Products Research: Mixtures of Solids as Extraction Solvents. *J. Nat. Prod.* **2013**, *76*, 2162–2173. [CrossRef]
77. Xiao, L.; Su, D.; Yue, C.; Wu, W. Protic ionic liquids: A highly efficient catalyst for synthesis of cyclic carbonate from carbon dioxide and epoxides. *J. CO_2 Util.* **2016**, *6*, 1–6. [CrossRef]
78. Wu, K.; Su, T.; Hao, D.; Liao, W.; Zhao, Y.; Ren, W.; Lü, H. (Choline chloride-based deep eutectic solvents for efficient cycloaddition of CO_2 with propylene oxide. *Chem. Commun.* **2018**, *54*, 9579–9582. [CrossRef]
79. Liu, F.; Gu, Y.; Xin, H.; Zhao, P.; Gao, J.; Liu, M. Multifunctional phosphonium-based deep eutectic ionic liquids: Insights into simultaneous activation of CO_2 and epoxide and their subsequent cycloaddition. *ACS Sustain. Chem. Eng.* **2019**, *7*, 16674–16681. [CrossRef]
80. Wilhelm, M.E.; Anthofer, M.H.; Cokoja, M.; Markovits, I.I.E.; Herrmann, W.A.; Kuhn, F.E. Cycloaddition of Carbon Dioxide and Epoxides using Pentaerythritol and Halides as Dual Catalyst System. *ChemSusChem* **2014**, *7*, 1357–1360. [CrossRef]
81. Liu, F.; Gu, Y.; Zhao, P.; Xin, H.; Gao, J.; Liu, M. N-hydroxysuccinimide based deep eutectic catalysts as a promising platform for conversion of CO_2 into cyclic carbonates at ambient temperature. *J. CO_2 Util.* **2019**, *33*, 419–426. [CrossRef]
82. Alves, M.; Grignard, B.; Gennen, S.; Méreau, R.; Detrembleur, C.; Jérôme, C.; Tassaing, T. Organocatalytic promoted coupling of carbon dioxide with epoxides: A rational investigation of the cocatalytic activity of various hydrogen bond donors. *Catal. Sci. Technol.* **2015**, *5*, 4636–4643. [CrossRef]
83. Liu, Y.; Cao, Z.; Zhou, Z.; Zhou, A. Imidazolium-based deep eutectic solvents as multifunctional catalysts for multisite synergistic activation of epoxides and ambient synthesis of cyclic carbonates. *J. CO_2 Util.* **2021**, *53*, 101717. [CrossRef]
84. Vagnoni, M.; Samori, C.; Galletti, P. Choline-based eutectic mixtures as catalysts for effective synthesis of cyclic carbonates from epoxides and CO_2. *J. CO_2 Util.* **2020**, *42*, 101302. [CrossRef]
85. Sopena, S.; Fiorani, G.; Martin, C.; Kleij, A.W. Highly Efficient Organocatalyzed Conversion of Oxiranes and CO_2 into Organic Carbonates. *ChemSusChem* **2015**, *8*, 3248–3254. [CrossRef] [PubMed]
86. He, L.; Zhang, W.; Yang, Y.; Ma, J.; Liu, F.; Liu, M. Novel biomass-derived deep eutectic solvents promoted cycloaddition of CO_2 with epoxides under mild and additive-free conditions. *J. CO_2 Util.* **2021**, *54*, 101750. [CrossRef]
87. Wang, S.; Zhu, Z.; Hao, D.; Su, T.; Len, C.; Ren, W.; Lü, H. Synthesis cyclic carbonates with BmimCl-based ternary deep eutectic solvents system. *J. CO_2 Util.* **2020**, *40*, 101250. [CrossRef]
88. Hardman-Baldwin, A.M.; Mattson, A.E. Silanediol-catalyzed carbon dioxide fixation. *ChemSusChem* **2014**, *7*, 3275–3278. [CrossRef]
89. Kaneko, S.; Shirakawa, S. Potassium iodide–tetraethylene glycol complex as a practical catalyst for CO_2 fixation reactions with epoxides under mild conditions. *ACS Sustain. Chem. Eng.* **2017**, *5*, 2836–2840. [CrossRef]
90. Zhou, H.; Wang, G.X.; Zhang, W.Z.; Lu, X.B. CO_2 adducts of phosphorus ylides: Highly active organocatalysts for carbon dioxide transformation. *ACS Catal.* **2015**, *5*, 6773–6779. [CrossRef]
91. Natongchai, W.; Pornpraprom, S.; D'Elia, V. Synthesis of Bio-Based Cyclic Carbonates Using a Bio-Based Hydrogen Bond Donor: Application of Ascorbic Acid to the Cycloaddition of CO_2 to Oleochemicals. *Asian J. Org. Chem.* **2020**, *9*, 801–810. [CrossRef]
92. Schäffner, B.; Blug, M.; Kruse, D.; Polyakov, M.; Köckritz, A.; Martin, A.; Rajagopalan, P.; Bentrup, U.; Brückner, A.; Jung, S.; et al. Synthesis and Application of Carbonated Fatty Acid Esters from Carbon Dioxide Including a Life Cycle Analysis. *ChemSusChem* **2014**, *7*, 1133–1139. [CrossRef]
93. Werner, T.; Buttner, H. Phosphorus-based Bifunctional Organocatalysts for the Addition of Carbon Dioxide and Epoxides. *ChemSusChem* **2014**, *7*, 3268–3271. [CrossRef] [PubMed]
94. Tsutsumi, Y.; Yamakawa, K.; Yoshida, M.; Ema, T.; Sakai, T. Bifunctional Organocatalyst for Activation of Carbon Dioxide and Epoxide To Produce Cyclic Carbonate: Betaine as a New Catalytic Motif. *Org. Lett.* **2010**, *12*, 5728–5731. [CrossRef] [PubMed]
95. Meng, X.; Ju, Z.; Zhang, S.; Liang, X.; von Solms, N.; Zhang, X.; Zhang, X. Efficient transformation of CO_2 to cyclic carbonates using bifunctional protic ionic liquids under mild conditions. *Green Chem.* **2019**, *21*, 3456–3463. [CrossRef]
96. Wang, T.; Ma, Y.; Jiang, J.; Zhu, X.; Fan, B.; Yu, G.; Zhang, J. Hydroxyl-functionalized pyrazolium ionic liquids to catalyze Chemical fixation of CO_2: Further benign reaction condition for the singlecomponent catalyst. *J. Mol. Liq.* **2019**, *293*, 111479. [CrossRef]
97. Zhou, Y.; Hu, S.; Ma, X.; Liang, S.; Jiang, T.; Han, B. Synthesis of cyclic carbonates from carbon dioxide and epoxides over betaine-based catalysts. *J. Mol. Catal. A Chem.* **2008**, *284*, 52–57. [CrossRef]
98. Xiao, L.F.; Lv, D.W.; Su, D.; Wu, W.; Li, H.F. Influence of acidic strength on the catalytic activity of Brønsted acidic ionic liquids on synthesizing cyclic carbonate from carbon dioxide and epoxide. *J. Clean. Prod.* **2014**, *67*, 285–290. [CrossRef]
99. Li, Y.; Dominelli, B.; Reich, R.M.; Liu, B.; Kühn, F.E. Bridge-functionalized bisimidazolium bromides as catalysts for the conversion of epoxides to cyclic carbonates with CO_2. *Catal. Commun.* **2019**, *124*, 118–122. [CrossRef]
100. Hu, J.; Ma, J.; Liu, H.; Qian, Q.; Xie, C.; Han, B. Dual-ionic liquid system: An efficient catalyst for chemical fixation of CO_2 to cyclic carbonates under mild conditions. *Green Chem.* **2018**, *20*, 2990–2994. [CrossRef]

101. Wang, T.; Zheng, D.; Hu, Y.; Zhou, J.; Liu, Y.; Zhang, J.; Wang, L. Efficient responsive ionic liquids with multiple active centers for the transformation of CO_2 under mild conditions: Integrated experimental and theoretical study. *J. CO_2 Util.* **2021**, *49*, 101573. [CrossRef]
102. Yue, S.; Wang, P.; Hao, X.; Zang, S. Dual amino-functionalized ionic liquids as efficient catalysts for carbonate synthesis from carbon dioxide and epoxide under solvent and cocatalyst-free conditions. *J. CO_2 Util.* **2017**, *21*, 238–246. [CrossRef]
103. Kumar, P.; Varyani, M.; Khatri, P.K.; Paul, S.; Jain, S.L. Post combustion capture and conversion of carbon dioxide using histidine derived ionic liquid at ambient conditions. *J. Ind. Eng. Chem.* **2017**, *49*, 152–157. [CrossRef]
104. Hui, W.; Wang, X.; Li, X.-N.; Wang, H.-J.; He, X.-M.; Xu, X.-Y. Protic ionic liquids tailored by different cationic structures for efficient chemical fixation of diluted and waste CO_2 into cyclic carbonates. *New J. Chem.* **2021**, *45*, 10741–10748. [CrossRef]
105. Yue, Z.X.; Pudukudy, M.; Chen, S.; Liu, Y.; Zhao, W.; Wang, J.; Jia, Q. A non-metal Acen-H catalyst for the chemical fixation of CO_2 into cyclic carbonates under solvent-and halide-free mild reaction conditions. *Appl. Catal. A Gen.* **2020**, *601*, 117646. [CrossRef]
106. Ravi, S.; Puthiaraj, P.; Ahn, W.S. Hydroxylamine-Anchored Covalent Aromatic Polymer for CO_2 Adsorption and Fixation into Cyclic Carbonates. *ACS Sustain. Chem. Eng.* **2018**, *6*, 9324–9332. [CrossRef]
107. Isik, M.; Ruiperez, F.; Sardon, H.; Gonzalez, A.; Zulfiqar, S.; Mecerreyes, D. Innovative poly (ionic liquid) s by the polymerization of deep eutectic monomers. *Macromol. Rapid Commun.* **2016**, *37*, 1135–1142. [CrossRef]
108. Fairuzov, D.; Gerzeliev, I.; Maximov, A.; Naranov, E. Catalytic Dehydrogenation of Ethane: A Mini Review of Recent Advances and Perspective of Chemical Looping Technology. *Catalysts* **2021**, *11*, 833. [CrossRef]
109. Khokarale, S.; Shelke, G.; Mikkola, J.P. Integrated and Metal Free Synthesis of Dimethyl Carbonate and Glycidol from Glycerol Derived 1,3-Dichloro-2-propanol via CO_2 Capture. *Clean Technol.* **2021**, *3*, 685–698. [CrossRef]

Review

Recent Advances in the Mitigation of the Catalyst Deactivation of CO_2 Hydrogenation to Light Olefins

Daniel Weber [1,†], Tina He [2,†], Matthew Wong [3], Christian Moon [1], Axel Zhang [4], Nicole Foley [1], Nicholas J. Ramer [1,*] and Cheng Zhang [1,*]

1. Chemistry Department, Long Island University (Post), Brookville, NY 11548, USA; Daniel.weber4@my.liu.edu (D.W.); Christian.moon@my.liu.edu (C.M.); nicole.foley2@my.liu.edu (N.F.)
2. College of Natural Science, The University of Texas at Austin, Austin, TX 78712, USA; tinah00@utexas.edu
3. College of Engineering, Cornell University, Ithaca, NY 14850, USA; mcw243@cornell.edu
4. John Jay College of Criminal Justice, New York, NY 10019, USA; axel.zhang@jjay.cuny.edu
* Correspondence: nicholas.ramer@liu.edu (N.J.R.); cheng.zhang@liu.edu (C.Z.); Tel.: +01-516-299-3034 (N.J.R.); +01-516-299-2013 (C.Z.); Fax: +01-516-299-3944 (C.Z.)
† These authors contributed equally.

Abstract: The catalytic conversion of CO_2 to value-added chemicals and fuels has been long regarded as a promising approach to the mitigation of CO_2 emissions if green hydrogen is used. Light olefins, particularly ethylene and propylene, as building blocks for polymers and plastics, are currently produced primarily from CO_2-generating fossil resources. The identification of highly efficient catalysts with selective pathways for light olefin production from CO_2 is a high-reward goal, but it has serious technical challenges, such as low selectivity and catalyst deactivation. In this review, we first provide a brief summary of the two dominant reaction pathways (CO_2-Fischer-Tropsch and MeOH-mediated pathways), mechanistic insights, and catalytic materials for CO_2 hydrogenation to light olefins. Then, we list the main deactivation mechanisms caused by carbon deposition, water formation, phase transformation and metal sintering/agglomeration. Finally, we detail the recent progress on catalyst development for enhanced olefin yields and catalyst stability by the following catalyst functionalities: (1) the promoter effect, (2) the support effect, (3) the bifunctional composite catalyst effect, and (4) the structure effect. The main focus of this review is to provide a useful resource for researchers to correlate catalyst deactivation and the recent research effort on catalyst development for enhanced olefin yields and catalyst stability.

Keywords: CO_2 hydrogenation; light olefins; catalyst deactivation; CO_2-Fischer-Tropsch (CO_2-FT); iron-based catalysts; methanol to olefins; bifunctional composite catalysts; SAPO-34

Citation: Weber, D.; He, T.; Wong, M.; Moon, C.; Zhang, A.; Foley, N.; Ramer, N.J.; Zhang, C. Recent Advances in the Mitigation of the Catalyst Deactivation of CO_2 Hydrogenation to Light Olefins. *Catalysts* **2021**, *11*, 1447. https://doi.org/10.3390/catal11121447

Academic Editors: Javier Ereña and Ainara Ateka

Received: 10 November 2021
Accepted: 25 November 2021
Published: 28 November 2021

Publisher's Note: MDPI stays neutral with regard to jurisdictional claims in published maps and institutional affiliations.

Copyright: © 2021 by the authors. Licensee MDPI, Basel, Switzerland. This article is an open access article distributed under the terms and conditions of the Creative Commons Attribution (CC BY) license (https://creativecommons.org/licenses/by/4.0/).

1. Introduction

1.1. General Aspects

While carbon-rich fossil fuels like coal, oil, and natural gas have powered human civilization, the massive emission of CO_2 as a greenhouse gas has caused severe and harmful effects on the ecological environment [1]. For example, the rise of sea levels is accelerating, the number of large hurricanes and wildfires is growing, and dangerous heat waves and more severe droughts are occurring in many areas. The CO_2 concentration in the atmosphere had climbed to 415 ppm by 2020 (Figure 1), an increase of more than 40% relative to the pre-industrial era [2]. The atmospheric CO_2 concentration will continue to rise to ~570 ppm by the end of the 21st century if no alleviation measures are taken [3]. Therefore, there is an urgent need to control CO_2 emissions in order to mitigate their negative impact on the environment. In recent years, capture and storage technologies for the CO_2 released from the burning of fossil fuels have emerged and developed in potential commercial scale applications [4–7]. In order to close the carbon gap, transforming the captured gas into value-added fuels and chemicals has become an urgent task for CO_2 remediation [8,9].

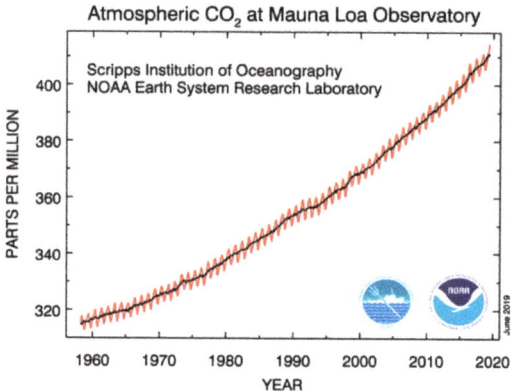

Figure 1. Trends in the atmospheric CO_2 concentration (ppm) [2].

The catalytic conversion of CO_2 is a favorable approach to the mitigation of CO_2 emissions by producing chemicals and fuels [8,10–17]. Light olefins such as ethylene, propylene and butylene ($C_2^=-C_4^=$), which are currently among the top petrochemicals, are the building blocks for the production of a wide variety of polymers, plastics, solvents, and cosmetics [8,13,18–21]. Moreover, light olefins can be oligomerized into long-chain hydrocarbons which can be used as fuels, making them a desirable product with high potential for the utilization—and therefore elimination—of up to 23% of CO_2 emissions [8]. A highly promising route is selective CO_2 hydrogenation to produce light olefins [10]. The huge market demand for the lower olefins offers a great opportunity for the target technology to profoundly impact the scale of CO_2 utilization once it is developed with renewable hydrogen. The current chemical industry relies heavily on petroleum (the steam cracking of naphtha) for the production of light olefins [22]. The depletion or movement away from the refining of petroleum and the gap between the supply and demand of light olefins call for a new strategy to synthesize light olefins from alternative carbon sources [18–21,23,24]. A one-step process for the conversion of CO_2 to light olefins is a highly desirable tactic to address the "3Rs" (reduce, reuse, and recycle) associated with ever-increasing CO_2 levels, and to solve the paradox between the supply and demand of light olefins [25].

Currently, there are two primary pathways, as shown in Scheme 1, to produce light olefins from CO_2 reduction by hydrogen (H_2) in a one-step process: (1) the CO_2 Fischer–Tropsch synthesis (CO_2–FTS) route consists of two consecutive processes, the reverse water–gas shift (RWGS) reaction (Equation (1)) and subsequent Fischer–Tropsch synthesis (FTS) (Equation (2)); (2) the methanol (MeOH) mediated route consists of two consecutive processes, i.e., CO_2-to-MeOH (Equation (3)) and a subsequent MeOH-to-olefins process (MTO) (Equation (4)). The complex reaction network in Scheme 2 indicates the competing reactions (i.e., Equation (5)) with the formation of light olefins. The control of the selectivity of the CO_2 hydrogenation to the desired olefin product requires the design of catalysts for reaction pathways that are compatible with favorable thermodynamics and a good understanding of the reaction kinetics [26]. The thermodynamic values in the equations (Equations (1)–(5)) indicate that lower temperatures favor FTS (Equation (2)), methanol (Equation (3)), and methane synthesis (Equation (5)), while higher temperatures are needed to activate CO_2 (Equation (1)) for rapid reaction rates [27]. The complex reaction network in Scheme 2 and thermodynamics suggest that the design and synthesis of catalysts for a one-step process to selectively produce olefins are challenging.

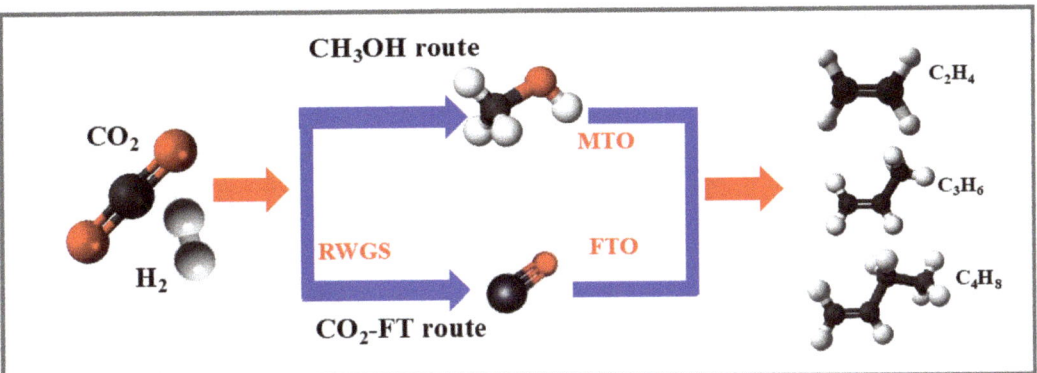

Scheme 1. Reaction route for CO_2 hydrogenation to light olefins.

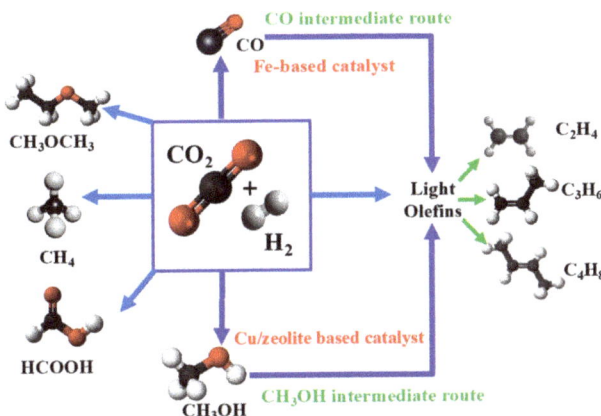

Scheme 2. Complex reaction network for CO_2 conversion to chemicals through hydrogenation.

CO_2–FTS reaction pathway:
Reverse water-gas shift reaction (RWGS):

$$CO_2 + H_2 \rightarrow CO + H_2O \qquad \triangle H_0^{298} = 41.1 \text{ kJ mol}^{-1} \qquad (1)$$

Fischer-Tropsch synthesis to olefins (FTS):

$$nCO + 2nH_2 \rightarrow (CH_2)_n + nH_2O \qquad \triangle H_0^{298} = -210.2 \text{ kJ mol}^{-1} \ (n=2) \qquad (2)$$

Methanol mediated reaction pathway:
Methanol synthesis:

$$CO_2 + 3H_2 \rightarrow CH_3OH + H_2O \qquad \triangle H_0^{298} = -49.3 \text{ kJ mol}^{-1} \qquad (3)$$

Methanol to olefins (MTO):

$$nCH_3OH \rightarrow (CH_2)_n + H_2O \qquad \triangle H_0^{298} = -29.3 \text{ kJ mol}^{-1} \ (n=2) \qquad (4)$$

CO_2 methanation:

$$CO_2 + 4H_2 \rightarrow CH_4 + 2H_2O \qquad \triangle H_0^{298} = -165.0 \text{ kJ mol}^{-1} \qquad (5)$$

1.2. Mechanistic Insights for CO_2 Conversion to Light Olefins

In reviewing the mechanistic details of the light olefin formation, it is clear that controlling the active H to C ratio is of primary importance. The presence of too much H* on the surface will result in excessive hydrogenation, and therefore methanation, while too little H* on the surface will restrict the hydrogenation ability of the catalyst and therefore reduce the CO_2 conversion activity. At its most fundamental, the pivotal steps of CO_2 conversion to light olefins are the cleavage of the C–O bonds and the formation of C–C bonds [25].

Iron-based catalysts have been extensively studied for use in the CO_2–FTS route due to their relatively high utility and activity for both the RWGS and FTS component reactions. When using Fe-based catalysts for CO_2–FTS, the initial Fe_2O_3 phase is reduced by hydrogen to Fe_3O_4 or a mixture of Fe_3O_4 and FeO. The resulting Fe_3O_4 is the active component for the RWGS reaction, and can be further reduced to form metallic Fe [27]. The reaction mechanism for the CO_2–FTS pathway is suggested as shown in Scheme 3a. CO_2 is first adsorbed and activated on the RWGS active phases (e.g., Fe_3O_4) to form a carboxylate (*CO_2, * representing the adsorption state). The *CO_2 can then be hydrogenated by adsorbed H to form an *HOCO intermediate. The intermediate then dissociates into *OH and *CO. The *OH is then hydrogenated into *H_2O. Then, *CO either desorbs as CO gas or reacts further via successive FTS. In order to form hydrocarbons, the *CO is first partially hydrogenated into *HCO and then undergoes complete hydrogenation, dissociation, and finally dehydration to form *CH_x species. The *CH_x species are precursors for the formation of olefins. In an alternative mechanism, *CO can dissociate into *C and *O. Some *C can diffuse into the Fe-metal lattice to form metal carbides as χ-Fe_5C_2, the active component for the FTS reaction [27]. The C* on the χ-Fe_5C_2 surface can then be hydrogenated to CH_x* species. C* + CH_x* and CH_x* + CH_x* were the most likely coupling pathways [25].

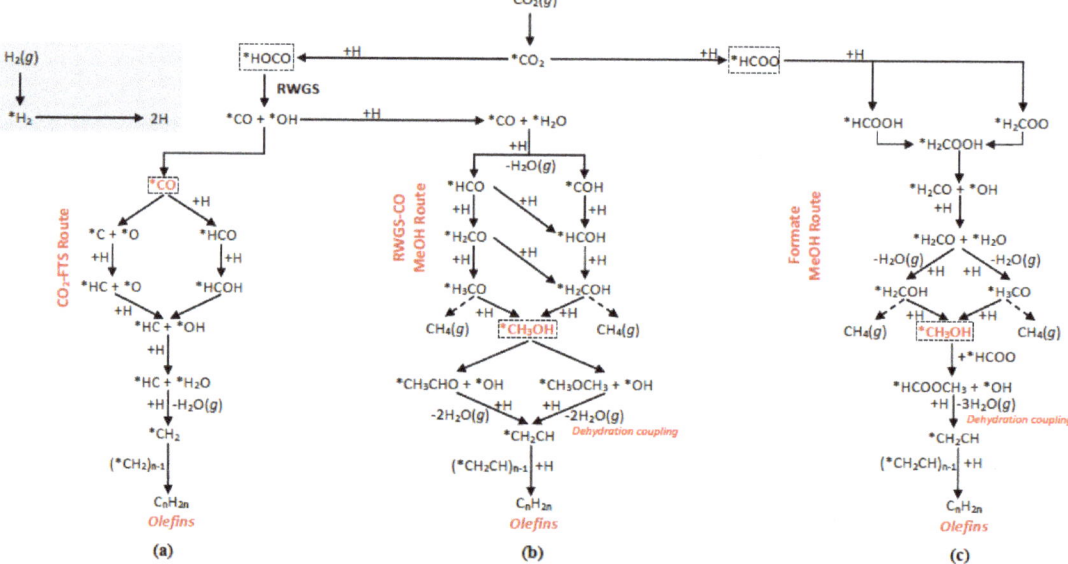

Scheme 3. (a–c) Reaction mechanism for CO_2 hydrogenation to light olefins (modified and adapted with permission from ref. [27]. Copyright 2021 Elsevier).

As indicated above, the *C from the dissociation of *CO during the FTS reaction may diffuse into the α-Fe metal lattice, resulting in the formation of Fe_7C_3, χ-Fe_5C_2, θ-Fe_3C, ε′-$Fe_{2.2}C$, and ε-Fe_2C phases, depending on reaction conditions [27]. Iron carbides play an essential role in CO hydrogenation/dissociation and C–C coupling. Some researchers have proposed that χ-Fe_5C_2 is the active phase, while θ-Fe_3C is less active and can cause catalyst deactivation due to production of graphite, which has increased stability under typical FTS reaction conditions and may block the production of other active phases [27,28].

Alternatively, the reaction mechanism for the MeOH pathway is suggested as shown in Scheme 3b,c. The synthesis of MeOH can proceed via two pathways: (1) CO-mediated, in which the *CO intermediate, which was produced from the RWGS reaction via the dissociation of the carboxyl (*HOCO) species, is hydrogenated to methanol via *HCO and *COH, and (2) formate-mediated, in which the formate (*HCOO) species results from the hydrogenation of the carboxylate intermediate (*CO$_2$), which is then reacted further to *H$_2$COOH, *H$_2$CO, *H$_2$COH, and *H$_3$CO. Through dehydration coupling, the methanol forms *CH$_2$CH, and then forms olefins via subsequent hydrogenation [27].

The factors that may affect the CO$_2$ conversion and light olefin selectivity are the catalyst composition (metals, supports, promotors, etc.), functionality (i.e., metal/zeolite bifunctionality), structure (i.e., layered metal oxide, core–shell, etc.), preparation methods (e.g., impregnation, hydrothermal, sol-gel, etc.) and testing conditions (e.g., temperature, pressure, CO$_2$/H$_2$ molar ratio, gas hourly space velocity, etc.). The focus of this review will be on the catalyst composition, functionality and structure. Other factors of catalyst preparation methods and testing conditions for CO$_2$ conversion to light olefins can be found elsewhere [15,20,27,29,30].

1.3. Catalysts for CO$_2$ Conversion to Light Olefins

As can be seen above, because each route has its own unique pathway of species and intermediates, different catalysts must be employed for the hydrogenation of CO$_2$ to olefins depending on the chosen route. In the CO$_2$–FTS path, Fe is one of the most widely used components in the catalysts, as catalysts containing Fe offer less methanation activity under higher reaction temperatures. As described above, it has been reported that Fe$_3$O$_4$ was the active phase responsible for RWGS; the metallic Fe and iron carbides could activate CO and produce hydrocarbons [31,32]. When incorporating alkali promoters, Fe-based catalysts showed greater olefin selectivity. The alkali metals, acting as electron donors to the Fe metal, facilitate the adsorption of CO$_2$ while lowering the affinity for H$_2$. The net result is a higher olefin yield [33–35]. There is also some indication that doping the catalyst with an additional metal may promote even higher olefin yields by forming a highly active interface. The second metal components allow for even greater adjustment of the CO$_2$ and H$_2$ adsorption and activation, shifting the distribution of the product more towards the desired hydrocarbons. By supporting the Fe-based catalysts on supports such as silica (SiO$_2$), alumina (Al$_2$O$_3$), titania (TiO$_2$), zirconia (ZrO$_2$) and carbon materials (i.e., carbon nanotubes (CNTs), carbon nanospheres (CNSs), graphene oxide (GO)), the catalytic performance may be further enhanced by improving the active metal dispersion and slowing down the sintering of the active particles [36,37]. Controlling hydrocarbon chain growth to achieve a desired carbon range (i.e., C$_2$–C$_4$) remains a challenge for CO$_2$ conversion due to the product selectivity limit governed by the Anderson–Schulz–Flory (ASF) distribution with a maximum achievable C$_2$–C$_4$ hydrocarbons selectivity of less than 60%, as shown in Figure 2 [30,38].

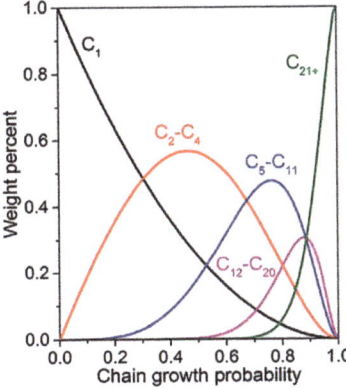

Figure 2. Product distribution as predicted by the Anderson–Schulz–Flory (ASF) model. Adapted with permission from ref. [30]. Copyright 2021 Elsevier.

Alternatively, in the MeOH path, light olefins can be synthesized with selectivity as high as 80–90% among hydrocarbons, exceeding the ASF product distribution limit for FTS reactions [18,39–41]. Some plausible reasons for the reported ASF distribution deviation are the blockage of surface polymerization by intermediates, (e.g., ketene (CH_2CO)), space confinement, or the use of catalysts with two types of active sites (i.e., bifunctional catalysts) [27]. Regardless of the reasons, the observed deviation from the ASF distribution offers opportunities to increase the selectivity to olefins [27]. Several recent studies have reported the results for the combination of MeOH synthesis catalysts (i.e., In_2O_3, In-Zr, $ZnGa_2O_4$, $MgGa_2O_4$, $ZnAl_2O_4$, $MgAl_2O_4$, ZnZrO and In_2O_3-$ZnZrO_2$) with an MTO catalyst (i.e., SAPO-34, SSZ-13 and ZSM-5), and their ability to produce light olefins with enhanced selectivity for CO_2 hydrogenation [13,42–44]. It has been proposed that the secondary functionality of acid–base sites on the catalytic support significantly impacts the light olefin selectivity. For example, by passivating the Brønsted acid sites of In_2O_3-ZnZrOx/SAPO-34, the secondary hydrogenation reaction is inhibited, thereby improving the olefin selectivity [27,30].

1.4. The Main Focus of This Review

Even though significant efforts have been made, considerable challenges remain in the development of highly efficient catalysts with selective pathways to light olefins due to the thermodynamically stable nature of the CO_2 molecule, the complexity of the reaction networks, and catalyst deactivation [8,45,46]. Several recent reviews have summarized CO_2 hydrogenation to value-added products, including light olefins [25,27,30,38]. However, it is necessary to present a review focused on the recent advances in the mitigation of the catalyst deactivation of CO_2 hydrogenation to light olefins, as catalyst deactivation has been a big challenge that provides economic hurdles to the adoption of the new technologies.

Because catalysts and mechanisms have been extensively reviewed in numerous review papers [25,27,30,38], the focuses of the current article are to identify possible causes that trigger catalyst deactivation and summarize recent advances on catalyst development with enhanced catalyst stability and light olefin selectivity for CO_2 hydrogenation. In this review, we first provide a brief summary of the two dominant reaction pathways (CO_2–FTS and MeOH-mediated), mechanistic insights and catalytic materials for CO_2 hydrogenation to light olefins. We then list the deactivation mechanism caused by carbon deposition, water formation, phase transformation and metal sintering/agglomeration. Finally, we summarize the recent progress published within five years on catalyst development that improves catalyst deactivation by the following catalyst functionalities: (1) the promoter effect, (2) the support effect, (3) the hybrid functional effect, and (4) the structure effect.

Each one of these aspects is accompanied by a suitable table in which the most significant literature findings are comparatively presented. To the best of our knowledge, no review has ever directly correlated the causes of catalyst deactivation and catalyst mitigation for CO_2 hydrogenation to light olefins. Herein, we attempt to provide a useful resource for researchers to correlate the catalyst deactivation and the recent research effort on catalyst development for enhanced olefin yield and catalyst stability.

2. Causes of Catalyst Deactivation

During CO_2 hydrogenation, catalyst deactivation can occur via several mechanisms, resulting in decreased activity and selectivity toward the desired olefins. The determination of the mechanism of deactivation is an important step toward mitigation. The primary causes of catalyst deactivation are the sintering (or agglomeration) of metal particles, phase transformation at the catalyst's surface, and catalyst poisoning by water or carbonaceous deposits (i.e., coke). An understanding of the deactivation causes is necessary to develop a mitigation strategy and sustain high selectivity toward the desired olefins during CO_2 hydrogenation. For context, we present brief descriptions of each of these causes with a few representative examples from the literature that demonstrate the necessity of robust and novel mitigation studies. More thorough reviews of the deactivation causes and their mechanisms can be found elsewhere [47–49].

2.1. Sintering

Catalyst sintering can occur through either Ostwald ripening or particle migration and coalescence, as shown in Figure 3 [50]. Through sintering, the agglomeration of smaller catalyst crystals into larger ones will bring about the loss of the pore structure, which lowers the internal surface area of the catalyst, leading to the deactivation. In the area of FT by cobalt catalysts, several groups have determined that the particle growth of cobalt is the largest factor causing deactivation [51,52].

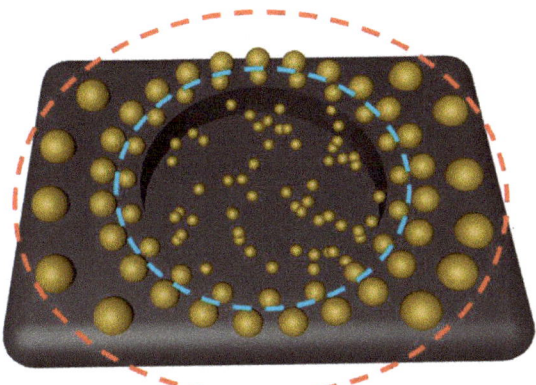

Figure 3. Diagram of active phase sintering occurring over a support material: the blue ring represents atomic migration to form larger crystallites; the red ring represents the coalescence of crystallites. Adapted with permission from ref. [50]. Copyright 2021 Elsevier.

Sun et al. [53] examined sintering in zinc- and alumina-supported copper catalysts ($Cu/ZnO/Al_2O_3$). It was found that the presence of CO in the process employed for CH_3OH synthesis strongly contributed to the deactivation of the catalysis over 0 to 50 h. Taken with corroborative evidence from the Cu surface area determination, the deactivation was likely attributed to the sintering of the Cu metal.

As mentioned above, sintering negatively affects the catalytic performance due to many reasons: for example, the overall catalytically active surface area is reduced due to the collapse of the structure and the chemical alteration of the catalytically active phases to non-active phases [50,54,55]. As this form of deactivation involves the coalescence of larger particles from smaller, it is extremely difficult to reverse. Sintering, therefore, is easier to prevent through careful catalyst design [50,56]. For example, Li et al. observed remarkable metal sintering on supported $FeCo/ZrO_2$ catalysts [56]. As shown in Figure 4A(a), for the $13Fe2Co/ZrO_2$ supported catalyst precursor prepared using the conventional impregnation method, the Co and Fe are distributed into separate oxide particles, which increased the possibility of sintering. As confirmed in Figure 4A(b,c), aggregates composed of Fe and Co oxide nanoparticles were observed on the ZrO_2 fibers, with an average diameter of ca. 15 nm before the reaction. The particle size increased to 48 nm after the reaction, which was responsible for the rapid deactivation of activity (Figure 4A(d–f)). By comparison, Fe-Co-Zr polymetallic fibers obtained via a one-step electrospinning technique showed that Fe and Co were dispersed in proximity to ZrO_2, as shown in Figure 4C(a), but separately from each other. In order to reduce the possibility of sintering, as demonstrated in Figure 4B(a–f), the Fe and Co oxides nanoparticles successfully dispersed with the ZrO_2 particles for the polymetallic oxide fibers, with an average size of roughly 1–2 nm before the reaction, and after the reaction, the particle size barely changed, which contributed to the stable catalytic activity after 500 mins on stream (Figure 4C(a,b)).

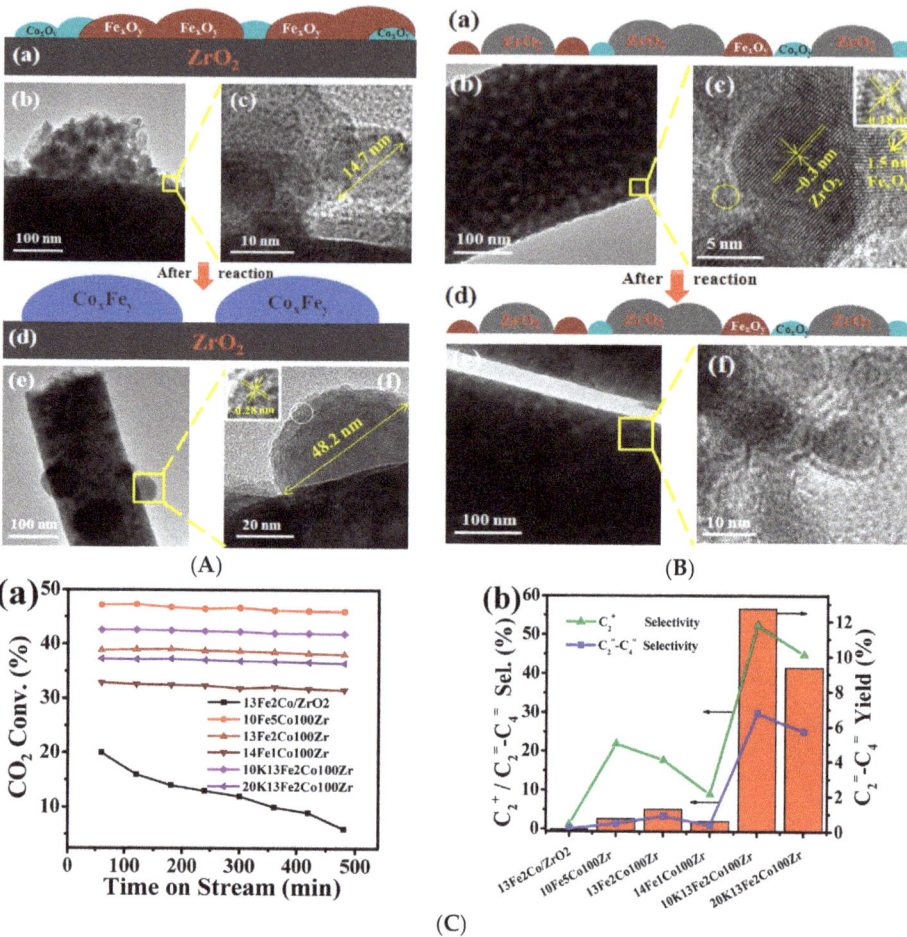

Figure 4. (**A**) (**a–f**) Schematic illustration of the metal distribution and TEM images of the 13Fe2Co/ZrO$_2$-supported catalyst precursor (**a–c**) and the spent catalyst (**d–f**). (**B**) (**a–f**) Schematic illustration of the metal distribution and TEM images of the 13Fe2Co100Zr polymetallic oxide fiber catalyst (**a–c**) and the spent catalyst (**d–f**). (**C**) CO$_2$ conversion (**a**) and the C$_{2+}$/C$_2^=$–C$_4^=$ selectivity and C$_2^=$–C$_4^=$ yield (**b**) over different catalysts after 8 h TOS (testing conditions: H$_2$/CO$_2$ molar ratio = 3/1, GHSV = 7200 mL g^{-1} h^{-1}, P = 3 MPa, T = 673 K). Adapted with permission from ref. [56]. Copyright 2019 Elsevier.

2.2. Phase Transformations

Phase transformations are processes of deactivation involving the conversion of an active crystalline phase of the catalyst (or one of its components) into a different inactive one. These transformations can involve both metal-supported and metal-oxide catalysts. In the former type of catalyst, atoms from the catalyst's support will diffuse into the catalyst's surface. A reaction at the surface can then result in an inactive phase, deactivating the catalyst.

Riedel et al. was able to demonstrate that the steady states of the synthesis of hydrocarbons using iron oxides could be separated into five episodes of distinct kinetic regimes. In episode I, the adsorption of the reactants takes place on the catalyst surface and carbonization occurs. During episodes II and III, products from the RWGS reaction dominate during ongoing carbon deposition. In episode IV, the rate of FT activity increases up to the steady state, and the maintenance of the steady state occurs in episode V. Prior to the reaction, the iron phases of the reduced catalyst are mainly α-Fe and Fe_3O_4, along with a small amount of Fe_2O_3. As the process proceeds, the Fe_3O_4 and Fe_2O_3 phases are consumed and a new oxidic iron amorphous phase is formed, which appears to be active for the RWGS reaction. Through a reaction of iron with carbon from the CO dissociation, FTS activity commences with the formation of iron carbide (Fe_5C_2). Upon the formation of the stable but inactive carbide (Fe_3C), which is the result of Fe_5C_2 carburization, the catalyst begins deactivating [57–59]. Lee et al. studied the causes of the deactivation of Fe–K/γ-Al_2O_3 for CO_2 hydrogenation to hydrocarbons, and found the causes for deactivation varied based on positioning inside the reactor. Over time, the Fe_2O_3 was reduced to active phase χ-Fe_5C_3, and then the χ-Fe_5C_3 was transformed to θ-FeC_3, a form which is not active for CO_2 hydrogenation. The primary reason for deactivation was the phase transformation at the top of the reactor. Conversely, at the bottom of the reactor, deactivation was largely the result of deposited coke generated by secondary reactions [57].

Zhang et al. reported the structure evolution of the iron catalyst during its full catalytic life cycle of CO_2 to olefins (CTO), including the catalyst activation, reaction/deactivation (120 h) and regeneration. The phase transition during the CO activation was observed to follow the sequence of $Fe_2O_3 \rightarrow Fe_3O_4 \rightarrow Fe \rightarrow Fe_5C_2$. The primary deactivation mechanism during CTO was identified as the irreversible transition of iron phases under reaction conditions. Two possible pathways of the phase transition of the iron catalyst under CTO conditions have been identified, i.e., $Fe_5C_2 \rightarrow Fe_3O_4$ and $Fe_5C_2 \rightarrow Fe_3C \rightarrow Fe_3O_4$. Moreover, carbon deposition and the agglomeration of the catalyst particle proves to have relatively minor impacts on the catalytic activity compared with phase transition during the 120 h of reaction [60].

It appears that transformation to iron oxides will destroy catalyst activity. There is some question as to whether the cementite phase itself is problematic to activity, as higher-surface-area cementite phases have been reported to perform CO_2 hydrogenation quite effectively [61].

2.3. Poisoning

Catalytic poisoning is a result of the strong chemisorption of reactants, products or impurities on sites that would otherwise be capable of catalysis. In essence, the poisoning ability of a particular species is related to the strength of its chemisorption to the catalysis relative to the other reactants that are competing for the catalytic active sites. The poisoning has two deactivating effects; a poison physically blocks the active sites from receiving additional reactants, and a poison can alter the electronic or structural properties of the catalytic surface, rendering it partially or completely ineffective toward catalysis [62,63].

While there are several different poisons which have been reported in the literature that have shown to deactivate CO_2 hydrogenation catalysts [64,65], we will focus on the two which are the most pervasive, namely water and carbonaceous deposits or coke.

2.3.1. Water Poisoning

As seen in the above CO_2 hydrogenation reactions, the dissociation of CO_2 produces oxygen atoms, which in turn results in the formation of water. This byproduct is necessary for the thermodynamic favorability of the entire process, but can be unfortunately detrimental to catalytic performance. It is because of this unavoidable mechanistic absolute that the mitigation of water poisoning must be part of all catalytic investigations [66].

Wu et al. [67,68] examined the effect of the produced water on the stability of Cu/ZnO-based catalysts in methanol synthesis from the high temperature hydrogenation of CO_2. Specifically, Cu/ZnO/ZrO_2/Al_2O_3 (40/30/25/5) was subjected to a CO_2-rich feed, which produces water, and a CO-rich feed, which does not produce water. The examination of the catalysts by X-ray powder diffraction (XRD) after 1 h and 500 h time-on-stream of a CO_2-rich feed containing steam showed that the Cu and ZnO crystallized more rapidly when compared to identical catalysts exposed to a CO-rich feed not containing steam.

In particular, the Cu particle size in the catalyst used with the CO_2-rich feed containing steam grew from 94 Å to 166 Å from 1 h to 500 h. The particle size growth under steam might be the key reason causing catalyst deactivation.

Huber et al. observed the rapid deactivation of Co/SiO_2 during an FTS reaction at high water partial pressure, and the loss of activity was attributed to the support breakdown byproduct water accompanied by the formation of stable, inactive cobalt-silicates and the loss of the BET surface area [69]. van Steen et al. stated that metallic cobalt crystallites with a diameter less than 4.4 nm are more susceptible to oxidation by water to form Co(II)O [70]. This is in agreement with Iglesia's work showing that small Co metal crystallites (<5–6 nm diameter) appear to re-oxidize and deactivate rapidly in the presence of a water reaction product in typical FTS conditions [71].

Water poisoning has the most dramatic effect on zeolite-based CO_2 hydrogenation catalysts for which the acidic sites of the zeolite are essential for catalysis. Recently, Zhang et al. investigated the water effect over zeolite-based catalysts at high temperatures, and found that water caused the loss of crystallinity and modified acid sites, thereby deactivating the catalyst [72]. Their studies show, by functionalization with organosilanes, that the tolerance of defective zeolites to hot liquid water can be greatly enhanced. This method renders the zeolite hydrophobic, which prevents the wetting of the surface. At the same time, the organosilanes act as a capping agent of Si−OH species, reducing their reactivity. Both aspects are important for the prevention of water attack [72].

It appears that there are several analogies of Fe catalysts for CO_2 hydrogenation and CO catalysts for conventional FT synthesis. Kliewer et al., for example, showed that for a supported CO catalyst, water can oxidize the surface of the CO to an inactive oxide phase, and it also plays a large role in sintering. With a high water partial pressure in the Fe system, it appears that this can also oxidize iron carbides to inactive surface oxide phases and also promotes particle growth sintering [51].

2.3.2. Carbonaceous Deposits (Coke)

Coke is produced by the decomposition or condensation of hydrocarbons on the surfaces of catalysts, and is primarily is comprised of polymerized hydrocarbons. There have been several books and reviews that describe the formation of coke on catalysts, and the resulting deactivation [73–78].

These deposits are most problematic for catalysis involving zeolites, because the active sites of the zeolites become blocked or fouled by the coke deposits. The deactivation of MTO reactions over zeolites due to coke deposition results in a reduction in both the catalyst activity and product selectivity [79–81].

Nishiyama et al. [82] studied the effect of the SAPO-34 crystal size on the catalyst lifetime, and found that the amount of coke deposited on the deactivated SAPO-34 catalyst increased with the decreasing crystal size, indicating that for larger crystals, the reactants were unable to penetrate further into the larger crystals to reach other acidic sites. Because MTO reactions and coke formation take place simultaneously in the same pores, it seems likely that the effectiveness of the catalyst increased with the decreasing crystal size. Their studies demonstrated that the coke formation was inhibited in small-crystal SAPO-34 due to reduced diffusive resistance.

The work of Wei et al. on CO_2 hydrogenation found that the deactivation of the zeolites HMCM-22 and HBeta was the result of coke formation, which deposited in the zeolites' cavities and channels. The deposition blocked the reactants' access to the zeolites' acid sites, leading to the deactivation [11]. Muller et al. investigated the MTO process on H-ZSM-5 catalysts in plug-flow (PFR) and fully back-mixed reactors (CSTR). They found that the catalysts deactivated under the homogeneous gas phase in the CSTR. It was shown unequivocally that, in the early stages of the reaction, the zeolite deactivates via Brønsted acid site blocking, and not by coke-induced deposition restricting the pore access. The deactivation of H-ZSM-5 in the CSTR occurred at first rapidly, and then at a much slower rate (Figure 5). The rapid deactivation was observed in a PFR due to the formation of a larger fraction of the oxygen-containing carbon species. The larger fraction of oxygen-containing carbon species increases the reaction with the desired olefins, which results in a strongly adsorbed aromatic molecule. The formation of aromatic coke proceeds mostly by hydride transfer between olefins and carbon growth via multiple methylations of such aromatic species [83].

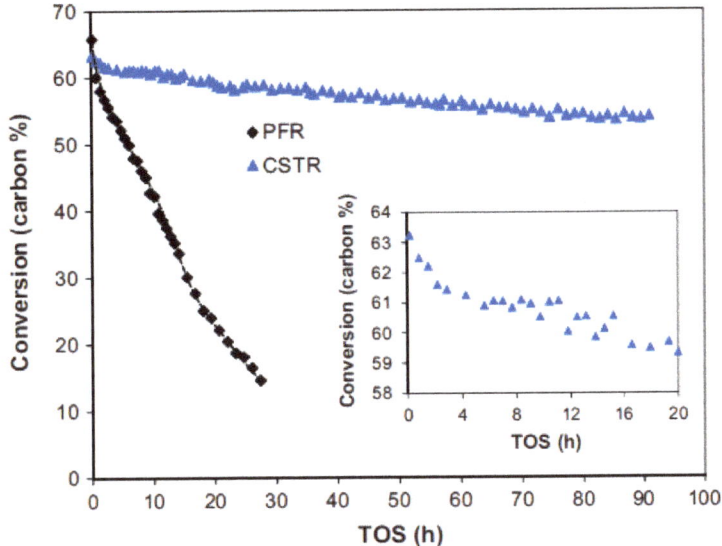

Figure 5. MTO reaction over the H-ZSM-5-S catalyst in the PFR at ambient pressure and the H-ZSM-5-S catalyst in the CSTR at 6.5 bar, T = 723 K and p_{MeOH} = 178 mbar. Adapted with permission from ref. [83]. Copyright 2015 Elsevier.

Zeolite-based catalysts that show promise for high olefin selectivity are unfortunately typically limited by mass transfer, suffering from rapid deactivation due to carbon deposition and water poisoning [83]. The issues with coke deactivation on the zeolite catalysts involved in MTO reactions are seen in classical MTO chemistry. The directed transformation of coke into active intermediates in a methanol-to-olefins catalyst was reported to boost the light olefin selectivity [84]. Another strategy to mitigate the deactivation was to synthesize nanozeolites, which have shortened diffusion paths, or mesoporous hierarchical zeolites, which exhibit longer catalyst lifetimes because of their larger pores and improved mass transfer [85–87].

With an Fe catalyst, the deactivation by coke is not related to the constriction of narrow pores, but several authors have reported the formation of carbonaceous residues on the active sites. Lee et al. investigated the deactivation behavior of an Fe–K/-Al$_2$O$_3$ catalyst, and found that the deactivation pathway was different according to the reaction position and reaction time. The main deactivation reason was the phase transformation at the top of the reactor. Conversely, the main factor at the bottom of the reactor was the deposited coke generated by secondary reactions. In particular, the produced olefins may have been adsorbed on acidic sites, and thus the olefins served as major precursors to coke. The SEM micrographs of the used catalysts clearly showed that most of the surface was covered by deposited graphite and graphite clusters protruding on the surface, mixed with some fine filamentous carbon (Figure 6) [57].

With these multiple deactivation pathways having been identified, it now becomes a critical issue to find ways of modifying the catalyst to become more stable. In the section directly below, we will describe the approaches that multiple researchers have examined in an attempt to mitigate deactivation.

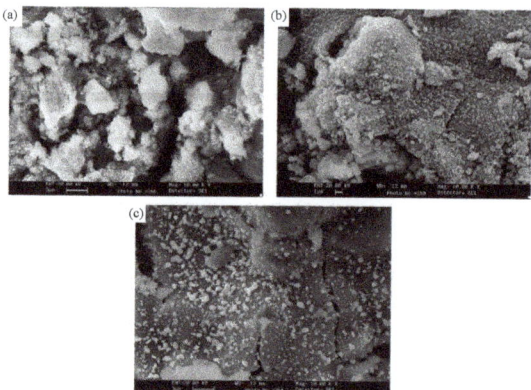

Figure 6. SEM image of the Fe–K/γ-Al$_2$O$_3$ catalysts after the CO$_2$ hydrogenation reaction: (**a**) 100 h, (**b**) 300 h, and (**c**) 500 h. Adapted with permission from ref. [57]. Copyright 2009 Elsevier.

3. Recent Progress on the Mitigation of the Catalyst Deactivation

We will discuss the effect of promotors, metal oxide support, bifunctional composition and structure on the catalyst design to minimize the catalyst deactivation. We will summarize reports published within the last five years showing that promoters, supports, and novel morphology designs have mitigated the deactivation effects.

3.1. Promoter Effect

Fe-based catalysts have been widely studied in CO$_2$ hydrogenation, and usually show unsatisfactory selectivity toward lower olefins. The addition of suitable promotors to increase the yield of light olefins and the stability of the catalysts by controlling the electronic and structural properties have been extensively studied. Alkali metals such as K and Na have been broadly used as promotors to control the electronic properties. Mn, Ce, Ca metals have been used as structural promotors. Transition metals such as Zn, Co, Cu, V, Zr, etc., have been used as both electronic and structural promotors. Some representative catalysts on the promoter effect for CO$_2$ hydrogenation to light olefins with improved catalyst stability are presented in Table 1.

Table 1. Some representative catalysts on promoter effect for CO$_2$ hydrogenation to light olefins.

Catalyst	CO$_2$ Conv., %	Selectivity, %				Yield, %	O/P Ratio	Stability	Ref.
		CO	CH$_4$	C$_2$–C$_4$	C$_2$–C$_4^0$	C$_2$–C$_4^=$			
10Mn-Fe$_3$O$_4$	44.7	9.4	22.0	46.2	7.1	18.7	6.5	24 h	[88]
0.58% Zn-Fe-Co/K-Al$_2$O$_3$	57.8	8.8	6.2	63.2	21.8	19.9	2.9	50 h	[89]
Na-Zn-Fe	38.0	15.0	13.0	42.0	4.9	16.0	8.5	100 h	[90]
Na-CoFe$_2$O$_4$	41.8	10.0	~18.0	37.2	~7.0	15.5	~5.3	100 h	[91]
Fe-Co/K-Al$_2$O$_3$	40.0	12.2	24.8	46.1	7.9	16.2	5.9	6 h	[33]
0.5%Na-Fe$_5$C$_2$	35.3	13.2	31.8	57.0	10.1	20.1	5.7	10 h	[92]
Fe-Zn-2Na	43.0	15.7	22.8	54.1	7.4	23.2	7.3	10 h	[93]
Fe/C-KHCO$_3$	33.0	20.8	12.7	59.8 [a]	27.3	9.0	2.2	100 h	[94]
5Mn-Na/Fe	38.6	11.7	11.8	30.2	4.0	11.7	11.0	10 h	[95]
FeNa(1.18)	40.5	13.5	15.8	46.6	7.5	15.7	6.2	60 h	[31]
Na/Fe-Zn	30.6	n/a	13.0	26.8	3.9	8.4	6.9	200 h	[96]
Fe/C-Bio	31.0	23.2	11.8	21.7	24.4	6.7	0.9	6 h	[97]
5%Na/Fe$_3$O$_4$	36.8	~11.0	~5.0	64.3	~13.0	23.7	~4.9	10 h	[98]
Fe/C+K(0.75)	40.0	~16.0	~22.0	~39.0	~12.0	~15.6	~3.3	50 h	[99]
35Fe-7Zr-1Ce-K	57.3	3.05	20.6	55.6	7.9	31.8	7.1	84 h	[100]
Fe-Mn/K-Al$_2$O$_3$	29.4	20.2	18.7	48.7	6.5	14.3	7.4	>6 h	[101]
Fe-Cu(0.17)/K(1.0)	29.3	17.0	7.0	63.8	12.2	19.1	5.2	50 h	[102]
Na-CoFe$_2$O$_4$/CNT	34.4	19.0	~5.0	38.8	18.0	13.3	12.9	>24 h	[103]
Fe-Co-K(0.3)/TiO$_2$	21.2	54.0	9.0	37.0 [b]	n/a	n/a	4.1	18 h	[104]
Fe$_2$Zn$_1$	35.0	~15.0	~20.0	57.8 [c]	7.2	20.2	8.0	200 h	[28]
ZnCo$_{0.5}$Fe$_{1.5}$O$_4$	49.6	~7.5	~17.5	36.1	~10.0	17.9	~3.6	80 h	[105]

[a] refers to high valued olefins (HVO); [b] includes C$_2$-C$_4$ and C$_{5+}$ hydrocarbons; [c] refer to C$_2$–C$_7^=$ olefins.

Adding alkali metals (i.e., Na, K) could increase the selectivity towards light olefins due to the enhanced CO_2 adsorption on the more electron-rich Fe phases and suppressed H_2 chemisorption, which inhibits olefin re-adsorption. Numpilai et al. reported on the effect of varying the content of the K promoter on the Fe-Co/K-Al$_2$O$_3$ catalysts via the CO_2–FTS reaction pathway. Unpromoted catalysts evidenced low-light olefin yields when compared to K-promoted ones with an ascending K/Fe ratio from 0 to 2.5. The maximum light olefin ($C_2^=$–$C_4^=$) distribution of 46.7% and O/P ratio of 7.6 were achieved over the catalyst promoted with a K/Fe atomic ratio of 2.5. The positive effect of K's addition is attributed to the strong interaction of H adsorbed with the catalyst surface caused by the electron donor from K to Fe species. This notion is also rationalized by the fact that the K promoter enhances the bond strength of absorbed CO_2 and H_2, retarding the hydrogenation of olefins to paraffins. In the same operating conditions, the catalyst promoted with a K/Fe atomic ratio of 0.5 provides the maximum light olefin ($C_2^=$–$C_4^=$) yield of 16.4%, which is significantly higher than that of 2.5 KFe catalysts (13.4%). This is explained by the K enriched surface of 2.5 KFe catalysts significantly reducing the BET surface area and generating a hydrogen-lean environment, ultimately lessening the catalytic activity [101].

A different promoter source plays an important role to affect catalytic CO_2 hydrogenation. Han et al. demonstrated that as the series of K-promoters changes from K_2CO_3, CH_3COOK, $KHCO_3$, and KOH, the electron transfer from potassium to iron species is facilitated, which forms a more active and distinct χ-Fe$_5$C$_2$-K$_2$CO$_3$ interface during CO_2 hydrogenation. This results in a higher selectivity to light olefins (75%) and a higher CO_2 conversion (32%). In contrast, the non-carbonaceous K-promoters do not facilitate iron species to form iron carbides, which causes an undesirable catalytic performance (Figure 7a). Additionally, the close proximity between carbonaceous K-promoters and Fe/C catalyst components produced high olefin yields and catalytic stability (Figure 7b) [94]. Guo et al. reported that K derived from biological rather than inorganic precursors showed a stronger migration ability during the CO_2 hydrogenation to light olefins. These surface-enriched K ions extracted from corncobs could promote the carburization of iron species to form more Fe$_5$C$_2$, promoting both the reverse water–gas shift reaction and subsequent C–C coupling [97].

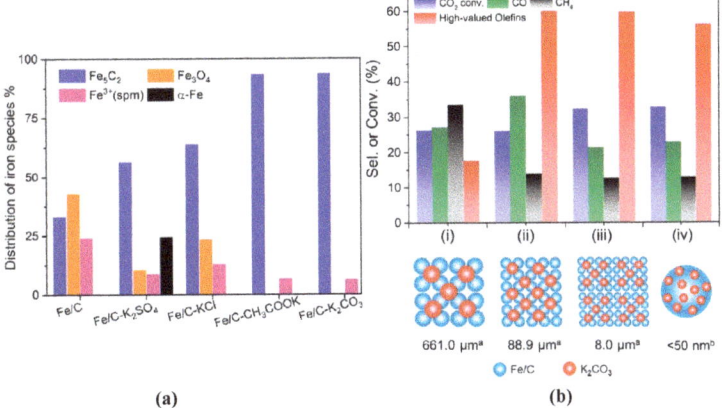

Figure 7. (a) Distribution of iron species content over different spent catalysts. (b) CO_2 hydrogenation over Fe/C–K$_2$CO$_3$ catalysts with varying proximity. Adapted with permission from ref. [94]. Copyright 2020 American Chemical Society.

Metal organic frameworks as precursors for the preparation of heterogeneous catalysts have been used recently [99,106,107]. Ramirez et al. used a metal organic framework as a catalyst precursor to synthesize a highly active, selective, and stable catalyst, as shown in Figure 8a–f for the hydrogenation of CO_2 to light olefins. Comparing the addition of Cu, Mo, Li, Na, K, Mg, Ca, Zn, Ni, Co, Mn, Fe, Pt, and Rh to an Fe/C composite, only K is able to enhance olefin selectivity, as shown in Figure 8c. The presence of K promoted the formation of Fe_5C_2 and Fe_7C_3 carbides, as confirmed by XRD (Figure 8e). K helped keep a good balance between the iron oxide for RWGS and iron carbide for FTS. The results presented in Figure 8f indicated a trend in which methane formation decreased and olefin selectivity increased as the K loading increased. The catalyst Fe/C+K(0.75) exhibited good stability (Figure 8d) and outstanding C_2–C_4 olefin space time yields of 33.6 mmol·g_{cat}^{-1}·h^{-1} at X_{CO2} = 40%, 593K, 30 bar, H_2/CO_2 = 3, and 24,000 mL·g^{-1}·h^{-1} [99].

Some work may shed light on the ways in which the alkali promoters affect the behavior of iron catalysts. By the precisely controlled addition of promoters to fine tune the catalytic performance for the hydrogenation of CO_2 to olefins, Yang et al. investigated how a zinc ferrite catalyst system could be affected by the addition of sodium and potassium promoters, specifically on the conversion of CO and CO_2 to olefins. It was found that the catalyst's composition of iron oxides and iron carbides was altered in the presence of the promoters, which affected the CO and CO_2 conversion. The production of C_{2+} olefins was greatly facilitated by the Na- and K-promoted catalysts. The Na/Fe-Zn catalyst was found to possess the optimal olefin productivity, and inhibited the competitive methanation reaction. It showed a total carbon conversion of 34.0%, which decreased by only 12.2% over 200 h [96]. Similarly, Wei et al. unraveled the effect of the Na promoter on the evolution of iron and carbon species, as well as the consequent tuning effect on the hydrogenation of CO_2 to olefins. With the contents of the Na promoter increasing from 0 wt% to 0.5 wt%, the ratio of olefins to paraffins (C_{2+}) rose markedly, from 0.70 to 5.67. The in situ XRD and temperature programed surface reaction (TPSR) confirmed that the introduction of the Na promoter decreased the particle size of Fe_5C_2 and regulated the distribution of surface carbon species. Furthermore, the in situ XRD and Raman demonstrated that the interaction between the Na promoter and the catalysts inhibited the hydrogenation of Fe_5C_2 and surface graphitic carbon species, consequently improving the stability of the Fe_5C_2 and enhancing the formation of olefins by inhibiting the hydrogenation of the intermediate carbon species [92]. Using a similar approach, Liang et al. modified the xNa/Fe-based catalysts with tunable amounts of sodium promoter for CO_2 hydrogenation to alkenes, with CO_2 conversion at 36.8% and a light olefin selectivity of 64.3%. It was found that the addition of the Na promoter into Fe-based catalysts boosted the adsorption of CO_2, facilitated the formation and stability of the active Fe_5C_2 phase, and inhibited the secondary hydrogenation of alkenes under the CO_2 hydrogenation reaction conditions (Figure 9a–c). The content of Fe_5C_2 correlated with the amount of Na is shown in Figure 9d [98].

Wei et al. synthesized a series of Fe_3O_4-based nanocatalysts with varying sodium contents. The residual sodium markedly influenced the textural properties of the Fe_3O_4-based catalysts, and faintly hampered the reduction of the catalysts. However, it discernibly promoted the surface basicity and prominently improved the carburization degrees of the iron catalysts, which is favored for olefin production. Compared with the sodium-free Fe_3O_4 catalysts, the sodium-promoted Fe_3O_4 catalysts displayed higher activity and selectivity for C_2–C_4 olefins. The FeNa catalyst (1.18) (Na/Fe weight ratio of 1.18/100) exhibited a high degree of catalytic activity with a high olefin/paraffin ratio (6.2) and selectivity to C_2–C_4 olefins (46.6%), and fairly low CO and CH_4 production at a CO_2 conversion of 40.5%. This catalyst also exhibited superb stability during the 100 h test at 593K. Comparing the scanning electron microscopy (SEM) image after reduction, there was no apparent indication of particle size growth after catalytic reaction for 100 h, further revealing the improved reaction stability of these iron nanoparticles [31]. Zhang et al. fabricated a Na- and Zn-promoted iron catalyst by a sol-gel method, and demonstrated its high activity, selectivity and stability towards the formation of C_{2+} olefins in the hydrogenation of CO_2 into C_{2+} olefins. The selectivity of the C_{2+} olefins reached 78%, and the space–time yield of olefins was as high as 3.4 g g_{cat}^{-1} h^{-1}. The catalyst was composed of ZnO and χ-Fe_5C_2 phases with Na^+ dispersed on both ZnO and χ-Fe_5C_2. Zhang et al. found that ZnO functions for the RWGS reaction of CO_2 to CO, while χ-Fe_5C_2 is responsible for CO hydrogenation to olefins. The presence of Na^+ enhanced the selectivity of C_{2+} olefins by regulating the hydrogenation ability and facilitating the desorption of olefins (Figure 10a). The presence of ZnO not only efficiently catalyzes the RWGS reaction but also improves the activity and stability of CO_2 hydrogenation by controlling the size of χ-Fe_5C_2 (Figure 10c,d). It was further discovered that the close proximity between ZnO and χ-Fe_5C_2 is beneficial for the conversion of CO_2 to olefins (Figure 10b). The larger interface could facilitate the diffusion and transfer of intermediate CO from ZnO to χ-Fe_5C_2, favoring CO_2 adsorption and subsequent CO hydrogenation to C_{2+} olefins [90]. Malhi et al. also investigated the effect of Na and Zn on iron-based catalysts, and found that the modified Fe-based catalyst exhibited a good performance for CO_2 hydrogenation to olefins, with a CO_2 conversion of 43%, a selectivity of 54.1% to $C_{2+}^=$ olefins, and a high olefins-to-paraffins ratio of 7.3 [93].

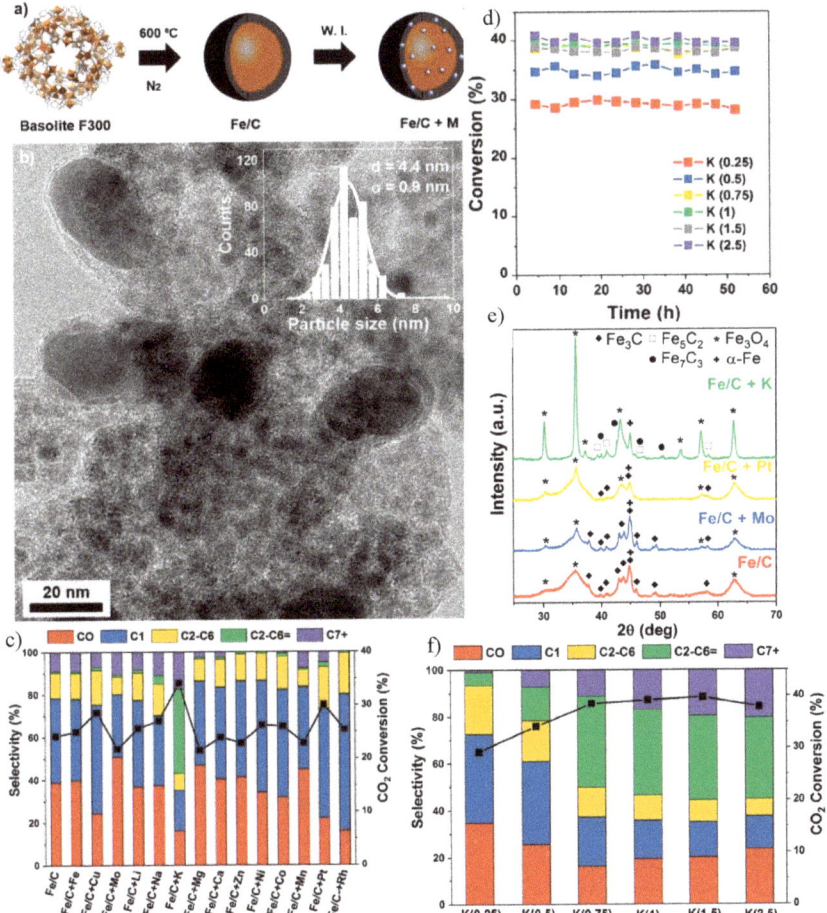

Figure 8. (a) Illustrated synthesis of Fe/C catalysts. (b) TEM image of Fe/C catalysts. (c) Catalytic performance over promoted and unpromoted Fe catalysts. (d) CO_2 conversion after 50 h of TOS. (e) XRD of promoted and unpromoted Fe catalysts. (f) Effect of K loading on the selectivity and CO_2 conversion after 50 h of TOS. Testing conditions: 593 K, 30 bar, H_2/CO_2 molar ratio = 3, and GHSV = 24,000 mL·g^{-1}·h^{-1}. Adapted with permission from ref. [99]. Copyright 2018 American Chemical Society.

Chaipraditgul et al. investigated the effect of transition metals (Cu, Co, Zn, Mn or V) on the Fe/K-Al_2O_3 catalyst and found that the inclusion of the transition metal remarkably affected the interaction between the catalysts' surface and the adsorptive CO_2 and H_2.

The Fe/K-Al$_2$O$_3$ promoted with Cu, Co or Zn showed a lower the olefin to paraffin ratio, owning to a markedly increased number of weakly adsorbed H atoms resulting from the enhanced hydrogenation ability of the promoted catalysts. On the contrary, the addition of a Mn promoter to Fe/K-Al$_2$O$_3$ reduced the number of weakly adsorbed H atoms, lowering the hydrogenation ability to result in a high olefin to paraffin ratio of 7.4. The presence of either Mn or V inhibited the CO hydrogenation to hydrocarbon, leading to the low CO$_2$ conversion, while the CO$_2$ conversion was enhanced by incorporating either Co or Cu onto the Fe/K-Al$_2$O$_3$ catalyst [33]. Gong et al. investigated the promoting effect that Cu had on Fe-Mn-based catalysts in the production of light olefins via the CO$_2$–FTS process. The Cu promoter was found to facilitate the reduction process and enhance CO dissociative adsorption by altering the interactions between Fe, Mn and the SiO$_2$ binder, which led to increased activity. The addition of Cu weakened the surface basicity, which in turn decreased the chain growth probability and yielded a higher selectivity of light olefins [108].

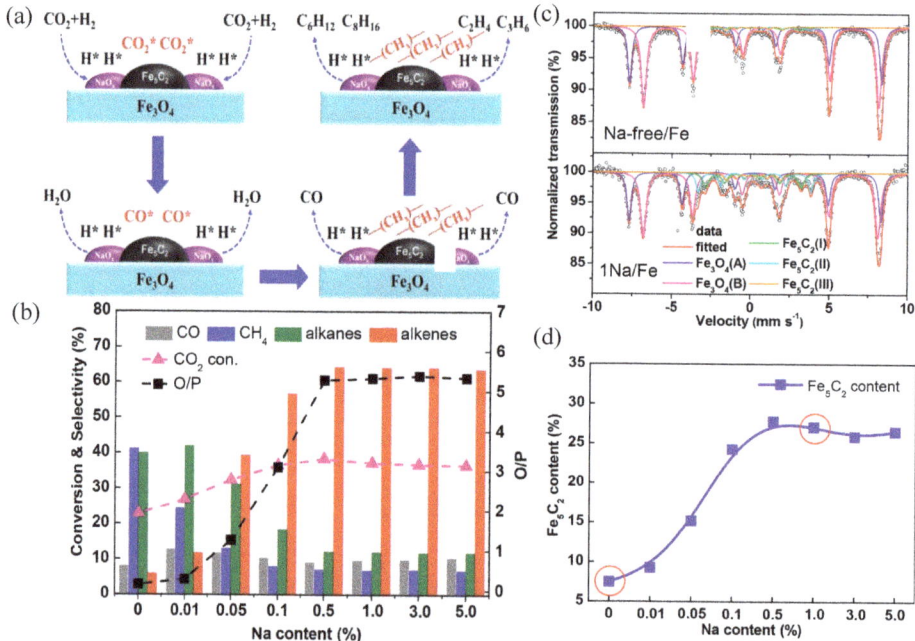

Figure 9. (**a**) Illustrated scheme of CO$_2$ hydrogenation. (**b**) Conversion and selectivity of CO$_2$ hydrogenation over an xNa/Fe catalyst (testing conditions: H$_2$/CO$_2$ molar ratio = 3; P = 3 MPa; T = 593 K; GHSV = 2040 mL h^{-1} g$_{cat}^{-1}$; TOS = 10 h). (**c**) Mössbauer spectra of the spent Na-free/Fe and 1Na/Fe catalysts. (**d**) Fe$_5$C$_2$ content of the spent xNa/Fe catalyst vs. the Na content. Adapted with permission from ref. [98]. Copyright 2019 American Chemical Society.

Jiang et al. reported the synthesis of Mn-modified Fe$_3$O$_4$ microsphere catalysts. These catalysts demonstrated excellent catalytic performance, with a 44.7% CO$_2$ conversion, 46.2% light olefin selectivity, and 18.7% light olefin yield over the 10 Mn−Fe$_3$O$_4$ catalyst. The O/P ratio increased from 3.7 for the unpromoted Fe$_3$O$_4$ catalyst to 6.5 for the Mn-promoted catalyst. An even distribution of manganese was found over the surface of the Fe$_3$O$_4$ microsphere. Such homogeneous dispersion allows for an increase in the basicity of the catalyst, which prevents the further hydrogenation of olefins into paraffins. It was noted that the synergistic effects between Fe and Mn improve the dissociation and conversion of CO$_2$ to hydrocarbons. The addition of Mn was found to promote the production of Fe carbides and enhance the active phases of CO$_2$ hydrogenation and the FTS reaction, as well as preventing the hydrogenation of light olefins into paraffins and chain growth into longer hydrocarbons [88]. A similar effect of the addition of Mn to Na/Fe catalysts was also observed by Liang et al. [95].

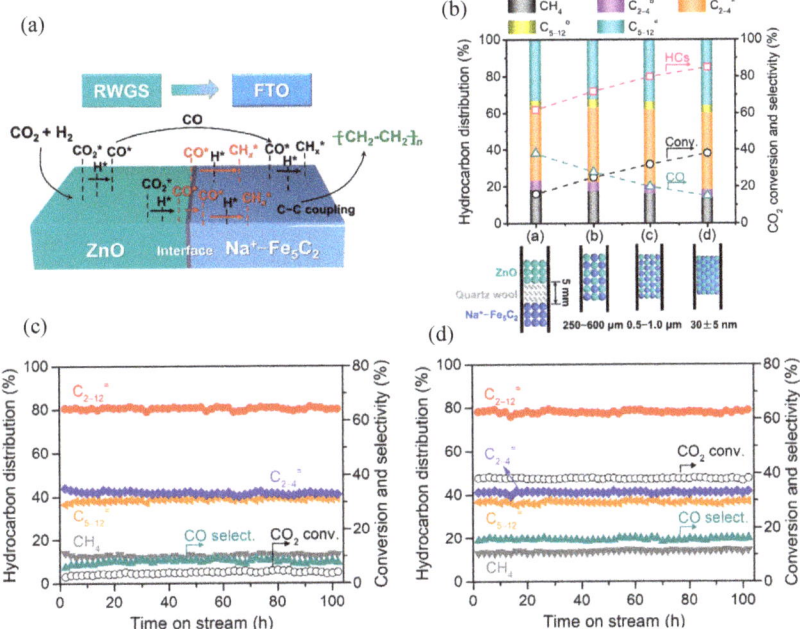

Figure 10. (**a**) Illustrated reaction mechanism for CO_2 hydrogenation over the catalyst Na-Zn-Fe. (**b**) Effect of the proximity between ZnO and $Na^+Fe_5C_2$ on catalytic behaviors for CO_2 hydrogenation. Catalyst stability in CO_2 hydrogenation over (**c**) an $Na^+Fe_5C_2$ catalyst and (**d**) a Na-Zn-Fe catalyst. Testing conditions: H_2/CO_2 molar ratio = 3, P = 2.5 MPa, W = 0.10 g, F = 25 mL min^{-1}, T = 613 K. Adapted with permission from ref. [90]. Copyright 2021 Elsevier.

Zhang et al. synthesized uniform microspheres of Fe-Zr-Ce-K catalysts by microwave-assisted homogeneous precipitation, and found that the reducibility, surface basicity and surface atom composition of the catalysts were greatly affected by varying the Ce content. CeO_2, as the structural promoter, restrained the growth of Fe_2O_3 crystallite, weakening the interaction between Fe species and zirconia, and enabling the easier reduction of Fe_2O_3. The best performance was obtained on a 35Fe-7Zr-1Ce-K catalyst at 593 K and 2 MPa, with a CO_2 conversion of 57.34%, a C_2–C_4 olefin selectivity of 55.67%, and a ratio of olefin/paraffin of 7 [100].

Extensive research efforts have been exerted on the development of bi-metallic catalysts for the conversion of CO_2 to light olefins. Yuan et al. demonstrated the influence of Na, Co and intimacy between Fe and Co on the catalytic performance of Fe-Co bimetallic catalysts for CO_2 hydrogenation that offers an olefin to paraffin ratio of 6 at a CO_2 conversion rate of 41%. With the introduction of Co into the Fe catalyst, the CO_2 conversion is significantly enhanced. The intimate contact between the Fe and Co sites favored the production of C_2–$C_4^=$. When Na was added to the system, the surface of the catalyst became carbon-rich and hydrogen-poor, allowing C–C coupling to form light olefins and suppress the methane formation. Moreover, the addition of a Na promoter facilitated the generation of χ-$(Fe_{1-x}Co_x)_5C_2$ under the CO_2 hydrogenation reaction conditions, and thus further improved the catalytic performances. A superb stability over 100 h was observed (Figure 11) [91]. Witoon et al. investigated the effect of Zn addition to Fe-Co/K-Al_2O_3 catalysts. The addition of Zn resulted in the improved dispersion and reducibility of iron oxides. For example, the 0.58 wt% Zn-promoted Fe-Co/K-Al_2O_3 catalyst afforded a large number of active sites for the adsorption of CO and H_2 due to higher dispersion and an eased reducibility (Figure 12a). The catalyst exhibited superior activity for light olefin formation with yield of 19.9% under the optimum testing conditions of 613 K, 25 bar, 9000 mL g_{cat}^{-1} h^{-1} and a H_2/CO_2 ratio of 4. Figure 12b also shows a gradual decrease in the olefin to paraffin ratio, with an almost constant CO_2 conversion as a function of the time-on-stream (TOS). The X-ray photoelectron spectroscopy (XPS) analysis of the spent catalyst showed the continuous growth of iron carbide with the time-on-stream, indicating that iron carbide may be the active component resulting in paraffin production (Figure 12c). XRD confirmed the formation of Fe-C phases over the spent 0.58 wt% Zn-promoted Fe-Co/K-Al_2O_3 catalyst at the time-on-stream (Figure 12d) [89].

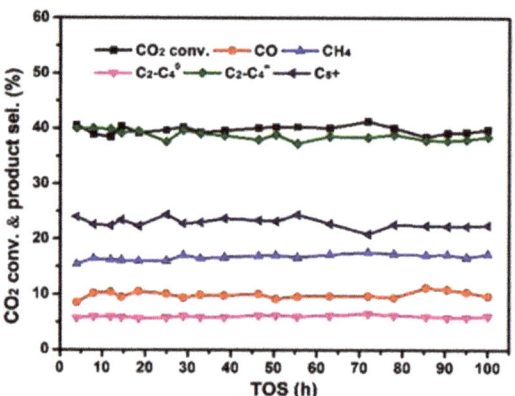

Figure 11. Catalytic performance of the CO_2 hydrogenation over the Na-$CoFe_2O_4$ catalyst at TOS (reaction conditions: H_2/CO_2 molar ratio = 3, T= 593 K, P = 3 MPa, GHSV = 7200 mL h^{-1} $gcat^{-1}$, TOS = 100 h). Adapted with permission from ref. [91]. Copyright 2021 Elsevier.

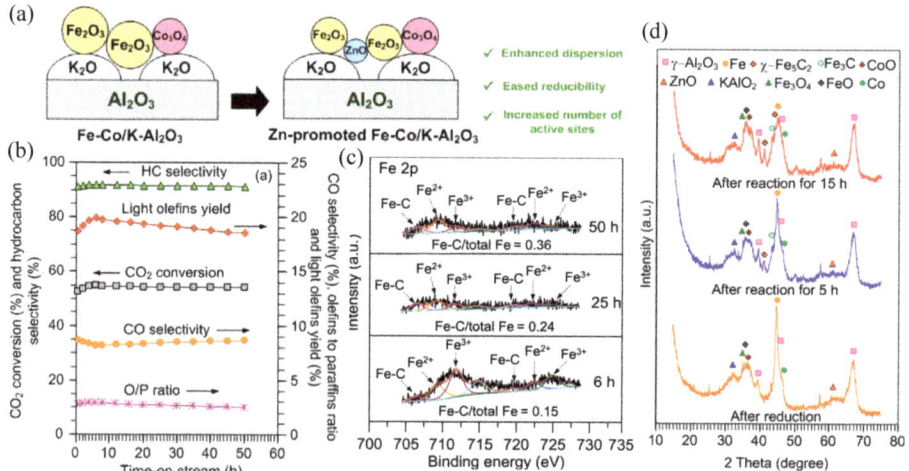

Figure 12. (**a**) Illustrated reaction mechanism. (**b**) Catalytic performance of the CO_2 hydrogenation over a Zn-promoted Fe-Co/K-Al_2O_3 catalyst. (**c**) XPS spectra (Fe 2p region) of the 0.58 wt% Zn-promoted Fe-Co/K-Al_2O_3 catalysts. (**d**) XRD pattern of the 0.58 wt% Zn-promoted Fe-Co/K-Al_2O_3 catalyst at varying TOS. Testing conditions: T = 613 K, P = 25 bar, GHSV = 9000 mL g_{cat}^{-1} h^{-1} and H_2/CO_2 molar ratio = 4. Adapted with permission from ref. [89]. Copyright 2021 Elsevier.

Wang et al. reported the synthesis of γ-alumina supported Fe-Cu bimetallic catalysts, and found a strong bimetallic promotion for selective CO_2 conversion to olefin-rich C_{2+} hydrocarbons resulting from the combination of Fe and Cu at a specific composition. The suppression of the undesired CH_4 formation was achieved by the addition of Cu to Fe while simultaneously enhancing the C–C coupling for C_{2+} hydrocarbon formation. The formation of the Fe-Cu alloy in the Fe-Cu(0.17)/Al_2O_3 catalyst is suggested by the XRD results. Furthermore, the addition of K into the Fe-Cu considerably enhanced the production of $C_2^=$–$C_4^=$ light olefins and the O/P ratio over Fe-Cu bimetallic catalysts. The Fe-Cu/K catalysts exhibited the superior selectivity of C_{2+} hydrocarbons compared to Fe-Co/K catalysts under the same reaction conditions [102]. Kim et al. synthesized monodisperse nanoparticles (NPs) of $CoFe_2O_4$ by the thermal decomposition of metal−oleate complexes, as shown in Scheme 4. The prepared NPs were supported on carbon nanotubes (CNTs), and Na was added to investigate the promoter and support effects on the catalyst for CO_2 hydrogenation to light olefins. The resulting Na-$CoFe_2O_4$/CNT exhibited a superior CO_2 conversion of 34% and a light olefin selectivity of 39% compared to other reported Fe-based catalysts under similar reaction conditions. The superb performance of Na-$CoFe_2O_4$/CNT was attributed to the formation of a bimetallic alloy carbide, $(Fe_{1-x}Co_x)_5C_2$. Higher CO_2 conversion and better light olefin selectivity were found in comparison with conventional Fe-only catalysts which possess χ-Fe_5C_2 active sites and drastically improved the C_{2+} hydrocarbon formation in comparison with Co-only catalysts which contain Co_2C sites [103].

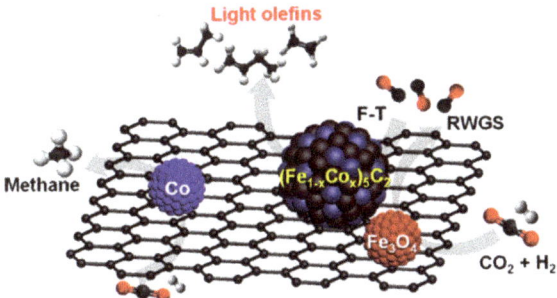

Scheme 4. Schematic demonstration of CO_2 hydrogenation over CNT supported bi-metallic catalyst $CoFe_2O_4$. Adapted with permission from ref. [103]. Copyright 2020 American Chemical Society.

Song et al. investigated titania-supported monometallic and bimetallic Fe-based catalysts for CO_2 conversion, and found that the mono-metallic catalyst (Fe-, Co-, Cu-) performed poorly for C–C coupling reactions. However, adding a small amount of a second metal (Co and Cu) to Fe revealed the synergetic promotion on the CO_2 conversion and the space–time yields (STY) of hydrocarbon products. The inclusion of K and La as promoters further improved the activity, giving a higher hydrocarbon selectivity and O/P ratio, indicating that the promotor facilitated the CO_2 activation and C–C couplings over bi-metallic catalysts [106]. Zhang et al. investigated Fe-Zn bimetallic catalysts for CO_2 hydrogenation to C_{2+} olefins. A high C_{2+} olefin selectivity of 57.8% after 200 h of time-on-stream at a CO_2 conversion of 35.0% was obtained over an Fe_2Zn_1 catalyst. In bimetallic Fe_5C_2-ZnO catalysts, the ZnO plays a crucial role in improving the performance by altering the structure of the Fe components. Without ZnO, the chief deactivation mechanism was attributed to a phase transition from FeC_x to FeO_x over Fe_2O_3. However, with the addition of Zn to Fe_2O_3, the phase transformation and the carbon deposits over Fe_2Zn_1 were greatly diminished. Furthermore, the addition of Na inhibited the oxidation of χ-Fe_5C_2 active species for Fe-Zn bimetallic catalysts. During activation, both Zn and Na were shown to migrate onto the catalysts' surfaces. The oxidation of FeC_x by H_2O and CO_2 was shown to be diminished by the interaction between Zn and Na [28].

Xu et al. investigated the roles of Fe-Co interactions over ternary spinel-type $ZnCo_xFe_{2-x}O_4$ catalysts for CO_2 hydrogenation to produce light olefins. As shown in Figure 13, a high light olefin selectivity of 36.1%, a low CO selectivity of 5.8% at a high CO_2 conversion of 49.6%, and an excellent catalyst stability were obtained over the $ZnCo_{0.5}Fe_{1.5}O_4$ via the RWGS–FTS reaction pathway. It was shown that during the CO_2 hydrogenation over ternary ZnCo$_{0.5}$Fe$_{1.5}$O$_4$ catalysts, the formation of electron-rich Fe^0 atoms in the CoFe alloy phase significantly boosted the generation of the active χ-Fe$_5$C$_2$, Co$_2$C, and θ-Fe$_3$C phases, in which the χ-Fe$_5$C$_2$ phase facilitated the C–C coupling, the Co$_2$C species suppressed the formation of CH$_4$, and the formation of the θ-Fe$_3$C phase with lower hydrogenation activity inhibited the second hydrogenation of light olefins [105].

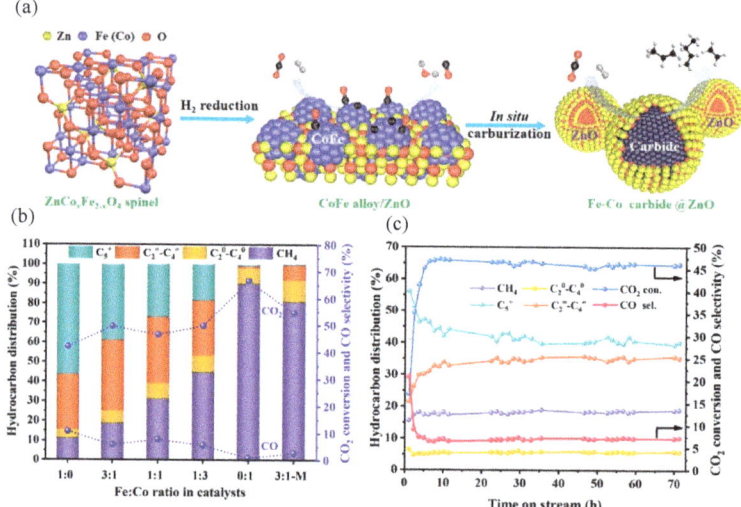

Figure 13. (**a**) Schematic illustration of the structural transformations of as-formed ZnCo$_x$Fe$_{2-x}$O$_4$ catalysts during the reduction and reaction steps. (**b**) CO$_2$ conversion and product distributions over K-containing ZnCo$_x$Fe$_{2-x}$O$_4$ catalysts with various Fe/Co molar ratios. (**c**) The stability of the K-containing ZnCo$_{0.5}$Fe$_{1.5}$O$_4$ catalyst in CO$_2$ hydrogenation (testing conditions: T = 583 K, P = 2.5 MPa, GHSV = 4800 mLh^{-1}g$_{cat}$$^{-1}$, CO$_2$/H$_2$ molar ratio = 1:3). Adapted with permission from ref. [105]. Copyright 2021 Elsevier.

In summary, the use of the appropriate K or Na promoter, the inclusion of Cu, Co, Zn, Mn or Ce in the Fe phase, and the bi-metallic formation played important roles for enhanced catalytic performance and stability.

3.2. Support Effect

Catalyst support plays an important role in the overall activity and selectivity due to the interactions between the active metal components and the support during CO$_2$–FTS. Some representative catalysts of the support effect for CO$_2$ hydrogenation to light olefins with improved catalyst stability are presented in Table 2.

Table 2. Some representative catalysts of the support effect for CO$_2$ hydrogenation to light olefins.

Catalyst	CO$_2$ Conv., %	Selectivity, %				Yield, %	O/P Ratio	Stability	Ref.
		CO	CH$_4$	C$_2$–C$_4$$^=$	C$_2$–C$_4$^{0}	C$_2$–C$_4$$^=$			
Fe-K/HPCMs-1	33.4	38.9	13.5	18.0	11.5	6.0	1.6	35 h	[109]
ZIF-8(a)/Fe$_2$O$_3$	~24.0	~24.0	~21.0	~20.0	~24.0	~4.8	0.83	n/a	[110]
Fe(0.5)-Mo$_2$C^c	9.8	0.5	2.1	92.0	3.5	9.0	26.3	2 h	[111]
K-Zr-Co/aTiO$_2$	70.0	n/a	n/a	17.0	n/a	11.9	n/a	8h	[10]
Fe-Cr-K/Nb$_2$O$_5$	31.0	57.0	32.0	10.0	1.0	3.1	3.1	n/a	[4]
15Fe-K/m-ZrO$_2$	38.8	19.9	30.1	42.8	12.8	16.6	3.3	12 h	[112]
20%Fe/CeO$_2$-NC	18.9	73.5	75.5	18.2	4.0	3.4	4.1	n/a	[113]
10Fe-1K/m-ZrO$_2$	40.5	n/a	n/a	15.0	n/a	6.1	n/a	100 h	[114]
Fe$_5$C$_2$-10K/a-Al$_2$O$_3$	40.9	n/a	n/a	73.5	n/a	30.1	n/a	100 h	[15]
Co-Na-Mo/CeO$_2$	15.1	70.2	22.1	10.7	36.0	1.6	0.03	n/a	[115]

Owen et al. investigated the effect of Co-Na-Mo on various supports (SiO$_2$, CeO$_2$, ZrO$_2$, γ-Al$_2$O$_3$, TiO$_2$, ZSM-5 (NH$_4{}^+$) and MgO) for CO$_2$ hydrogenation. It was found that the surface area of the support and the metal–support interaction played a key role in the determination of the cobalt crystallite size, which strongly affected the catalytic activity. Cobalt particles with sizes < 2 nm supported on MgO showed low RWGS conversion with negligible FT activity, which is in agreement with the work of de Jong et al. [51]. When the cobalt particle size increased to 15 nm supported on SiO$_2$ and ZSM-5, both the CO$_2$ conversion and C$_{2+}$ hydrocarbon selectivity increased markedly. When the cobalt particle size further increased to 25–30 nm, a lower CO$_2$ conversion but higher C$_{2+}$ light olefin selectivity was obtained. The authors reported that the higher the metal–support interaction, the higher the growth chain probability of the hydrocarbons. By altering the TiO$_2$/SiO$_2$ ratio in the support, the CO$_2$ conversion and C$_{2+}$ light olefin selectivity could be tuned [115]. Li et al. evaluated cobalt catalysts supported on TiO$_2$ with different crystal forms of anatase (a-TiO$_2$) and rutile (r-TiO$_2$), and it was found that the addition of Zr, K, and Cs improved the CO, CO$_2$, and H$_2$ adsorption in both the capacity and strength over a-TiO$_2$- and r-TiO$_2$-supported catalysts. The surface C/H ratio increased drastically in the presence of promoters, leading to a high C$_{2+}$ selectivity of 17% with 70% CO$_2$ conversion over a K-Zr-Co/a-TiO$_2$ catalyst. As a result, the product distribution could be tuned by adjusting the metal–support interaction and surface C/H ratio through Zr, K, and Cs modification over Co-based catalysts for CO$_2$ hydrogenation, as shown in Scheme 5 [10].

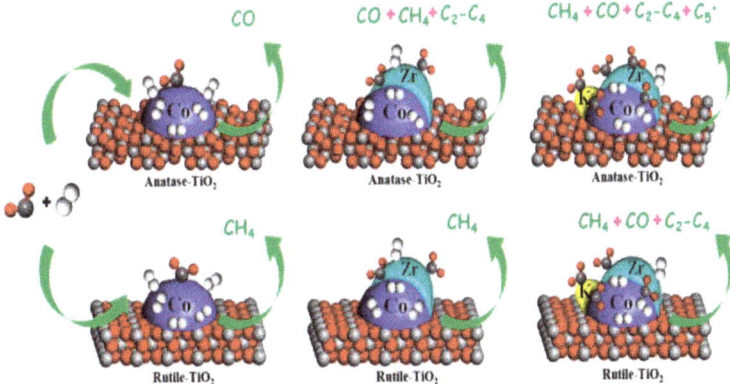

Scheme 5. Schematic illustration of CO$_2$ hydrogenation over unpromoted and Zr- and K-promoted cobalt catalysts supported on a-TiO$_2$ and r-TiO$_2$. Adapted with permission from ref. [10]. Copyright 2013 American Chemical Society.

Da Silva et al. found the Fe-Cr catalyst, promoted with K and supported on niobium oxide, was more active (CO$_2$ conversion = 20%) and selective to light olefins (25%) compared to the same composition supported on silica (CO$_2$ conversion = 11%, light olefin selectivity = 18%) under the same testing conditions. Alkali metal promotion increased the selectivity of olefins, probably due to electron-donor effects and the basicity of niobium oxide. A niobium oxide-supported Fe-Cr catalyst presented higher activity and selectivity to olefins, which is probably due to strong metal–support interactions when compared with traditional SiO$_2$ [4]. Very recently, Huang et al. revealed the dynamic evolution of the active Fe and carbon species over different phases of zirconia (m-ZrO$_2$ and t-ZrO$_2$) on CO$_2$ hydrogenation to light olefins, as shown in Scheme 6. Fe-K/m-ZrO$_2$ catalysts performed better than the corresponding Fe-K/t-ZrO$_2$ catalysts under the optimal reaction conditions. Among them, the 15Fe-K/m-ZrO$_2$ catalyst showed remarkable catalytic activity, with a CO$_2$ conversion of 38.8% and a C$_2$–C$_4{}^=$ selectivity of 42.8%. More active species (Fe$_3$O$_4$ and χ-Fe$_5$C$_2$) with smaller particle sizes were obtained for the Fe-K/m-ZrO$_2$ catalysts. The larger specific surface area facilitated the highly dispersed Fe species on the surface of the m-ZrO$_2$ support when compared to the t-ZrO$_2$ support. In addition, the monoclinic phase m-ZrO$_2$ support provided more strong basic sites, effectively decreasing the deposited carbon species and coke generation. Moreover, the electron-donating ability of iron elements and more oxygen vacancies (Ov) improved the charge transfer between ZrO$_2$ and Fe. The synergy effect between K$_2$O and ZrO$_2$ fostered the generation of active carbide species. The formation of more χ-Fe$_5$C$_2$ species contributed to the high yield of light olefins [112]. Similarly, Gu et al. investigated Fe-K supported on ZrO$_2$ with different crystal phases, revealing 40.5% CO$_2$ conversion, 15.0% light olefin selectivity, and excellent stability (Figure 14) over 10Fe-1K/m-ZrO$_2$ (10 wt% Fe and 1 wt% K) at 2.0 MPa and 613 K. The CO$_2$ conversion was almost 200% higher than that of 10Fe-1K/t-ZrO$_2$ [114].

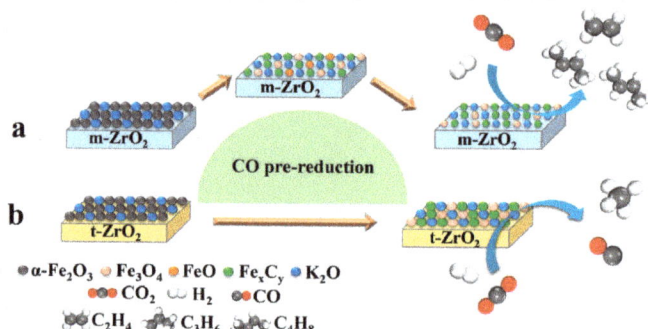

Scheme 6. CO pre-reduction and CO$_2$ hydrogenation process on (**a**) m-ZrO$_2$- and (**b**) t-ZrO$_2$-supported Fe-Zr catalysts. Adapted with permission from ref. [112]. Copyright 2021 American Chemical Society.

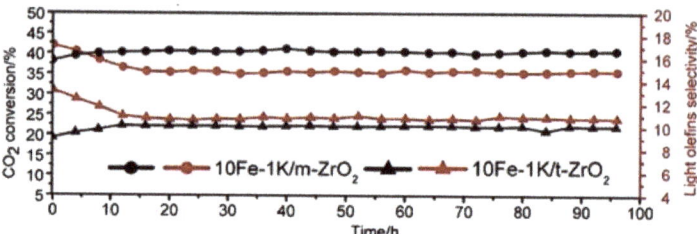

Figure 14. The stability of 10Fe1K/m-ZrO$_2$ and 10Fe1K/t-ZrO$_2$ for CO$_2$ conversion, and the light olefin selectivity at a TOS of 100 h at 613K. Adapted with permission from ref. [114]. Copyright 2019 Elsevier.

Torrente-Murciano demonstrated that iron-based catalysts could be improved not only through the inclusion of promoters but also by the judicious control of the morphology of the ceria support (nanoparticle, nanorods, nanocubes) for CO$_2$ hydrogenation to light olefins. For example, 20 wt% Fe/CeO$_2$ cubes provided better catalytic performance (CO$_2$ conversion = 15.2%, C$_2$–C$_4$$^=$ selectivity = 20.2%) when compared with nanorods and their nanoparticle counterparts. TPR showed that the ceria reducibility decreased in the order of rods > particles > cubes, suggesting that the catalytic effect had a direct dependence on the reducibility of the different nanostructured ceria supports and their interaction with the iron particles [113]. By the physical mixing of Fe$_5$C$_2$ and K-modified Al$_2$O$_3$, Liu et al. discovered that Fe$_5$C$_2$-10K/a-Al$_2$O$_3$ exhibited a CO$_2$ conversion of 40.9% and C$_{2+}$ selectivity of 73.5%, containing 37.3% C$_2$–C$_4$$^=$ and 31.1% C$_{5+}$ (Figure 15). The superior catalytic performance was due to the potassium which migrated into the Fe$_5$C$_2$ during the reaction, and the intimate contact between the Fe$_5$C$_2$ and K/a-Al$_2$O$_3$. Among the various supports tested, as shown in Figure 15, alkaline Al$_2$O$_3$ is the best support for the high selectivity of value-added hydrocarbons [15].

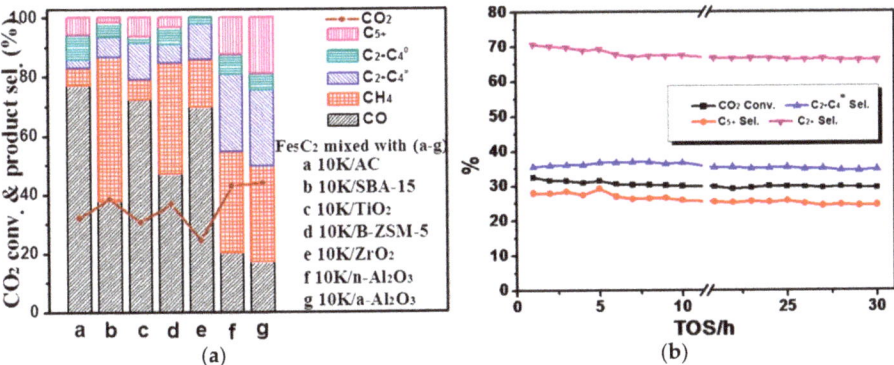

Figure 15. (a) Catalytic performance of CO_2 hydrogenation over Fe_5C_2-based catalysts on various supports. (b) Catalytic performance and stability over an Fe_5C_2-10K/a-Al_2O_3 catalyst (testing conditions: T = 593 K, P = 3.0 MPa, GHSV = 3600 mLg^{-1}h^{-1}, H_2/CO_2 molar ratio = 3). Adapted with permission from ref. [15]. Copyright 2018 American Chemical Society.

Dai et al. synthesized hierarchical porous carbon monoliths (HPCMs) by an adaptable strategy employing a one-step desilication process for a coke-deposited spent zeolite catalyst. This hierarchical porous carbon was shown to be a better support for the reduction of the nanoparticle size and heightening the synergism of the Fe–K catalyst for CO_2 hydrogenation, with a CO_2 conversion of 33.4% and a $C_2^=$–$C_4^=$ selectivity of 18.0% [109].

Metal organic frameworks (MOFs) as novel porous materials had a considerable effect on the activity and selectivity of Fe-based catalysts. Hu et al. synthesized a type of hydrothermally stable MOF, zeolitic imidazolate frameworks (ZIF-8) with different sizes and morphologies, which were used as supports for CO_2 hydrogenation. The acidity, internal diffusion process and crystal size enabled the ZIF-8 supports to show different levels of substantial light olefin selectivity [110]. Raghav et al. developed a simple method for the synthesis of hierarchical molybdenum carbide (β-Mo_2C). The β-Mo_2C phase exhibited the strongest metallic and some ionic character, and it behaved as both a support and co-catalyst for CO_2 hydrogenation to light olefins. The Fe(0.5)-Mo_2C catalyst exhibited a conversion of CO_2 of 7.3% and a $C_2^=$ olefin selectivity of 79.4% at 300 °C and 4.0 mPa. The XRD patterns of the fresh and used Fe(0.5)-Mo_2C catalyst did not show a noticeable difference, indicating the stability of the catalysts to achieve high olefin selectivity [111].

In summary, various supports (SiO_2, CeO_2, m-ZrO_2, γ-Al_2O_3, TiO_2, ZSM-5, MgO, NbO HPCMs, MOFs, β-Mo_2C) have been used for the dispersal of active species. The surface area, basicity, reducibility, oxygen vacancies, and morphology of the support played important roles, in most cases with the presence of promoters (K, Zr, Cs), in affecting the amount and particle size of the active carbide species; the synergy effect; the metal–support interaction; the strength and capacity of the CO, CO_2, and H_2 adsorption on support; and the surface C/H ratio for CO_2 hydrogenation. By tuning the above-mentioned characteristics properly, the physically deposited carbon species, coke generation and metal sintering could be mitigated as reported.

3.3. Bifunctional Composite Catalyst Effect

The zeolite–methanol composite catalyst can also be improved by compositional modifications. The composite catalyst is composed of two functional components: one is the target for methanol synthesis, mainly Cu, Zn, and In metal oxide catalysts; the other one is for the MTO process, mainly zeolite catalysts. Here, in this section, the recent progress on composite catalysts for improved catalytic performance and stability are described accordingly. Some representative catalysts for the bifunctional composite catalyst effect for CO_2 hydrogenation to light olefins with improved catalyst stability are presented in Table 3.

Table 3. Some representative catalysts for the bifunctional composite catalyst effect for CO_2 hydrogenation to light olefins.

Catalyst	CO_2 Conv., %	Selectivity, %				Yield, % C_2–$C_4^=$	O/P Ratio	Stability	Ref.
		CO	CH_4	C_2–$C_4^=$	C_2–C_4^0				
CuO-ZnO & SAPO-34	41.3	9.3	11.8	63.4	15.5	26.2	4.1	13 h	[116]
(CuO-ZnO)-kaolin & SAPO-34	57.6	9.6	11.4	63.8	15.2	36.7	4.2	20 h	[116]
In_2O_3/ZrO_2 & SAPO	19.0	87.0	~17.0	90.0 [a]	n/a	17.1	n/a	>50 h	[43]
In_2O_3/ZrO_2 & SAPO	~14.0	<5.0	<5.0	70.0	n/a	9.8	n/a	>100 h	[44]
In−Zr/SAPO-34	26.7	n/a	4.3	76.4 [a]	~14.0 [a]	20.4	5.5	>150 h	[19]
$Zn_{0.5}Ce_{0.2}Zr_{1.8}O_4$ & H-RUB-13 (200)	10.7	28.3	2.9	83.4	5.4	8.9	15.4	>30 h	[117]
$ZnZrO_x$ & bio-ZSM-Si	10.0	~80.0	5.5	64.4	30.1	6.4	2.1	60 h	[118]
InCrOx(0.13) & SAPO	33.6	55.0	35.0	75.0 [a]	20.0 [a]	11.3	3.8	>120 h	[119]
ZnZrO & SAPO-34	12.6	47.0	3.0	80.0 [a]	14.0 [a]	10.1	5.7	>100 h	[14]
CZZ@Zn & SAPO-34	~7.0	n/a	~18.0	72.0	8.0	1.3	8.6	>120 h	[120]
In_2O_3-$ZnZrO_x$ & SAPO-34-S-a	17.0	55.8	1.6	85.0 [a]	11.1 [a]	14.5	7.7	>90 h	[42]
In_2O_3-$ZnZrO_x$ & SAPO-34-H-a	17.0	53.4	1.2	84.5 [a]	11.0 [a]	14.4	7.7	>90 h	[42]
$ZnAl_2O_4$ & SAPO-34	15.0	49.0	0.7	87.0 [a]	10.0 [a]	13.1	8.7	10 h	[121]
$ZnGa_2O_4$ & SAPO-34	13.0	46.0	1.0	86.0 [a]	11.0 [a]	11.2	7.8	10 h	[121]
ZnO-ZrO_2 & $Mn_{0.1}$SAPO-34	24.4	42.2	3.7	61.7	33.6	15.1	1.8	10 h	[122]
In-Zr (4:1) & SAPO-34	26.2	63.9	2.0	74.5 [a]	21.5 [a]	19.5	3.5	>140 h	[18]

[a] CO is not considered when calculating selectivity.

Wang et al. prepared kaolin-supported CuO-ZnO/SAPO-34 catalysts using kaolin as the support and raw material to prepare SAPO-34 molecular sieves. It was found that the resultant SAPO-34 molecular sieves showed a lamellar structure, relatively high crystallinity, and a larger specific surface area, which enabled the good dispersion of CuO-ZnO on the surface of the kaolin, and exposed more active sites for CO_2 conversion. The confinement effect of (CuO-ZnO)-kaolin/SAPO-34 catalysts could prevent methanol dissipation, and provided an increased driving force for the conversion of CO_2. Furthermore, the lamellar structure of SAPO-34 molecular sieves shortened the diffusion path of the intermediate product, and therefore enhanced the catalytic lifetime [116].

Gao et al. shown a selective hydrogenation process to directly convert CO_2 to light olefins via a bifunctional catalyst composed of a methanol synthesis catalyst (In_2O_3-ZrO_2) and a MTO catalyst (SAPO-34) by simple physical mixing. This bifunctional process exhibited an outstanding light olefin (C_2–$C_3^=$) selectivity of 80–90% with a CO_2 conversion of ~20% and superior catalyst stability, running 50 h without obvious deactivation. The excellent catalytic performance was ascribed to the hybrid catalyst that suppressed the usually uncontrollable surface polymerization of CH_x in conventional CO_2–FTS. This was the highest selectivity reported to date, which dramatically surpassed the value obtained from traditional Fe or Co CO_2–FTS catalysts (typically less than 50%) [43].

Similarly, Tan et al. evaluated CO_2 conversion to light olefins over an In_2O_3-ZrO_2/SAPO-34 hybrid catalyst. This hybrid catalyst combined a In_2O_3-ZrO_2 component, which would provide the benefit of oxygen vacancy to foster CO_2 activation for hydrogenation into methanol, and a SAPO-34 component, to provide sites for the dehydration of the formed methanol into light olefins (Figure 16a). The light olefin selectivity reached 77.6% with less than 5% CO formation, which was ascribed to the strong adsorption of CO_2 to defects in the In_2O_3 and ZrO_2 components, creating a large energy barrier that suppressed CO_2 dissociation into CO. The weaker acidity from In_2O_3-ZrO_2 suppressed the further hydrogenation of the generated light olefins to paraffins. The catalyst displayed excellent stability, running for 100 h without obvious deactivation (Figure 16b) [44].

Furthermore, Gao et al. discovered that a bifunctional catalyst with an appropriate proximity containing In−Zr oxide, which was responsible for the CO_2 activation, and SAPO-34, which was responsible for the selective C−C coupling, could greatly improve the CO_2 hydrogenation to lower olefins with excellent selectivity (80%) and high activity (35% CO_2 conversion) (Figure 17a). They showed that the incorporation of zirconium significantly improved the catalytic stability by preventing the sintering of the oxide nanoparticles caused by the increase in surface oxygen vacancies. No obvious deactivation was observed over 150 h (Figure 17b) [19].

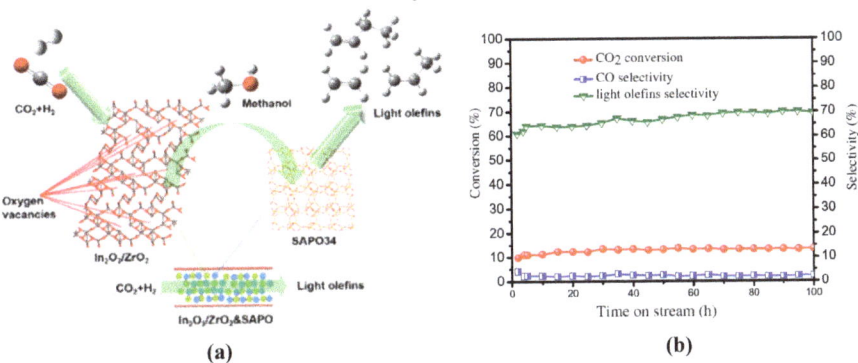

Figure 16. (a) Illustrated reaction mechanism over the bifunctional composite catalyst In$_2$O$_3$-ZrO$_2$/SAPO-34, and (b) the stability of the In$_2$O$_3$-ZrO$_2$/SAPO-34 composite catalyst for CO$_2$ hydrogenation to light olefins (testing conditions: P = 2.0 MPa, T = 573 K, GHSV = 2160 cm^{3}h^{-1}g$_{cat}$$^{-1}$). Adapted with permission from ref. [44]. Copyright 2019 Elsevier.

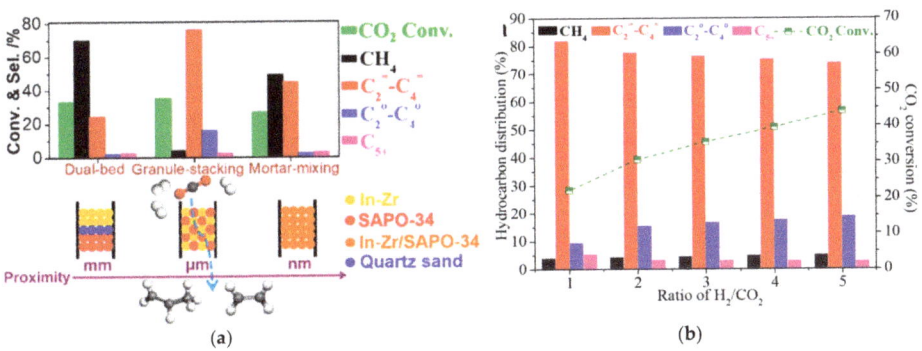

Figure 17. (a) Effect of the proximity of the active components on the CO$_2$ conversion and product selectivity, and (b) the catalytic stability of the composite catalyst In-Zr/SAPO-34 (testing conditions: T = 673 K, P = 3.0 MPa, GHSV = 9000 mL g$_{cat}$$^{-1}$ h^{-1}, molar ratio of H$_2$/CO$_2$/N$_2$ = 73/24/3, and mass ratio of oxide/zeolite = 2). Adapted with permission from ref. [19]. Copyright 2018 American Chemical Society.

Wang et al. developed a new catalyst system composed of a Zn$_{0.5}$Ce$_{0.2}$Zr$_{1.8}$O$_4$ solid solution and H-RUB-13 zeolite. This composite exhibited a remarkable C$_2$$^=$–C$_4$$^=$ yield as high as 16.1%, with a CO selectivity of only 26.5% due to the hindering of the RWGS reaction. It was demonstrated that methanol was first generated on the Zn$_{0.5}$Ce$_{0.2}$Zr$_{1.8}$O$_4$ solid solution via the formate–methoxyl intermediate mechanism, and was then converted into light olefins on H-RUB-13. By adjusting the H-RUB-13 acidity, the light olefin distribution can be effectively regulated, with propene and butene accounting for 90% of the light olefins [117].

Li et al. proposed a new synthetic strategy to prepare the bifunctional catalysts ZnZrO$_x$/bio-ZSM-5. Hierarchically porous structured bio-ZSM-5 was prepared by using a natural rice husk as a template, which was then integrated with the ZnZrO$_x$ solid solution nanoparticles by physical mixing. The derived bifunctional catalysts ZnZrO$_x$ and bio-ZSM-5 exhibited superior light olefin selectivity and stability due to their unique pore structure, which was advantageous for mass transport and coke formation inhibition. *CH$_x$O was identified to be the key intermediate formed on the ZnZrO$_x$ surface, and was transferred to the Brønsted acid sites in the bio-ZSM-5 for the subsequent conversion to light olefins. The addition of a Si promoter to the ZnZrO$_x$/bio-ZSM-5 catalyst prominently enhanced the light olefin selectivity. The ZnZrO$_x$/bio-ZSM-5—Si catalyst exhibited an outstanding light olefin selectivity of 64.4%, with a CO$_2$ conversion of 10% and an excellent stability without noticeable deactivation during 60 h on stream (Figure 18a). In addition, the proximity of the catalyst components plays a key role in light olefin selectivity. As seen in Figure 18b, increasing the proximity resulted in a greater olefin selectivity [118]. By incorporating proper amounts of Ce or Cr ions into indium oxides, the methanol selectivity is increased, along with a reduction in the CH$_4$ amount, as shown in Figure 19. Upon complexing with SAPO-34, a CO$_2$ conversion of 33.6% and a C$_2^=$–C$_4^=$ selectivity of 75.0% were achieved over InCrO$_x$(0.13)/SAPO-34, which was about 1.5–2.0 times those obtained on In$_2$O$_3$/SAPO-34 and In–Zr/SAPO-34. This is because the incorporation of Ce or Cr ions into In$_2$O$_3$ lattice sites promoted the generation of more surface oxygen vacancies, as shown in Figure 19a, and enhanced the electronic interaction of HCOO* with InCeO$_x$(0.13) and InCrO$_x$(0.13) surfaces, which decreased the free energy barrier and enthalpy barrier for the formation of HCOO* and CH$_3$OH. The composite catalysts also displayed excellent stability after 120 h on stream (Figure 19b) [119].

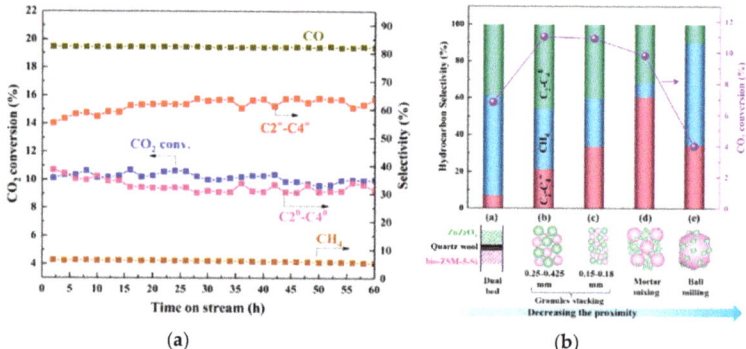

Figure 18. (**a**) Catalytic performance over the bifunctional composite catalysts ZnZrO$_x$/bio-ZSM-5—Si at TOS (testing conditions: mass of catalyst = 0.6 g, T = 653 K, P = 3 MPa, gas flow rate = 20 mL min^{-1}). (**b**) Effect of the proximity of the active components of ZnZrO$_x$/bio-ZSM-5-Si on the catalytic performance. Adapted with permission from ref. [118]. Copyright 2021 American Chemical Society.

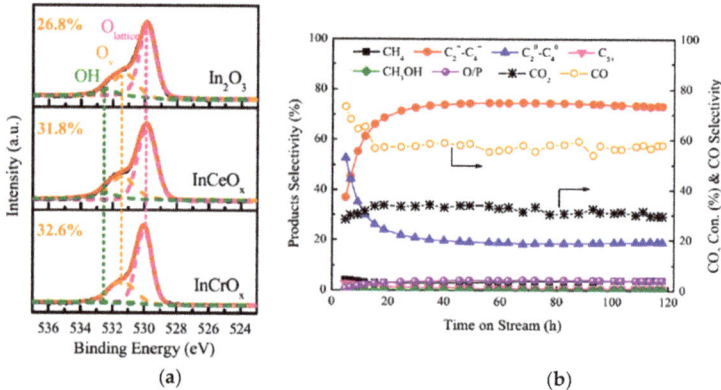

Figure 19. (**a**) The content of the surface oxygen vacancies (O$_v$) from O (1s) XPS spectra for the catalysts In$_2$O$_3$, InCeO$_x$(0.13), and InCrO$_x$(0.13). (**b**) The catalytic stability of the bifunctional composite catalysts InCrO$_x$(0.13) and SAPO-34 for CO$_2$ hydrogenation (testing conditions: H$_2$/CO$_2$ molar ratio = 3/1, T = 623 K, P = 3.5 MPa, and GHSV = 1140 mLg$_{cat}^{-1}$h^{-1}). Adapted with permission from ref. [119]. Copyright 2020 Elsevier.

Similarly, Li et al. developed a bifunctional composite catalyst ZnZrO/SAPO-34 containing a ZnOZrO$_2$ component to activate CO$_2$ and H$_2$ to form methanol, and a SAPO-34 component to perform C–C bond formation for the conversion of the produced methanol to light olefins. The derived dual function tandem catalyst exhibited an outstanding light olefin selectivity of 80% with good stability, and a CO$_2$ conversion of 12.6% (Figure 20a,b). The kinetic and thermodynamic coupling between the tandem reactions enabled the highly efficient conversion of CO$_2$ to lower olefins through the transfer and migration of CH$_x$O intermediate species [13].

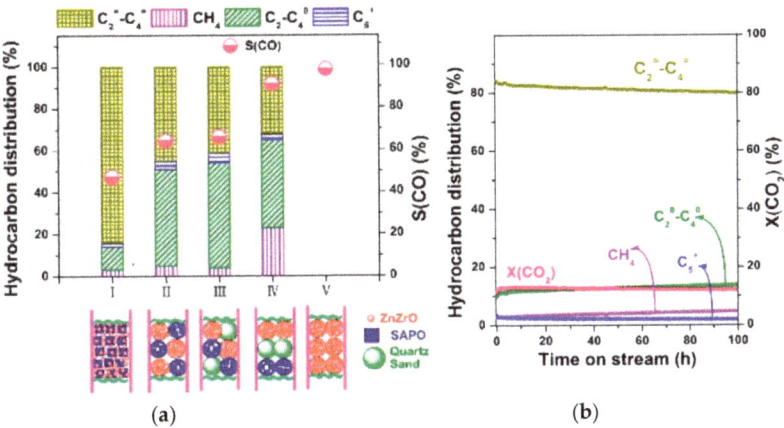

Figure 20. (a) CO$_2$ hydrogenation over the bifunctional composite catalyst ZnZrO/SAPO-34, with the effect of the proximity of the active components of ZnZrO and SAPO-34 on the catalytic performance. (b) The catalytic stability of the catalyst ZnZrO/SAPO-34 (testing conditions: T = 653K, P = 2 MPa, and GHSV = 3600 mL g$_{cat}^{-1}$ h^{-1}). Adapted with permission from ref. [13]. Copyright 2017 American Chemical Society.

Dang et al. advanced a series of dual function tandem catalysts containing In$_2$O$_3$-ZnZrO$_x$ oxides and various SAPO-34 zeolites with varying crystal sizes (0.4–1.5 mm) and pore structures. It was found that decreasing the crystal size of SAPO-34 could shorten the diffusion path from the surface to the acid sites inside the zeolite pores, thus favoring the mass transfer of intermediate species for efficient C–C coupling to produce lower olefins and enhance the selectivity of C$_2^=$–C$_4^=$. Interestingly, further HNO$_3$ post-treatment caused the formation of the SAPO-34 zeolites with a hierarchical structure comprised of micro-/meso-/macropores, and reduced the amount of the Brønsted acid sites, both of which led to a significant increase in the catalytic performance, with the C$_2^=$–C$_4^=$ selectivity reaching as high as 85% among all of the hydrocarbons (Figure 21a), a very low CH$_4$ selectivity of only 1%, and an O/P ratio of 7.7 at a CO$_2$ conversion of 17%. The C$_2^=$–C$_4^=$ selectivity is much higher than the maximum predicted by the Anderson–Schulz–Flory distribution over modified FTS catalysts. The composite catalysts also exhibited excellent stability after 90 h on stream (Figure 21b) [42].

Liu et al. synthesized bifunctional composite catalysts composed of a spinel binary metal oxide ZnAl$_2$O$_4$/ZnGa$_2$O$_4$ and SAPO-34, with the selectivity of C$_2$–C$_4$ olefins reaching 87% at CO$_2$ conversions of 15%. This study revealed that the oxygen vacancy site on metal oxides played a crucial role in the adsorption and activation of CO$_2$, while the -Zn-O- domain accounted for H$_2$ activation. It was demonstrated that the methanol reaction intermediates formed on the metal oxide, then converted to lower olefins at the Brønsted acid sites in SAPO-34 zeolite [121]. Tong et al. developed a dual-function composite catalyst, 13%ZnO-ZrO$_2$/Mn$_{0.1}$SAPO-34, and attained a high CO$_2$ conversion of 21.3% with a light olefin selectivity of 61.7%, and suppressed the selectivity of CO below 43% and the CH$_4$ selectivity below 4%. The fine-tuned acidity of zeolite by the addition of Mn and the granule stacking arrangement contributed to the excellent catalytic performance. Mn was embedded into the zeolite ionic structure to tune the acidity of the molecular sieve and limit secondary hydrogenation reactions. The granule stacking arrangement facilitated the tandem catalysis [122]. Dang et al. presented a series of bifunctional catalysts containing In-Zr composite oxides with different In/Zr atomic ratios and SAPO-34 zeolite for CO_2 conversion to light olefins. It was demonstrated that the inclusion of a certain amount of ZrO_2 could provide more oxygen vacancy sites (Figure 22a), stabilize the intermediates in the CO_2 hydrogenation, and prevent the sintering of the active nanoparticles. This, in turn, would lead to significantly enhanced catalytic activity, selectivity of hydrocarbons and stability for direct CO_2 hydrogenation to lower olefins at the relatively high reaction temperature of 653K. A light olefin selectivity as high as 80% at a CO_2 conversion rate of 27% and less than 2.5% methane selectivity was obtained over the optimized indium-zirconium/SAPO-34 bifunctional catalyst. The catalyst exhibited excellent stability for over 140 h without showing obvious deactivation (Figure 22b) [18].

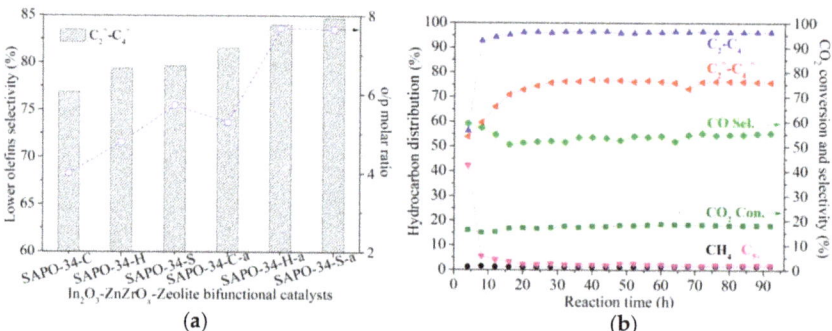

Figure 21. (a) Catalytic performance for CO_2 hydrogenation over a In_2O_3-$ZnZrO_x$ catalyst with different types of SAPO-34. (b) The stability of the bifunctional composite catalysts In_2O_3-$ZnZrO_x$/SAPO-34-H-a (testing conditions: T = 653 K, P = 3.0 MPa, GHSV = 9000 mLg$_{cat}^{-1}$h^{-1}, molar ratio of $H_2/CO_2/N_2$ = 73:24:3, mass ratio of oxide/zeolite = 0.5). Adapted with permission from ref. [42]. Copyright 2019 Wiley-VCH.

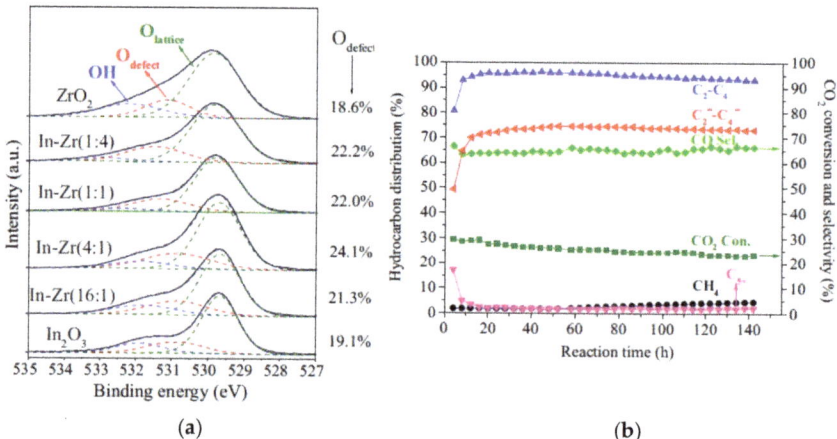

Figure 22. (a) XPS spectra (O1s) of various oxides and the content of surface oxygen vacancies (O_v). (b) The stability tests of the bifunctional composite catalysts In-Zr(4:1)/SAPO-34 (testing conditions: T = 653 K, P = 3.0 MPa, GHSV = 9000 mL gcat^{-1} h^{-1}, molar ratio of $H_2/CO_2/N_2$ = 73/24/3, and mass ratio of oxide/zeolite = 0.5). Adapted with permission from ref. [18]. Copyright 2018 Elsevier.

In summary, the majority of the catalysts tested for CO_2 hydrogenation to light olefins via the MeOH-mediated route involve two active components (metal oxides and zeolite), which are so-called bifunctional composite catalysts. In this section, multiple variations (acidity, particle size, proximity, oxygen vacancy) in the combination of methanol synthesis catalysts (Cu, Zn, In, Ce, Zr, etc. metal oxides) with various zeolites (SAPO-34 and ZSM5) have been reported to give improved olefin selectivity and catalyst stability by mitigating coke formation, reducing the particle size growth of active carbide species, and inhibiting inactive species formation for CO_2 hydrogenation to light olefins.

3.4. Structure Effect

The structure of the catalysts plays an important role in converting CO_2 to light olefins. In this section, we will report the recent progress on the ways in which morphology changes in both the Fe-based and methanol zeolite composite catalysts can improve the catalytic performance. Some representative catalysts on the structure effect for CO_2 hydrogenation to light olefins with improved catalyst stability are presented in Table 4.

Table 4. Some representative catalysts on the structure effect for CO_2 hydrogenation to light olefins.

Catalyst	CO_2 Conv., %	Selectivity, %				Yield, % C_2–$C_4^=$	O/P Ratio	Stability	Ref.
		CO	CH_4	C_2–$C_4^=$	C_2–C_4^0				
0.8K-2.4Fe-1.3Ti	35.0	36.3	22.0	60.0	8.0	21.0	7.5	200 h	[123]
Fe@NC-400	29.0	17.5	27.0	21.0	12.0	6.1	1.7	>15 h	[36]
K/Fe-Al-O Spinel E.1 nanobelts	48.0	16.0	10.0	52.0	5.0	24.0	3.1	120 h	[124]
MgH_2/Cu_xO	20.7	n/a	40.0	54.8	7.0	11.3	7.8	210 h	[125]
CZA/SAPO-34	50.0	3.0	10.0	62.0	25.0	33.0	2.5	12 h	[126]
FeK1.5/HSG	50	39	31	56	9.9	28	5.7	>120 h	[127]
Carbon-confined MgH_2 nano-lamellae	10.5	27.6	17.5	50.9	4.0	5.3	12.7	>2 h	[128]
Fe-Co/K-(CM-Al_2O_3)	41.0	12.4	33.7	41.1	6.4	14.4	6.4	50 h	[37]
ZnO-Y_2O_3 & SAPO-34	27.6	85.0	1.8	83.9 [a]	12.9 [a]	23.2	6.5	n/a	[40]
Cu-Zn-Al (6:3:1) oxide & HB zeolite	27.6	53.4	0.7	45.5 [b]	n/a	12.6 [b]	n/a	9 h	[129]

[a] Refers to hydrocarbon distribution %; [b] refers to C_2-C_{5+} hydrocarbons.

Wang et al. developed a layered metal oxides (LMO) structure, K-Fe-Ti, that displayed high catalytic activity, olefin selectivity and decent stability toward CO_2–FTS. The light olefin selectivity achieved approximately 60% with an olefin/paraffin ratio of 7.3 over the catalyst 0.8K-2.4Fe-1.3Ti (Figure 23). The LMO structure exfoliated through the acid treatment was found to weaken the interaction between Fe and Ti, which made it easier for the reduction and activation of iron oxides to form active iron carbide species that favored a shift from the RWGS to the FTS reaction. Meantime, C_2H_4 adsorption was hindered due to the low surface area of the LMO structure, contributing to higher olefin selectivity by inhibiting the secondary hydrogenation of primary olefins. The acid treatment played a key role in the formation of a slice structure that favored CO_2 conversion to light olefins with lower CO selectivity [123]. Fujiwara et al. found the composite catalysts obtained from the simple mixing of Cu–Zn–Al oxide together with HB zeolite, which was modified with 1,4-bis(hydroxydimethylsilyl) benzene, to be very effective for CO_2 hydrogenation to C_{2+} hydrocarbons. The modification of zeolite with the disilane compound made the catalysts' surface hydrophobic, a characteristic which was effective in preventing catalyst deactivation by the formation of water during CO_2 hydrogenation. The highest yield of C_{2+} hydrocarbons over the modified composite catalysts reached about 12.6 C-mol% at 573 K under a pressure of 0.98 Mpa. The diminishing of the deactivation of the strong acid sites of HB zeolite with the hydrophobic surface is the source of the enhanced catalytic activity [129].

Liu et al. synthesized a unique structure with ZnO and nitrogen-doped carbon (NC)-overcoated Fe-based catalysts (Fe@NC) (Figure 24), and found that the reaction rate increased by ~25%, while the O/P ratio increased from 0.07 to 1.68 when compared with the benchmark Fe_3O_4 catalyst. The inactive θ-Fe_3C phase disappeared, and the active phases (Fe_3O_4 and Fe_5C_2) formed for CO_2 hydrogenation. The introduction of NC to the surface of the Fe catalysts significantly boosted the catalyst activity, the selectivity toward light olefins, and the stability due to the enhanced metal–support-reactant interaction and interfacial charge transfer [36].

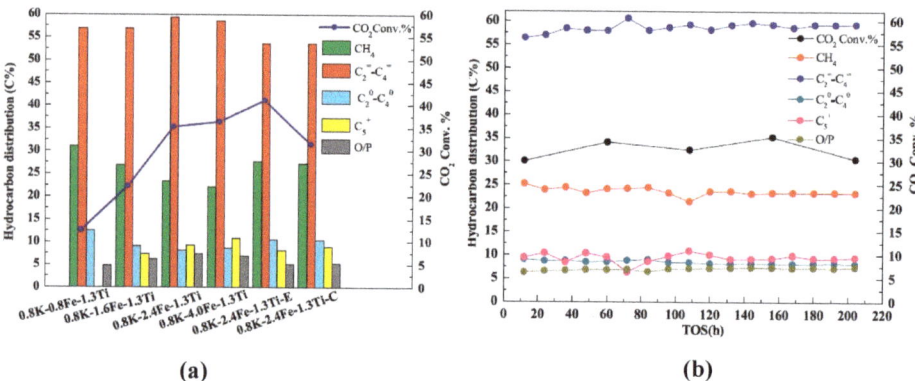

Figure 23. (**a**) Catalytic performance over different catalysts. (**b**) The catalytic stability of 0.8K-2.4Fe-1.3Ti at TOS (testing conditions: H_2/CO_2 molar ratio = 3/1, T = 593 K, P = 2.0 MPa and GHSV = 10,000 mL gcat^{-1}h^{-1}). Adapted with permission from ref. [123]. Copyright 2019 Elsevier.

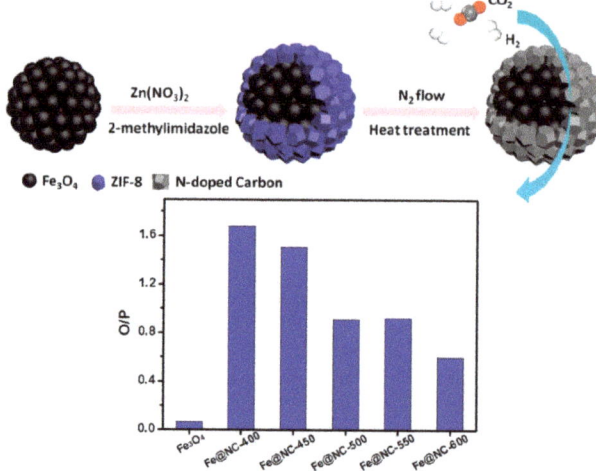

Figure 24. Schematic illustration of the formation of Fe@NC catalysts and the reaction for CO_2 hydrogenation. Adapted with permission from ref. [36]. Copyright 2019 American Chemical Society.

Numpilai et al. studied the hydrogenation of CO_2 to light olefins over Fe-Co/K-Al$_2$O$_3$ catalysts, and discovered that the pore sizes of the Al$_2$O$_3$ support had profound effects on the Fe$_2$O$_3$ crystallite size, the reducibility, the adsorption–desorption of CO_2 and H_2, and the catalytic performances. The highest olefins to paraffins ratio of 6.82 was obtained from the largest pore catalyst (CL-Al$_2$O$_3$) due to the suppression of the hydrogenation of olefins to paraffins by increasing the pore sizes of Al$_2$O$_3$ to eliminate diffusion limitation. The maximum light olefin yield of 14.38% was obtained over the catalyst with an appropriated Al$_2$O$_3$ pore size (49.7 nm) owing to the suppression of the olefins' hydrogenation and chain growth reaction [37].

The electrospun ceramic K/Fe-Al-O nanobelt catalysts synthesized by Elishav et al. showed a much higher CO_2 conversion of 48%, a C$_2$-C$_5$ olefin selectivity of 52%, and a high olefin/paraffin ratio of 10.4, while the K/Fe-Al-O spinel powder catalyst produced mainly

C_{6+} hydrocarbons. The enhanced olefin selectivity of the electrospun materials is related to a high degree of reduction of the surface Fe atoms due to the more efficient interaction with the K promoter [124].

A defect-rich MgH_2/Cu_xO hydrogen storage composite might inspire the catalysts' design for the hydrogenation of CO_2 to lower olefins. Chen et al. presented a defect-rich MgH_2/Cu_xO composite catalyst that achieved a $C_2^=-C_4^=$ selectivity of 54.8% and a CO_2 conversion of 20.7% at 623 K under a low H_2/CO_2 ratio of 1:5. It is the defective structure of MgH_2/Cu_xO that promotes CO_2 molecule adsorption and activation, while the electronic structure of MgH_2 was more conducive to the provision of lattice H^- for the hydrogenation of the CO_2 molecule. The lattice H^- could combine with the C site of the CO_2 molecule to promote the formation of Mg formate, which was further hydrogenated to lower olefins under a low H^- concentration [125]. The same group reported carbon-confined MgH_2 nano-lamellae which stored solid hydrogen for the hydrogenation of CO_2 to lower olefins and demonstrated a high selectivity under low H_2/CO_2 ratios. The high selectivity of lower olefins was attributed to the low concentration of solid hydrogen under low H_2/CO_2 ratios that suppressed the further hydrogenation of light olefins from Mg formate [128].

SAPO-34 molecular sieves were considered to be the best catalysts due to their excellent structure selectivity, suitable acidity, favorable thermal stability, and hydrothermal stability, as well as their high selectivity for light olefins. Tian et al. used Palygorskite as a silicon and partial aluminum source, and DEA, TEA, MOR and TEAOH as template agents to prepare SAPO-34 molecular sieves with higher purity. Composite catalysts of $CuO-ZnO-Al_2O_3$/SAPO-34 were prepared by mechanically mixing SAPO-34 molecular sieves with $CuO-ZnO-Al_2O_3$ (CZA), and a superb CO_2 conversion of 53.5%, a light olefin selectivity of 62.1% and a yield of 33.2% were obtained over the CZA/SAPO-34(TEAOH)HCl composite catalyst [126]. CO_2 conversion and product distribution are strongly dependent on the oxide composition and structure. Li et al. developed a bifunctional catalyst composed of $ZnO-Y_2O_3$ oxide and SAPO-34 zeolite that offered a CO_2 conversion of 27.6% and a light olefin selectivity of 83.6% [40].

Some Fe-containing catalysts can also be improved by creating unique architectures. Wei et al. created Fe-based catalysts with honeycomb-structured graphene (HSG) as the catalyst support and K as the promoter, and achieved the 59% selectivity of light olefins over a $FeK_{1.5}$/HSG catalyst. No obvious deactivation was observed within 120 h on stream (Figure 25). The excellent catalytic performance was ascribed to the confinement effect of HSG and the K promotion effect on the activation of inert CO_2 and the formation of iron carbide. The complex three-dimensional (3D) architecture of the porous HSG effectively impeded the sintering of the active sites' iron carbide nanoparticles (NPs). Meanwhile, CO_2 and H_2 could more easily permeate the mesoporous–macroporous framework of HSG and access the catalysts' active sites. Similarly, the generated light olefins could more easily emerge from the catalyst so as to avoid further unwanted hydrogenation [127].

Consequently, multiple reports indicate that the modification of the morphology of zeolite–methanol synthesis composites by creating core–shell configurations can have a beneficial effect [16,120,130]. For example, dual-function composite catalysts containing CuZnZr (CZZ) and SAPO-34 were synthesized by Chen et al. for the tandem reactions of CO_2 to methanol and methanol to olefins. The assembled core–shell CZZ@SAPO-34 catalyst, as shown in Figure 26, exhibited an enhanced light olefin selectivity of 72% and inhibited CH_4 formation due to reduced contact interface between CZZ and SAPO-34 and weakened hydrogenation ability at the metal sites. Furthermore, the addition of Zn reduced the acidity of SAPO-34; as a result, the secondary reactions of the primary olefins were significantly diminished (Figure 26) [120].

In summary of this section, the structure and the properties associated with the structure of the catalysts are pivotal for CO_2 hydrogenation to light olefins. The low surface area of the LMO structure could hinder the C_2H_4 secondary reaction, contributing to higher olefin selectivity. The surface modification of zeolite from hydrophilic to hydrophobic could prevent the catalyst deactivation caused by the formation of water. The unique structure of

Fe@NC enables phase transformation from the inactive (θ-Fe$_3$C) phase to active species (Fe$_3$O$_4$ and Fe$_5$C$_2$). Increasing the pore sizes of Al$_2$O$_3$ could eliminate the diffusion limitation for CO$_2$ and H$_2$. The electrospun ceramic K/Fe-Al-O nanobelt catalysts led to a high degree of the reduction of surface iron atoms. The defective structure of MgH$_2$/Cu$_x$O and carbon-confined MgH$_2$/C nano-lamellae could promote CO$_2$ adsorption and activation, with the electronic structure of MgH$_2$ offering lattice H$^-$ for CO$_2$ hydrogenation. The 3D architecture of the porous HSG could impede the sintering of the active sites' iron carbide NPs. The confinement of core–shell CZZ@SAPO-34 structure could increase the access frequency of the methanol intermediate to the active zeolite sites, consequently improving the light olefin selectivity.

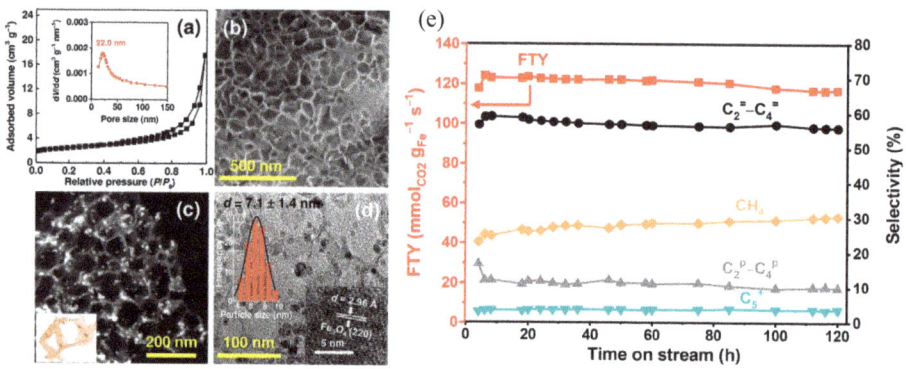

Figure 25. (**a**) N$_2$ physisorption isotherms, (**b**) SEM image, (**c**) HAADF–STEM image, and (**d**) TEM image and particle distribution of the FeK1.5/HSG catalyst. (**e**) CO$_2$ hydrogenation over the catalyst FeK1.5/HSG during a TOS of 120 h (testing conditions: mass of catalyst = 0.15 g, T = 613 K, P = 20 bar, H$_2$/CO$_2$ molar ratio = 3, and GHSV = 26 L h^{-1}g^{-1}). Adapted with permission from ref. [127]. Copyright 2018 American Chemical Society.

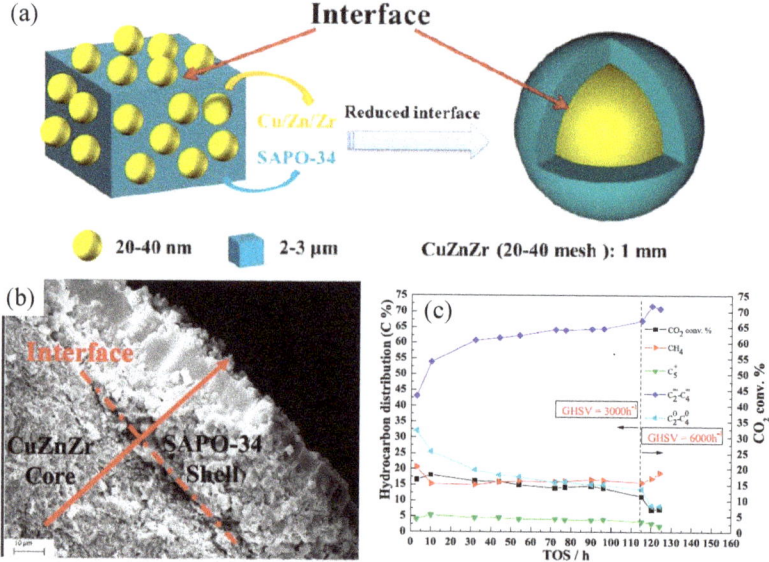

Figure 26. (**a**) Schematic illustration of the interface between CZZ and SAPO-34. (**b**) Core–shell interface of CZZ and SAPO-34. (**c**) Stability of the composite catalyst CZZ@Zn-SAPO-34 at TOS (testing conditions: H$_2$/CO$_2$ molar ratio = 3, T = 673 K). Adapted with permission from ref. [120]. Copyright 2019 Elsevier.

4. Conclusions

There is an urgent need to control CO_2 emissions in order to mitigate their negative impact on the environment. The catalytic conversion of CO_2 is an encouraging approach to mitigate CO_2 emissions by producing chemicals and fuels. A highly promising route is selective CO_2 hydrogenation to produce light olefins. The huge market demand for the lower olefins offers a great opportunity for the target technology to profoundly impact the scale of CO_2 utilization once it is developed with renewable hydrogen. Currently, there are two primary pathways (the CO_2–FTS route and the MeOH-mediated route) to produce light olefins from CO_2 hydrogenation in a one-step process. In the CO_2–FTS path, Fe is one of the most widely used components, while in the MeOH path, Cu/zeolite has been used the most. Even though significant efforts have been made, considerable challenges remain in the development of highly efficient catalysts with selective pathways to light olefins due to the thermodynamically stable nature of the CO_2 molecule, the complexity of the reaction networks, and catalyst deactivation. During CO_2 hydrogenation, the primary causes for catalyst deactivation are the sintering (or agglomeration) of metal particles, phase transformation at the catalyst's surface, and catalyst poisoning by water or carbonaceous deposits (i.e., coke). A firm grasp of the causes for deactivation is essential in order to develop a mitigation strategy and sustain a high selectivity toward the desired olefins during CO_2 hydrogenation. In this review, we summarized the reports published within five years on the effect of the promotors, metal oxide support, bifunctional composites and structure on the catalyst design in order to minimize catalyst deactivation.

Promoter effect: Alkali metals such as K and Na have been broadly used as promotors to control the electronic properties. Mn, Ce, and Ca metals have been used as structural promotors. Transition metals such as Zn, Co, Cu, V, Zr, etc., have been used as both electronic and structural promotors. With the inclusion of alkali promoters, Fe-based catalysts can possess higher olefin selectivity. The alkali metals act as electron donors to Fe metal centers, fostering CO_2 adsorption while decreasing their affinity with H_2, and consequently leading to a higher olefin yield. Some studies show that doping the catalyst with a second metal improves the olefin yield by forming a highly active interface. The second metal promoters may provide a way to tune the CO_2 and H_2 adsorption and activation, shifting the product distribution towards the desired hydrocarbons.

Support effect: Supporting the Fe-based species on supports such as SiO_2, CeO_2, m-ZrO_2, γ-Al_2O_3, TiO_2, ZSM-5, MgO, NbO HPCMs, MOFs, and β-Mo_2C may enhance the catalytic performance by improving the active metal dispersion and retarding the sintering of the active particles. The surface area, basicity, reducibility, oxygen vacancies, and morphology of the support played important roles—in most cases, with the presence of promoters (K, Zr, Cs)—in affecting the amount and particle size of the active carbide species; the synergy effect; the metal–support interaction; the strength and capacity of CO, CO_2, and H_2 adsorption on support; and the surface C/H ratio for CO_2 hydrogenation. By tuning the above-mentioned characteristics properly, the physically deposited carbon species, coke generation and metal sintering could be mitigated.

Bifunctional composite catalyst effect: The catalysts tested for CO_2 hydrogenation to light olefins via the MeOH-mediated route mainly involve two active components (metal oxides and zeolite), and so are called bifunctional composite catalysts. In this review, multiple variations (acidity, particle size, proximity, oxygen vacancy) of the combination of methanol synthesis catalysts (Cu, Zn, In, Ce, Zr, etc. metal oxides) with various zeolites (SAPO-34 and ZSM5) have been reported for enhanced olefin selectivity and catalyst stability by mitigating coke formation, reducing the particle size growth of active carbide species, and inhibiting inactive species formation for CO_2 hydrogenation to light olefins.

Structure effect: The structure of the catalysts plays a pivotal role in CO_2 hydrogenation to light olefins. The structures and properties (for example, LMO, the surface modification of zeolite from hydrophilic to hydrophobic, Fe@NC, the pore sizes of Al_2O_3, the defective structure of MgH_2/Cu_xO and carbon-confined MgH_2/C nano-lamellae, the 3D architecture of the porous HSG, and core–shell CZZ@SAPO-34) could be tuned to mitigate catalyst deactivation by retarding the sintering of active species and coke deposition, tolerating water formation and enabling favorable phase transformation for an enhanced light olefin yield and catalyst stability.

Despite the many advances made in catalytic development, especially with light olefin yield and stability, a novel catalytic system that is both economically viable and resistant to deactivation has not yet been achieved. Most research efforts have focused on the development of catalytic materials and the adjustment of properties and metal interactions for the desired catalyst activity and long-term stability. Future research directions for CO_2 hydrogenation should consider: (1) the further modification of the catalytic surface H/C molar ratio and the fostering of C-C coupling; (2) tuning the basicity and oxygen vacancies of the catalyst support to facilitate the CO_2 adsorption and activation; (3) examining more novel catalytic materials/structures to boost the catalyst stability; and (4) exploring more energy-saving catalysts for CO_2 hydrogenation to light olefins. In addition, in situ measurements using synchrotron-based techniques, such as X-ray adsorption spectroscopy (XAS), should be performed in order to understand the ways in which the local environment of the catalysts affects their activity, stability and efficient mitigation.

Author Contributions: Conceptualization, C.Z.; writing—original draft, C.Z., N.J.R., T.H., D.W., M.W., C.M., A.Z. and N.F.; writing—review and editing, C.Z. and N.J.R.; funding acquisition, C.Z.; resources, C.Z.; supervision, C.Z. All authors have read and agreed to the published version of the manuscript.

Funding: This work was supported by the National Science Foundation under Grant No. 1955521 (C.Z.).

Acknowledgments: The authors are grateful for the U.S. Department of Energy, the Office of Science, and the Office of Workforce Development for Teachers and Scientists under the Science Undergraduate Laboratory Internships Program (T.H. and A.Z.) and Visiting Faculty Program (C.Z.).

Conflicts of Interest: The authors declare no conflict of interest.

Abbreviations

FTS	Fisch–Tropsch Synthesis
MeOH	methanol
RWGS	reverse water–gas shift
MTO	MeOH-to-olefins
ASF	Anderson–Schulz–Flory
CTO	CO_2 to olefins
XRD	X-ray powder diffraction
PFR	plug-flow
CSTR	fully back-mixed reactors
TPSR	Temperature-programed surface reaction
XPS	X-ray photoelectron spectroscopy
SEM	scanning electron microscopy
TEM	transmission electron microscopy
HAADF–STEM	high-angle annular dark-field-scanning transmission electron microscopy
O/P ratio	olefins/paraffin ratio
NPs	nanoparticles
CNT	carbon nanotubes
FTY	Fe time yield
STY	space–time yields
HPCMs	hierarchical porous carbon monoliths
LMO	layered metal oxides
MOF	metal organic framework

References

1. Zhang, X.; Zhang, A.; Jiang, X.; Zhu, J.; Liu, J.; Li, J.; Zhang, G.; Song, C.; Guo, X. Utilization of CO_2 for aro-matics production over $ZnO/ZrO2$-ZSM-5 tandem catalyst. *J. CO_2 Util.* **2019**, *29*, 140–145. [CrossRef]
2. Atmospheric CO_2 Levels Defy the Pandemic to Hit Record High. Available online: https://newatlas.com/environment/atmospheric-co2-pandemic-record-concentrations/ (accessed on 15 October 2021).
3. Monastersky, R. Global carbon dioxide levels near worrisome milestone. *Nature* **2013**, *497*, 13–14. [CrossRef] [PubMed]
4. Da Silva, I.A.; Mota, C. Conversion of CO_2 to light olefins over iron-based catalysts supported on niobium oxide. *Front. Energy Res.* **2019**, *7*, 49. [CrossRef]
5. Keith, D.; Holmes, G.; Angelo, D.S.; Heidel, K. A process for capturing CO_2 from the atmosphere. *Joule* **2018**, *2*, 1573–1594. [CrossRef]
6. Centi, G.; Quadrelli, E.A.; Perathoner, S. Catalysis for CO_2 conversion: A key technology for rapid introduc-tion of renewable energy in the value chain of chemical industries. *Energy Environ. Sci.* **2013**, *6*, 1711. [CrossRef]
7. Dutta, A.; Farooq, S.; Karimi, I.A.; Khan, S.A. Assessing the potential of CO_2 utilization with an integrated framework for producing power and chemicals. *J. CO_2 Util.* **2017**, *19*, 49–57. [CrossRef]
8. Ma, Z.; Porosoff, M. Development of tandem catalysts for CO_2 hydrogenation to olefins. *ACS Catal.* **2019**, *9*, 2639–2656. [CrossRef]
9. Science Daily. Available online: https://www.sciencedaily.com/releases/2018/11/181108130533.htm (accessed on 15 August 2021).
10. Li, W.; Zhang, G.; Jiang, X.; Liu, Y.; Zhu, J.; Ding, F.; Liu, Z.; Guo, X.; Song, C. CO_2 hydrogenation on un-promoted and M-promoted Co/TiO_2 catalysts (M = Zr, K, Cs): Effects of crystal phase of supports and met-al–support interaction on tuning product distribution. *ACS Catal.* **2019**, *9*, 2739–2751. [CrossRef]
11. Wei, J.; Yao, R.W.; Ge, Q.J.; Wen, Z.Y.; Ji, X.W.; Fang, C.Y.; Zhang, J.X.; Xu, H.Y.; Sun, J. Catalytic hydrogena-tion of CO_2 to isoparaffins over Fe-based multifunctional catalysts. *ACS Catal.* **2018**, *8*, 9958–9967. [CrossRef]
12. Wei, J.; Ge, Q.; Yao, R.; Wen, Z.; Fang, C.; Guo, L.; Xu, H.; Sun, J. Directly converting CO_2 into a gaso-line fuel. *Nat. Commun.* **2018**, *8*, 15174. [CrossRef]
13. Li, Z.; Wang, J.; Qu, Y.; Liu, H.; Tang, C.; Miao, S.; Feng, Z.; An, H.; Li, C. Highly selective con-version of carbon dioxide to lower olefins. *ACS Catal.* **2017**, *7*, 8544–8548. [CrossRef]
14. Li, Z.L.; Qu, Y.Z.; Wang, J.J.; Liu, H.L.; Li, M.R.; Miao, S.; Li, C. Highly selective conversion of carbon dioxide to aromatics over tandem catalysts. *Joule* **2019**, *3*, 570–583. [CrossRef]
15. Liu, J.H.; Zhang, A.F.; Jiang, X.; Liu, M.; Zhu, J.; Song, C.S.; Guo, X.W. Direct transformation of carbon dioxide to value-added hydrocarbons by physical mixtures of Fe_5C_2 and K-modified Al_2O_3. *Ind. Eng. Chem. Res.* **2018**, *57*, 9120–9126. [CrossRef]
16. Xie, C.L.; Chen, C.; Yu, Y.; Su, J.; Li, Y.F.; Somorjai, G.A.; Yang, P.D. Tandem catalysis for CO_2 hydrogenation to C_2–C_4 hydrocarbons. *Nano Lett.* **2017**, *17*, 3798–3802. [CrossRef] [PubMed]
17. Liu, M.; Yi, Y.H.; Wang, L.; Guo, H.C.; Bogaerts, A. Hydrogenation of carbon dioxide to value-added chemi-cals by heterogeneous catalysis and plasma catalysis. *Catalysts* **2019**, *9*, 275. [CrossRef]
18. Dang, S.S.; Gao, P.; Liu, Z.Y.; Chen, X.Q.; Yang, C.G.; Wang, H.; Zhong, L.S.; Li, S.G.; Sun, Y.H. Role of zirconium in direct CO_2 hydrogenation to lower olefins on oxide/zeolite bifunctional catalysts. *J. Catal.* **2018**, *364*, 382–393. [CrossRef]
19. Gao, P.; Dang, S.S.; Li, S.G.; Bu, X.N.; Liu, Z.Y.; Qiu, M.H.; Yang, C.G.; Wang, H.; Zhong, L.S.; Han, Y.; et al. Direct production of lower olefins from CO_2 conversion via bifunctional catalysis. *ACS Catal.* **2018**, *8*, 571–578. [CrossRef]
20. Saeidi, S.; Najari, S.; Hessel, V.; Wilson, K.; Keil, F.J.; Concepción, P.; Suib, S.L.; Rodrigues, A.E. Recent ad-vances in CO_2 hydrogenation to value-added products-Current challenges and future directions. *Prog. Energy Combust. Sci.* **2021**, *85*, 100905. [CrossRef]
21. Gnanamani, M.K.; Jacobs, G.; Hamdeh, H.H.; Shafer, W.D.; Liu, F.; Hopps, S.D.; Thomas, G.A.; Davis, B.H. Hydrogenation of carbon dioxide over Co−Fe bimetallic catalysts. *ACS Catal.* **2016**, *6*, 913–927. [CrossRef]
22. Ren, T.; Patel, M.; Blok, K. Olefins from conventional and heavy feedstocks: Energy use in steam cracking and alternative processes. *Energy* **2006**, *31*, 425–451. [CrossRef]
23. Amghizar, I.; Vandewalle, L.A.; Geem, K.M.V.; Marin, G.B. New trends in olefin production. *Engineering* **2017**, *3*, 171–178. [CrossRef]
24. One, O.; Niziolek, A.M.; Floudas, C.A. Optimal production of light olefins from natural gas via the methanol intermediate. *Ind. Eng. Chem. Res.* **2016**, *5511*, 3043–3063.
25. Li, W.H.; Wang, H.Z.; Jiang, X.; Zhu, J.; Liu, Z.M.; Guo, X.W.; Song, C.S. A short review of recent advances in CO_2 hydrogenation to hydrocarbons over heterogeneous catalysts. *RCS Adv.* **2018**, *8*, 7651–7669. [CrossRef]
26. Porosoff, M.; Yan, B.; Chen, J. Catalytic reduction of CO_2 by H_2 for synthesis of CO, methanol and hydrocarbons: Challenges and opportunities. *Energy Environ. Sci.* **2016**, *9*, 62–73. [CrossRef]
27. Wang, D.; Xie, Z.; Porosoff, M.D.; Chen, J. Recent advances in carbon dioxide hydrogenation to produce ole-fins and aromatics. *Chem* **2021**, *7*, 1–35. [CrossRef]
28. Zhang, C.; Cao, C.; Zhang, Y.; Liu, X.; Xu, J.; Zhu, M.; Tu, W.; Han, Y. Unraveling the role of zinc on bimetal-lic Fe_5C_2–ZnO catalysts for highly selective carbon dioxide hydrogenation to high carbon α-olefins. *ACS Catal.* **2021**, *11*, 2121–2133. [CrossRef]
29. Wang, X.; Zhang, J.; Chen, J.; Ma, Q.; Fan, S.; Zhao, T.S. Effect of preparation methods on the structure and catalytic performance of Fe–Zn/K catalysts for CO_2 hydrogenation to light olefins. *Chin. J. Chem. Eng.* **2018**, *26*, 761–767. [CrossRef]

30. Numpilai, T.; Cheng, C.K.; Limtrakul, J.; Witoon, T. Recent advances in light olefins production from cata-lytic hydrogenation of carbon dioxide. *Proc. Saf. Environ. Prot.* **2021**, *151*, 401–427. [CrossRef]
31. Wei, J.; Sun, J.; Wen, Z.; Fang, C.; Ge, Q.; Xu, H. New insights into the effect of sodium on Fe_3O_4-based nano-catalysts for CO_2 hydrogenation to light olefins. *Catal. Sci. Technol.* **2016**, *6*, 4786. [CrossRef]
32. Wezendonk, T.A.; Sun, X.; Duglan, A.I.; van Hoof, A.J.F.; Hensen, E.J.M.; Kapteijn, F.; Gascon, J. Controlled formation of iron carbides and their performance in Fischer-Tropsch synthesis. *J. Catal.* **2018**, *362*, 106–117. [CrossRef]
33. Chaipraditgul, N.; Numpilai, T.; Cheng, C.K.; Siri-Nguan, N.; Sornchamni, T.; Wattanakit, C.; Limtrakul, J.; Witoon, T. Tuning interaction of surface-adsorbed species over $Fe/K-Al_2O_3$ modified with transition metals (Cu, Mn, V, Zn or Co) on light olefins production from CO_2 hydrogenation. *Fuel* **2021**, *283*, 119248. [CrossRef]
34. Gnanamani, M.K.; Jacobs, G.; Hamdeh, H.H.; Shafer, W.D.; Liu, F.; Hopps, S.D.; Thomas, G.A.; Davis, B.H. Hydrogenation of carbon dioxide over iron carbide prepared from alkali metal promoted iron oxalate. *Appl. Catal. A Gen.* **2018**, *564*, 243–249. [CrossRef]
35. Guo, L.; Sun, J.; Ge, Q.; Tsubaki, N. Recent advances in direct catalytic hydro-genation of carbon dioxide to valuable C_{2+} hydrocarbons. *J. Mater. Chem. A* **2018**, *6*, 23244–23262. [CrossRef]
36. Liu, J.; Zhang, A.; Jiang, X.; Zhang, G.; Sun, Y.; Liu, M.; Ding, F.; Song, C.; Guo, X. Overcoating the surface of Fe-based catalyst with ZnO and nitrogen-doped carbon toward high selectivity of light olefins in CO_2 Hy-drogenation. *Ind. Eng. Chem. Res.* **2019**, *58*, 4017–4023. [CrossRef]
37. Numpilai, T.; Chanel, N.; Poo-Arporn, Y.; Cheng, C.; Siri-Nguan, N.; Sornchamni, T.; Chareonpanich, M.; Kongkachuichay, P.; Yigit, N.; Rupprechter, G.; et al. Pore size effects on physicochemical properties of $Fe-Co/K-Al_2O_3$ catalysts and their catalytic activity in CO_2 hydrogenation to light olefins. *Appl. Surf. Sci.* **2019**, *483*, 581–592. [CrossRef]
38. Yang, H.; Zhang, C.; Gao, P.; Wang, H.; Li, X.; Zhong, L.; Wei, W.; Sun, Y. A review of the catalytic hydro-genation of carbon dioxide into value-added hydrocarbons. *Catal. Sci. Technol.* **2017**, *7*, 4580. [CrossRef]
39. Jiao, F.; Li, J.; Pan, X.; Xiao, J.; Li, H.; Ma, H.; Wei, M.; Pan, Y.; Zhou, Z.; Li, M.; et al. Selective conversion of syngas to light olefins. *Science* **2016**, *351*, 1065–1068. [CrossRef] [PubMed]
40. Li, J.; Yu, T.; Miao, D.; Pan, X.; Bao, X. Carbon dioxide hydrogenation to light olefins over ZnO/Y_2O_3 and SAPO-34 bifunctional catalysts. *Catal. Commun.* **2019**, *129*, 105711. [CrossRef]
41. Liu, X.; Wang, M.; Zhou, C.; Zhou, W.; Cheng, K.; Kang, J.; Zhang, Q.; Deng, W.; Wang, Y. Selective transfor-mation of carbon dioxide into lower olefins with a bifunctional catalyst composed of $ZnGa_2O_4$ and SAPO-34. *Chem. Commun.* **2018**, *54*, 140–143. [CrossRef]
42. Dang, S.; Li, S.; Yang, C.; Chen, X.; Li, X.; Zhong, L.; Gao, P.; Sun, Y. Selective transformation of CO_2 and H_2 into lower olefins over $In2O_3$-ZnZrOx/SAPO-34 bifunctional catalysts. *ChemSusChem* **2019**, *12*, 3582–3591. [CrossRef]
43. Gao, J.; Jia, C.; Liu, B. Direct and selective hydrogenation of CO_2 to ethylene and propene by bifunctional catalysts. *Catal. Sci. Technol.* **2017**, *7*, 5602–5607. [CrossRef]
44. Tan, L.; Zhang, P.; Cui, Y.; Suzuki, Y.; Li, H.; Guo, L.; Yang, G.; Tsubaki, N. Direct CO_2 hydrogenation to light olefins by suppressing CO by-product formation. *Fuel Process. Technol.* **2019**, *196*, 106174–106178. [CrossRef]
45. Porosoff, M.D.; Kattel, S.; Li, W.; Liu, P.; Chen, J.G. Identifying trends and descriptors for selective CO_2 conversion to CO over transition metal carbides. *Chem. Commun.* **2015**, *51*, 6988–6991. [CrossRef] [PubMed]
46. Porosoff, M.D.; Chen, J.G. Trends in the catalytic reduction of CO_2 by hydrogen over supported monometal-lic and bimetallic catalysts. *J. Catal.* **2013**, *301*, 30–37. [CrossRef]
47. Butt, J.B.; Petersen, E. *Activation, Deactivation and Poisoning of Catalysts*; Academic Press: Cambridge, MA, USA, 1988.
48. Bartholomew, C. Mechanisms of catalyst deactivation. *Appl. Catal. A Gen.* **2001**, *212*, 17–60. [CrossRef]
49. Moulijn, J.A.; van Diepen, A.E.; Kapteijn, F. Catalyst deactivation: Is it predictable? What to do? *Appl. Catal. A Gen.* **2001**, *212*, 3–16. [CrossRef]
50. Price, C.A.H.; Reina, T.R.; Liu, J. Engineering heterogenous catalysts for chemical CO_2 utilization: Lessons from thermal catalysis and advantages of yolk@shell structured nanoreactors. *J. Energy Chem.* **2021**, *57*, 304–324. [CrossRef]
51. Kliewer, C.E.; Soled, S.L.; Kiss, G. Morphological transformations during Fischer-Tropsch synthesis on a titania-supported cobalt catalyst. *Catal. Today* **2019**, *323*, 233–256. [CrossRef]
52. Eschemann, T.O.; de Jong, K.P. Deactivation behavior of Co/TiO_2 catalysts during Fischer–Tropsch synthesis. *ACS Catal.* **2015**, *5*, 3181–3188. [CrossRef]
53. Sun, J.T.; Metcalfe, I.S.; Sahibzada, M. Deactivation of $Cu/ZnO/Al_2O_3$ methanol synthesis catalyst by sintering. *Ind. Eng. Chem. Res.* **1999**, *38*, 3868–3872. [CrossRef]
54. Argyle, M.; Bartholomew, C. Heterogeneous catalyst deactivation and regeneration: A review. *Catalysts* **2015**, *5*, 145–269. [CrossRef]
55. Hansen, T.W.; DeLaRiva, A.T.; Challa, S.R.; Datye, A.K. Sintering of catalytic nanoparticles: Particle migration or Ostwald ripening. *Acc. Chem. Res.* **2013**, *46*, 1720–1730. [CrossRef] [PubMed]
56. Li, W.; Zhang, A.; Jiang, X.; Janik, M.J.; Qiu, J.; Liu, Z.; Guo, X.; Song, C. The anti-sintering catalysts: Fe–Co–Zr polymetallic fibers for CO_2 hydrogenation to C_2=–C_4=—Rich hydrocarbons. *J. CO_2 Util.* **2018**, *23*, 219–225. [CrossRef]
57. Lee, S.-C.; Kim, J.-S.; Shin, W.C.; Choi, M.-J.; Choung, S.-J. Catalyst deactivation during hydrogenation of carbon dioxide: Effect of catalyst position in the packed bed reactor. *J. Mol. Catal. A Chem.* **2009**, *301*, 98–105. [CrossRef]

58. Riedel, T.; Schulz, H.; Schaub, G.; Jun, K.W.; Hwang, J.S.; Lee, K.W. Fischer–Tropsch on iron with H_2-CO and H_2-CO_2 as synthesis gases: The episodes of formation of the Fischer–Tropsch regime and construction of the catalyst. *Top. Catal.* **2003**, *26*, 41–54. [CrossRef]
59. Wang, W.; Wang, S.P.; Ma, X.B.; Gong, J.L. Recent advances in catalytic hydrogenation of carbon dioxide. *Chem. Soc. Rev.* **2011**, *40*, 3703–3727. [CrossRef]
60. Zhang, Y.; Cao, C.; Zhang, C.; Zhang, Z.; Liu, X.; Yng, Z.; Zhu, M.; Meng, B.; Jing, X.; Han, Y.-F. The study of structure-performance relationship of iron catalyst during a full life cycle for CO_2 hydrogenation. *J. Catal.* **2019**, *378*, 51–62. [CrossRef]
61. Fiato, R.A.; Rice, G.W.; Miseo, S.; Soled, S.L. Laser Produced Iron Carbide-Based Catalysts. U.S. Patent 4,687,753, 18 August 1987.
62. Hegedus, L.L.; McCabe, R.W. Catalyst Poisoning. In *Catalyst Deactivation 1980 (Studies in Surface Science and Catalysis)*; Delmon, B., Froment, G.F., Eds.; Elsevier: Amsterdam, The Netherlands, 1980; Volume 6, pp. 471–505.
63. Bartholomew, C.H. Mechanisms of nickel catalyst poisoning. In *Catalyst Deactivation 1987 (Studies in Surface Science and Catalysis)*; Delmon, B., Froment, G.F., Eds.; Elsevier: Amsterdam, The Netherlands, 1987; Volume 34, pp. 81–104.
64. Schühle, P.; Schmidt, M.; Schill, L.; Rilsager, A.; Wasserscheid, P.; Albert, J. Influence of gas impurities on the hydrogenation of CO_2 to methanol using indium-based catalysts. *Catal. Sci. Technol.* **2020**, *10*, 7309–7322. [CrossRef]
65. Szailer, T.; Novák, É.; Oszkó, A.; Erdőhelyi, A. Effect of H_2S on the hydrogenation of carbon dioxide over supported Rh catalysts. *Top. Catal.* **2007**, *46*, 79–86. [CrossRef]
66. Rytter, E.; Holmen, A. Perspectives on the effect of water in cobalt Fischer–Tropsch synthesis. *ACS Catal.* **2017**, *7*, 5321–5328. [CrossRef]
67. Wu, J.; Luo, S.; Toyir, J.; Saito, M.; Takeuchi, M.; Watanabe, T. Optimization of preparation conditions and improvement of stability of Cu/ZnO-based multicomponent catalysts for methanol synthesis from CO_2 and H_2. *Catal. Today* **1998**, *45*, 215–220. [CrossRef]
68. Wu, J.; Saito, M.; Takeuchi, M.; Watanabe, T. The stability of Cu/ZnO-based catalysts in methanol synthesis from a CO_2-rich feed and from a CO-rich feed. *Appl. Catal. A Gen.* **2001**, *218*, 235–240. [CrossRef]
69. Huber, G.W.; Guymon, C.G.; Conrad, T.L.; Stephenson, B.C.; Bartholomew, C.H. Hydrothermal stability of Co/SiO_2 Fischer-Tropsch Synthesis Catalysts. *Stud. Surf. Sci. Catal.* **2001**, *139*, 423–429.
70. Van Steen, E.; Claeys, M.; Dry, M.E.; van de Loosdrecht, J.; Viljoen, E.L.; Visagie, J.L. Stability of nanocrystals: Thermodynamic analysis of oxidation and re-reduction of cobalt in water/hydrogen mixtures. *J. Phys. Chem. B* **2005**, *109*, 3575–3577. [CrossRef] [PubMed]
71. Iglesia, E. Design, synthesis, and use of cobalt-based Fischer-Tropsch synthesis catalysts. *Appl. Catal. A Gen.* **1997**, *161*, 59–78. [CrossRef]
72. Zhang, L.; Chen, K.; Chen, B.; White, J.L.; Resasco, D.E. Factors that determine zeolite stability in hot liquid water. *J. Am. Chem. Soc.* **2015**, *137*, 11810–11819. [CrossRef]
73. Rostrup-Neilson, J.; Trimm, D.L. Mechanisms of carbon formation on nickel-containing catalysts. *J. Catal.* **1977**, *48*, 155–165. [CrossRef]
74. Trimm, D.L. The formation and removal of coke from nickel catalyst. *Catal. Rev. Sci. Eng.* **1977**, *16*, 155–189. [CrossRef]
75. Trimm, D.L. Catalyst design for reduced coking (review). *Appl. Catal.* **1983**, *5*, 263–290. [CrossRef]
76. Bartholomew, C.H. Carbon deposition in steam reforming and methanation. *Catal. Rev. Sci. Eng.* **1982**, *24*, 67–112. [CrossRef]
77. Albright, L.F.; Baker, R.T.K. (Eds.) *Coke Formation on Metal Surfaces*; ACS Symposium Series; American Chemical Society: Washington, DC, USA, 1982; Volume 202, pp. 1–200.
78. Menon, P.G. Coke on catalysts-harmful, harmless, invisible and beneficial types. *J. Mol. Catal.* **1990**, *59*, 207–220. [CrossRef]
79. Chen, D.; Moljord, K.; Holmen, A. Methanol to Olefins: Coke Formation and Deactivation. In *Deactivation and Regeneration of Zeolite Catalysts*; Guisnet, M., Ribeiro, F., Eds.; Imperial College Press: London, UK, 2011; Volume 9, pp. 269–292.
80. Chen, D.; Rebo, H.P.; Grønvold, A.; Moljord, K.; Holmen, A. Methanol conversion to light olefins over SAPO-34: Kinetic modeling of coke formation. *Microporous Mesoporous Mater.* **2000**, *35–36*, 121–135. [CrossRef]
81. Chen, D.; Moljord, K.; Fuglerud, T.; Holmen, A. The effect of crystal size of SAPO-34 on the selectivity and deactivation of the MTO reaction. *Microporous Mesoporous Mater.* **1999**, *29*, 191–203. [CrossRef]
82. Nishiyama, N.; Kawaguchi, M.; Hirota, Y.; Van Vu, D.; Egashira, Y.; Ueyama, K. Size control of SAPO-34 crystals and their catalyst lifetime in the methanol-to-olefin reaction. *Appl. Catal. A Gen.* **2009**, *362*, 193–199. [CrossRef]
83. Müller, S.; Liu, Y.; Vishnuvarthan, M.; Sun, X.; van Veen, A.C.; Haller, G.L.; Sanchez-Sanchez, M.; Lercher, J.A. Coke formation and deactivation pathways on H-ZSM-5 in the conversion of methanol to olefins. *J. Catal.* **2015**, *325*, 48–59. [CrossRef]
84. Zhou, J.; Gao, M.; Zhang, J.; Liu, W.; Zhang, T.; Li, H.; Xu, Z.; Ye, M.; Liu, Z. Directed transforming of coke to active intermediates in methanol-to-olefins catalyst to boost light olefins selectivity. *Nat. Commun.* **2021**, *12*, 17. [CrossRef]
85. Yang, M.; Tian, P.; Wang, C.; Yuan, Y.; Yang, Y.; Xu, S.; He, Y.; Liu, Z. A top-down approach to prepare sili-coaluminophosphate molecular sieve nanocrystals with improved catalytic activity. *Chem. Commun.* **2014**, *50*, 1845–1847. [CrossRef]
86. Xi, D.; Sun, Q.; Chen, X.; Wang, N.; Yu, J. The recyclable synthesis of hierarchical zeolite SAPO-34 with ex-cellent MTO catalytic performance. *Chem. Commun.* **2015**, *51*, 11987–11989. [CrossRef]
87. Guo, G.; Sun, Q.; Wang, N.; Bai, R.; Yu, J. Cost-effective synthesis of hierarchical SAPO-34 zeolites with abundant intracrystalline mesopores and excellent MTO performance. *Chem. Commun.* **2018**, *54*, 3697–3700. [CrossRef]
88. Jiang, J.; Wen, C.; Tian, Z.; Wang, Y.; Zhai, Y.; Chen, L.; Li, Y.; Liu, Q.; Wang, C.; Ma, L. Manga-nese-promoted Fe_3O_4 microsphere for efficient conversion of CO_2 to light olefins. *Ind. Eng. Chem. Res.* **2020**, *59*, 2155–2162. [CrossRef]

89. Witoon, T.; Chaipraditgul, N.; Numpilai, T.; Lapkeatseree, V.; Ayodele, B.; Cheng, C.; Siri-Nguan, N.; Sorn-chamni, T.; Limtrakul, J. Highly active Fe-Co-Zn/K-Al$_2$O$_3$ catalysts for CO$_2$ hydrogenation to light olefins. *Chem. Eng. Sci.* **2021**, *233*, 116428. [CrossRef]
90. Zhang, Z.; Yin, H.; Yu, G.; He, S.; Kang, J.; Liu, Z.; Cheng, K.; Zhang, Q.; Wang, Y. Selective hydrogenation of CO$_2$ and CO into olefins over sodium- and zinc-promoted iron carbide catalysts. *J. Catal.* **2021**, *395*, 350–361. [CrossRef]
91. Yuan, F.; Zhang, G.; Zhu, J.; Ding, F.; Zhang, A.; Song, C.; Guo, X. Boosting light olefin selectivity in CO$_2$ hy-drogenation by adding Co to Fe catalysts within close proximity. *Catal. Today* **2021**, *371*, 142–149. [CrossRef]
92. Wei, C.; Tu, W.; Jia, L.; Liu, Y.; Lian, H.; Wang, P.; Zhang, Z. The evolutions of carbon and iron species modi-fied by Na and their tuning effect on the hydrogenation of CO$_2$ to olefins. *Appl. Surf. Sci.* **2020**, *525*, 146622. [CrossRef]
93. Malhi, H.S.; Sun, C.; Zhang, Z.; Liu, Y.; Liu, W.; Ren, P.; Tu, W.; Han, Y.-F. Catalytic consequences of the dec-oration of sodium and zinc atoms during CO$_2$ hydrogenation to olefins over iron-based catalyst. *Catal. Today* **2021**, *4*. [CrossRef]
94. Han, Y.; Fang, C.; Ji, X.; Wei, J.; Ge, Q.; Sun, J. Interfacing with carbonaceous potassium promoters boosts catalytic CO$_2$ hydrogenation of iron. *ACS Catal.* **2020**, *10*, 12098–12108. [CrossRef]
95. Liang, B.; Sun, T.; Ma, J.; Duan, H.; Li, L.; Yang, X.; Zhang, Y.; Su, X.; Huang, Y.; Zhang, T. Mn decorated Na/Fe catalysts for CO$_2$ hydrogenation to light olefins. *Catal. Sci. Technol.* **2019**, *9*, 456–464. [CrossRef]
96. Yang, S.; Chun, H.; Lee, S.; Han, S.J.; Lee, K.; Kim, Y.T. Comparative study of olefin production from CO and CO$_2$ Using Na-and K-promoted zinc ferrite. *ACS Catal.* **2020**, *10*, 10742–10759. [CrossRef]
97. Guo, L.; Sun, J.; Ji, X.; Wei, J.; Wen, Z.; Yao, R.; Xu, H.; Ge, Q. Directly converting carbon dioxide to linear α-olefins on bio-promoted catalysts. *Commun. Chem.* **2018**, *1*, 11. [CrossRef]
98. Liang, B.; Duan, H.; Sun, T.; Ma, T.; Liu, X.; Xu, J.; Su, X.; Huang, Y.; Zhang, T. Effect of Na promoter on Fe-based catalyst for CO$_2$ hydrogenation to alkenes. *ACS Sustain. Chem. Eng.* **2019**, *7*, 925–932. [CrossRef]
99. Ramirez, A.; Gevers, L.; Bavykina, A.; Ould-Chikh, S.; Gascon, J. Metal organic framework-derived iron cat-alysts for the direct hydrogenation of CO$_2$ to short chain olefins. *ACS Catal.* **2018**, *8*, 9174–9182. [CrossRef]
100. Zhang, J.; Su, X.; Wang, X.; Ma, Q.; Fan, S.; Zhao, T. Promotion effects of Ce added Fe–Zr–K on CO$_2$ hydro-genation to light olefins. *React. Kinet. Mech. Catal.* **2018**, *124*, 575–585. [CrossRef]
101. Numpilai, T.; Chanlek, N.; Poo-Arporn, Y.; Cheng, C.K.; Siri-Nguan, N.; Sornchamni, T.; Chareonpanich, M.; Kongkachuichay, P.; Yigit, N.; Rupprechter, G.; et al. Tuning interactions of sur-face-adsorbed species over Fe-Co/K-Al$_2$O$_3$ catalyst by different K content: Selective CO$_2$ hydrogenation to light olefins. *ChemCatChem* **2020**, *12*, 3306–3320. [CrossRef]
102. Wang, W.; Jiang, X.; Wang, X.; Song, C. Fe−Cu bimetallic catalysts for selective CO$_2$ hydrogenation to ole-fin-rich C$_2$+ hydrocarbons. *Ind. Eng. Chem. Res.* **2018**, *57*, 4535–4542. [CrossRef]
103. Kim, K.Y.; Lee, H.; Noh, W.Y.; Shin, J.; Han, S.J.; Kim, S.K.; An, K.; Lee, J.S. Cobalt ferrite nanoparticles to form a catalytic Co−Fe alloy carbide phase for selective CO$_2$ hydrogenation to light olefins. *ACS Catal.* **2020**, *10*, 8660–8671. [CrossRef]
104. Boreriboon, N.; Jiang, X.; Song, C.; Prasassarakich, P. Fe-based bimetallic catalysts supported on TiO$_2$ for se-lective CO$_2$ hydrogenation to hydrocarbons. *J. CO$_2$ Util.* **2018**, *25*, 330–337. [CrossRef]
105. Xu, Q.; Xu, X.; Fan, G.; Yang, L.; Li, F. Unveiling the roles of Fe-Co interactions over ternary spinel-type ZnCo$_x$Fe$_2$-$_x$O$_4$ catalysts for highly efficient CO$_2$ hydrogenation to produce light olefins. *J. Catal.* **2021**, *400*, 355–366. [CrossRef]
106. Numpilai, T.; Witoon, T.; Chanlek, N.; Limphirat, W.; Bonura, G.; Chareonpanich, M.; Limtrakul, J. Struc-ture–activity relationships of Fe-Co/K-Al$_2$O$_3$ catalysts calcined at different temperatures for CO$_2$ hydro-genation to light olefins. *Appl. Catal. A Gen.* **2017**, *547*, 219–229. [CrossRef]
107. Rodemerck, U.; Holeňa, M.; Wagner, E.; Smejkal, Q.; Barkschat, A.; Baerns, M. Catalyst development for CO$_2$ hydrogenation to fuels. *ChemCatChem* **2013**, *5*, 1948–1955. [CrossRef]
108. Gong, W.; Ye, R.-P.; Ding, J.; Wang, T.; Shi, X.; Russell, C.K.; Tang, J.; Eddings, E.G.; Zhang, Y.; Fan, M. Effect of copper on highly effective Fe-Mn based catalysts during production of light olefins via Fischer-Tropsch process with low CO$_2$ emission. *Appl. Catal.* **2020**, *278*, 119302. [CrossRef]
109. Dai, C.; Zhang, A.; Liu, M.; Song, F.; Song, C.; Guo, X. Facile one-step synthesis of hierarchical porous carbon monoliths as superior supports of Fe-based catalysts for CO$_2$ hydrogenation. *RSC Adv.* **2016**, *6*, 10831–10836. [CrossRef]
110. Hu, S.; Liu, M.; Ding, F.; Song, C.; Zhang, G.; Guo, X. Hydrothermally stable MOFs for CO$_2$ hydrogenation over iron-based catalyst to light olefins. *J. CO$_2$ Util.* **2016**, *15*, 89–95. [CrossRef]
111. Raghav, H.; Siva Kumar Konathala, L.N.; Mishra, N.; Joshi, B.; Goyal, R.; Agrawal, A.; Sarkar, B. Fe-decorated hierarchical molybdenum carbide for direct conversion of CO$_2$ into ethylene: Tailoring activity and stability. *J. CO$_2$ Util.* **2021**, *50*, 101607. [CrossRef]
112. Huang, J.; Jiang, S.; Wang, M.; Wang, X.; Gao, J.; Song, C. Dynamic evolution of Fe and carbon species over different ZrO$_2$ supports during CO prereduction and their effects on CO$_2$ hydrogenation to light olefins. *ACS Sustain. Chem. Eng.* **2021**, *9*, 7891–7903. [CrossRef]
113. Torrente-Murciano, L.; Chapman, R.S.L.; Narvaez-Dinamarca, A.; Mattia, D.; Jones, M.D. Effect of nanostructured ceria as support for the iron catalysed hydrogenation of CO$_2$ into hydrocarbons. *Phys. Chem. Chem. Phys.* **2016**, *18*, 15496–15500. [CrossRef] [PubMed]
114. Gu, H.; Ding, J.; Zhong, Q.; Zeng, Y.; Song, F. Promotion of surface oxygen vacancies on the light olefins synthesis from catalytic CO$_2$ hydrogenation over Fe-K/ZrO$_2$ catalysts. *Int. J. Hydrogen Energy* **2019**, *44*, 11808–11816. [CrossRef]

115. Owen, R.E.; Pluckinski, P.; Mattia, D.; Torrente-Murciano, L.; Ting, V.; Jones, M.D. Effect of support of Co-Na-Mo catalysts on the direct conversion of CO_2 to hydrocarbons. *J. CO_2 Util.* **2016**, *16*, 97–103. [CrossRef]
116. Wang, P.; Zha, F.; Yao, L.; Chang, Y. Synthesis of light olefins from CO_2 hydrogenation over (CuO-ZnO)-kaolin/SAPO-34 molecular sieves. *Appl. Clay Sci.* **2018**, *163*, 249–256. [CrossRef]
117. Wang, S.; Zhang, L.; Zhang, W.; Wang, P.; Qin, Z.; Yan, W.; Dong, M.; Li, J.; Wang, J.; He, L.; et al. Selective conversion of CO_2 into propene and butene. *Chem* **2020**, *6*, 3344–3363. [CrossRef]
118. Li, W.; Wang, K.; Zhan, G.; Huang, J.; Li, Q. Design and synthesis of bioinspired ZnZrOx & bio-ZSM-5 inte-grated nanocatalysts to boost CO_2 hydrogenation to light olefins. *ACS Sustain. Chem. Eng.* **2021**, *9*, 6446–6458.
119. Wang, S.; Wang, P.; Qin, Z.; Yan, W.; Dong, M.; Li, J.; Wang, J.; Fan, W. Enhancement of light olefin produc-tion in CO_2 hydrogenation over In2O3-based oxide and SAPO-34 composite. *J. Catal.* **2020**, *391*, 459–470. [CrossRef]
120. Chen, J.; Wang, X.; Wu, D.; Zhang, J.; Ma, Q.; Gao, X.; Lai, X.; Xia, H.; Fan, S.; Zhao, T. Hydrogenation of CO_2 to light olefins on CuZnZr@(Zn-)SAPO-34 catalysts: Strategy for product distribution. *Fuel* **2019**, *239*, 44–52. [CrossRef]
121. Liu, X.; Wang, M.; Yin, H.; Hu, J.; Cheng, K.; Kang, J.; Zhang, Q.; Wang, Y. Tandem catalysis for hydrogena-tion of CO and CO_2 to lower olefins with bifunctional catalysts composed of spinel oxide and SAPO-34. *ACS Catal.* **2020**, *10*, 8303–8314. [CrossRef]
122. Tong, M.; Chizema, L.G.; Chang, X.; Hondo, E.; Dai, L.; Zeng, Y.; Zeng, C.; Ahmad, H.; Yang, R.; Lu, P. Tan-dem catalysis over tailored ZnO-ZrO2/MnSAPO-34 composite catalyst for enhanced light olefins selectivity in CO_2 hydrogenation. *Microporous Mesoporous Mater.* **2021**, *320*, 111105. [CrossRef]
123. Wang, X.; Wu, D.; Zhang, J.; Gao, X.; Ma, Q.; Fan, S.; Zhao, T. Highly selective conversion of CO_2 to light ole-fins via Fischer-Tropsch synthesis over stable layered K-Fe-Ti catalysts. *Appl. Catal. A Gen.* **2019**, *573*, 32–40. [CrossRef]
124. Elishav, O.; Shener, Y.; Beilin, V.; Langau, M.V.; Herskowitz, M.; Shter, G.E.; Grader, G.S. Electrospun Fe−Al−O nanobelts for selective CO_2 hydrogenation to light olefins. *ACS Appl. Mater. Interfaces* **2020**, *12*, 24855–24867. [CrossRef] [PubMed]
125. Chen, H.; Liu, P.; Li, J.; Wang, Y.; She, C.; Liu, J.; Zhang, L.; Yang, Q.; Zhou, S.; Feng, X. MgH2/CuxO hydrogen storage composite with defect-rich surfaces for carbon dioxide hydrogenation. *ACS Appl. Mater. Interfaces* **2019**, *11*, 31009–31017. [CrossRef] [PubMed]
126. Tian, H.; Yao, J.; Zha, F.; Yao, L.; Chang, Y. Catalytic activity of SAPO-34 molecular sieves prepared by using palygorskite in the synthesis of light olefins via CO_2 hydrogenation. *Appl. Clay Sci.* **2020**, *184*, 105392. [CrossRef]
127. Wu, T.; Lin, J.; Cheng, Y.; Tian, J.; Wang, S.; Xie, S.; Pei, Y.; Yan, S.; Qiao, M.; Xu, H.; et al. Porous gra-phene-confined Fe-K as highly efficient catalyst for CO_2 direct hydrogenation to light olefins. *ACS Appl. Mater. Interfaces* **2018**, *10*, 23439–23443. [CrossRef]
128. Chen, H.; Liu, J.; Liu, P.; Wang, Y.; Xiao, H.; Yang, Q.; Feng, X.; Zhou, S. Carbon-confined magnesium hydride nano-lamellae for catalytic hydrogenation of carbon dioxide to lower olefins. *J. Catal.* **2019**, *379*, 121–128. [CrossRef]
129. Fujiwara, M.; Satake, T.; Shiokawa, K.; Sakurai, H. CO_2 hydrogenation for C2+ hydrocarbon synthesis over composite catalyst using surface modified HB zeolite. *Appl. Catal. B Environ.* **2015**, *179*, 37–43. [CrossRef]
130. Tan, L.; Wang, F.; Zhang, P.P.; Suzuki, Y.; Wu, Y.; Chen, J.; Yang, G.; Tsubaki, N. Design of a core–shell catalyst: An effective strategy for suppressing side reactions in syngas for direct selective conversion to light olefins. *Chem. Sci.* **2020**, *11*, 4097–4105. [CrossRef] [PubMed]

Review

The Role of CO₂ as a Mild Oxidant in Oxidation and Dehydrogenation over Catalysts: A Review

Sheikh Tareq Rahman [1], Jang-Rak Choi [1,2], Jong-Hoon Lee [1] and Soo-Jin Park [1,*]

[1] Department of Chemistry, Inha University, 100 Inharo, Incheon 22212, Korea; rahman19@inha.edu (S.T.R.); 22161120@inha.edu (J.-R.C.); boy834@naver.com (J.-H.L.)
[2] Evertech Enterprise Co. Ltd., Dongtansandan 2 gil, Hwaseong 18487, Korea
* Correspondence: sjpark@inha.ac.kr; Tel.: +82-32-876-7234

Received: 28 July 2020; Accepted: 8 September 2020; Published: 17 September 2020

Abstract: Carbon dioxide (CO_2) is widely used as an enhancer for industrial applications, enabling the economical and energy-efficient synthesis of a wide variety of chemicals and reducing the CO_2 levels in the environment. CO_2 has been used as an enhancer in a catalytic system which has revived the exploitation of energy-extensive reactions and carry chemical products. CO_2 oxidative dehydrogenation is a greener alternative to the classical dehydrogenation method. The availability, cost, safety, and soft oxidizing properties of CO_2, with the assistance of appropriate catalysts at an industrial scale, can lead to breakthroughs in the pharmaceutical, polymer, and fuel industries. Thus, in this review, we focus on several applications of CO_2 in oxidation and oxidative dehydrogenation systems. These processes and catalytic technologies can reduce the cost of utilizing CO_2 in chemical and fuel production, which may lead to commercial applications in the imminent future.

Keywords: carbon dioxide; soft oxidant; oxidation; dehydrogenation; nano-catalyst

---

1. Introduction

Global warming is an imminent threat to our planet. It is essential to diminish the emission of greenhouse gases, especially carbon dioxide (CO_2), to slow global warming. Different sources of CO_2 emissions are a significant part to dictate by the ignition of liquid, solid, and gaseous chemicals. Rising atmospheric CO_2 concentrations and the increasing temperature of the planet's surface have increased public awareness of this problem [1]. CO_2 is utilized in the in the manufacturing industries, which is mostly released by the combustion of fossil fuels [2]. Among different products, methanol and formic acid can be synthesized from CO_2 which is used directly as fuels or to generate H_2 on demand at low temperatures (<100 °C) [1]. However, CO_2 can be used efficiently in various value-adding strategies and research pursuits which are converted waste emissions into valuable chemicals products, such as hydrocarbons and oxygenates [3]. Electrochemical activation technologies and conversion of CO_2 and H_2O into hydrocarbons has seen a marked increase in research activity over the past few years [4]. The impressive separation and utilization of CO_2 technologies in a higher challenge of organizing than other gases [5]. The development of CO_2 for novel approaches can add value to CO_2 recycling as it may result in commercially useful carbon-based products. Today, CO_2 is used commercially in the production of pharmaceuticals, air-conditioning systems, beverages, fertilizers, inert agents for food packaging, the water treatment process, fire extinguishers, and other applications. To achieve sustainable economic growth, it is crucial to study the conversion of CO_2 into carbon-based chemicals and materials. Industrial companies use massive amounts of CO_2 to enhance oil restoration. Biomass conversion to fuels also utilizes CO_2 [6]. Recently, Drisdell et al. [7] reported that oxide-derived copper catalysts are better at making fuel products from CO_2. According to Drisdell group, CO_2 is initially converted into carbon monoxide under first conditions for producing fuel and then hydrocarbon chains are developed. Oxide-derived catalysts are better, not because they have oxygen remaining while they reduce carbon monoxide, but because the process of removing the oxygen creates a metallic copper structure that is better at forming ethylene. Using solar energy to convert CO_2 into most needed fuels has the potential to decrease global warming impact (GWI) and produce sustainable fuels at large scale [8]. A great deal of research has focused on combining heteroatoms in the carbon structure to improve the exchangeable action of CO_2 along with the adsorbent surfaces over the past few years [6].

CO_2 utilization has recently become an alluring sector of research, as it will help to alleviate climate change and reduce industrial operating costs. Globally, CO_2 capture and utilization are significant goals for chemicals and materials scientists [9]. Researchers are working to diminish the negative effects of CO_2 by adsorption [10,11], reduction, and fixation as well as through the development of metal-organic frameworks (MOFs), zeolites, polymers and micro-porous carbons [12]. Currently, CO_2 is used in an impenetrable phase under harsh conditions as an active promoter, making it a green substitute for organic compounds [13]. There are several limitations of dense phase CO_2 media, including the high pressures required to assure sufficient solubility of various transition metal catalysts and low reaction rates [14]. Jessop et al. proposed, as a solution to the solubility issue, an exchangeable process using 1,8-diazabicyclo-[5.4.0]-undec-7-ene. Additionally, they were able to eliminate partition steps by adjusting polarities with the use of CO_2 [15]. Another way to utilize CO_2 is to use it as an oxygen source. Park et al. demonstrated the mild oxidant character of CO_2 in the oxidative dehydrogenation of various types of alkyl benzene in both liquid and gaseous phases [16,17]. Using CO_2 in catalytic reactions offers other advantages; for instance, absorption of hydrogen from alkanes, alkyl aromatics, and alcohols using CO_2 as a reactant to create CO and oxygen species results in an expedited reaction rate, increased conversion, higher yield, and suppression of oxidation [16]. The presence of both CO_2 and O_2 increases the reaction rates as well as the conversion and selectivity. This process is performed under subcritical pressures of CO_2 and involves CO_2-promoted systems (CPS) instead of a CO_2-expanded system, as evidenced by the low-pressure approach as well as catalytic CO_2 activation. Recently developed CO_2 use technologies require the utilization of high-energy initiators [18]. Although great progress has been made in the carbon dioxide sector, there remain innate limitations, such as high-energy requirements, and the hydrogen recession. CO_2 has various benefits as a mild oxidant over several oxidizing promoters tested for oxidative dehydrogenation reaction, such as dry air, SO_2, and N_2O [19]. C1 products such as methanol, formic acid has become possible to produce with high initial selectivity by using CO_2 over simple metal-based catalysts [4]. CO_2 promotes selectivity by contaminating the non-selective species of several catalysts, preventing the production of several by-products [20]. Additionally, CO_2 is used as a carbon source in the decoking process ($C + CO_2 = 2CO$) which sustains catalytic activity [21]. Therefore, the oxidative dehydrogenation (ODH) reaction with CO_2 primarily considered to be a gas-interposed adaptation of the catalyst surface. This affects the diffusion, adsorption, and red-ox characteristics of the catalyst [22]. In recent years, the CO_2 conversion process has been utilized in various sectors, including thermo-chemical [23], photochemical [24], solar-chemical [25], electrochemical [26], biochemical [27] and homogenous catalysis [28] (Scheme 1).

In this review, we discuss a way to improve various technologies using CO_2 as a mild oxidant and enhancer for the production of essential chemicals. The purpose of this review is to illustrate the limitations and scope of CO_2 utilization and to highlight the advantages and challenges of carbon management. The use of CO_2 as a feedstock is a major goal, which could have a modest impact in practice, but may impart a significant symbolic effect on worldwide carbon stability. The further impact would result from the use of CO_2 as a soft oxidant and for oxidative dehydrogenation in catalytic reactions. Bartholomew et al. [29] studied the oxidizing capability of different gases in the gasification of coke. Their activities were ranked as follows: O_2 (105) > H_2O (3) > CO_2 (1) > H_2 (0.003). This demonstrates that carbon dioxide is less active than molecular oxygen and water, but still offers high oxidative capacity. However, carbon dioxide has the greatest heat capability among the commonly used alternative gases. Furthermore, CO_2 can reduce the occurrence of hotspots, which cause problems, such as catalyst deactivation, runaway temperature, and undesirable product oxidation.

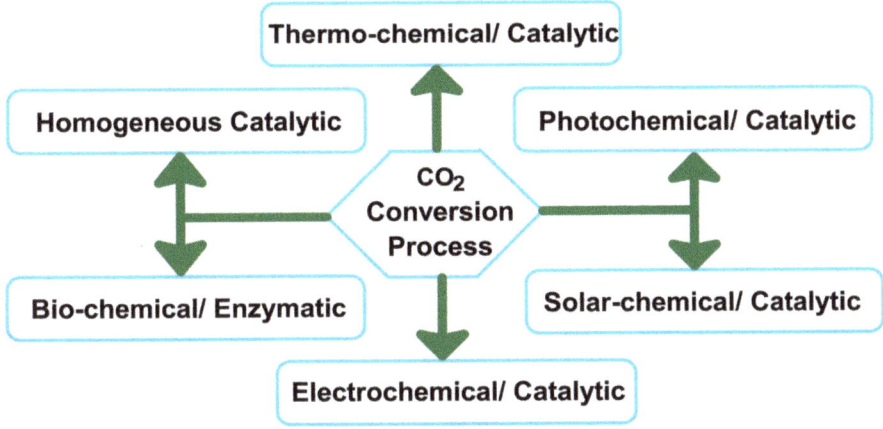

Scheme 1. The various chemical processes for CO₂ conversion.

2. Effect of CO$_2$ in Oxidation

2.1. Influence of CO$_2$ on Oxidation of Cyclohexene

The impact of CO$_2$, at various concentrations, was investigated on the oxidation of cyclohexene which is a small and symmetric molecule, similar to many starting compounds in chemical synthesis (Scheme 2) [30]. The results revealed that O$_2$/CO$_2$ conversion (%) was higher than O$_2$/N$_2$ conversion (%) rate. However, at a gas ratio of 0.066 O$_2$:CO$_2$:N$_2$ (Table 1, entry 1), cyclohexene was not converted. Park et al. revealed the positive impact of carbon dioxide on mesoporous metal-free oxidation carbon nitride (MCN) catalysts [31]. These mesoporous MCN elements exhibit oxygen-carrying capabilities which are effective sites for oxidation. Additionally, the large nitrogen quantity in the CN matrix acts as a CO$_2$-philic exterior for the incitation of CO$_2$. Molecular oxygen promotes this synergy, allowing for the oxidation of cyclic olefins and improving the conversion of cyclic olefins with better selectivity. In-between the conversion of the O$_2$/CO$_2$ and the O$_2$/N$_2$, Park et al. observed the enhancive performance as a premier time, which can be expressed as ΔC (%) and can be calculated using the Equation (1):

$$\Delta C(\%) = \frac{\left(C_{O_2/CO_2}\right) - \left(C_{O_2/N_2}\right)}{\left(C_{O_2/CO_2}\right) + \left(C_{O_2/N_2}\right)} \times 100 \tag{1}$$

where,

$\left(C_{O_2/CO_2}\right)$ = Conversion in O$_2$/CO$_2$ and $\left(C_{O_2/N_2}\right)$ = Conversion in O$_2$/N$_2$

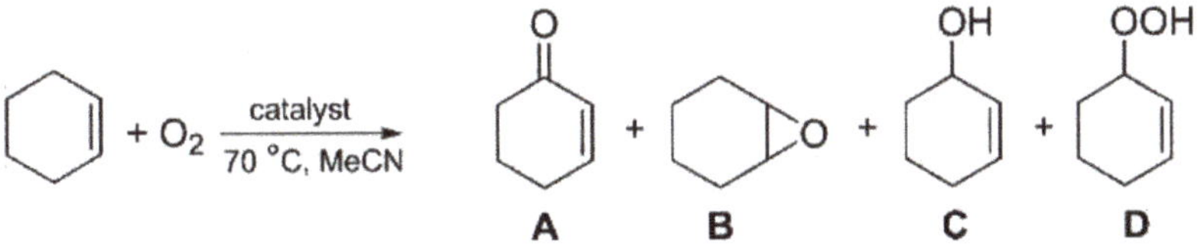

Scheme 2. Cyclohexene oxidation reaction over catalyst. (Redrawn from [30]; copyright (2018), WILEY-VCH). (**A**) = 2-cyclohexene-1-one, (**B**) = cyclohexene oxide, (**C**) = 2-cyclohexene-1-ol, (**D**) = 2-cyclohexene-1-hydroperoxide). Reaction conditions: 10 bar O$_2$; 2.5 mL cyclohexene; 0.5 mL cyclohexane(IS); 10 mg catalyst; 15 mL MeCN; stirred in an autoclave (1000 rpm); 70 °C; 16 h.

Table 1. Effect of the CO_2 on oxidation of cyclohexene over MCN Ref [31] (Reproduced from [31]; copyright (2011), Royal Society of Chemistry).

Entry	Gas Ratio (PSI) [a]	Conversion (%) O_2/CO_2	Conversion (%) O_2/N_2	$\Delta C(\%)$ [b]
1	0.066	0	0	0
2	0.142	16	9	28
3	0.230	25	18	16.3
4	0.333	33	24	15.7
5	0.454	34	24	15.7

Reaction conditions: 20 mg Melamine mesoporous carbon nitride (M-MCN), 10 mL Dimethylformamide (DMF), temperature 373 K, Pressure 80 PSI, time 10 h; Estimated by Gas Chromatography (GC) analysis. [a] PSI = Pounds per Square Inch, [b] Conversion (%) of cyclic olefin.

The efficiency of CO_2 in the oxidation of cyclohexene at varying CO_2 concentrations is shown in Table 1. Higher conversions were achieved by the O_2/CO_2 system. The results showed that the conversion of cyclohexene was nothing at a content of 0.066 O_2 (entry 1). This is Possibly due to the low frictional pressure of O_2, which is deficient to drive the reaction. Further, the $\Delta C\%$ value was higher for higher concentrations of CO_2. No meaningful change of $\Delta C\%$ was demonstrated for gas ratios beyond 0.333 in the catalytic process, demonstrating the impregnation of activity.

CO_2 has been used with metal-supported systems that were observed to produce a per-oxycarbonate species which are highly active in oxidation reactions. Aresta et al. were reported the composition of a metal per-oxycarbonate species, as determined by spectroscopic analysis [32]. A process for the production of per ox-carbonate has acceded in Scheme 3. Park et al. investigated the oxidation of alkyl aromatics via an EPR analysis using a metal carbonate catalyst. They demonstrated the production of metal per-oxycarbonate groups in the presence of carbon dioxide by the hyperfine cracking of manganese. Yoo et al. [20] observed the production of per-oxycarbonate on Fe/Mo/DBH (deboronated borosilicate molecular sieve); the production of per-oxycarbonate is illustrated in Figure 1. All of the catalytic schemes discussed above involve transitional metal catalysts and CO_2 coupled with oxygen. The resulting enhancement over traditional metal oxide systems in O_2/CO_2 mixtures may occur because of an oxygen exchange between O_2 and CO_2, which would increase the rate of the reaction. During isotope-labeling studies, these types of exchanges have been detected by Iwata et al. [33] using different metal oxide structures.

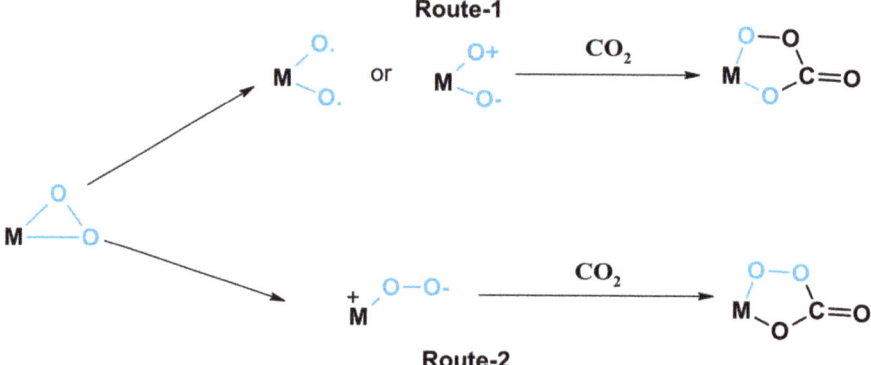

Scheme 3. Per-oxycarbonate production reaction mechanisms (Redrawn from [32]; copyright (1996), American Chemical Society).

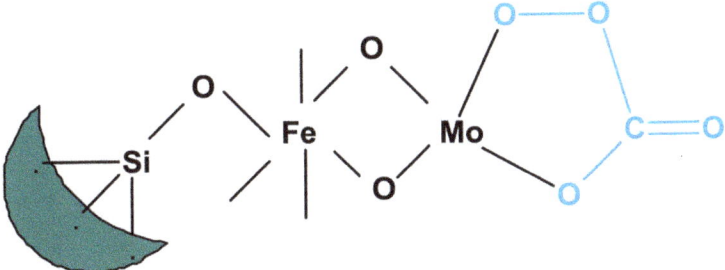

Figure 1. Per ox-carbonate over Fe/Mo/DBH in the O_2/CO_2 system (Reproduced from [20]; copyright (1993), Elsevier (Amsterdam, The Netherlands)).

2.2. Promotional Effect of CO_2 on Oxidation of Cyclic Olefins

Park et al. demonstrated the use of CO_2 as a promoter for the oxidation of cyclic olefins with mesoporous carbon nitrides (CN) as a metal-free catalyst in the presence of molecular oxygen. Analysis of the surface characteristics of the catalyst after the reaction revealed the presence of carbamate, confirmed by a new band in the FTIR spectrum at 1419 cm^{-1}. This measurement illustrated the incitation of CO_2 owing to the accumulation of surface carbamate. This surface carbamate can then react with the cyclic olefins, assisted by the catalyst. After the reaction, the IR spectra showed the presence of extra bands at 2174 and 2115 cm^{-1}, possibly due to a gaseous CO doublet. However, these absorption bands were not present before the reaction. This analysis exposed the production of CO, which is revealed to the increased catalytic activity to credit to carbon dioxide sharing as an 'oxygen atom' onset [31,32]. The production of CO was previously observed in nitrogen including heterocyclic systems [34–36]. The positive impact of CO_2 in the oxidation of cyclic olefin was quantified by measuring the catalytic performance using various reactants, cyclopentene ($n = 1$), cyclohexene ($n = 2$), cyclooctene ($n = 4$), and cyclododecene ($n = 8$) (Table 2). The epoxide selectivity was greater in O_2/CO_2 than O_2/N_2, suggesting that in the presence of CO_2, the mechanism may be altered to improve the conversion and selectivity. The blend of gaseous from the autoclave was studied by IR spectroscopy to better understand the positive impact of CO_2. In the reaction with no oxidant and source oxygen, it was presumed that CO_2 is reduced to CO and aldehyde is oxidized to carboxylic acid in the same process. The reaction may have occurred via the addition of carbon dioxide to the quickly produced Breslow intermediate A to produce the hydroxy carboxylate B and the tautomer C (Scheme 4) [34]. Possibly, the following intermediate can lose CO and hydroxide to support benzoic acid. Additionally, it was observed that intermediate D is supplicated in the oxidative esterification of aldehydes with CO. Interestingly, phenylglyoxylic acid was revealed to nucleophilic heterocyclic carbenes (NHC) under similar experimental conditions wherein phenylglyoxylic acid was switched to benzoic acid. (Scheme 5). Under mild experimental conditions, CO_2 was utilized in an NHC-intermediated conversion of the aldehyde to the carboxylic acid.

Table 2. Enhancive role of CO_2 on cyclic olefins oxidation (Reproduced from [31]; copyright (2011), Royal Society of Chemistry).

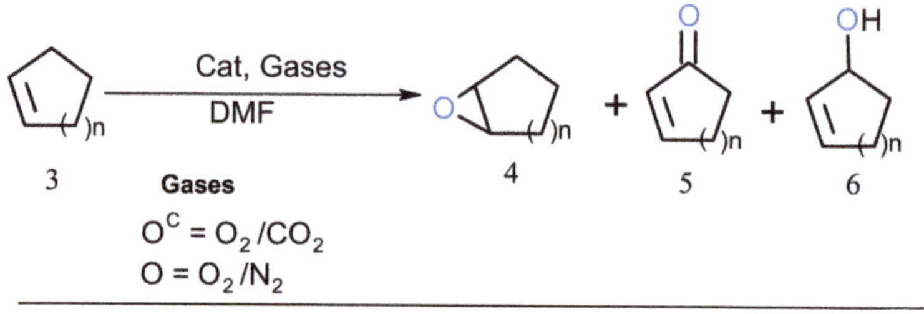
Cat, Gases
DMF
3
4
5
6
Gases
O^C = O2/CO2
O = O2/N2

Entry	n	Gas	Conversion of 3 (%)	Selectivity (%)			
				4	5	6	ΔC (%)
1	1	O^C	40	37	24	29	12.6
		O	31	30	22	40	-
2	2	O^C	33	30	21	49	15.8
		O	24	25	16	53	-
3	4	O^C	21	> 99	-	-	27.0
-	-	O	12	> 99	-	-	-
4	8	O^C	17	> 99	-	-	30.7
-	-	O	9	> 99	-	-	-

Reaction conditions: 20 mg Melamine mesoporous carbon nitride (M-MCN), 10 mL Dimethylformamide (DMF), temperature 373 K, Pressure 80 PSI, gas ratio 0.333, time 10 h; Produced analyzed by GC and GC-MS.

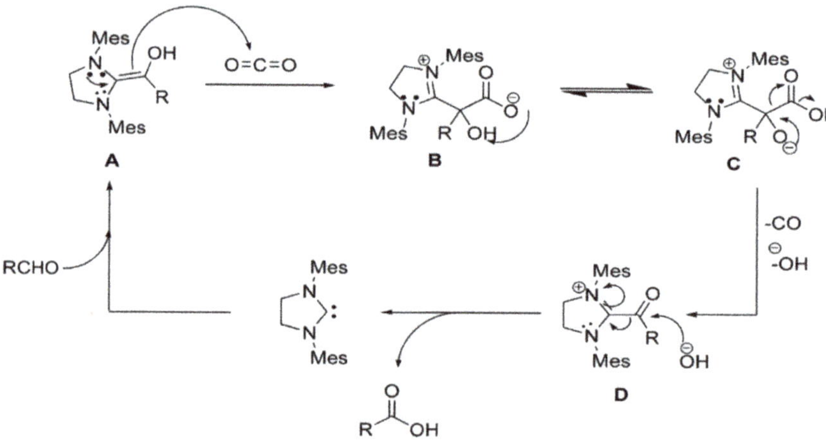
O=C=O
A
B
C
D
-CO
-OH
RCHO
Mes

Scheme 4. Proposed Mechanism for aldehyde assisted CO_2 to carboxylic acid process. (Reprinted from [34]; copyright (2010), American Chemical Society).

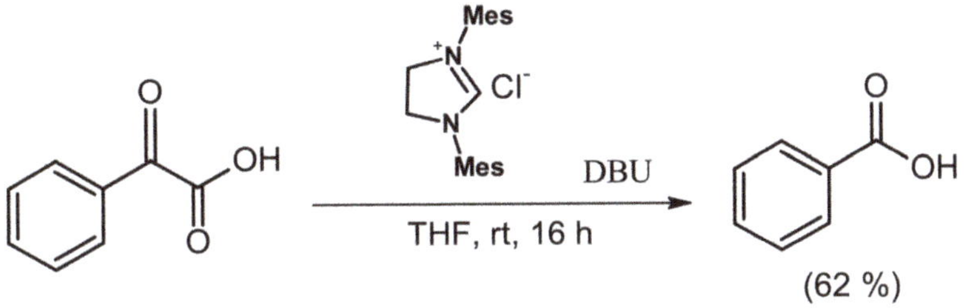

Scheme 5. Phenylglyoxylic acid to benzoic acid reaction with NHC-intermediate. (Reprinted from [34]; copyright (2010), American Chemical Society).

2.3. Influence of CO_2 on Oxidation of p-Xylene

It was proposed that in the O_2-CO_2 system, metal peroxy-carbonate groups assist as oxygen transfer promoters to the oxyphilic substrate. Aresta et al. [32] also reported that the presence of O-O bonds in Rh (η^2-O_2) complexes imply the accumulation of metal per-oxycarbonate during the other oxidation reaction. They demonstrated that CO_2 promotes the oxidative ability of O_2 over the RhCl(Pet$_2$-Ph)$_3$ catalyst. In the presence of CO_2 over the metal-based structure was found to be formation of peroxycarbonate species which are more active than hydrogen peroxide in oxidation reaction [37]. Park et al. [22] reported the performance of carbon dioxide in the liquid-phase oxidation reaction of toluene, p-tolu-aldehyde, and p-xylene with O_2 over an MC-based catalyst (Co/Mn/Br). The reaction rate, selectivity, and the conversion were all enhanced by the co-presence of CO_2. This enhancement was attributed to the creation of per-oxycarbonate species, as determined by electron paramagnetic resonance (EPR) analysis of the reaction with and without carbon dioxide. A hyperfine manganese arrangement was noticed in the existence of CO_2, confirming the formation of a per-oxycarbonate species.

Additionally, Park et al. observed the oxidation of different alkyl aromatics applying MC-supported catalysts [22]. Oxidations were carried out using O_2 as the oxidant (with N_2) and compared to reactions in the presence of both O_2 and CO_2. The conversion of p-xylene without CO_2 (Table 3) was 57.2%, whereas the conversion of p-xylene was increased to 66.8% in the presence of CO_2. Moreover, in O_2/CO_2, the yield of terephthalic acid was improved. The Amoco Chemical Research Laboratory studied the activation of CO_2 in the gas-state of p-xylene oxidation to p-tolualdehyde and terephthaldehyde over the chemical vapor deposition (CVD) of Fe/Mo/DBH [20]. The oxidation reaction was performed in two feed streams varying compositions, including p-xylene with O_2/N_2/He and p-xylene with O_2/N_2/CO_2. The catalytic activity is shown in both the feeds at various temperatures in Figure 2, as shown in the figure, p-xylene conversion in the existence of CO_2 in O_2 was greater than the absence of CO_2 in O_2. This improved conversion was connected to the production of per-oxycarbonate groups over the catalyst surface. Furthermore, in the existence of CO_2, the secondary reactions also emerged more remarkable, possibly due to the acidity of the CO_2 molecules adsorbed onto the DBH matrix. In comparison with O_2 alone, the conversion of p-xylene was higher in the co-presence of CO_2 at all temperatures (Figure 2). The O_2/N_2/CO_2 feed system, resulted in a higher conversion of p-xylene and greater selectivity towards benzaldehyde at temperatures from 300 °C to 375 °C (Table 4). It was observed that no carbon dioxide was formed by the burning of p-xylene over the catalyst at 375 °C; however, in the O_2/N_2/He feed system, the formation of CO_2 started (10.7%) at 300 °C and significantly increased (20.2%) at 375 °C. Thus, CO_2 performed as a co-oxidant for the gas-phase p-xylene oxidation reaction with oxygen. Yoo et al. [20] also reported a significant enhancement in the conversion of p-xylene, p-ethyl toluene, and o-xylene in the presence of CO_2 at varying temperatures.

Table 3. Effect of CO_2 on the MC-type catalyst for oxidation of *p*-xylene (Reproduced from [22]; copyright (2012), Royal Society of Chemistry).

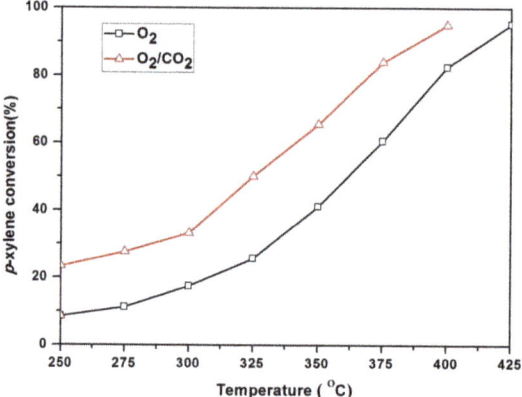

Gas	Conversion of 1 (%)	Yield mol (%)				
		2	3	4	5	6
O_2	57.2	17.7	47.9	2.8	1.7	29.2
O_2/CO_2	66.8	34.8	36.9	1.7	2.4	24.2

Reaction conditions: Temperature 170 °C, time 3 h, Mesoporus carbon (MC) type catalyst with transition metal additive (Co/Mn/Br), Co 100 ppm, Mn 200 ppm, Br 300 ppm [38].

Figure 2. Promotional role of CO_2 on Fe/Mo/DBH for the oxidation of *p*-xylene.

Table 4. Enhancive effect of CO_2 on oxidation of *p*-xylene (Reproduced from [20]; copyright (1993), Elsevier). Reaction conditions: WHSV: 0.22 h^{-1}, contact time: 0.21 s, gas flowrate: 400 sccm, Feed gas 1: 0.1% *p*-xylene, 1% O_2, 1% N_2 in He. Feed gas 2: 0.1% *p*-xylene, 1% O_2, 1% N_2 in commercial grade CO_2.

Temperature (°C) Feed [a]	300		350		375	
	1	2	1	2	1	2
p-Xylene (Con.%)	17.6	33.3	41.2	65.5	60.7	84.1
Product selectivity (mol%)	-	-	-	-	-	-
p-Tolu-aldehyde	57.9	57.2	50.2	40.9	40.6	40.6
Terephthaldehyde	16.4	27.5	23.5	33.6	32.6	30.2
Benzaldehyde	1.3	1.5	2.4	2.7	2.7	3.1
Maleic anhydride	0.0	0.0	2.4	6.0	5.8	13.7
Toluene	6.2	6.9	3.5	4.8	3.1	4.8
Trimethyl biphenyl methane	7.5	6.8	1.7	0.6	0.4	0.0
CO	0.0	0.0	0.6	3.4	4.7	7.5
CO_2	10.7	0.0	15.6	0.0	20.2	0.0

[a] Feed gas 1: O_2/N_2/He, Feed gas 2: $O_2/N_2/CO_2$.

2.4. Oxidation of p-Toluic Acid and p-Methyl-Anisole

CO_2 acts as a promoter in catalytic systems and as a co-oxidant with O_2 resulting in improved reaction kinetics, more desirable product distributions, better selectivity, and higher conversion. Initially, Aresta et al. [32] reported that carbon dioxide enhanced the oxidative characteristics of dioxide in transition metal systems. Park et al. [38] studied the use of Co/Mn/Br catalysts in the fluid-phase oxidation of olefins. Interestingly, they observed the expansion effect of carbon dioxide on mesoporous carbon nitride (MCN) catalytic systems, whereas the CO_2-promoted system was fabricated by them on the oxidation of alkyl-aromatics. In the presence of CO_2, the conversion of p-toluic acid over the metal carbonate (MC) catalyst was increased by 12% (Table 5) compared to oxidation in O_2 alone. Furthermore, the yield of terephthalic acid increased from 58.2% to 64.9%. These data demonstrate that the catalytic activity is significantly enhanced by CO_2. Interestingly, over an MC-supported catalytic system, the main product of the oxidation of p-methyl-anisole is p-methoxy phenol (Table 6) along with a limited number of other products, such as p-anisaldehyde and p-anisic acid. However, the yield of p-anisaldehyde has increased the presence of CO_2, again demonstrating the capacity of CO_2 to sustain mono-oxygen transfer.

Table 5. p-toluic acid oxidation with CO_2 on an MC-type catalyst (Reproduced from [22]; copyright (2012), Royal Society of Chemistry).

Gas	Conversion of 1 (1%)	Yield mol (%)	
		2	3
O_2	60.9	58.2	3.7
O_2/CO_2	72.7	64.9	10.6

Reaction conditions: 6 mL p-toluic acid, 0.1183 g $CoBr_2$, 0.1587 g $Mn(OAc)_2 \cdot 4H_2O$ in 24 mL HOAc, temperature 190 °C, time 3 h; P_{CO2} = 0-6 atm, P_{O2} = 2 atm [38].

Table 6. Performance of CO_2 on oxidation of p-methylanisole (Reproduced from [22]; copyright (2012), Royal Society of Chemistry).

Gas	Conversion of 1 (%)	Yield mol (%)		
		2	3	4
O_2	94.9	2.05	0.83	92
O_2/CO_2	98	7.7	0.38	90

Reaction conditions: 43.5 mmol p-methylanisole, 0.6 mmol $CoBr_2$, 0.6 mmol $Co(OAc)_2$, 0.6 mmol $Mn(OAc)_2 \cdot 4H_2O$ in 30 g HOAc, total pressure 12 atm (P_{CO2} = 0-2 atm, P_{O2} = 2,3,6 atm, P_{N2} balance). temperature 120 °C, time 3 h [38].

3. Performance of CO_2 in Oxidative Dehydrogenation

3.1. Influence of CO_2 on Dehydrogenation of Ethyl Benzene

Styrene is typically formed by the dehydrogenation of ethyl benzene under the steam on a metal oxide catalyst in an adiabatic reactor [39]. There are several limitations to this process, including thermodynamic drawbacks, low conversion rates, high endothermic energy ($\Delta H°_{298}$ = 123.6 kJ mol^{-1}), huge energy destruction, and catalyst deactivation by coke production [40]. An alternative method of styrene production is the oxidative dehydrogenation reaction with O_2; however, this results in the burning of large quantities of valuable hydrocarbons. In this context, the use of CO_2 in the oxidative dehydrogenation of ethyl benzene may prove useful [39–61]. Zhang et al. [43] confirmed coke deposition using spectroscopy and reported the deactivation of a ceria catalyst without CO_2 present. In a two-step, reaction mechanism for the dehydrogenation of ethyl benzene to produce styrene with H_2 in the initial step and in the presence of CO_2, ejection of H_2 through a reverse water-gas shift (RWGS) reaction was also demonstrated [41]. Kovacevic et al. revealed the results of CeO_2 catalyst morphology (i.e., rods vs. cubes vs. particles) in the presence and absence of CO_2 [42]. They reported that in the presence of CO_2 cubic catalysts showed higher initial benzene selectivity, and about two times more activity per m^2 compared to the reaction without CO_2. Interestingly, the number of oxygen species was increased by the presence of CO_2. They also observed that these additional oxygen molecules were expended in the ethyl benzene conversion, demonstrating their performance as active sites for styrene formation. Periyasamy et al. reported that in the ODH reaction the conversion of ethyl benzene (EB) was 50% and the selectivity for styrene was 93% at gas hourly space velocity (GHSV) 2400 h^{-1}. They also observed that the conversion and selectivity increased with enhancing oxidant flow ratio, up to GHSV 2400 h^{-1}.

Park et al. [49] reported on the use of SBA-15 as a beneficial backing for a ceria-zirconium (25:75) combined oxide catalyst for oxidative dehydrogenation of ethyl benzene utilizing carbon dioxide. Ce-Mn oxide nanoparticles enclosed inside carbon nanotubes (CNTs) were used for the oxidative dehydrogenation of ethyl benzene with CO_2 acting as a soft oxidant. The high diffusion and the encapsulation effect of CNTs resulted in excellent performance of the entrapped catalysts. Correlated to CeO_2 support CNTs, the restriction result of CNT pathways enhanced the communication between carbon nanotube (CNT) inner walls and CeO_2 particles, which is orderly, convinced the misrepresentation of CeO_2 crystal lattice which is advertised CeO_2 reduction and invigoration of CeO_2 surface lattice oxygen. The unique process of promoting oxidative catalytic activity the addition of CO_2 was reported by Zhang et al. [44]. They observed that multi-walled carbon nanotubes (MWCNTs) have a significant quantity of surface hydroxyl groups which are produced by an alkali-supported hydrothermal method after ball milling. The MWCNTs can mostly arrange the active sites for the oxidative dehydrogenation of ethyl benzene (EB) in the existence of CO_2. Figure 3a shows the conversion of ethyl benzene over various types of MWCNTs at 3 hr. The HMWCNTs-OH exhibits significant catalytic activity, indicating that the surface hydroxyl groups are the active sites for the oxidative dehydrogenation of ethyl benzene. The O1s spectra of HMWCNTs-B-OH identified by XPS is shown in Figure 3b. Figure 3c demonstrates the production of carbonyl groups in the reaction. The results indicate that CO_2 acts effectively as a soft oxidant, directly oxidizing -OH groups into carbonyl groups. As shown in Figure 3d, CO and H_2 were also identified as byproducts for the reaction, indicating that CO_2 is reduced in the RWGS reaction. CO_2 activation occurs via electron donation from the surface of the catalyst to the anti-bonding orbital of CO_2 [62]. However, ethyl benzene (EB) can be activated for oxidative dehydrogenation (ODH) by donating an electron to the acidic portion of the catalyst surface.

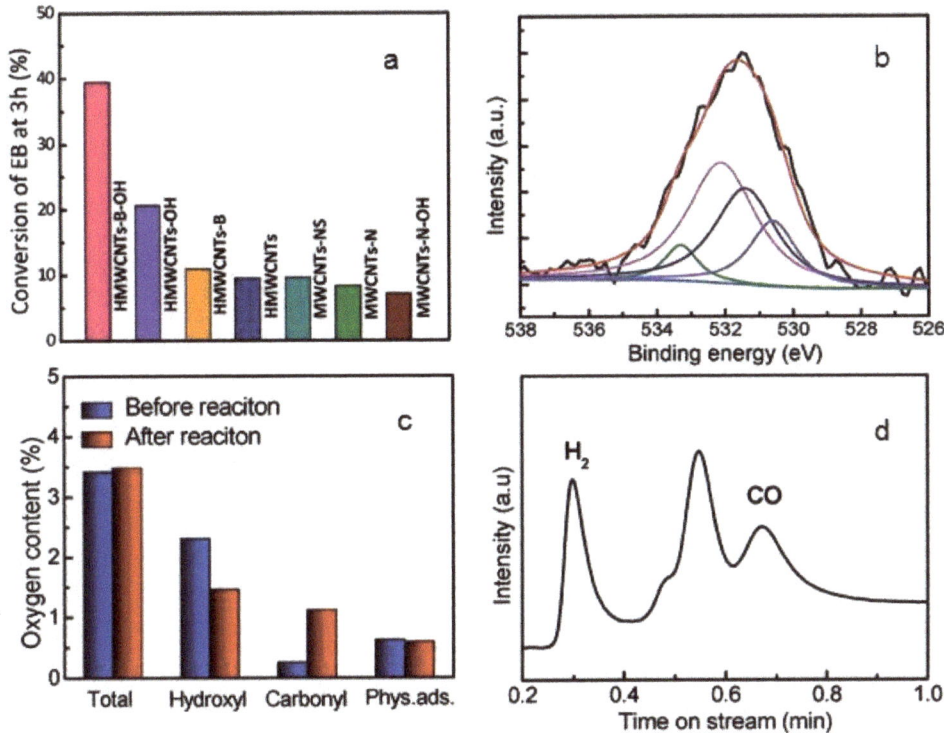

Figure 3. (a) Conversion of EB using CO_2 as a soft oxidant; (b) O1s spectra of HMWCNTs-B-OH after oxidative dehydrogenation by XPS for 3 hr; (c) Oxygen substance (mol%) of HMWCNTs-B-OH earlier and later oxidative dehydrogenation reaction at 3 h; (d) Gas derivatives of HMWCNTs-B-OH after oxidative dehydrogenation at 3 h. (Reprinted from [44]; copyright (2013), Royal Society of Chemistry).

Additionally, basic sites abstract hydrogen from ethyl benzene. Thus, the aggregated effect of the basic and acidic sites of the catalyst face is the oxidative dehydrogenation reaction, resulting in high catalytic efficiency in the existence of CO_2 [63]. Sato et al. reported on the use of CO_2 as a mild oxidant in the oxidative dehydrogenation reaction as well as the typical dehydrogenation process in the absence of CO_2. Two mechanisms utilizing acidic and basic sites were proposed, as depicted in Figure 4. Vanadium-embed catalysts also used in CO_2 based oxidative dehydrogenation of oxidative dehydrogenation of ethyl benzene (ODHEB) reactions [21,45]. CO_2, being a mild oxidant, cannot reproduce the active sites on the V_2O_5 (001) surface of the catalyst quickly enough due to the large activation energy (3.16 eV) [46]. A ceria-supported vanadium catalyst floated on a titania-zirconia combined oxide (TiO_2-ZrO_2) has moderate constancy which was reported by Reddy et al. [47]. XPS analysis of Ce 3d indicated the presence of Ce^{4+} and Ce^{3+} on the Ti-Zr catalyst. They also reported that CeO_2-V_2O_5/TiO_2-ZrO_2 (TZ) catalysts resulted in 56% conversion of ethylbenzene and 98% selectivity of styrene. Liu et al. [48] illustrated the red-ox mechanism for the CO_2-oxidative dehydrogenation of ethyl benzene (CO_2-ODEB) using a ceria promoted vanadium catalyst, as shown in Figure 5. In the CO_2-ODEB case, CO_2 directly oxidizes Ce^{3+} to Ce^{4+}, and ethylbenzene reduces of V^{5+} to V^{4+}. Then, the reduction of Ce^{4+} to Ce^{3+} and the oxidation of V^{4+} to V^{5+} completes the full cycle. In the existence of CO_2, modified vanadium catalysts are effective, selective, and stable for the ODEB, as reported by Park et al. [49] Rapid regeneration of active sites on a silica-assisted vanadium catalyst along in the presence of CO_2 has also been reported [64]. 10% La_2O_3-15%V_2O_5/SBA-15 (wt.%)

catalyst resulted in a 74% styrene yield, with La^{3+} resisting coke ejection [50]. The use of supporting materials, such as Aluminum mesoporous cylindrical molecular sieve (Al MCM-41) also resulted in substantial EB conversion in the ODEB using a VO_x/Al MCM-41 catalyst in the presence of CO_2 [51]. ZrO_2-containing combined oxide catalysts for oxidative dehydrogenation of ethylbenzene with CO_2 in the presence of MnO_2, CeO_2 and TiO_2 have exhibited high activity. The styrene yield was also increased over the MnO_2-ZrO_2 dual oxide catalyst at a high temperature. Significant enhancement of catalytic activity was checked with increasing CO_2/EB ratios [52]. A TiO_2-ZrO_2 catalyst was used, and the proportion of TiO_2/ZrO_2 determined the catalytic activity [53–55,65]. A 60% titania content resulted the best performance for the ODEB [65,66]. Commercial Fe-supported catalysts are unsuitable for the oxidative dehydrogenation of ethyl benzene in the existence of carbon dioxide due to the atomization of the active catalytic site [56]. However, the use of appropriate dopants' support materials might enhance the activity by promoting re-oxidation of Fe^{2+} and preventing coke deposition [57,58]. High product yield stability was observed in a mesoporous silica COK12-assisted CoO_3 catalyst [59]. The performance of several effective catalysts in the oxidative dehydrogenation of ethyl benzene to styrene in the existence of CO_2 is shown in Table 7.

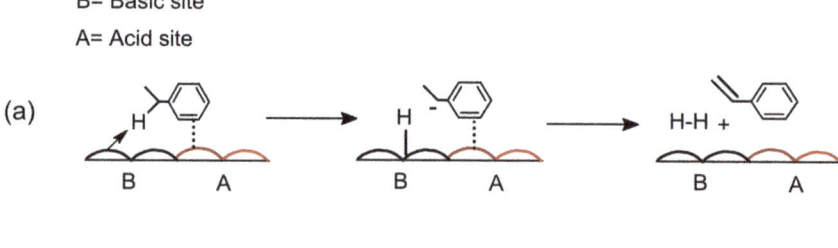

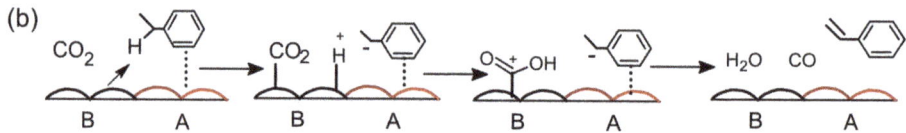

Figure 4. The procedure of oxidative dehydrogenation of ethylbenzene to styrene (**a**) without CO_2 and (**b**) with CO_2. (Reproduced with permission from [62]; copyright (2016), Elsevier).

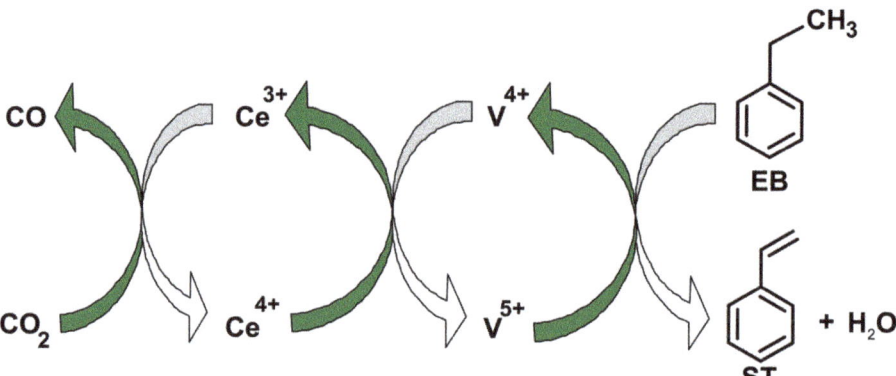

Figure 5. Red-ox cycle for CO_2-ODEB over the ceria-promoted vanadium catalyst. (Reproduced from [48]; copyright (2011), WILEY-VCH).

Table 7. Performance of CO_2 on oxidative dehydrogenation of ethylbenzene to styrene.

Catalyst	Reaction Temperature (°C)	EB Conversion (%)	ST Selectivity (%)	ST Yield (%)	Ref.
Co_3O_4/COK-12	600	57.5	95.5	54.9	[59]
$CeZrO_{4-\delta}$	550	7	97	6.8	[61]
Na X zeolite	545	9.4	89.6	8.4	[60]
K X zeolite	545	10.5	92.1	9.6	[60]
VO_x/Al MCM-41	550	52.3	96.7	50.6	[51]
TiO_2-ZrO_2	600	69.3	96.2	66.6	[67]
V_2O_5/SiO_2	550	50.5	96.8	48.8	[64]
SnO_2-ZrO_2	600	61.1	97.6	59.6	[68]
MnO_2-ZrO_2	600	51.1	99.1	50.9	[69]

3.2. Performance of CO_2 on Dehydrogenation of Ethane

Ethylene is one of the most prominent raw materials in the chemical industry. Presently, it is used to produce industrial products such as PVC, ethylene glycol, ethylbenzene, ethylene oxide, and vinyl acetate. Commercially, ethylene is formed by steam cracking dehydrogenation of hydrocarbons and fluid catalytic cracking (FCC). These conventional methods have several major limitations including the reaction endothermicity, thermodynamic drawbacks, rapid coke formation, and high energy consumption. The oxidative dehydrogenation of ethane (ODHE) to ethylene in the presence of CO_2 as a mild oxidant is an environmentally friendly alternative method for the production of ethylene. A Cr-oxide catalyst with zeolite support was successfully used for the oxidative dehydrogenation of ethane in the presence of CO_2 as a soft oxidant. A novel Clinoptilolite-based Cr-oxide (Cr/CLT-IA) catalyst for the ODHE in the existence of CO_2 was investigated by Rahamani et al. [70]. This Cr-supported catalyst exhibits high selectivity and catalytic activity which was expected due to its acidity. Homogeneous, tunable smaller Clinoptilolite-based Cr catalyst particles with higher surface area can be generated. Thus, using a Cr/CLT-IA nano-catalyst may be feasible and favorable for the oxidative dehydrogenation of ethane to ethylene in the existence of CO_2. Cr/H-ZSM-5 (SiO_2/Al_2O_3 ≥ 190) outperformed the SiO_2-based catalyst in the oxidative dehydrogenation of ethane to ethylene with CO_2 [71]. CO_2 is a promising soft oxidant for the ODHE reaction acting as a channel for transporting heat to the endothermic dehydrogenation. Further, CO_2 improves the conversion by modifying alkanes and maintains the catalytic activity by eliminating coke from the catalyst surface. the texture of the Cr active sites and the catalyst activity are determined by the SiO_2/Al_2O_3 ratio. The presence of more alumina amount in the zeolite negatively affected the activity of the catalyst, due to the incorporation of alumina with the Cr into the catalyst structure, affecting the red-ox properties of Cr. Mimura et al. [71] reported on the dehydrogenation of ethane on a Cr-doped HZSM-5 catalyst which is established on the redox phase of the eminent oxidation type Cr species. In their work, C_2H_6 was absorbed on the acidic site of CrO_x and H-ZSM-5. Then, the activated C_2H_6 reacted with CrO_x (active O species) to produce ethylene. The CrO_{x-1} species is then re-oxidized by the soft oxidant CO_2 regenerating the active O species and eliminating coke from the surface of the catalyst. The catalytic performance of the Cr-supported mesoporous catalyst, as well as a Cr-doped silicate MSU-1 catalyst, in the ethane oxidative dehydrogenation to ethylene in the presence of CO_2 was reported on by Liu et al. [72]. They initially observed high catalytic activity due to the Cr(VI) active species. However, even in the existence of CO_2, the reduction of Cr(VI) to Cr(III) occurred, resulting in the deactivation of the catalyst during the dehydrogenation reaction. Shi et al. [73] reported that Cr-supported Ce/SBA-15 catalysts were comprised of hexagonally ordered mesoporous frameworks and exhibited high catalytic activity in the oxidative dehydrogenation of ethane in the existence of CO_2. They confirmed the addition of Ce species using high-angle XRD, which increased the Cr species distribution in the Cr-Ce based SBA-15 zeolite. TPR results determined that Cr species in SBA-15-type zeolites are Cr^{6+} and Cr^{3+} groups. Among those two ions, Cr^{6+} exhibited significant activity for the oxidative dehydrogenation reaction in the existence of CO_2. Including a Ce-supported in 5Cr/SBA-15 catalysts modified the red-ox properties and enhanced the activity of the catalyst. Ethane conversion was 55% and ethylene selectivity was 96% on the 5Cr-10Ce/SBA-15 catalyst in the existence of CO_2 (Table 8). Cr^{6+} is reduced to Cr^{3+} during the oxidative dehydrogenation method reaction, however, in the presence of CO_2, Cr^{3+} is re-oxidized to Cr^{6+}. Cr_2O_3/ZrO_2 supported catalysts with Fe, Co, Mn was also investigated in an effort to fully understand the excellent catalytic activity for the ethane dehydrogenation reaction to ethylene under CO_2 treatment [64,65,74]. The Cr^{6+}/Cr^{3+} red-ox cycle is crucial in the oxidative dehydrogenation reaction, as is a Fe^{3+}/Fe^{2+} red-ox cycle which was removes H_2 from the lattice oxygen. An SBA-15-based, Cr-modified catalyst using a Fe-Cr-Al alloy [75] also exhibited remarkable selectivity of ethylene and high ethane conversion in the oxidative dehydrogenation reaction with CO_2. Wang et al. [64] observed the red-ox properties and the acidity/basicity of the Cr-supported catalyst in the oxidative dehydrogenation of ethane to ethylene with CO_2. They found that Cr-supported catalysts exhibited different activities in the ODHE with CO_2. Cr_2O_3/SiO_2 showed higher ethane conversion and ethylene selectivity. The catalytic activities were ranked as follows $Cr/SiO_2 > Cr/ZrO_2 > Cr/Al_2O_3 > Cr/TiO_2$ [76,77]. Notably, Cr_2O_3 interacted more with Al_2O_3 than with SiO_2, resulting in tetrahedral Cr^{6+} sites and declining activity [78]. Cr is one of the vital elements of various types of nano-catalysts (Table 9). The active site of these catalysts contains both Cr^{3+} and Cr^{6+}. The Cr^{6+}/Cr^{3+} ratio strongly influences the reducibility of Cr/H-ZSM-5 catalysts. The red-ox performance of Cr-supported catalysts is crucial for the oxidative dehydrogenation of ethane to ethylene in the presence of CO_2 as a soft oxidant. Cr^{6+} (or Cr^{5+}) sites are reduced to Cr^{3+} as ethane is dehydrogenated. Then, the reduced Cr^{3+} sites are re-oxidized by carbon dioxide treatment. Mimura et al. reported that the highly active Cr-based catalysts had Cr^{6+} or Cr^{5+} species on the surface of the catalyst [79]. Apart from Cr-supported catalysts, several other effective catalysts have been used in research on ethane oxidative dehydrogenation. Among these, the Ni-Nb-mixed oxide catalyst performed very well at relatively low temperatures [80–82]. Additionally, a TiO_2-based Ga catalyst proved applicable for oxidative dehydrogenation with CO_2 [83].

Table 8. Catalytic activity for the dehydrogenation of ethane (Reproduced from [73]; copyright (2008), Springer (Berlin, Germany)).

Catalyst	In the Presence of CO_2				In the Presence of Ar		
	Conv. (%)		Selectivity (%)		Conv. (%)	Selectivity (%)	
	C_2H_6	CO_2	C_2H_4	CH_4	C_2H_6	C_2H_4	CH_4
SBA-15	2.7	0.04	93.5	6.5	2.4	93.0	7.0
2.5Cr/SBA-15	39.6	15.9	95.5	4.5	30.2	89.7	10.3
5.0Cr/SBA-15	46.3	16.6	94.7	5.3	34.1	91.4	8.6
7.5Cr/SBA-15	45.3	18.8	92.2	7.8	33.9	92.8	7.2
10Cr/SBA-15	44.2	18.9	92.0	8.0	31.2	90.9	9.1
5Cr-5Ce/SBA-15	48.4	17.9	96.4	4.6	35.8	87.6	12.4
5Cr-7.5Ce/SBA-15	50.0	20.9	96.0	4.0	37.9	88.2	11.8
5Cr-10Ce/SBA-15	55.0	21.9	96.0	4.0	40.8	83.1	16.9
5Cr-15Ce/SBA-15	52.2	21.2	95.5	4.5	40.1	82.4	17.6

Reaction conditions: GHSV = 3600 mL/g h, T = 700 °C.

Table 9. Influence of CO_2 on oxidative dehydrogenation of ethane.

Catalyst	Ethane Conversion (%)	Ethylene Selectivity (%)	Ethylene Yield (%)	Ref.
Cr_2O_3 (5 wt.%) CLT-IA	39.7	98.8	39.2	[70]
3Cr/NaZSM-5-160	65.5	75.4	49.3	[84]
Cr_2O_3/Al_2O_3-ZrO_2	36.0	56.2	20.2	[85]
Cr/MSU-1	68.1	81.6	55.6	[72]
Cr_2O_3/ZrO_2	77.4	46.3	35.8	[86]
2.5 Cr/SBA-15	46.3	94.7	43.8	[75]
5 Cr-10Ce/SBA-15	55.0	96.0	52.8	[73]
5% Cr_2O_3/Al_2O_3	19.2	56.5	10.8	[77]

Step-1

$$H_3C\text{---}CH_3 + CrO_x \rightleftarrows H_2C=CH_2 + H_2O + CrO_{x-1} \text{ (oxidative dehydrogenation)}$$

Step-2

$$H_3C\text{---}CH_3 \rightleftarrows H_2C=CH_2 + H_2 \text{ (Simple dehydrogenation)}$$

$$CrO_x + H_3C\text{---}CH_3 \rightleftarrows CH_4 + C + H_2O + CrO_{x-1} \text{ (methane and coke formation)}$$

$$H_3C\text{---}CH_3 + H_2 \rightleftarrows 2CH_4 \text{ (hydrocracking)}$$

Step-3

$$CrO_{x-1} + CO_2 \rightleftarrows CrO_x + CO \text{ (reoxidizing)}$$

$$C + CO_2 \rightleftarrows 2CO$$

3.3. Influence of CO_2 on the Alkylation of Toluene Side-Chain

The dehydrogenation of ethylbenzene produces the most styrene using the Friedel-Crafts alkylation reaction [87]. However, the ethylbenzene dehydrogenation method has some limitations such as catalyst deactivation and, high energy consumption [88,89]. Alkylation of the toluene side-chain is a promising alternative process that uses basic catalysts for the formation of styrene in the existence of CO_2. Another process was reported by Sindorenko et al. [90] utilizing K^+ and Rb^+ ion transposing Faujasite supported catalysts in 1967. However, the catalytic conversion of toluene and styrene monomer (SM) selectivity was low (Table 10) [91]. The side-chain alkylation is primarily carried out on solid base catalysts [92–96]. Toluene side-chain alkylation with methanol enhanced by the promotional use of alkali metal oxides. Greater catalyst acidity accelerates methanol dehydration, [97] while low concentrations of alkali metal ions prevent the decomposition of formaldehyde produced from methanol [98]. Thus, catalysts for this reaction must be optimized for their acidity and basicity [99]. Generally, catalyst sites for the side-chain alkylation are limited to alkali metal-altered zeolites [100]. One reliable, widely studied catalyst is the cesium ion-exchanged or Ce_2O-impregnated zeolite-X. The advantages of a MgO-supported mesoporous catalyst for this reaction has also been reported by Park el al. [101]. Hattori et al. observed that the impregnation of Cs_2O in ion-exchanged zeolite-X results in high conversion of toluene, owing to the strongly basic sites [102]. Carbon dioxide has been under consideration as a renewable, low-cost, safe, and environmentally beneficial feedstock in current years. CO_2 utilization is difficult for commercial applications, owing to its high thermal stability as well as the solid oxidation phase [103]. Hence, remarkable research efforts are being directed to detect innovative technologies for the utilization of CO_2. Toluene side-chain alkylation was performed to assess the efficacy of the catalytic approach with methanol over cesium-supported catalysts. Toluene and methanol conversion over the Cs-X and Cs-modified zeolites in the presence of He and CO_2 are shown in Table 10. In these reactions, styrene and ethylbenzene were formed as main products. Other side-chain alkylated components, including cumin and α-methyl styrene, as well as other xylenes, tri-methylbenzene, and benzene were identified as by-products. When the catalytic reaction was carried out in the existence of CO_2, methanol and toluene conversion increased. Though the styrene selectivity decreased, there was a significant increase in the conversion as well as product selectivity in the presence of He and CO_2 streams. TG/DTA analysis of the used Cs-X catalyst in the presence of CO_2 and He streams is shown in Figure 6. In the range of 25–200 °C, weight loss occurred owing to the desorption of adsorbed water [94]. The continued weight loss in the 200–450 °C region occurred due to the deposition of coke on the surface of the catalyst. Relatively high quantities of coke were deposited on the Cs-X catalyst in the existence of the CO_2. This suggests greater deactivation of the catalyst in the presence of carbon dioxide owing to coke deposition [89]. Still the Cesium-supported catalysts performed better in the presence of CO_2 than under He in terms of toluene and methanol conversion. CO_2 acted as a significant performance in hydrogen skulking and enhanced the reaction rate in the decisive route. Additionally, CO_2 increases alkylation to produce cumin and α-methyl styrene, which are side-chain alkylation products. Further, the increased toluene conversion enhances the aromatic yields.

Table 10. Performance of CO_2 in the toluene side-chain alkylation (Reproduced with permission from [91]; copyright (2018), Elsevier).

Catalyst	Carrier Gas	MeOH Conv. (%)	Toluene Conv. (%)	Selectivity (%)		
				SM	EB	Others
Ce-X	He	12.54	1.42	78.61	15.32	6.07
	CO_2	35.35	3.48	45.83	33.36	20.81
Cs_2O/ Cs-X	He	46.48	3.59	28.76	68.02	3.22
	CO_2	39.16	2.52	36.02	43.02	20.78

Reaction conditions: WHSV = 2.1 h^{-1}, Reaction temperature = 425 °C, Toluene/MeOH molar ratio = 2, SM = Styrene Monomer and other byproducts = Cumene, Xylenes, TMB and Benzene.

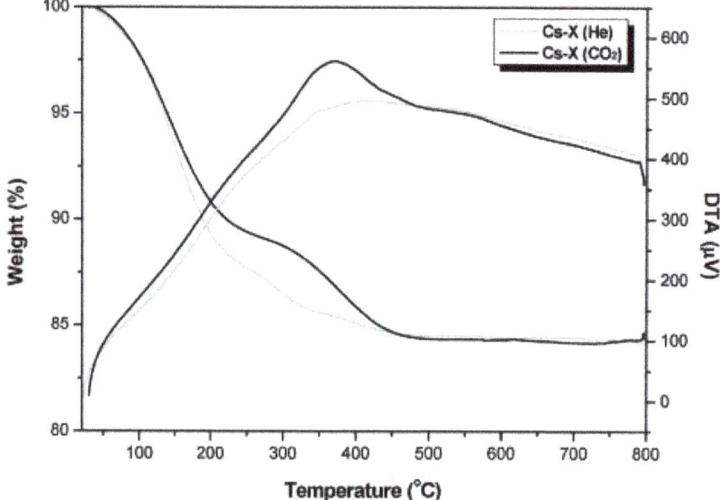

Figure 6. TG/DTA results obtained for used Cs-X catalyst in the presence of CO_2 and He streams (Redrawn with permission from [91]; copyright (2018), Elsevier).

3.4. Role of CO_2 on Dehydrogenation of Propane

Propylene is the most prominent raw material in the chemical industries. It is primarily manufactured by steam cracking and propane oxidative dehydrogenation [104–106]. Oxidative dehydrogenation (ODH) is preferred due to its low energy requirements and lack of thermodynamic limitations [107,108]. However, the ODH reaction with O_2 occurs under potentially flammable conditions and forms of carbon oxides due to over-oxidation with low selectivity [109,110]. This complication can be resolved using CO_2 as a mild, safer oxidant. Thus, this reaction is a favorable example of CO_2 utilization. Interestingly, CO_2 was used as a mild oxidant to shift the equilibrium more toward the products, as well as enhance the dehydrogenation over the coupling between propane oxidative dehydrogenation to propylene and the reverse water gas shift (RWGS) reaction [111–113]. Dehydrogenation of propane occurred on the acid site of the catalyst. The SiO_2/Al_2O_3 proportion is critical in determining both the catalyst physicochemical properties and its reactivity characteristics [114–117]. The HZSM-5, SBA-15, MCM-41, SBA-1 catalyst which is a two-dimensional microchannel system, has been used in the oxidative dehydrogenation of alkanes especially for the conversion of methane to propane in the existence of CO_2. Various research groups have reported on the influence of catalyst acidity in the oxidative dehydrogenation reaction with CO_2. The activity of the zeolites decreased with increasing Si/Al proportion in HZSM-5 based Ga_2O_3, although the selectivity increased, as shown is in Figure 7 [118]. Lewis acidity is present in the metal oxide (Ga_2O_3) catalyst, while Bronsted acidity is present in HZSM-5. Thus, extracting the aluminum from HZSM-5 declines the Bronsted acidity more than it decreases the Lewis acidity. Several transition metals, such as vanadium, molybdenum, and chromium, have been used to support catalysts for ODH of light alkanes including propane [105,112,119,120]. Among these, chromium oxide provided high catalytic performance with CO_2, despite fractional deactivation by coke production. Chromium oxide enhanced propane conversion and the propylene selectivity by expelling H_2 produced in the ODH reaction [112]. The catalytic performance of Cr-supported catalysts was observed by the character of chromium categories on the support surface of the catalysts [121–125] Park et al. found that different Cr doping of Cr-TUD-1 catalysts (3, 5, 7 and 9 wt.%) with soft oxidant (CO_2) were formed by MW irradiation and investigated the propane oxidative dehydrogenation [126]. The effect of reaction temperature on the oxidative dehydrogenation of propane in the existence of CO_2 as a mild oxidant over the Cr-TUD-1 catalyst (7 wt.%) was investigated thoroughly to improve the catalytic activity. The conversion of CO_2 was 3.5% at 550 °C and improved to 5.5% at 650 °C. To demonstrate the importance of CO_2 in the propane oxidative dehydrogenation on Cr-TUD-1 catalysts, the process was carried out at 550 °C on 7 wt.% catalyst under the same conditions in the presence of CO_2 as well as He. The decline in the catalytic activity of the catalyst with helium may be due to coke production and the reduction of the Cr groups on the surface of the zeolite. The proposed mechanism of propane oxidative dehydrogenation over metal oxide surfaces with the CO_2 stream is shown below [112]:

A weak exclusive propane adsorption on the lattice oxygen

$$C_3H_8 + O^* \rightarrow C_3H_8O^* \tag{2}$$

C-H schism via H-abstraction from propane utilizing an abutting lattice oxygen

$$C_3H_8O^* + O^* \rightarrow C_3H_7O^* \tag{3}$$

Propylene desorption by hybrid expulsion from adsorbed alkoxide groups

$$C_3H_7O^* \rightarrow C_3H_6 + OH^* \tag{4}$$

Reconsolidation of OH groups to produce H_2O, reduced metal center (*)

$$OH^* + OH^* \rightarrow H_2O + O^* + * \tag{5}$$

Re-oxidation of abridged M-centers by separating chemisorptions of CO_2

$$2CO_2 + * + * \rightarrow 2CO + 2O^* \tag{6}$$

To evaluate the deactivation of the catalyst by coke creation and the enhancement of CO_2, (Equation (7)) can be used as the deactivation parameter:

Deactivation parameter (%) = Conversion of propane (initial amount − final amount)/ (initial amount) * 100 (7)

The rate of Cr degradation by H_2 liberated from dehydrogenation is faster than the rate at which CO_2 re-oxidizes the degraded Cr species, resulting in catalytic deactivation. Selective adsorption properties can be improved by surface functional groups on activated carbons. Thus, surface treatment of activated carbon may result in more selective and efficient adsorption of the gas, liquid and the alleviation of pollution [127].

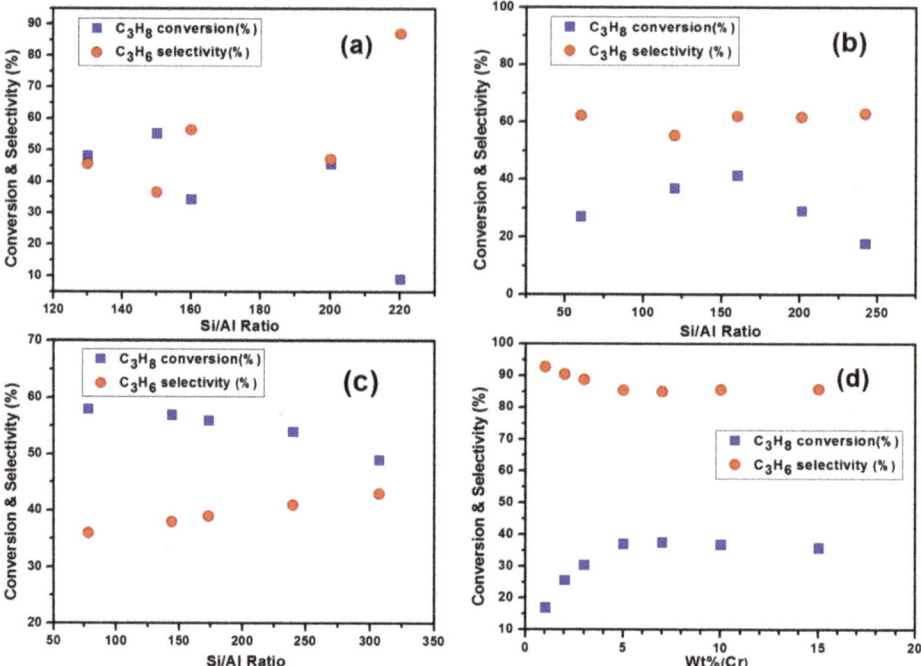

Figure 7. Influence of Si/Al proportion on the efficiency of (**a**) Ga$_2$O$_3$/ZSM-48 zeolites (Reproduced from [118]; copyright (2012), Elsevier), (**b**) ZnO-HZSM-5 zeolites in the oxidative dehydrogenation of propane along with CO$_2$ (Reproduced from [105]; copyright (2009), Elsevier), (**c**) Ga$_2$O$_3$/M-HZSM-5 zeolites in the absence of CO$_2$ (Reproduced from [118]; copyright (2012), Elsevier), (**d**) Influence of Cr substance on the effectiveness of Cr/SBA-15 in the carriage of CO$_2$ (Reproduced from [128]; copyright (2012), Elsevier).

4. Conclusions

This review article has comprised a number of CO$_2$ conversions, which are still in the research scale. These promising technologies are mitigating the continuously increasing atmospheric CO$_2$ concentration. Among the methods employing CO$_2$, the ethyl benzene ODH process has seen significant progress. Currently, most of the ethylbenzene dehydrogenation plants apply the oxidative dehydrogenation method, which leads to large heat losses upon compression at the gas–liquid separator. Further, this reaction is thermodynamically restrictive and energy intensive. Several industrial companies such as SABIC (Saudi Basic Industry Corporation, Saudi Arabia), Samsung General Co. in south Korea have tested the catalytic consummation for this method. The commercial implementation of such a process may support the economics of styrene monomer production. According to European Rubber Journal (ERJ), Asahi Kasei Chemical Company 's (Japan) 6th generation SBR (Styrene-butadiene rubber) is currently being tested by many customers in the world with positive feedback and company is planning to commercialize some grades in 2021. Moreover, Trinseo's highly functionalized SPRINTAN^{TM} 918S Solution-Styrene Butadiene Rubber (S-SBR) has awarded second position in the elastomers for sustainability initiative of the European Rubber Journal. Based on lab indicator data confirmed by tire customers, grade 918S (compared to non-functionalized high-grip SSBR) improves fuel efficiency of the whole car approximately 1.5%. Considering in Europe alone, the benefit of this increased fuel efficiency would translate in approximately 540 tons less fuel consumed or a reduction of CO$_2$ emissions by 1.3 million tons.

Several methods using CO_2 as a mild oxidant have appeared in the technology sector. It is a long-term goal and alluring dream for chemical engineers to establish commercial industries based on the utilization of CO_2. Challenges for the commercial utilization of this technology include the process rate required to ensure CO_2 conversion with low coke deposition, the need to decrease energy expenditure, and the need for improved catalysts offering higher conversion. Despite the challenges, there is great room for catalyst improvement in these sectors. Recently, the carbon XPRIZE is a $20 million competition to capture and CO_2 conversion which is jointly funded by COSIA (Canada's Oil Sends Innovation Alliance) [129]. Most of the countries' governments are concern about climate changes with a high priority. China, the world's largest energy consumer and carbon emitter, announced USD 360 billion in renewable energy investments by 2020 to reduce carbon emissions [130]. Canada has implemented federally a carbon pricing policy with a current tax of USD 10/ton CO_2 and a steady rise to USD 50/ton CO_2 nationwide by 2022. However, the positive effects of CO_2 in benzene hydroxylation over commercial and hierarchical zeolites in the liquid phase as well as the gas phase are under investigation by our group, wherein the byproducts are various aromatic compounds. The recycling of CO_2 from the atmosphere to fuels, chemicals will lead to a real sustainable future for humanity. We expect that the use of CO_2 as a promoter and as a mild oxidant at the laboratory level can be translated to the industrial scale in the future, thus contributing also to the world economy.

Author Contributions: S.T.R.: Writing original draft; J.-R.C.: Editing; J.-H.L.: Editing; S.-J.P.: Writing review & editing. All authors have read and agreed to the published version of the manuscript.

Funding: This work was supported by the Technology Innovation Program (or Industrial Strategic Technology Development Program-Development of technology on materials and components) (20010881, Development of ACF for rigid (COG)/ flexible (COP) and secured mass production by developing core material technology for localizing latent hardener for low temperature fast curing) funded By the Ministry of Trade, Industry & Energy (MOTIE, Korea) and supported by Korea Evaluation institute of Industrial Technology (KEIT) through the Carbon Cluster Construction project [10083586, Development of petroleum based graphite fibers with ultra-high thermal conductivity] funded by the Ministry of Trade, Industry & Energy (MOTIE, Korea).

Conflicts of Interest: The authors declare that they have no conflicts of interest.

References

1. Kar, S.; Goeppert, A.; Prakash, G.K.S. Integrated CO_2 Capture and Conversion to Formate and Methanol: Connecting Two Threads. *Acc. Chem. Res.* **2019**, *52*, 2892–2903. [CrossRef] [PubMed]
2. Smol, J.P. Climate Change: A Planet in flux. *Nature* **2012**, *483*, 12–15. [CrossRef] [PubMed]
3. Ross, M.B.; De Luna, P.; Li, Y.; Dinh, C.-T.; Kim, D.; Yang, P.; Sargent, E.H. Designing materials for electrochemical carbon dioxide recycling. *Nat. Catal.* **2019**, *2*, 648–658. [CrossRef]
4. De Luna, P.; Hahnet, C.; Higgins, D.; Jaffer, S.A.; Jaramillo, T.F.; Sargent, E.H. What would it take for renewably powered electrosynthesis to displace petrochemical processes? *Science* **2019**, *364*. [CrossRef]
5. Nugent, P.; Belmabkhout, Y.; Burd, S.D.; Cairns, A.J.; Luebke, R.; Forrest, K.; Pham, T.; Ma, S.; Space, B.; Wojtas, L.; et al. Porous materials with optimal adsorption thermodynamics and kinetics for CO_2 separation. *Nature* **2013**, *495*, 80–84. [CrossRef]
6. Rehman, A.; Park, S.-J. From chitosan to urea-modified carbons: Tailoring the ultra-microporosity for enhanced CO_2 adsorption. *Carbon* **2020**, *159*, 625–637. [CrossRef]
7. Lee, S.H.; Sullivan, I.; Larson, D.M.; Liu, G.; Toma, F.M.; Xiang, C.; Drisdell, W.S. Correlating Oxidation State and Surface Area to Activity from Operando Studies of Copper CO Electroreduction Catalysts in a Gas Fed Device. *ACS Catal.* **2020**, *10*, 8000–8011. [CrossRef]
8. Han, L.; Zhou, W.; Xiang, C. High-Rate Electrochemical Reduction of Carbon Monoxide to Ethylene Using Cu-Nanoparticles Based Gas Diffusion Electrodes. *ACS Energy Lett.* **2018**, *3*, 855–860. [CrossRef]
9. Heo, Y.J.; Park, S.-J. Facile Synthesis of MgO-Modified Carbon Adsorbents with Microwave-Assisted Methods: Effect of MgO Particles and Porosities on CO_2 Capture. *Sci. Rep.* **2017**, *7*, 1–9. [CrossRef] [PubMed]
10. Qi, S.-C.; Wu, J.-K.; Lu, J.; Yu, G.-X.; Zhu, R.R.; Liu, Y.; Liu, X.-Q.; Sun, L.-B. Underlying mechanism of CO_2 adsorption onto conjugated azacyclo-copolymers: N-doped adsorbents capture CO_2 chiefly through acid-base interaction? *J. Mater. Chem. A* **2019**, *7*, 17842–17853. [CrossRef]

11. Qi, S.C.; Liu, Y.; Peng, A.Z.; Xue, D.M.; Liu, X.; Liu, X.Q.; Sun, L.B. Fabrication of porous carbons from mesitylene for highly efficient CO_2 capture: A rational choice improving the carbon loop. *Chem. Eng. J.* **2019**, *361*, 945–952. [CrossRef]
12. Rehman, A.; Park, S.-J. Tunable nitrogen-doped microporous carbons: Delineating the role of optimum pore size for enhanced CO_2 adsorption. *Chem. Eng. J.* **2019**, *362*, 731–742. [CrossRef]
13. Jessop, P.G.; Ikariya, T.; Noyori, R. Homogeneous catalysis in supercritical fluids. *Science* **1995**, *269*, 1065–1069. [CrossRef]
14. Musie, G.; Wei, M.; Subramaniam, B.; Busch, D.H. Catalytic oxidations in carbon dioxide-based reaction media, including novel CO_2-expanded phases. *Coord. Chem. Rev.* **2001**, *219–221*, 789–820. [CrossRef]
15. Heldebrant, D.J.; Jessop, P.G.; Thomas, C.A.; Eckert, C.A.; Liotta, C.L. The reaction of 1,8-diazabicyclo [5.4.0]undec-7-ene (DBU) with carbon dioxide. *J. Org. Chem.* **2005**, *70*, 5335–5338. [CrossRef]
16. Chang, J.S.; Vislovskiy, V.P.; Park, M.S.; Hong, D.Y.; Yoo, J.S.; Park, S.E. Utilization of carbon dioxide as soft oxidant in the dehydrogenation of ethyl benzene over supported vanadium-antimony oxide catalysts. *Green Chem.* **2003**, *5*, 587–590. [CrossRef]
17. Park, M.S.; Chang, J.S.; Kim, D.S.; Park, S.-E. Oxidative dehydrogenation of ethyl benzene with carbon dioxide over zeolite-supported iron oxide catalysts. *Res. Chem. Intermed.* **2002**, *28*, 461–469. [CrossRef]
18. Sakakura, T.; Choi, J.C.; Yasuda, H. Transformation of carbon dioxide. *Chem. Rev.* **2007**, *107*, 2365–2387. [CrossRef]
19. Reddy, B.M.; Lakshmanan, P.; Loridant, S.; Yamada, Y.; Kobayashi, T.; López-Cartes, C.; Rojas, T.C.; Fernández, A. Structural characterization and oxidative dehydrogenation Activity of v2O5/CexZr1-xO2/SiO2 catalysts. *J. Phys. Chem. B* **2006**, *110*, 9140–9147. [CrossRef]
20. Yoo, J.S.; Lin, P.S.; Elfline, S.D. Gas-phase oxygen oxidations of alkyl aromatics over CVD Fe/Mo/borosilicate molecular sieve. II. The role of carbon dioxide as a co-oxidant. *Appl. Catal. A Gen.* **1993**, *106*, 259–273. [CrossRef]
21. Sun, A.; Qin, Z.; Wang, J. Reaction coupling of ethylbenzene dehydrogenation with water-gas shift. *Appl. Catal. A Gen.* **2002**, *234*, 179–189. [CrossRef]
22. Ansari, M.B.; Park, S.E. Carbon dioxide utilization as a soft oxidant and promoter in catalysis. *Energy Environ. Sci.* **2012**, *5*, 9419–9437. [CrossRef]
23. Abanades, S.; Le Gal, A. CO_2 splitting by thermo-chemical looping based on ZrxCe1-xO2 oxygen carriers for synthetic fuel generation. *Fuel* **2012**, *102*, 180–186. [CrossRef]
24. Wang, S.; Wang, X. Imidazolium ionic liquids, imidazolylidene heterocyclic carbenes, and zeolitic imidazolate frameworks for CO2 capture and photochemical reduction. *Angew. Chemie Int. Ed.* **2016**, *55*, 2308–2320. [CrossRef] [PubMed]
25. Nikulshina, V.; Hirsch, D.; Mazzotti, M.; Steinfeld, A. CO_2 capture from air and co-production of H2 via the Ca(OH)2-CaCO3 cycle using concentrated solar power-Thermodynamic analysis. *Energy* **2006**, *31*, 1715–1725. [CrossRef]
26. Spinner, N.S.; Vega, J.A.; Mustain, W.E. Recent progress in the electrochemical conversion and utilization of CO_2. *Catal. Sci. Technol.* **2012**, *2*, 19–28. [CrossRef]
27. Singh, G.; Lakhi, K.S.; Ramadass, K.; Sathish, C.I.; Vinu, A. High-Performance Biomass-Derived Activated Porous Biocarbons for Combined Pre- and Post-Combustion CO_2 Capture. *ACS Sustain. Chem. Eng.* **2019**, *7*, 7412–7420. [CrossRef]
28. Grice, K.A. Carbon dioxide reduction with homogenous early transition metal complexes: Opportunities and challenges for developing CO_2 catalysis. *Coord. Chem. Rev.* **2017**, *336*, 78–95. [CrossRef]
29. Bartholomew, C.H. *Catalyst Deactivation 1991*; Elsevier: Amsterdam, The Netherlands, 1991; pp. 96–112.
30. Denekamp, I.M.; Antens, M.; Slot, T.K.; Rothenberg, G. Selective Catalytic Oxidation of Cyclohexene with Molecular Oxygen: Radical Versus Nonradical Pathways. *ChemCatChem* **2018**, *10*, 1035–1041. [CrossRef]
31. Ansari, M.B.; Min, B.H.; Mo, Y.H.; Park, S.-E. CO_2 activation and promotional effect in the oxidation of cyclic olefins over mesoporous carbon nitrides. *Green Chem.* **2011**, *13*, 1416–1421. [CrossRef]
32. Aresta, M.; Tommasi, I.; Quaranta, E.; Fragale, C.; Tranquille, M.; Galan, F.; Fouassier, M. Mechanism of formation of peroxycarbonates RhOOC(O)O(Cl)(P)3 and Their Reactivity as Oxygen Transfer Agents Mimicking Monooxygenases. The First Evidence of CO_2 Insertion into the O-O Bond of Rh(η2-O2) Complexes. *Inorg. Chem.* **1996**, *35*, 4254–4260. [CrossRef] [PubMed]

33. Iwata, R.; Ido, T.; Fujisawa, Y.; Yamazaki, S. On-line interconversion of [^{15}O]O$_2$ and [^{15}O]CO$_2$ via metal oxide by isotopic exchange. *Int. J. Radiat. Appl. Instrum. Part A.* **1988**, *39*, 1207–1211. [CrossRef]
34. Nair, V.; Varghese, V.; Paul, R.R.; Jose, A.; Sinu, C.R.; Menon, R.S. NHC catalyzed transformation of aromatic aldehydes to acids by carbon dioxide: An unexpected reaction. *Org. Lett.* **2010**, *12*, 2653–2655. [CrossRef]
35. Chiang, P.C.; Bode, J.W. On the role of CO$_2$ in NHC-catalyzed oxidation of aldehydes. *Org. Lett.* **2011**, *13*, 2422–2425. [CrossRef] [PubMed]
36. Gu, L.; Zhang, Y. Unexpected CO$_2$ splitting reactions to form CO with N-heterocyclic carbenes as organocatalysts and aromatic aldehydes as oxygen acceptors. *J. Am. Chem. Soc.* **2010**, *132*, 914–915. [CrossRef]
37. Lane, B.S.; Vogt, M.; De Rose, V.J.; Kevin, B. Manganese-Catalyzed Epoxidations of Alkenes in Bicarbonate Solutions. *J. Am. Chem. Soc.* **2002**, *124*, 11946–11954. [CrossRef]
38. Park, S.-E.; Yoo, J.S. *Studies in Surface Science and Catalysis 153*; Elsevier: Amsterdam, The Netherlands, 2004; pp. 303–314.
39. Cavani, F.; Trifiro, F. Alternative Processes for the Production of Sponge Iron. *Appl. Catal. A Gen.* **1995**, *133*, 219–239. [CrossRef]
40. Li, X.; Li, W. Effect of TiO$_2$ loading on the activity of V/TiO2-Al2O3 in the catalytic oxidehydrogenation of ethylbenzene with carbon dioxide. *Front. Chem. Eng. China* **2010**, *4*, 142–146. [CrossRef]
41. Nowicka, E.; Reece, C.; Althahban, S.M.; Mohammed, K.M.H.; Kondrat, S.A.; Morgan, D.J.; He, Q.; Willock, D.J.; Golunski, S.; Kiely, C.J.; et al. Elucidating the Role of CO$_2$ in the Soft Oxidative Dehydrogenation of Propane over Ceria-Based Catalysts. *ACS Catal.* **2018**, *8*, 3454–3468. [CrossRef]
42. Kovacevic, M.; Agarwal, S.; Mojet, B.L.; Van Ommen, J.G.; Lefferts, L. The effects of morphology of cerium oxide catalysts for dehydrogenation of ethylbenzene to styrene. *Appl. Catal. A Gen.* **2015**, *505*, 354–364. [CrossRef]
43. Rao, R.; Zhang, Q.; Liu, H.; Yang, H.; Ling, Q.; Yang, M.; Zhang, A.; Chen, W. Enhanced catalytic performance of CeO$_2$ confined inside carbon nanotubes for dehydrogenation of ethylbenzene in the presence of CO$_2$. *J. Mol. Catal. A Chem.* **2012**, *363–364*, 283–290. [CrossRef]
44. Rao, R.; Yang, M.; Ling, Q.; Li, C.; Zhang, Q.; Yang, H.; Zhang, A. A novel route of enhancing oxidative catalytic activity: Hydroxylation of MWCNTs induced by sectional defects. *Catal. Sci. Technol.* **2014**, *4*, 665–671. [CrossRef]
45. Sakurai, Y.; Suzaki, T.; Ikenaga, N.O.; Suzuki, T. Dehydrogenation of ethylbenzene with an activated carbon-supported vanadium catalyst. *Appl. Catal. A Gen.* **2000**, *192*, 281–288. [CrossRef]
46. Fan, H.; Feng, J.; Li, X.; Guo, Y.; Li, W.; Xie, K. Ethylbenzene dehydrogenation to styrene with CO$_2$ over V2O5(001): A periodic density functional theory study. *Chem. Eng. Sci.* **2015**, *135*, 403–411. [CrossRef]
47. Rao, K.N.; Reddy, B.M.; Abhishek, B.; Seo, Y.H.; Jiang, N.; Park, S.E. Effect of ceria on the structure and catalytic activity of V$_2$O$_5$/TiO$_2$-ZrO$_2$ for oxidehydrogenation of ethylbenzene to styrene utilizing CO$_2$ as soft oxidant. *Appl. Catal. B Environ.* **2009**, *91*, 649–656. [CrossRef]
48. Liu, Z.W.; Wang, C.; Fan, W.B.; Liu, Z.T.; Hao, Q.Q.; Long, X.; Lu, J.; Wang, J.G.; Qin, Z.F.; Su, D.S. V2O5/Ce0.6Zr0.4O2-Al2O3 as an efficient catalyst for the oxidative dehydrogenation of ethylbenzene with carbon dioxide. *ChemSusChem* **2011**, *4*, 341–345. [CrossRef]
49. Burri, A.; Jiang, N.; Ji, M.; Park, S.-E.; Khalid, Y. Oxidative dehydrogenation of ethylbenzene to styrene with CO$_2$ over V$_2$O$_5$-Sb$_2$O$_5$-CeO$_2$/TiO$_2$-ZrO$_2$ catalysts. *Top. Catal.* **2013**, *56*, 1724–1730. [CrossRef]
50. Liu, B.S.; Rui, G.; Chang, R.Z.; Au, C.T. Dehydrogenation of ethylbenzene to styrene over LaVOx/SBA-15 catalysts in the presence of carbon dioxide. *Appl. Catal. A Gen.* **2008**, *335*, 88–94. [CrossRef]
51. Li, Z.; Su, K.; Cheng, B.; Shen, D.; Zhou, Y. Effects of VOx /AlMCM-41 surface structure on ethyl benzene oxydehydrogenation in the presence of CO$_2$. *Catal. Lett.* **2010**, *135*, 135–140. [CrossRef]
52. Jiang, N.; Burri, A.; Park, S.-E. Ethylbenzene to styrene over ZrO$_2$-based mixed metal oxide catalysts with CO$_2$ as soft oxidant. *Chin. J. Catal.* **2016**, *37*, 3–15. [CrossRef]
53. Reddy, B.M.; Khan, A. Recent advances on TiO2-ZrO$_2$ mixed oxides as catalysts and catalyst supports. *Catal. Rev. Sci. Eng.* **2005**, *47*, 257–296. [CrossRef]
54. Jiang, N.; Han, D.S.; Park, S.-E. Direct synthesis of mesoporous silicalite-1 supported TiO$_2$-ZrO$_2$ for the dehydrogenation of EB to styrene with CO$_2$. *Catal. Today* **2009**, *141*, 344–348. [CrossRef]
55. Manríquez, M.E.; López, T.; Gómez, R.; Navarrete, J. Preparation of TiO$_2$-ZrO$_2$ mixed oxides with controlled acid-basic properties. *J. Mol. Catal. A Chem.* **2004**, *220*, 229–237. [CrossRef]

56. Zangeneh, F.T.; Sahebdelfar, S.; Ravanchi, M.T. Conversion of carbon dioxide to valuable petrochemicals: An approach to clean development mechanism. *J. Nat. Gas Chem.* **2011**, *20*, 219–231. [CrossRef]
57. Balasamy, R.J.; Tope, B.B.; Khurshid, A.; Al-Ali, A.A.S.; Atanda, L.A.; Sagata, K.; Asamoto, M.; Yahiro, H.; Nomura, K.; Sano, T.; et al. Ethylbenzene dehydrogenation over FeOx/(Mg,Zn)(Al)O catalysts derived from hydrotalcites: Role of MgO as basic sites. *Appl. Catal. A Gen.* **2011**, *398*, 113–122. [CrossRef]
58. Braga, T.P.; Pinheiro, A.N.; Teixeira, C.V.; Valentini, A. Dehydrogenation of ethylbenzene in the presence of CO_2 using a catalyst synthesized by polymeric precursor method. *Appl. Catal. A Gen.* **2009**, *366*, 193–200. [CrossRef]
59. Pochamoni, R.; Narani, A.; Varkolu, M.; Dhar Gudimella, M.; Prasad Potharaju, S.S.; Burri, D.R.; Rao Kamaraju, S.R. Studies on ethylbenzene dehydrogenation with CO_2 as soft oxidant over Co_3O_4/COK-12 catalysts. *J. Chem. Sci.* **2015**, *127*, 701–709. [CrossRef]
60. Zhao, G.; Chen, H.; Li, J.; Wang, Q.; Wang, Y.; Ma, S.; Zhu, Z. Acid-based co-catalysis for oxidative dehydrogenation of ethylbenzene to styrene with CO_2 over X zeolite modified by alkali metal cation exchange. *RSC Adv.* **2015**, *5*, 75787–75793. [CrossRef]
61. Periyasamy, K.; Aswathy, V.T.; Ashok Kumar, V.; Manikandan, M.; Shukla, R.; Tyagi, A.K.; Raja, T. An efficient robust fluorite $CeZrO_4$-δ oxide catalyst for the eco-benign synthesis of styrene. *RSC Adv.* **2015**, *5*, 3619–3626. [CrossRef]
62. Mukherjee, D.; Park, S.-E.; Reddy, B.M. CO2 as a soft oxidant for oxidative dehydrogenation reaction: An eco benign process for industry. *J. CO2 Util.* **2016**, *16*, 301–312. [CrossRef]
63. Sato, S.; Ohhara, M.; Sodesawa, T.; Nozaki, F. Combination of ethylbenzene dehydrogenation and carbon dioxide shift-reaction over a sodium oxide/alumina catalyst. *Appl. Catal.* **1988**, *37*, 207–215. [CrossRef]
64. Chen, S.; Qin, Z.; Wang, G.; Dong, M.; Wang, J. Promoting effect of carbon dioxide on the dehydrogenation of ethylbenzene over silica-supported vanadium catalysts. *Fuel* **2013**, *109*, 43–48. [CrossRef]
65. Burri, D.R.; Choi, K.M.; Han, S.C.; Burri, A.; Park, S.-E. Dehydrogenation of ethylbenzene to styrene with CO2 over TiO_2-ZrO_2 bifunctional catalyst. *Bull. Korean Chem. Soc.* **2007**, *28*, 53–58.
66. Burri, D.R.; Choi, K.M.; Han, S.C.; Burri, A.; Park, S.-E. Selective conversion of ethylbenzene into styrene over K_2O/TiO_2-ZrO_2 catalysts: Unified effects of K_2O and CO_2. *J. Mol. Catal. A Chem.* **2007**, *269*, 58–63. [CrossRef]
67. Burri, A.; Jiang, N.; Park, S.-E. High surface area TiO_2-ZrO_2 prepared by caustic solution treatment, and its catalytic efficiency in the oxidehydrogenation of para-ethyl toluene by CO_2. *Catal. Sci. Technol.* **2012**, *2*, 514–520. [CrossRef]
68. Burri, D.R.; Choi, K.M.; Han, D.S.; Sujandi; Jiang, N.; Burri, A.; Park, S.-E. Oxidative dehydrogenation of ethylbenzene to styrene with CO_2 over SnO_2-ZrO_2 mixed oxide nanocomposite catalysts. *Catal. Today* **2008**, *131*, 173–178. [CrossRef]
69. Burri, D.R.; Choi, K.M.; Han, D.S.; Koo, J.B.; Park, S.-E. CO2 utilization as an oxidant in the dehydrogenation of ethylbenzene to styrene over MnO2-ZrO2 catalysts. *Catal. Today* **2006**, *115*, 242–247. [CrossRef]
70. Rahmani, F.; Haghighi, M.; Amini, M. The beneficial utilization of natural zeolite in preparation of Cr/clinoptilolite nanocatalyst used in CO2-oxidative dehydrogenation of ethane to ethylene. *J. Ind. Eng. Chem.* **2015**, *31*, 142–155. [CrossRef]
71. Mimura, N.; Takahara, I.; Inaba, M.; Okamoto, M.; Murata, K. High-performance Cr/H-ZSM-5 catalysts for oxidative dehydrogenation of ethane to ethylene with CO_2 as an oxidant. *Catal. Commun.* **2002**, *3*, 257–262. [CrossRef]
72. Liu, L.; Li, H.; Zhang, Y. A comparative study on catalytic performances of chromium incorporated and supported mesoporous MSU-x catalysts for the oxidehydrogenation of ethane to ethylene with carbon dioxide. *Catal. Today* **2006**, *115*, 235–241. [CrossRef]
73. Shi, X.; Ji, S.; Wang, K. Oxidative dehydrogenation of ethane to ethylene with carbon dioxide over Cr-Ce/SBA-15 catalysts. *Catal. Lett.* **2008**, *125*, 331–339. [CrossRef]
74. Deng, S.; Li, S.; Li, H.; Zhang, Y. Oxidative dehydrogenation of ethane to ethylene with CO_2 over Fe-Cr/ZrO2 catalysts. *Ind. Eng. Chem. Res.* **2009**, *48*, 7561–7566. [CrossRef]
75. Shi, X.; Ji, S.; Li, C. Oxidative dehydrogenation of ethane with CO_2 over novel Cr/SBA-15/Al_2O_3/FeCrAl monolithic catalysts. *Energy Fuels* **2008**, *22*, 3631–3638. [CrossRef]
76. Tedeeva, M.A.; Kustov, A.L.; Pribytkov, P.V.; Leonov, A.V.; Dunaev, S.F. Dehydrogenation of Propane with CO_2 on Supported CrO_x/SiO_2 Catalysts. *Russ. J. Phys. Chem. A* **2018**, *92*, 2403–2407. [CrossRef]

77. Wang, S.; Murata, K.; Hayakawa, T.; Hamakawa, S.; Suzuki, K. Dehydrogenation of ethane with carbon dioxide over supported chromium oxide catalysts. *Appl. Catal. A Gen.* **2000**, *196*, 1–8. [CrossRef]
78. Weckhuysen, B.M.; Schoonheydt, R.A.; Jehng, J.M.; Wachs, I.E.; Cho, S.J.; Ryoo, R.; Kijlstra, S.; Poels, E. Combined DRS-RS-EXAFS-XANES-TPR study of supported chromium catalysts. *J. Chem. Soc. Faraday Trans.* **1995**, *91*, 3245–3253. [CrossRef]
79. Mimura, N.; Okamoto, M.; Yamashita, H.; Oyama, S.T.; Murata, K. Oxidative dehydrogenation of ethane over Cr/ZSM-5 catalysts using CO_2 as an oxidant. *J. Phys. Chem. B* **2006**, *110*, 21764–21770. [CrossRef]
80. Heracleous, E.; Lemonidou, A.A. Ni-Nb-O mixed oxides as highly active and selective catalysts for ethene production via ethane oxidative dehydrogenation. Part I: Characterization and catalytic performance. *J. Catal.* **2006**, *237*, 162–174. [CrossRef]
81. Heracleous, E.; Lemonidou, A.A. Ni-Nb-O mixed oxides as highly active and selective catalysts for ethene production via ethane oxidative dehydrogenation. Part II: Mechanistic aspects and kinetic modeling. *J. Catal.* **2006**, *237*, 175–189.
82. Heracleous, E.; Delimitis, A.; Nalbandian, L.; Lemonidou, A.A. HRTEM characterization of the nanostructural features formed in highly active Ni-Nb-O catalysts for ethane ODH. *Appl. Catal. A Gen.* **2007**, *325*, 220–226. [CrossRef]
83. Koirala, R.; Buechel, R.; Krumeich, F.; Pratsinis, S.E.; Baiker, A. Oxidative dehydrogenation of ethane with CO2 over flame-made Ga-loaded TiO2. *ACS Catal.* **2015**, *5*, 690–702. [CrossRef]
84. Cheng, Y.; Zhang, F.; Zhang, Y.; Miao, C.; Hua, W.; Yue, Y.; Gao, Z. Oxidative dehydrogenation of ethane with CO2 over Cr supported on submicron ZSM-5 zeolite. *Chinese J. Catal.* **2015**, *36*, 1242–1248. [CrossRef]
85. Ramesh, Y.; Thirumala Bai, P.; Hari Babu, B.; Lingaiah, N.; Rama Rao, K.S.; Prasad, P.S.S. Oxidative dehydrogenation of ethane to ethylene on Cr2O3/Al2O3–ZrO2 catalysts: The influence of oxidizing agent on ethylene selectivity. *Appl. Petrochem. Res.* **2014**, *4*, 247–252. [CrossRef]
86. Deng, S.; Li, H.; Li, S.; Zhang, Y. Activity and characterization of modified Cr_2O_3/ZrO_2 nano-composite catalysts for oxidative dehydrogenation of ethane to ethylene with CO_2. *J. Mol. Catal. A Chem.* **2007**, *268*, 169–175. [CrossRef]
87. Yashima, T.; Sato, K.; Hayasaka, T.; Hara, N. Alkylation on synthetic zeolites. III. Alkylation of toluene with methanol and formaldehyde on alkali cation exchanged zeolites. *J. Catal.* **1972**, *26*, 303–312. [CrossRef]
88. Rossetti, I.; Bencini, E.; Trentini, L.; Forni, L. Study of the deactivation of a commercial catalyst for ethylbenzene dehydrogenation to styrene. *Appl. Catal. A Gen.* **2005**, *292*, 118–123. [CrossRef]
89. Meima, G.R.; Menon, P.G. Catalyst deactivation phenomena in styrene production. *Appl. Catal. A Gen.* **2001**, *212*, 239–245. [CrossRef]
90. Sindorenko, L.N.; Galich, P.N.; Gutirya, V.S. Condensation of toluene and methanol upon synthetic zeolites containing ion-exchange cations of alkali metals. *Dokl. Akad. Nauk. SSSR* **1967**, *173*, 132–133.
91. Seo, D.W.; Rahman, S.T.; Reddy, B.M.; Park, S.-E. Carbon dioxide assisted toluene side-chain alkylation with methanol over Cs-X zeolite catalyst. *J. CO2 Util.* **2018**, *26*, 254–261. [CrossRef]
92. Hattori, H. Solid base catalysts: Fundamentals and their applications in organic reactions. *Appl. Catal. A Gen.* **2015**, *504*, 103–109. [CrossRef]
93. Yoo, K.S.; Smirniotis, P.G. Zeolites-catalyzed alkylation of isobutane with 2-butene: Influence of acidic properties. *Catal. Lett.* **2005**, *103*, 249–255. [CrossRef]
94. Alabi, W.O.; Tope, B.B.; Jermy, R.B.; Aitani, A.M.; Hattori, H.; Al-Khattaf, S.S. Modification of Cs-X for styrene production by side-chain alkylation of toluene with methanol. *Catal. Today* **2014**, *226*, 117–123. [CrossRef]
95. Yoo, K.; Smirniotis, P.G. The deactivation pathway of one-dimensional zeolites, LTL and ZSM-12, for alkylation of isobutane with 2-butene. *Appl. Catal. A Gen.* **2003**, *246*, 243–251. [CrossRef]
96. Yoo, K.; Burckle, E.C.; Smirniotis, P.G. Isobutane/2-butene alkylation using large-pore zeolites: Influence of pore structure on activity and selectivity. *J. Catal.* **2002**, *211*, 6–18. [CrossRef]
97. Han, H.; Liu, M.; Nie, X.; Ding, F.; Wang, Y.; Li, J.; Guo, X.; Song, C. The promoting effects of alkali metal oxide in side-chain alkylation of toluene with methanol over basic zeolite X. *Microporous Mesoporous Mater.* **2016**, *234*, 61–72. [CrossRef]
98. Itoh, H.; Hattori, T.; Suzuki, K.; Murakami, Y. Role of acid and base sites in the side-chain alkylation of alkylbenzenes with methanol on two-ion-exchanged zeolites. *J. Catal.* **1983**, *79*, 21–33. [CrossRef]

99. Tope, B.B.; Alabi, W.O.; Aitani, A.M.; Hattori, H.; Al-Khattaf, S.S. Side-chain alkylation of toluene with methanol to styrene over cesium ion-exchanged zeolite X modified with metal borates. *Appl. Catal. A Gen.* **2012**, *443–444*, 214–220. [CrossRef]
100. Philippou, A.; Anderson, M.W. Solid-State NMR Investigation of the Alkylation of Toluene with Methanol over Basic Zeolite X. *J. Am. Chem. Soc.* **1994**, *116*, 5774–5783. [CrossRef]
101. Jiang, N.; Jin, H.; Jeong, E.-Y.; Park, S.-E. Mgo Encapsulated Mesoporous Zeolite for the Side Chain Alkylation of Toluene with Methanol. *J. Nanosci. Nanotechnol.* **2010**, *10*, 227–232. [CrossRef]
102. Hattori, H.; Alabi, W.O.; Jermy, B.R.; Aitani, A.M.; Al-Khattaf, S.S. Pathway to ethylbenzene formation in side-chain alkylation of toluene with methanol over cesium ion-exchanged zeolite X. *Catal. Lett.* **2013**, *143*, 1025–1029. [CrossRef]
103. Mikkelsen, M.; Jørgensen, M.; Krebs, F.C. The teraton challenge. A review of fixation and transformation of carbon dioxide. *Energy Environ. Sci.* **2010**, *3*, 43–81. [CrossRef]
104. Cavani, F.; Ballarini, N.; Cericola, A. Oxidative dehydrogenation of ethane and propane: How far from commercial implementation? *Catal. Today* **2007**, *127*, 113–131. [CrossRef]
105. Ren, Y.; Zhang, F.; Hua, W.; Yue, Y.; Gao, Z. ZnO supported on high silica HZSM-5 as new catalysts for dehydrogenation of propane to propene in the presence of CO_2. *Catal. Today* **2009**, *148*, 316–322. [CrossRef]
106. Chen, M.; Xu, J.; Liu, Y.M.; Cao, Y.; He, H.Y.; Zhuang, J.H. Supported indium oxide as novel efficient catalysts for dehydrogenation of propane with carbon dioxide. *Appl. Catal. A Gen.* **2010**, *377*, 35–41. [CrossRef]
107. Schimmoeller, B.; Jiang, Y.; Pratsinis, S.E.; Baiker, A. Structure of flame-made vanadia/silica and catalytic behavior in the oxidative dehydrogenation of propane. *J. Catal.* **2010**, *274*, 64–75. [CrossRef]
108. Liu, Y.M.; Cao, Y.; Yan, S.R.; Dai, W.L.; Fan, K.N. Highly effective oxidative dehydrogenation of propane over vanadia supported on mesoporous SBA-15 silica. *Catal. Lett.* **2003**, *88*, 61–67. [CrossRef]
109. Santamaría-González, J.; Mérida-Robles, J.; Alcántara-Rodríguez, M.; Maireles-Torres, P.; Rodríguez-Castellón, E.; Jiménez-López, A. Catalytic behaviour of chromium supported mesoporous MCM-41 silica in the oxidative dehydrogenation of propane. *Catal. Lett.* **2000**, *64*, 209–214. [CrossRef]
110. Davies, T.; Taylor, S.H. The oxidative dehydrogenation of propane using gallium-molybdenum oxide-based catalysts. *J. Mol. Catal. A Chem.* **2004**, *220*, 77–84. [CrossRef]
111. Raju, G.; Reddy, B.M.; Abhishek, B.; Mo, Y.H.; Park, S.-E. Synthesis of C4 olefins from n-butane over a novel VOx/SnO2-ZrO2 catalyst using CO2 as soft oxidant. *Appl. Catal. A Gen.* **2012**, *423–424*, 168–175. [CrossRef]
112. Atanga, M.A.; Rezaei, F.; Jawad, A.; Fitch, M.; Rownaghi, A.A. Oxidative dehydrogenation of propane to propylene with carbon dioxide. *Appl. Catal. B Environ.* **2018**, *220*, 429–445. [CrossRef]
113. Reddy, B.M.; Lee, S.C.; Han, D.S.; Park, S.-E. Utilization of carbon dioxide as soft oxidant for oxydehydrogenation of ethylbenzene to styrene over V2O5-CeO2/TiO2-ZrO2 catalyst. *Appl. Catal. B Environ.* **2009**, *87*, 230–238. [CrossRef]
114. Uy, D.; O'Neill, A.E.; Xu, L.; Weber, W.H.; McCabe, R.W. Observation of cerium phosphate in aged automotive catalysts using Raman spectroscopy. *Appl. Catal. B Environ.* **2003**, *41*, 269–278. [CrossRef]
115. Armaroli, T.; Simon, L.J.; Digne, M.; Montanari, T.; Bevilacqua, M.; Valtchev, V.; Patarin, J.; Busca, G. Effects of crystal size and Si/Al ratio on the surface properties of H-ZSM-5 zeolites. *Appl. Catal. A Gen.* **2006**, *306*, 78–84. [CrossRef]
116. Thakkar, H.; Eastman, S.; Hajari, A.; Rownaghi, A.A.; Knox, J.C.; Rezaei, F. 3D-Printed Zeolite Monoliths for CO2 Removal from Enclosed Environments. *ACS Appl. Mater. Interfaces* **2016**, *8*, 27753–27761. [CrossRef]
117. Thakkar, H.; Eastman, S.; Al-Mamoori, A.; Hajari, A.; Rownaghi, A.A.; Rezaei, F. Formulation of Aminosilica Adsorbents into 3D-Printed Monoliths and Evaluation of Their CO2 Capture Performance. *ACS Appl. Mater. Interfaces* **2017**, *9*, 7489–7498. [CrossRef]
118. Ren, Y.; Wang, J.; Hua, W.; Yue, Y.; Gao, Z. Ga2O3/HZSM-48 for dehydrogenation of propane: Effect of acidity and pore geometry of support. *J. Ind. Eng. Chem.* **2012**, *18*, 731–736. [CrossRef]
119. Abello, M.C.; Gomez, M.F.; Ferretti, O. Oxidative conversion of propane over Al2O3-supported molybdenum and chromium oxides. *Catal. Lett.* **2003**, *87*, 43–49. [CrossRef]
120. Cherian, M.; Rao, M.S.; Hirt, A.M.; Wachs, I.E.; Deo, G. Oxidative dehydrogenation of propane over supported chromia catalysts: Influence of oxide supports and chromia loading. *J. Catal.* **2002**, *211*, 482–495. [CrossRef]
121. Rao, T.V.M.; Zahidi, E.M.; Sayari, A. Ethane dehydrogenation over pore-expanded mesoporous silica-supported chromium oxide: 2. Catalytic properties and nature of active sites. *J. Mol. Catal. A Chem.* **2009**, *301*, 159–165. [CrossRef]

122. Hakuli, A.; Kytökivi, A.; Krause, A.O.I. Dehydrogenation of i-butane on CrOx/Al2O3 catalysts prepared by ALE and impregnation techniques. *Appl. Catal. A Gen.* **2000**, *190*, 219–232. [CrossRef]
123. Liu, L.; Li, H.; Zhang, Y. Effect of synthesis parameters on the chromium content and catalytic activities of mesoporous Cr-MSU-x prepared under acidic conditions. *J. Phys. Chem. B* **2006**, *110*, 15478–15485. [CrossRef] [PubMed]
124. Santhosh Kumar, M.; Hammer, N.; Rønning, M.; Holmen, A.; Chen, D.; Walmsley, J.C.; Øye, G. The nature of active chromium species in Cr-catalysts for dehydrogenation of propane: New insights by a comprehensive spectroscopic study. *J. Catal.* **2009**, *261*, 116–128. [CrossRef]
125. Weckhuysen, B.M.; Wachs, I.E.; Schoonheydt, R.A. Surface chemistry and spectroscopy of chromium in inorganic oxides. *Chem. Rev.* **1996**, *96*, 3327–3349. [CrossRef]
126. Burri, A.; Hasib, M.A.; Mo, Y.H.; Reddy, B.M.; Park, S.-E. An Efficient Cr-TUD-1 Catalyst for Oxidative Dehydrogenation of Propane to Propylene with CO2 as Soft Oxidant. *Catal. Lett.* **2018**, *148*, 576–585. [CrossRef]
127. Park, S.-J.; Kim, K.D. Adsorption behaviors of CO2 and NH3 on chemically surface-treated activated carbons. *J. Colloid Interface Sci.* **1999**, *212*, 186–189. [CrossRef]
128. Michorczyk, P.; Pietrzyk, P.; Ogonowski, J. Preparation and characterization of SBA-1-supported chromium oxide catalysts for CO2 assisted dehydrogenation of propane. *Microporous Mesoporous Mater.* **2012**, *161*, 56–66. [CrossRef]
129. COSIA. $20M NRG COSIA Carbon XPRIZE Finalists Announced; Teams Ready to Test Transformative CO2 Technologies at Alberta's Carbon Conversion Centre. Available online: www.cosia.ca/resources/newsreleases/20m-nrg-cosia-carbon-xprize-finalists-announcedteams-ready-test (accessed on 9 April 2018).
130. Zhang, D.; Wang, J.; Lin, Y.; Si, Y.; Huang, C.; Yang, J.; Huang, B.; Li, W. Present situation and future prospect of renewable energy in China. *Renew. Sustain. Energy Rev.* **2017**, *76*, 865–871. [CrossRef]

© 2020 by the authors. Licensee MDPI, Basel, Switzerland. This article is an open access article distributed under the terms and conditions of the Creative Commons Attribution (CC BY) license (http://creativecommons.org/licenses/by/4.0/).

MDPI
St. Alban-Anlage 66
4052 Basel
Switzerland
Tel. +41 61 683 77 34
Fax +41 61 302 89 18
www.mdpi.com

Catalysts Editorial Office
E-mail: catalysts@mdpi.com
www.mdpi.com/journal/catalysts

www.ingramcontent.com/pod-product-compliance
Lightning Source LLC
LaVergne TN
LVHW070149100526
838202LV00015B/1915